# PSYCHOLOGICAL STATISTICS

# PSYCHOLOGICAL STATISTICS

## FOURTH EDITION

QUINN McNEMAR

JOHN WILEY AND SONS, INC.

*New York · London · Sydney · Toronto*

*Library of Congress Catalog Card Number: 79-76057*

*SBN 471 58708 7*

Printed in the United States of America

# PREFACE

As in the preceding three editions, the aim of the present effort is to provide a concise, balanced, and integrated coverage of the statistical techniques most frequently used in psychology and associated sciences. The stress is on interpretations and assumptions and the logical place of statistical inference as an inescapable part of scientific methodology. The near-ubiquitous electronic computer renders an ever lessening need for textbook display of computational details, but the researcher must somehow acquire sufficient "savvy" to be sure a selected computer program is appropriate.

Edition watchers will find the following major additions: further interpretations of correlation; more on sampling aspects of multiple regression; differences in linear and quadratic trends as part of inter-action: extension of covariance to two-way analysis of variance designs; introduction to four-way designs; nonorthogonality of factors and contrast of multiple regression and factorial analysis of variance; variance components estimation; control by "levels by treatments" design. Minor extensions on the following topics: joint probabilities; type II error and power; confidence intervals; effect of skewness on correlation; illustra-tions of biserial and fourfold correlation; interaction as differences.

The failure to present additional nonparametric techniques simply reflects my lack of enthusiasm for this area. And those who had hoped for a chapter on factor analysis should realize that I have negative enthusiasm for factor techniques.

Chapter 16 has been reorganized in order to achieve a better sequence. A new, and I hope, simpler notation is used in Chapters 17 and 18. At least the notation in these two chapters is now consistent!

Since my initial exposure to statistics in courses taught by the late Truman L. Kelley and by Harold Hotelling, there have been many direct and indirect factors that have contributed to the content and writing of this book. Former students of mine may be surprised to learn that they have had an influence. As usual, my greatest personal gratitude goes to Olga W. McNemar whose keen editorial eye has done much to clarify the exposition.

Acknowledgement is hereby made to authors (R. A. Fisher and F. Yates) and their publisher (Oliver and Boyd Limited, Edinburgh) for permission to reprint Tables III, IV, V, and VII from the book *Statistical Tables for Biological, Agricultural and Medical Research*. These tables appear in our appendix as E, D, F, and C, respectively.

*Austin, Texas*                                         Quinn McNemar
*January 1969*

# CONTENTS

# PSYCHOLOGICAL STATISTICS

# Chapter 1

# INTRODUCTION

Statistical methods are concerned with the reducing of either large or small masses of data to a few convenient descriptive terms and with the drawing of inferences therefrom. The data are collected by any of several methods of research with the aid of measuring devices appropriate to a given area of investigation. The research methods are variously named and classified. Thus in psychology we have methods which are labeled experimental, clinical, observational, etc. The devices for measuring or securing responses vary from those which involve delicate apparatus through paper-and-pencil schemes to controlled observations and interviews. Statistical techniques are not to be considered as coordinate either with research methods or with devices for obtaining and recording responses, but rather as tools for analyzing data collected by whatever means.

The reduction of a batch of data to a few descriptive measures is the part of statistical analysis which should lead to a better over-all comprehension of the data. All readers will be more or less familiar with the concept of *average*. An average is a measure which describes what is typical of a group with respect to some trait, characteristic, or variable. If we are comparing two or more groups, the determination of an average for each group permits a better appraisal of possible group differences than would be obtained by casual examination of the data. There are various statistical measures, or types of averages, which have proven useful as descriptive terms for a variety of data. One aim of this book is to present and discuss the descriptive statistical measures most frequently needed in psychological research. Proper usage and interpretation of these terms and evaluation of their use by others are not possible without knowledge of their meaning and their limiting assumptions. Incidentally, the user of statistical measures must give some thought to computational procedures.

As we proceed, it will be necessary not only to define descriptive measures but also to distinguish between the usage of a given measure as being descriptive of a *sample* as opposed to a *population*. Since sample descriptive statistics are knowns (i.e., computable) whereas the corresponding population values are unknowns (but estimable), we will in this book define and discuss the descriptive measures in terms of samples and subsequently consider the problem of drawing inferences about, or estimating, population values. Sample values are frequently referred to as *statistics* and population values are called *parameters*.

That part of statistical analysis which has to do with the drawing of inferences is imposed on us because of certain inadequacies of research data. For instance, an investigator who wishes to know the average height of adult women in the United States will never have facilities for measuring every woman. Accordingly, he is compelled to measure a sample of women; then on the basis of information yielded by the sample he can make an inference concerning the average height of the population of women. Another investigator, wishing to evaluate the relative merits of two learning methods, tries out the methods with two small groups of students, and from the results, makes an inference concerning what might be expected if he had facilities for working with very large groups. An opinion poller may seek information about the reactions of Republicans and Democrats to some world event. By questioning a sample of each group he can secure sufficient data for drawing an inference regarding a possible difference between the population of Republicans and the population of Democrats.

The problem of statistical inference is usually that of determining whether statistical significance can be attached to results after due allowance is made for known sources of error. There are many and varied situations for which we need tests of significance, and accordingly several tests are available. Intelligent and critical inferences cannot be made by those who do not understand the purposes, assumptions, and applicability of the various techniques for judging significance.

It is in connection with the problem of drawing inferences that a knowledge of statistical methods is most helpful. A research should be planned in such a way that the resulting data are amenable to treatment by the available statistical techniques. With sufficient information concerning these techniques of analysis, one should be able to lay out in advance of data collecting the main types of statistical analysis to be used. If a proposed experimental setup precludes the possibility of adequate analysis, it may be found that a slight alteration in the plan will remedy the situation. All too frequently the statistician is called in to help with data which have not been collected in such a manner as to permit efficient analysis. Only by knowing the available methods of analysis can one plan a research with assurance that the results can be handled statistically.

Another reason for keeping in mind statistical considerations while planning a research is the fact that some experimental designs are preferable because they permit, with small additional cost, or even at a saving, better control of error than other plans. Indeed, certain designs lead to a marked reduction in known sources of error.

A third reason for planning with foresight regarding the statistical analysis is that a set of data can sometimes be made to serve for checking several different hypotheses.

The student should be warned that he cannot expect miracles to be wrought by the use of statistical tools. Although statistical methods have an important place in present-day psychological research, it does not follow that they can be utilized to salvage data that result from a haphazardly planned and sloppily executed investigation. No amount of statistical juggling can transfigure bad data into acceptable form. It is doubtful whether the student who comes to the statistician with a batch of data and the question, "Can I compute a correlation coefficient . . . ?" will make a scientific contribution, but such a student deserves sympathy, especially if his major advisor has suggested that he need not worry about statistics until he has collected data.

The purpose of the present book is to acquaint the student with the statistical techniques commonly used, to suggest economical computational procedures, and to state the assumptions and limitations of the various techniques. Whenever the understanding of a particular technique can be clarified by a simple derivation, such a derivation will be given. Unfortunately, many of the derivations are too complicated mathematically to permit consideration in an elementary or intermediate treatment. The qualified and interested student will find some of these derivations in more advanced textbooks and others in original sources.

Statistical methods belong in the realm of applied mathematics, and consequently extensive scholarship in mathematics is required of those who choose to specialize in statistics. It is possible, however, to secure a practical working knowledge of statistical techniques without first becoming a mathematician, provided the deficiency in mathematics is not accompanied by an emotional reaction to symbols.

Within the realm of psychological research there is wide variation in the need for statistical procedures. We can find current research reports which involve no use of statistics, some which involve very simple statistical treatment, still others which lean heavily on the tools of statistics, and a few which are highly statistical. We need not shift from one area of investigation to another to find this variation, but it is true that certain areas of research in psychology have less dependency than others on statistical procedures. The area of psychology which seems the most dependent on statistics is psychological measurement. This dependency is due mainly to the very nature of psychological measurement, the theory of which is largely statistical.

The presence or absence of statistical analysis *per se* is not a safe criterion for judging the worth of a study—some studies would have been improved by the utilization of statistics, whereas others would be better if they had been so designed as to depend less on statistical analysis. Except for the requirement that the statistical analysis be adequate, there are no general rules as to how statistical a research should be. Of two experimental plans, either of which would provide appropriate data for checking a given hypothesis or sets of hypotheses, that plan which calls for simple statistical analysis is certainly preferable to the one which requires elaborate analysis. Experimental control of errors is far better than statistical adjustments.

# Chapter 2

# TABULAR AND GRAPHIC METHODS

When we are faced with a mass of data, the first manipulative step is tabulation or classification. If we are dealing with the number of children per family, the tabulation is equivalent to counting the number of one-child families, two-child families, etc.; or if we have information on 1000 persons regarding their national origin, we can tabulate, or count, the number of those of German, French, Italian, etc., origin; or these same individuals can be classified as to eye color. If we have their heights, we can also classify (or tabulate) them as being 58, 59, 60, etc., inches in height, and if the shortest person is 58 and the tallest is 78 inches, we would tabulate our 1000 into 21 different inch groups. If we also know the weights of these individuals, we can classify again, this time as 100, 101, up to (say) 229 pounds, and thereby have 130 groups. In all these situations we can classify with respect to the given characteristics, but the resulting tabulations will show marked differences as we pass from trait to trait. For instance, we may have only six national groups, and it will make little difference whether Germans or Russians are first on the tabulation sheet. Such a characteristic as nationality or eye color is said to be *unordered* (and somewhat *discrete*). The number of children per family is discrete but can be ordered, from least to greatest number. Such a trait as height can also be ordered, but it is said to be *continuous* (non-discrete) because it is possible to have an infinite number of in-between values very closely spaced. Such a series is sometimes called *graduated*. It will of course be obvious that a discrete series does not permit of in-between values, e.g., no family can have $2\frac{1}{4}$ children.

For most purposes it is adequate if we tabulate, or classify, individuals

5

into certain large groups. For example, instead of classifying our 1000 persons into pound groups (130 such groups) it is usually sufficient to classify them into broader groups, say 100–109, 110–119, etc., thereby obtaining 13 large groups. As a matter of fact, the use of fewer groups has a distinct advantage in that the labor of tabulating and computing descriptive terms is greatly lessened. The factors influencing the choice of the grouping interval are two: first, its size should be such as to permit at least 10 or 12, but not more than 20, classes or groups; and second, it should promote tabulating convenience. Suggestions for choosing tabulating intervals are: (1) determine the *range* of measures or scores, i.e.,

**Table 2.1. Frequency distribution of IQs for 161 five-year-old boys**

| Interval | $f$ | Smoothed $f$ | Cumulative $f$ |
|---|---|---|---|
| 160–169 | 1 | .3 | 161 |
| 150–159 |   | 1.3 | 160 |
| 140–149 | 3 | 4.0 | 160 |
| 130–139 | 9 | 13.7 | 157 |
| 120–129 | 29 | 25.7 | 148 |
| 110–119 | 39 | 34.3 | 119 |
| 100–109 | 35 | 35.3 | 80 |
| 90–99 | 32 | 25.0 | 45 |
| 80–89 | 8 | 14.0 | 13 |
| 70–79 | 2 | 3.7 | 5 |
| 60–69 | 1 | 1.3 | 3 |
| 50–59 | 1 | 1.0 | 2 |
| 40–49 | 1 | .7 | 1 |

the difference between the lowest and highest; (2) by inspection determine whether the range can be divided into 12 to 20 equal intervals of some convenient size, say 5 or 10; and (3) let the lower number of each interval be a multiple of the size of the interval. It is customary to arrange the tabulation sheet with the highest or largest values of the variable at the top and to use either dots or tally marks when tabulating. The tallies per interval can be counted and recorded to the right of the tally marks. This column is usually labeled $f$, and the sum of the $f$s will be $N$, or the total number of individuals in all the grouping intervals. Tabulation results in a *frequency table* or *frequency distribution*, such as that shown in the first two columns of Table 2.1.

It should be noted that the expressed interval limits in a frequency table are not necessarily the actual limits. Thus, if weight has been taken to the nearest pound, the actual limits of the interval 130–139 would be 129.5 and 139.5; but if the ages of individuals have been taken as at the last birthday, the interval 20–24 would have actual limits of 20 and 24.999+. Obviously for purposes of tabulation we need not use the implied actual limits, and for computational purposes we usually need either the lower limit or the midpoint of certain intervals, so there is nothing to be gained by meticulously labeling the intervals with actual limits.

## GRAPHIC PRESENTATION

If we scrutinize the tally marks or the frequency table, we can obtain some notion as to how the individual values are distributed. A number of pictorial schemes have been suggested as aids in the study of frequency distributions. It is possible to lay off the various values (or intervals) of the variable on the horizontal or $x$ axis, and to let the vertical or $y$ axis represent the frequency per value or interval. The frequencies of the several intervals can be represented by drawing a horizontal line across each interval at the height corresponding to the number of cases in that interval, and then connecting these horizontals with verticals erected at the interval limits. This yields a *histogram* (Fig. 2.1). Using the same arrangement of the vertical and horizontal scales, we can merely indicate the frequency with a dot or cross placed directly above the midpoint of the interval, and then connect the adjacent points with straight lines. This results in a *frequency polygon* (Fig. 2.2). Such a polygon or the corresponding histogram will usually show irregularities; on the assumption that these are due to the operation of chance, we can draw a smooth curve, cutting as near the points as possible, and this curve can be thought of as giving a better picture than the original polygon. A curve which is

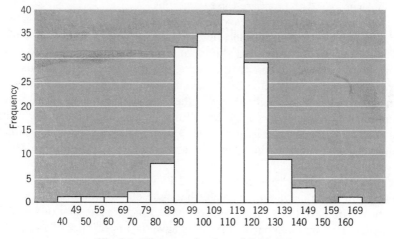

**Fig. 2.1.** Histogram for data of Table 2.1.

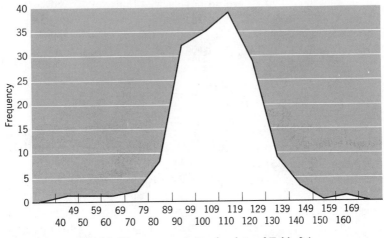

**Fig. 2.2.** Frequency polygon for data of Table 2.1.

obtained by freehand drawing or by graphic smoothing schemes or by repeated smoothing of the frequencies by a method of moving averages is known as a *frequency curve*. One method of moving averages is illustrated in Table 2.1, in which an average is taken over three intervals. The smoothed value for an interval is obtained by summing the frequencies in that interval and the two adjacent intervals and dividing by 3. Thus the smoothed value for the interval 80–90 is equal to the sum of the frequencies 2, 8, and 32, divided by 3. For the 90 interval, 8, 32, and 35 are summed and divided by 3. The student should plot both the original and smoothed frequencies so as to compare the two graphs.

Although it is relatively easy to depict a frequency distribution by a histogram, by a frequency polygon, or by a smoothed frequency curve, it is necessary that we note a shift in interpretation as we pass from the histogram to the polygon to the curve. In drawing the histogram, we are in effect drawing a series of vertical bars with a common boundary for any two that are adjacent to each other. Since the height of each bar represents a frequency, we may, by arbitrarily assigning unity as the width of each bar, say that the *area* of a bar also represents a frequency. Then the sum of the areas of the several bars will be the total number of cases, or $N$.

If we think of the polygon in Fig. 2.2 as being superimposed on the histogram of Fig. 2.1 and imagine that the common boundaries of the vertical bars have been erased, we will have a picture like that in Fig. 2.3, in which the remaining parts of the bars have the appearance of an up and then down irregular staircase. A little thought should convince the reader that the total area under this staircase is $N$, or precisely the same as the sum of the areas of all the bars.

Next consider the polygon. Note that as we pass from interval to interval, the polygon in conjunction with the staircase histogram forms a series of pairs of equal-area triangles. One of each pair is an area included under the polygon but not under the histogram, whereas the other is an area included under the histogram but not under the polygon. The net effect of this balancing of areas, in and out, is that the total areas under the polygon and histogram are equal; each total area represents $N$.

Now it should not stretch our imagination too much to regard the total area under a smoothed polygon or under a frequency curve as being equal to $N$. With this notion that area, not height, represents frequency, we can readily speak of the area under the curve between ordinates erected at any two score values on the base line ($x$ axis) as the number of cases between the two score points. And of course the area under any part of the curve could be expressed as a proportion or a percentage of the total area.

This concept of area as frequency will have considerable value for us as a basis for interpreting certain statistical measures, and the concept will be indispensable to our understanding of certain "ideal," or mathematical, frequency curves, as yet undefined.

Another type of graph can be obtained by the use of *cumulative* frequencies. In Table 2.1 is a column headed "Cumulative $f$." These values are obtained by successive adding of the frequencies, beginning with the lowest interval. Adding 1 and 1 gives 2, adding to this the next frequency gives 3, to which in turn is added the next, giving 5, and so on until we have 160 plus 1 for the last cumulative value, which is the total number of cases.

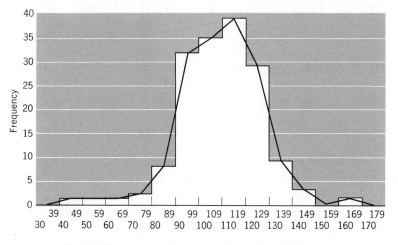

**Fig. 2.3.** Frequency polygon superimposed on histogram.

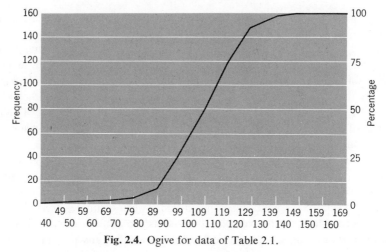

**Fig. 2.4.** Ogive for data of Table 2.1.

Obviously, from the cumulative table we can tell how many individuals fall below a given point. If we plot the cumulative values and connect the plotted points, an *ogive* curve results (Fig. 2.4). Note that, in plotting the cumulative frequencies, we do not use the midpoint of the interval, but rather the upper boundary. Why?

The use of frequency polygons in the comparison of two groups is quite simple and often very enlightening. All that is necessary is to plot the data for both groups on the same sheet and with reference to the same axes. If the number of cases in the two groups differs markedly, a better comparison can be obtained by converting the frequencies for each group to percentages of the total number in each group. Polygons based on percentage frequencies will not portray differences which are merely a reflection of differing $N$s and therefore are more comparable. A glance at two such frequency polygons will reveal whether the two groups show marked differences in the trait in question or to what extent the two distributions overlap. More refined methods for comparing groups are discussed later.

When we wish to picture a discrete series, it is customary to use either horizontal or vertical bars, separated from each other, to represent the several frequencies. As in the case of frequency polygons and histograms, there are no hard and fast rules regarding the heights (or lengths) of the bars relative to the horizontal (or vertical) base. The student should attempt to avoid extreme lack of proportion. Newspapers and magazines often represent frequencies as areas or solids. A circular diagram, or pie chart, in which the sizes of the separate sectors represent the percentage falling into given groups or classes is sometimes used to picture relative frequencies. There is some evidence, and a general consensus of opinion, that some type of linear graph is less likely to be misinterpreted than one that depends on areas or solids.

Another type of graphical representation is used to picture the relationship· between two variables, e.g., growth in stature and age, or price change with year. To make such a line graph, we can lay off time or age or trials on the horizontal axis, choose a convenient scale on the $y$ axis for the other variable, and then plot the observational values. The line graph should be arranged so that the graph is read from left to right and from the bottom to the top, and the scales on the two axes should allow the inclusion of all observed values of the two variables and at the same time permit of a well-balanced or well-proportioned picture. A line graph can be made misleading by the choice of the scales on the two axes. For instance, if we are plotting the practice curve for card sorting (number of cards sorted on $y$ axis, trial number on $x$ axis), it is possible to make a tremendous difference in the appearance of the graph simply by altering the scale on the $y$ axis. Of two curves which represent the same relationship, one (Fig. 2.5) would give the impression that the learning had progressed quite rapidly, whereas the other (Fig. 2.6) would lead us to think that progress was slow. The student will do well to develop a healthy scepticism of all graphs he encounters for the simple reason that either scale can be so selected as to lead to gross misinterpretation.

It should be noted that smoothing may be applied to line graphs as well as to frequency polygons. Often, if a line graph is smoothed, the relationship between the two variables can be more adequately characterized.

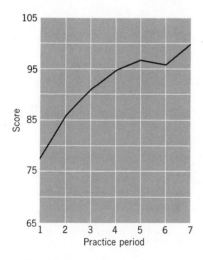

Fig. 2.5. Learning curve (same data as Fig. 2.6).

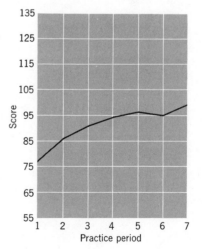

Fig. 2.6. Learning curve (same data as Fig. 2.5).

Smoothing out the irregularities helps us to see whether the relationship is linear or logarithmic or parabolic or of some other common type. Frequently a verbal description of a curve will aid in understanding something of the functional relatedness of the two variables. To state a relationship in more exact mathematical language involves the application of some form of curve fitting by which the constants of the equation can be determined.

The student who is interested in a complete discussion and treatment of graphic methods is referred to books on the subject by Brinton and by Arkin and Colton.*

* Brinton, W. C., *Graphic presentation*, New York: Brinton Associates, 1939; Arkin, Herbert, and Colton, R. R., *Graphs, how to make and use them*, New York: Harper, 1936.

# Chapter 3

# DESCRIBING FREQUENCY
# DISTRIBUTIONS

It has been implied in Chapter 2 that a variable, such as height, IQ, or reading ability, can be represented by $X$, where $X$ takes on various values, i.e., varies from individual to individual. Obviously, $X$ is not used here to represent an unknown but rather as a symbol for any of several known quantities. When a frequency polygon is drawn and smoothed, it is often found to be a curve which has a peak or maximum near the center of the $X$s and drops off gradually toward the base line or $x$ axis on either side of the point of maximum value. In other words, a typical frequency curve (or polygon) or a frequency distribution can be roughly characterized as one which shows four chief features: a clustering of individuals toward some central value, dispersion about this value, symmetry or lack of symmetry, and flatness or steepness. Many variables or traits yield distributions which are said to be approximately bell-shaped, but such a description is not adequate for scientific purposes. We want to know about what particular value and with how much scatter the individual scores are distributed, to what extent the distribution is symmetrical, and to what degree it is peaked or flat. That is, we need measures of central value or tendency, measures of scatter or dispersion or variability, and measures of skewness (lack of symmetry) and of kurtosis (peakedness or flatness). With such measures, we can describe the distribution mathematically, and in such a way that a statistically trained contemporary, say in Melbourne, can picture to himself the frequency distribution.

Thus we are led to a consideration of the various measures of central value, dispersion, skewness, and kurtosis. It is adequate and usually more economical of time to determine these measures from frequency distributions

13

rather than from the original undistributed scores. Since the computation of the descriptive terms frequently involves a determination of the lower limit or midpoint of a class interval, the student should recall what has been said about actual and expressed class limits. Obviously, if we need the midpoint of an interval, it is necessary only to add one-half the size of the interval to the actual lower limit, which must be determined by a consideration of the nature of the scores or measures which constitute the variable. Psychological measurements and test scores are usually treated as though rounded to the nearest value.

## MEASURES OF CENTRAL VALUE

**The mode.** A glance at a typical frequency distribution will indicate to us the most frequently occurring $X$ value, or for grouped data the group of $X$ values which has the greatest frequency. This maximal frequency roughly defines the *mode*. For nongrouped data the mode is the $X$ value having the greatest frequency, whereas for grouped data the mode is taken as the midpoint of the interval which has the greatest frequency. For a smoothed frequency curve, the mode is the $X$ value at which the curve reaches its maximum height. The mode is one indicator of central value, but as a descriptive statistic it has serious limitations. If a different size interval is used, the mode may be decidedly different. Furthermore, it occasionally happens that two nonadjacent intervals have the same maximal frequency, thereby yielding two modal values. Such a distribution is said to be bimodal, but it should be noted that the bimodality may not be real but merely accidental, the resultant of the particular grouping interval chosen. In dealing with certain discrete series, like size of family, the modal value is apt to be more typical than some other measure of central value and therefore should be used, even though as a measure it is subject to greater sampling fluctuations than either the mean or the median. (The question of sampling cannot be discussed at this time; the student is asked to take on faith statements regarding the efficiency of a given statistic.)

**The median.** As a measure of central value, the *median* is defined in two ways: (1) if the individual scores are arranged in order with respect to some trait, the median is the value of the midmost individual if $N$ is odd, or lies midway between the two middle individuals when $N$ is even; (2) when a distribution has been made, the median is defined as the point on the scale such that the frequency above or below the point is 50 per cent of the total frequency. For grouped data, the median may be determined by the following steps:

1. Find one-half of $N$.

2. Count the frequencies in a cumulative manner from the bottom up to that interval, say the $s$th, the frequency of which if included would give more than, if not included less than, $N/2$ cases. Obviously the median will fall somewhere in this interval unless exactly half the values fall below the lower limit of an interval, in which case this lower limit is the median. Let $f_c$ equal the total frequency up to the $s$th interval, and let $f_s$ equal the frequency in the $s$th interval.

### Table 3.1. The calculation of the median

| Score | $f$ | |
|---|---|---|
| 310–319 | 1 | |
| 300–309 | 2 | |
| 290–299 | 4 | $N/2 = 25$ |
| 280–289 | 1 | $s$th interval is 260–269 |
| 270–279 | 6 | $f_c = 24 \quad f_s = 12$ |
| 260–269 | 12 | $i = 10$ |
| 250–259 | 11 | $LLs = 259.5$ |
| 240–249 | 8 | $\text{Mdn} = 259.5 + 10\,\dfrac{25 - 24}{12} = 260.33$ |
| 230–239 | 2 | |
| 220–229 | 0 | |
| 210–219 | 3 | |
| | 50 | |

3. $(N/2 - f_c)/f_s$ will be the proportional distance required in the $s$th interval to locate the median.

4. Letting $i$ equal the size of the interval and $LLs$ the lower limit of the $s$th interval, the median will be given by

$$\text{Mdn} = LLs + i\,\frac{N/2 - f_c}{f_s} \tag{3.1}$$

This involves the defensible assumption that the scores for the cases falling in the $s$th interval are distributed fairly evenly over the possible score values in the interval.

The calculation of the median is illustrated in Table 3.1, in which is given the distribution of scores made by 50 college men on the Brown spool packer. The score is the number of spools packed in four 1-minute trials.

The chief merits of the median are its ease of computation, its independence of extremes (it can be computed even if a known number of extremes have not been measured), and the fact that it is not affected by the size of extremes. This last point will be clearer after a discussion of the mean.

**The mean.** This arithmetic average will already be familiar to most readers. The *mean* is defined simply as the sum of all the scores or measures divided by their number or

$$M = \frac{\Sigma X}{N} \qquad (3.2)$$

where $X$ represents any score, the symbol $\Sigma$ means "the sum of," and $N$ is the total number of cases. When $N$ is small, this definition form can be used to compute the mean, but when $N$ is large, say 50, 100, or more, such a method is not economical of time. Ordinarily, when $N$ is large, we make a frequency distribution from which it is possible to compute the mean and median and other statistical measures. Assuming that the midpoint of an interval is typical of all the individuals in the interval, we can obtain the mean by summing the products of the several midpoints times their respective frequencies and dividing this sum by $N$. The error introduced by the use of midpoints is nonsystematic, i.e., tends to be ironed out so far as the computed mean is concerned.

The computation of the mean can be shortened further by use of an arbitrary origin and deviations therefrom. The reasonableness of such a procedure can be readily grasped by considering the problem of determining the mean height of a group of men. We could measure each man's height from the floor or as so much in excess of a stationary bar 5 feet from the floor. The sum of the excesses divided by $N$ will be the mean excess, and obviously we must add 5 feet to this to obtain the mean height of the group.

When we have a frequency distribution the arithmetic can be shortened still further by expressing the deviation from an arbitrary origin in terms of step intervals, that is, as the number of intervals that a given interval deviates from the arbitrary origin. The arbitrary origin is taken as the midpoint of any interval, and it is assumed that the midpoint of each interval may be taken as representing the scores in that interval.

The procedure can be developed by simple algebra. Let $AO$ be the arbitrary origin, $i$ be the interval size, and $d$ be the deviation in step intervals of the midpoint of any interval from $AO$. Then each score can be expressed as $X = AO + id$ in which $AO$ and $i$ are constant and $d$ varies. From the definition formula for the mean we have

$$M = \frac{\Sigma X}{N} = \frac{\Sigma(AO + id)}{N} = \frac{\Sigma(AO) + \Sigma id}{N}$$

Now $\Sigma(AO)$ will equal $N(AO)$ because summing a constant $N$ times is the same as multiplying it by $N$. As an exercise, the student should demonstrate, by taking varying numbers each multiplied by a constant, that

$\Sigma id = i\Sigma d$; a constant can be brought out from under the summation sign. Hence we have

$$M = \frac{N(AO)}{N} + \frac{i\Sigma d}{N} = AO + i\frac{\Sigma d}{N}$$

Since we started by summing $N$ $X$s and since each $X$ is associated with a $d$ value, we should be summing $N$ $d$s. That is, the $d$ value for a particular interval needs to be summed $f$ times ($f$ being the frequency for the interval), but the sum for a particular interval is simply $f$ times its $d$. If we replace

Table 3.2. Calculation of the mean

| Score | $f$ | $d$ | $fd$ | |
|---|---|---|---|---|
| 310–319 | 1 | 10 | 10 | |
| 300–309 | 2 | 9 | 18 | |
| 290–299 | 4 | 8 | 32 | |
| 280–289 | 1 | 7 | 7 | $\Sigma fd = 235$ |
| 270–279 | 6 | 6 | 36 | |
| 260–269 | 12 | 5 | 60 | $i\dfrac{\Sigma fd}{N} = 47.00$ |
| 250–259 | 11 | 4 | 44 | |
| 240–249 | 8 | 3 | 24 | |
| 230–239 | 2 | 2 | 4 | $M = 214.5 + 47.00 = 261.50$ |
| 220–229 | 0 | 1 | 0 | |
| 210–219 | 3 | 0 | 0 | |
| | 50 | | 235 | |

$\Sigma d$ by $\Sigma fd$ we explicitly indicate that each $d$ is to be summed as often as it occurs. Accordingly, our computational formula for the mean is written as

$$M = AO + i\frac{\Sigma fd}{N} \tag{3.3}$$

In our algebraic derivation of formula (3.3) the only restriction placed on $AO$ was that it be the midpoint of an interval; hence we are free to choose arbitrarily the midpoint of any interval as $AO$. In order to avoid negative $d$s, $AO$ is ordinarily taken as the midpoint of the lowest interval. Table 3.2 indicates the computation of the mean from grouped data by use of an arbitrary origin and deviations therefrom in terms of step intervals.

If we had taken $AO$ near the center of the distribution we would be following the so-called *guessed average* method, a method which has the advantage of smaller $d$ values but has the disadvantage of both negative and positive $d$s.

Parenthetically, it might be pointed out that the use of the arbitrary origin, step-interval scheme is analogous to using *coded* scores. If we regard $d$ as a coded value, we see from $X = AO + id$ that $d = (X - AO)/i$, or that in general we have a coded score $X_c = (X - K)/k$, with $K$ and $k$ so chosen as to give coded values ranging from zero to between 10 and 20. Then the mean of the original scores is given by $M = K + k$ times the mean of the coded scores.

The beginning student who is puzzled about which measure to use, the median or the mean, should remember that the purpose of measures of central value is description. When we attempt to reduce a mass of scores or a distribution of measures to a few descriptive constants, the mean and median are both descriptive terms which more or less adequately depict the "average" or typical score, and the choice between the two is frequently determined on the basis of which is more typical. Thus, if six men run 100 yards in 9.6, 9.7, 9.8, 9.9, 10.0, and 14.0 seconds, the mean value of 10.5 is not as typical as the median value of 9.85. In general, the mean is not as typical as the median when there are extreme measures in one direction. However, when the scores are distributed in an approximately symmetrical fashion, the mean and median will be equal or nearly so, and either will be as typical as the other. The mean in this case has two distinct advantages over the median. (1) It is usually a more stable measure in the sampling sense, i.e., if we regard our scores as based on a sample of $N$ individuals and then take another sample, the means of the two samples will in general show closer agreement than the two medians. This point will be discussed in more detail in the chapter on sampling errors. (2) It can be handled arithmetically and algebraically. The student should prove that, if the mean of $N_1$ cases is $M_1$, and of $N_2$ cases is $M_2$, the mean of the two groups combined will be given by

$$M_c = \frac{N_1 M_1 + N_2 M_2}{N_1 + N_2}$$

The median cannot be handled in such a fashion. Furthermore, the mean is used in connection with more advanced topics in statistics, whereas the median is seldom mentioned. Thus, unless the distribution is markedly skewed, the mean should be used. The problem of describing skewness will receive consideration after measures of variation have been discussed.

As exercises, the student should show algebraically or to his own satisfaction by numerical examples that: (1) if a constant is added to or subtracted from the scores of a group, the new mean will be $M + C$ or $M - C$, where $C$ is the given constant and $M$ the mean of the original scores; (2) if all the scores are multiplied by a constant, $C$, the new mean will be $CM$, whereas dividing by a constant will lead to $M/C$ as the new mean.

## MEASURES OF VARIATION

The description of the extent of scatter (or cluster) about the central value may be obtained by any one of several measures. These measures differ somewhat in interpretation and usefulness. One may doubt whether the *range* (highest to lowest score) is of sufficient value in psychological research to justify its use as a measure of variation. It is, obviously, determined by the location of just two individual measures or scores, and consequently tells us nothing about the general clustering of the scores about a central value.

**Quartile deviation.** An easily computed description of dispersion is the *quartile deviation* ($Q$), defined as $(Q_3 - Q_1)/2$, in which $Q_3$ (or the third quartile) is the point above which one-fourth of the cases fall and $Q_1$ (or the first quartile) is the point with three-fourths of the cases above. $Q_2$ (or the median) has already been defined as the point above which one-half of the cases fall. The computation of the two quartiles $Q_3$ and $Q_1$ from grouped data is essentially the same as that of the median. For instance, in determining the third quartile we count up to the interval in which the point falls which divides the number of cases into two parts: three-fourths below and one-fourth above. The distance into this interval is found in exactly the same manner as in computing the median. Since the quartiles are not influenced by extremes, it is customary to use them along with the median. By definition, 50 per cent of the cases fall between the first and third quartiles, but in nonsymmetrical distributions it is not likely that the limits indicated by the median plus and minus $Q$ will include 50 per cent. It would seem better to report both the first and third quartiles, instead of $Q$, since these values along with the median make it possible to picture whether or not the clustering above the median is different from that below the median.

**Percentiles.** Closely allied to the quartiles are the percentiles. The *P*th *percentile* is defined as a point below which $P$ per cent of the cases fall. Thus the median is the 50th, the third quartile the 75th, and the first quartile the 25th percentile. The 10th, 20th, $\cdots$ 90th percentiles are sometimes called deciles. The computation of the percentiles from grouped data is accomplished in the manner indicated for computing the quartiles. The location of the zeroth and 100th percentiles is always perplexing. Since these two points are dependent upon the location of just two scores (i.e., are greatly influenced by chance), they are difficult to interpret. Common sense would suggest that the concept of these two percentiles be dropped.

Percentiles may readily be associated with the cumulative frequency distribution, and with the ogive curve if cumulative percentage frequencies (obtained by dividing the $f$s by $N$) are used along the ordinate when plotting the ogive. In fact, the ogive may be used as a graphic scheme for determining score values corresponding to given percentiles. For instance, if we wish to obtain the 25th percentile point, we find 25 on the ordinate scale, proceed horizontally to the ogive curve, then vertically to the $x$ axis, and read off the score corresponding to the 25th percentile. Scrutiny of Fig. 2.4 will help the student understand the process. Could we also use the ogive as a basis for determining the percentile value of a given score?

The use of the difference between percentiles as an indication of dispersion should be obvious. In fact, the 10th–90th percentile range is a somewhat better (more stable from sample to sample) measure of dispersion than the quartile deviation. Percentiles, however, are chiefly of value in reporting the scores of individuals on psychological and educational tests. Ordinarily a raw score gives no inkling of what it means, whereas when it is said that an individual scores at or near the 85th percentile, the implication is that 15 per cent of his fellows score higher or better than he. Thus a percentile score carries with it some idea of the location of the individual with reference to the group. Furthermore, percentile scores for entirely different tests are comparable if derived from the same group or sample. The original raw scores might be different units, e.g., number of additions per minute and time to read a page of prose, and consequently not at all comparable.

**The average deviation.** Sometimes called the mean deviation or mean variation, the *average deviation* ($AD$) is defined as the average of the deviations of the several scores from the mean. Thus, if $x = X - M$, then $AD = \Sigma |x|/N$, where $|x|$ is the absolute value of $x$, i.e., the negative deviations are treated as though positive. Currently the average deviation is seldom used; the student, however, needs to know something about it if he reads the earlier research literature in psychology.

Contrasted with the quartile deviation, the average deviation gives weight to extremes, and for the usual bell-shaped distribution the limits $M$ plus and minus $AD$ will include about 57.5 per cent of the cases; the average deviation is larger than $Q$ but not so large as the standard deviation to which we now turn.

**The standard deviation.** A third measure of variation, the *standard deviation*, $S$, is defined as

$$S = \sqrt{\Sigma x^2/N} \qquad (3.4)$$

where $x = X - M$. To compute the standard deviation directly from this formula would be very cumbersome and uneconomical, since $x$ will usually involve decimals. A computational formula involving deviations from an arbitrary origin ($AO$) can be easily derived by algebra. Such a derivation is included here in order further to familiarize the student with the method of handling summation signs. The derivation will be carried through for $S^2$, technically known as the *variance;* then at the end we can take the square root to obtain $S$.

From formula (3.4) we have

$$S^2 = \frac{\Sigma x^2}{N}$$

in which $x = X - M$.

As in deriving formula (3.3), we can set

$$X = AO + id$$

and since $M = AO + i(\Sigma d/N)$, we have, substituting in $x = X - M$,

$$x = AO + id - \left(AO + i\frac{\Sigma d}{N}\right)$$

$$= id - ic$$

where for convenience we let $c$ stand for $\Sigma d/N$.

$$x^2 = (id - ic)^2 = i^2(d - c)^2$$

$$\Sigma x^2 = i^2\Sigma(d - c)^2$$

$$= i^2(\Sigma d^2 - 2c\Sigma d + Nc^2)$$

Dividing both sides by $N$, we have,

$$S^2 = \frac{\Sigma x^2}{N} = i^2\left(\frac{\Sigma d^2}{N} - 2c\frac{\Sigma d}{N} + N\frac{c^2}{N}\right)$$

$$= i^2\left[\frac{\Sigma d^2}{N} - 2\left(\frac{\Sigma d}{N}\right)^2 + \left(\frac{\Sigma d}{N}\right)^2\right]$$

$$= \frac{i^2}{N^2}[N\Sigma d^2 - (\Sigma d)^2]$$

hence

$$S = \frac{i}{N}\sqrt{N\Sigma d^2 - (\Sigma d)^2}$$

But since this form does not make explicit the fact that each $d$, and each $d^2$, must be summed as often as it occurs, we will insert $f$ for the frequency of occurrence. Thus our computational formula becomes

$$S = \frac{i}{N}\sqrt{N\Sigma fd^2 - (\Sigma fd)^2} \qquad (3.5)$$

where $\Sigma fd =$ the algebraic sum of deviations (in step intervals) from an arbitrary origin, and $\Sigma fd^2 =$ the sum of the squares of the deviations (in step units). The arbitrary origin may be taken as the midpoint of the lowest interval or as a guessed average near the center of the distribution. The advantage of the latter procedure is that the $d$s will be relatively small and consequently will not lead to the handling of large numbers, whereas the first procedure avoids the use of negative numbers and is more readily adaptable to machine computation.

The computation of $S$ for grouped scores is illustrated in Table 3.3, which is identical to Table 3.2 except that we now have an $fd^2$ column. It is

**Table 3.3. Computation of $S$ by use of an arbitrary origin**

| Score | $f$ | $d$ | $fd$ | $fd^2$ | |
|-------|-----|-----|------|--------|---|
| 310–319 | 1 | 10 | 10 | 100 | |
| 300–309 | 2 | 9 | 18 | 162 | |
| 290–299 | 4 | 8 | 32 | 256 | |
| 280–289 | 1 | 7 | 7 | 49 | By formula (3.5): |
| 270–279 | 6 | 6 | 36 | 216 | |
| 260–269 | 12 | 5 | 60 | 300 | |
| 250–259 | 11 | 4 | 44 | 176 | $S = \dfrac{10}{50} \sqrt{50(1339) - (235)^2}$ |
| 240–249 | 8 | 3 | 24 | 72 | |
| 230–239 | 2 | 2 | 4 | 8 | $= 21.66$ |
| 220–229 | 0 | 1 | 0 | 0 | |
| 210–219 | 3 | 0 | 0 | 0 | |
| | 50 | | 235 | 1339 | |

easily seen that the $fd^2$ values can be obtained by multiplying the $fd$ values by the corresponding $d$s. If we regard $d$ as a coded score ($= X_c$) with $i$ as the constant $k$, we see that (3.5) is appropriate for computing $S$ by way of coded scores.

The $fd$ and $fd^2$ columns need not appear on the work sheet when we are computing the mean and standard deviation by a Monroe or Marchant or Friden type calculating machine. The two required sums can be obtained by punching in the lowest $d$ in the right-hand part of the keyboard and the corresponding $d^2$ just left of the center of the keyboard, multiplying both simultaneously by the given frequency, and then, without clearing the lower dial, punching in the next larger $d$ and its square, and so on. The successive products so obtained will be accumulated by the machine so that $\Sigma fd$ is read directly from the right-hand side of the lower dial, and $\Sigma fd^2$ is read from near the center of the same dial. If either an 8- or 10-bank machine is used, the $d$s of 9 and less are punched in the right-hand column of the keyboard, and higher values will of course require the first two columns. The squares of the $d$s will ordinarily be less than 400, rarely greater than 961, so that their values can be punched in columns 6, 7, and 8. The student should note that the squares of 1, 2, and 3 are to be punched in column 6, the squares of 4 to 9 in columns 6 and 7, and the squares of 10 to 31 in columns 6, 7, and 8. The sum of the squares will appear in the lower dial from window 6 to the left. With a little practice the two required sums for a distribution of 15 intervals and 200 cases can be obtained in less than a minute. It should not be necessary to say that the computation should be done twice as a check.

For use with a calculator, formula (3.5) has an advantage over formulas which involve two divisions under the radical. Thus we place the sum of the squares in the right-hand side of the keyboard, multiply by $N$, and leaving the product in the lower dial, punch the sum of the $d$s in the keyboard and subtract it $\Sigma fd$ times, and then from the dial copy the value of $N\Sigma fd^2 - (\Sigma fd)^2$.

Briefly summarizing, it will be noted that (1) with a machine, $\Sigma fd$ and $\Sigma fd^2$ taken from an arbitrary origin at the bottom of the distribution are no more difficult to compute than when taken from a guessed average, (2) all sums are positive, and (3) the two sums necessary for determining both the mean and standard deviation can be obtained in the same operation. It is helpful to write the $d$ column in red on the work sheet, thereby throwing it into contrast with the $f$ column.

When $N$ is small and the scores are not too large, $S$ can be computed economically by way of the original (raw) scores. The definition formula, (3.4), calls for $\Sigma x^2$. Note that since each $x = X - M$, we have

$$\Sigma x^2 = \Sigma(X - M)^2 = \Sigma X^2 - 2M\Sigma X + \Sigma M^2$$

Replacing the last $\Sigma$ by $N$ (we are summing $M^2 N$ times) and replacing $M$ by $\Sigma X/N$, we have

$$\Sigma x^2 = \Sigma X^2 - 2\frac{\Sigma X}{N}\Sigma X + N\left(\frac{\Sigma X}{N}\right)^2$$

$$= \frac{N\Sigma X^2 - 2(\Sigma X)^2 + (\Sigma X)^2}{N}$$

$$\Sigma x^2 = \frac{1}{N}[N\Sigma X^2 - (\Sigma X)^2] \qquad (3.6)$$

Substituting in formula (3.4) leads to an $N^2$ in the denominator, which can be brought out as $1/N$. Hence we have

$$S = \frac{1}{N}\sqrt{N\Sigma X^2 - (\Sigma X)^2} \qquad (3.7)$$

All the scores are simply squared and then summed to get $\Sigma X^2$, and $\Sigma X$ has the same meaning as in formula (3.2).

Although a mean computed by formula (3.3) from grouped data will not err systematically from the value obtained by formula (3.2), the use of formula (3.5) for calculating $S$ tends to give a value which is too large when compared with the nonapproximate value yielded either by (3.4) or by (3.7). The reason for this is easily explained at the blackboard—we give here a hint. In general for an interval below the mean there will be more scores above than below the midpoint of the interval, whereas for an interval above the mean there will be more scores below than above the midpoint. Thus in taking the several midpoints as representing the scores within the several intervals, we are in effect using values which deviate too far from the mean.

We may correct for the systematic error involved in using formula (3.5) by substituting in

$$S_{cor} = \sqrt{S^2 - (i^2/12)} \tag{3.8}$$

The $i^2/12$ is known as Sheppard's correction for grouping. The uncorrected and corrected values differ but little when 12 or 15 intervals have been used, and as the number of intervals is increased, the difference becomes smaller and smaller. If less than 10 intervals have been used, the error may be appreciable and the correction should be applied. These considerations form the basis for the suggested rule that at least 10 or 12, and not more than 20, intervals be used.

Regarding the interpretation of the standard deviation, it can be said that, when we have the usual symmetrical bell-shaped distribution, about 68 per cent of the cases will fall between the limits plus and minus $1S$ from the mean, about 95 per cent between plus and minus $2S$, and nearly all the cases (99.73 per cent) between plus and minus $3S$. The standard deviation, even more than the average deviation, gives weight to extremes and therefore may not be as good as the quartiles for describing the dispersion. The standard deviation has decided advantages over other measures of dispersion. (1) Typically, it is more stable from the sampling point of view. (2) It can be handled algebraically, e.g., if we have two groups of $N_1$ and $N_2$ cases, with $M_1$ and $M_2$, and $S_1$ and $S_2$, as the respective means and standard deviations, we can obtain the standard deviation for two groups combined by

$$S_c = \sqrt{\frac{N_1(M^2_1 + S^2_1) + N_2(M^2_2 + S^2_2)}{N_1 + N_2} - M^2_c} \tag{3.9}$$

where the subscript $c$ refers to the combined group. The mean for the combined group can be obtained by a formula given on p. 18. Formula

(3.9) can be extended for determining the standard deviation for three or more groups combined. (3) The standard deviation is a mathematical term which has considerable importance in more advanced statistical work. It is usually involved in the determination of sampling errors and is the measure of variation used in the analysis of variation and in connection with correlational analysis. Therefore, unless there are definite reasons for *not* using it, the standard deviation, instead of the average deviation or $Q$, should be used as a description of the amount of dispersion.

As an exercise, show that, if a constant is added to or subtracted from each of a set of scores, the standard deviation does not change, and that multiplying or dividing each by a positive constant, will lead to $CS$ or $S/C$, respectively, as the new standard deviation, where $S$ holds for the original scores and $C$ is the constant.

## MEASURES OF SKEWNESS AND KURTOSIS

If a distribution is not of the symmetrical bell-shaped type, it is not sufficient for descriptive purposes to report only the mean and standard deviation. We also need a measure of the lack of symmetry, i.e., of *skewness*, and frequently it is desirable to describe the distribution still further by giving a measure which indicates whether the distribution is relatively peaked or flat-topped, i.e., a measure of *kurtosis*.

Skewness can be described roughly by a number of measures, such as the difference between the mean and median divided by the standard deviation, or in terms of quartiles or percentiles. If an adequate and stable description of skewness is desired and if a measure of kurtosis is also needed, a method based on moments is to be preferred.

The first four *moments* about the mean are defined as follows:

$$\left. \begin{aligned} u_1 &= \frac{\Sigma x}{N} = 0 \\ u_2 &= \frac{\Sigma x^2}{N} = S^2 \\ u_3 &= \frac{\Sigma x^3}{N} \\ u_4 &= \frac{\Sigma x^4}{N} \end{aligned} \right\} \tag{3.10}$$

where $x$ represents the deviation of each score from the mean of all the scores. For purposes of computation, we can determine the moments about an arbitrary origin, and then from these values we can obtain the moments about the mean. This procedure has already been employed in computing the standard deviation; i.e., we took deviations from an arbitrary origin. (The definition of the standard deviation was in terms of deviations from the mean.) If we use $v$ to represent moments about an arbitrary origin, the first four moments about $AO$ can be defined as follows, where $d$ is the score deviation from $AO$ in step units:

$$
\left.
\begin{aligned}
v_1 &= \frac{\Sigma fd}{N} \\[1em]
v_2 &= \frac{\Sigma fd^2}{N} \\[1em]
v_3 &= \frac{\Sigma fd^3}{N} \\[1em]
v_4 &= \frac{\Sigma fd^4}{N}
\end{aligned}
\right\}
\tag{3.11}
$$

When the $v$s have been calculated, the $u$s can be readily determined from the following relationships:

$$
\left.
\begin{aligned}
u_1 &= 0 \\
u_2 &= i^2(v_2 - v^2_1) = S^2 \\
u_3 &= i^3(v_3 - 3v_2 v_1 + 2v^3_1) \\
u_4 &= i^4(v_4 - 4v_3 v_1 + 6v_2 v^2_1 - 3v^4_1)
\end{aligned}
\right\}
\tag{3.12}
$$

The student should note the similarity of the formula in (3.12) for the second moment to that given for the standard deviation [formula (3.5)].

A *measure of skewness* defined in terms of moments is

$$
g_1 = \sqrt{\beta_1} = \frac{u_3}{u_2 \sqrt{u_2}}
\tag{3.13}
$$

For symmetrical distributions the value of $g_1$ will be zero; hence the departure of $g_1$ from zero can be taken as a measure of skewness. The deviation of $g_1$ from zero, however, must be considered in light of the operation of chance or in terms of sampling errors (to be discussed later). The skewness is said to be positive when $g_1$ is positive and negative when $g_1$ is negative.

The *degree of kurtosis* can be described by

$$
g_2 = (\beta_2 - 3) = \frac{u_4}{u^2_2} - 3
\tag{3.14}
$$

When $g_2$ is less than zero, the distribution tends to be flat-topped (platykurtic) whereas for $g_2$ greater than zero it is relatively peaked with somewhat higher tails (leptokurtic). When both $g_1$ and $g_2$ are zero or near zero, the distribution is of the usual symmetrical bell-shaped type, which is referred to as the "normal" frequency distribution.

Formulas (3.13) and (3.14) also define $\beta_1$ and $\beta_2$, which have been and are still used as measures of skewness and kurtosis. Recently, the $g$ measures have come into use because of certain advantages that need not be discussed here.

It will be noted that the measure of skewness involves taking the third moment relative to $S^3$ (since $u_2 = S^2$), and that the measure of kurtosis depends on the fourth moment relative to $S^4$. For a given distribution, all the values of $u_2$, $u_3$, and $u_4$ are in terms of the same measurement unit, say inches or pounds or IQs or minutes; hence the ratios in formulas (3.13) and (3.14) are pure numbers, i.e., are not inches or pounds or IQs or minutes. If we have the distribution of the weights and of the heights for 1000 individuals, the measure of skewness for the height distribution may be compared directly with that for the weight distribution. This is true by virtue of the fact that for each we are expressing the third moment relative to the amount of variability, both in inches for one distribution, both in pounds for the other. Likewise, it can be reasoned that the measures of kurtosis for different distributions are comparable, although the distributions involve different measurement units.

In order to help the reader visualize the meaning of different values for $g_1$ as associated with different degrees of asymmetry, Fig. 3.1 has been prepared.

When we have determined the mean and the second, third, and fourth moments, and from the moments have derived expressions which tell us the degree of dispersion, skewness, and kurtosis, we have a description adequate for most distributions. These measures can be used to determine the type of mathematical equation which will fit an observed frequency polygon; i.e., we can write the equation of a frequency curve which fits the observed frequency distribution. A distribution frequently found in psychological research is of the "normal" type, which is sufficiently described by the mean and standard deviation. Ordinarily it is not necessary to compute $g_1$ unless the distribution "appears" to be skewed or to compute $g_2$ unless the distribution seems peaked or flat. The nature of the research, the type of variable being studied, and also the size of the sample are factors which need to be considered in making a decision as to the necessity for computing measures of skewness and kurtosis. It is seldom advisable to compute these measures when $N$ is less than 100.

The student should be apprised of the fact that the rather frequent occurrence of symmetrical distributions for psychological variables may

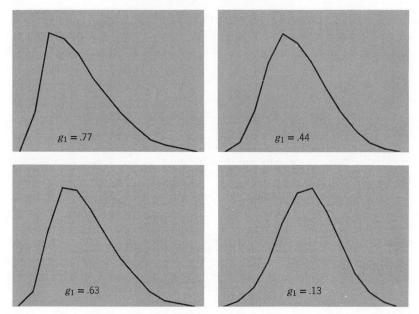

**Fig. 3.1.** Polygons with different degrees of skewness.

result from an artifact, and also that the occurrence of a skewed distribution may likewise be artifactual. This is true because very few of the instruments used in psychological "measurement" involve equal unit scales—the measuring units are frequently arbitrary or even accidental. Many of the variables are measured simply in terms of the number of items checked or the number of items correct. The shape of the resulting distributions is largely determined by the percentage checking the items or by the difficulty of the items. If the items are of medium difficulty for a group, it can be expected that the scale will yield a symmetrical distribution when applied to the group; if the items are easy, the scores will pile up toward the top (give negative skewness); if difficult, a piling up toward the bottom will occur. In the absence of equal scale units for the measuring devices, it cannot really be said whether the distribution of, for example, arithmetic ability for a given group is symmetrical or skewed—all that can be said is that in terms of the units used the distribution has a particular shape.

From the foregoing it would seem that, since skewness (and kurtosis too) is partly a function of the accidental nature of the measuring units, the descriptive measures of shape would have little value in psychology. The fact remains, however, that sometimes it is desirable to specify the skewness and kurtosis of a distribution of scores merely as a part of the description of what happens when a scale of measurement, however arbitrary the units, is applied to a given group. Furthermore, it is to the student's advantage to know something of measures of skewness and kurtosis, because we shall later have occasion to refer to them, and because he is apt to encounter them in more mathematical treatments of statistics.

## FURTHER CHARACTERIZATION OF THE MOMENTS

In formulas (3.10) it was stated without proof that $u = \Sigma x/N = 0$. Now $x = X - M$ is sometimes referred to as a *deviation score* or a *deviation unit*. Such scores, or units, simply use the mean as a reference point, or origin, and express each person's $X$ as deviating so many score points above or below the mean. Frequently in this book we will make use of the fact that the mean of the deviation scores is zero or that $\Sigma x = 0$, a fact that is easily proven:

$$\Sigma x = \Sigma(X - M) = \Sigma X - \Sigma M = NM - NM = 0$$

This is always true irrespective of distribution shape. The reader may recall the concept of moment of force as a weight operating through a specified distance. Consider $x$ as a distance and assume an equal weight for each deviation; then $\Sigma x = 0$ implies that a physical model of a frequency polygon will balance on a knife edge located at the mean. It is no accident that the quantities defined in formulas (3.10) are called moments: Karl Pearson, who originated these definitions, was a physicist before becoming interested in statistics.

The second moment, which involves the squares of the deviations, can be used to specify an important property of the mean: the sum of the squares of the deviations about the mean is smaller than the sum of squared deviations about any other value, or point. Suppose we take deviations from a point which is $D$ distance from the mean. Let

$$x' = X - (M + D) = (X - M) - D = x - D$$

Then

$$\Sigma(x')^2 = \Sigma(x - D)^2 = \Sigma x^2 - 2D\Sigma x + \Sigma D^2$$

Since the $2D\Sigma x$ term vanishes because $\Sigma x = 0$, we have

$$\Sigma(x')^2 = \Sigma x^2 + ND^2$$

which indicates clearly that the sum of squares of deviations about any point other than the mean is greater than the sum of the squares of deviations about the mean. In this sense the mean is regarded as a "least squares" statistic.

The student who did not immediately see that skewness will influence the third moment might consider two situations. First, suppose a perfectly symmetrical distribution; for each $+x$, a $-x$ of the same numerical value can always be found. The cubes, $x^3$ and $-x^3$, will also be of the same numerical value. This will be true for all such pairs, hence $\Sigma x^3 = 0$ for every symmetrical distribution. Second, suppose a distribution of 100 scores ranging from 6 to 19, with a mean of 9. The cube of the largest positive deviation, 10, is 1000. It would take 37 of the greatest negative deviations of $-3$ when cubed to balance off that 1000. If the next greatest positive deviation is 9, we would have to find 27 $-3$s to balance its cube. Thus we would have cubed values for $37 + 27 + 2$, or 66 scores with 64 of them piled up on an $X$ score of 6. Such a pile-up would certainly be indicative of skewness, but so far we have a sum of cubes very nearly equal to zero (actually, $+1$), which seems to say that the third moment can be zero. What of the remaining 34 scores? The sum of the deviations for the 66 scores is $-173$, which means that the sum of the deviations for the remaining 34 must be $+173$; i.e., the average of their deviations would need to be $+5.09$, which in turn means either that practically all of the 34 would be positive deviations or if some were negative, the rest would be large positive. The net result would be that the sum of the cubes of the 34 deviations would most certainly be a positive number. Hence the third moment would not be zero. In general, large deviations in *one* direction will lead to $\Sigma x^3$ not zero.

# Chapter 4
# DISTRIBUTION CURVES

By successive smoothing of a polygon (or distribution), we can iron out irregularities until the polygon becomes a "smooth" or regular and uniform curve. We can think of this curve as being similar or nearly identical to what we would obtain were we to increase indefinitely the size of our sample and at the same time use smaller and smaller grouping intervals. That is, the limit of a polygon, as we allow $N$ to approach infinity and the interval size to approach zero, is conceived to be a curve which is smooth and regular. Now such a uniform curve can usually be described in terms of a mathematical equation. The student may recall that the general equation for a straight line is $y = ax + b$, and that $y = 2x + 3$ is the equation for a particular line, that $x^2 + y^2 = a^2$ is the equation for a circle of radius $a$ with the origin or intersection of the abscissa and ordinate at the center, also that $y = a + bx + cx^2$ is the general equation for a parabola. It is not until we give specific numerical values to the constants that we have equations for particular curves.

Frequency curves can be thought of as representing the relationship between two variables: $h$, or the height of the curve, and $x$, the variate or variable under consideration. Frequency polygons or distributions, even when smoothed, may be of various shapes: symmetrical or skewed, flat-topped or steep, humped near the center or at one end, bimodal or unimodal, J-shaped or U-shaped, falling off gradually or suddenly, etc. A complete description of a frequency distribution is obtained when we have succeeded in writing the equation of the curve which "fits" the distribution. The type of curve to be fitted is chosen on the basis of certain criteria derived from the moments and the interrelations among the moments. The late Professor Karl Pearson developed the mathematics of a system of frequency curves and classified distributions according to several "types" of curves, but a complete exposition of these types is beyond the scope of this text.

**Normal curve.** A bell-shaped curve which is often approximated closely by frequency distributions and which is intimately involved in much of statistical inference is known as the *normal curve*. We need to know in detail the properties of this curve.

At this point we need to digress briefly to discuss a problem of notation. The mean and standard deviation have been defined in terms of an observed batch of scores for $N$ persons, presumably selected or drawn as a sample from some defined population of persons. The symbols, $M$ and $S$, stand for the sample mean and standard deviation. It is convenient to have symbols for the corresponding population values (parameters). Let us let $\mu$ (mu) stand for the population mean and $\sigma$ (sigma) symbolize the population standard deviation. Rarely will we have numerical values for $\mu$ and $\sigma$; $M$ and $S$ may be regarded as estimators of $\mu$ and $\sigma$.

The general equation for the normal distribution may be written as

$$h = \frac{N}{\sigma\sqrt{2\pi}} e^{-(X-\mu)^2/2\sigma^2} \tag{4.1}$$

for a population of $N$ scores or observations, or as

$$h = \frac{N}{S\sqrt{2\pi}} e^{-(X-M)^2/2S^2} \tag{4.2}$$

for a sample of $N$ scores having $g_1$ and $g_2$ values so near zero that one may regard the distribution as normal in form (within the limits of chance, or sampling, error—yet to be discussed). Equations (4.1) and (4.2) involve two well-known mathematical constants, $\pi$ (3.1416) and $e$ (2.7183). In each equation, $h$ is the height of the curve for any value of the variable $X$. In order to write the equation for a particular normal curve, i.e., one which corresponds to a particular distribution, we need $N$, $\mu$ or $M$, and $\sigma$ or $S$. This is the basis for saying that when we have the usual bell-shaped (normal) distribution, we need only the mean and standard deviation along with $N$ to describe it adequately. Referring again to equations (4.1) and (4.2), we note that the numerator part of the exponent could be written in terms of deviation units, i.e., with $x$ instead of $X - \mu$ in (4.1) or $X - M$ in (4.2). The $h$ for a positive deviation of, say, 10 will be exactly the same as that for a negative deviation of 10 for the simple reason that the deviation is squared. This indicates that the normal distribution is symmetrical about the mean, and therefore the mean and median coincide. When $x = 0$, i.e., an $X$ falls at the mean, $h$ has its maximal value, and therefore the mean and mode also coincide. For values of $x$ other than zero, the height of the curve will be less than that at the mean. This is evident if it is noted that the exponent of *e* is negative. As we go farther in either direction from the mean, the height of the curve becomes less and less (see Fig. 4.1). The dropping off is slow at first, then rapid, and then slow again. If we take the maximum value of *h* as unity, the ordinate at the point .5$\sigma$ above the mean is about .883; at 1$\sigma$, about .606; at 2$\sigma$, .135; and at 3$\sigma$, .011. As we go still farther from the mean, the value of *h* becomes smaller and smaller, and as *x* approaches infinity, *h* approaches zero (asymptotic). Theoretically, the curve never reaches the base line.

For both the frequency polygon and the histogram, the frequency for a given interval is represented along the *y* axis or ordinate, but for smoothed curves and for mathematical curves such as that defined by equation (4.1), it is advantageous to regard the area under the curve for a particular grouping interval on the *x* axis as indicating the frequency for the interval. Accordingly, the total area under the curve corresponds to the total frequency, or *N*, and the area under any given part of the curve, i.e., the area between any two *X* values, can be expressed as a percentage of the total. For example, the area included between the mean and the point on the base line 1$\sigma$ above the mean is 34.13 per cent of the total, and the area between plus and minus 1$\sigma$ is 68.26 per cent. The latter percentage has already been given on p. 24 as one way of interpreting the standard deviation. The limits plus and minus 2$\sigma$ will include 95.45 per cent; plus and minus 3$\sigma$, 99.73 per cent; and plus and minus 4$\sigma$, 99.9936 per cent.

The foregoing percentages hold for the theoretical curve, and will tend to be approximated in the distribution of a sample that tends to follow the normal distribution. Strictly speaking, no distribution of scores or observations in psychology can ever follow the normal curve insofar as the extremities of the distribution are concerned.

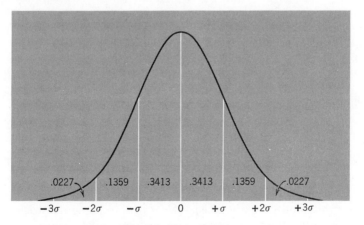

**Fig. 4.1.** Normal curve.

When we transform a set of scores into *relative deviates*, or so-called *standard scores* (*z* scores), by

$$z = \frac{X - \mu}{\sigma} \tag{4.3}$$

or by

$$z = \frac{X - M}{S} = \frac{x}{S} \tag{4.4}$$

we have each score expressed as a deviation from the mean in terms of fractions and/or multiples of the standard deviation of the distribution of the scores expressed as $X$s. Such a score transformation is ordinarily based on a sample, hence is accomplished by using (4.4); i.e., sample values are used because the parameters called for in (4.3) are unknowns.

The standard scores obtained by (4.4), or by (4.3) when possible, will have a mean of zero and a standard deviation of unity, as can be easily shown. Since the mean of any set of scores is their sum divided by their number, we have

$$M_z = \frac{\Sigma z}{N} = \frac{\Sigma(x/S)}{N} = \frac{1}{S} \frac{\Sigma x}{N}$$

It was shown (p. 29) that $\Sigma x = 0$, or that the mean of deviation scores must be zero. So we have $1/S$ times zero, or $M_z = 0$ always, a proposition that holds regardless of how the $X$s are distributed.

Since $M_z = 0$, each $z$ (or $x/S$) is a deviation from the mean of all the $z$ values. If these deviations are squared, summed, and divided by $N$ we have their variance, the square root of which gives their standard deviation. Thus,

$$S^2_z = \frac{\Sigma(x/S)^2}{N} = \frac{1}{S^2} \frac{\Sigma x^2}{N} = \frac{1}{S^2} S^2 = 1$$

The variance of standard scores is 1, hence the standard deviation is 1. The change to standard scores is a *linear* transformation because a manipulation of (4.4) leads to the (more recognizable) equation for a straight line. That is

$$z = \frac{X - M}{S} = \frac{1}{S} X - \frac{M}{S}$$

is the equation for a line representing the relation between $z$ and $X$; $1/S$ is the slope and $-M/S$ the intercept. Such a linear transformation will not, of course, alter the shape of the frequency polygon—the polygon for standard scores will have the same shape as that for the original, or $X$, scores. This transformation is equivalent to translating the origin along the $x$ axis to the point corresponding to the mean and changing the scale so as to make the standard deviation equal to unity. The student who is skeptical about this transformation business should be reminded that change of scale is commonplace. We change inches to feet, feet to miles; we change from the Fahrenheit to the centigrade temperature scale; etc.

The $N$ in the equation for the normal curve may be regarded as the total area under the curve. (It will be recalled that the area under a histogram or under a frequency polygon may be regarded as $N$.) It will be of considerable convenience to regard the total area under a normal curve as unity. With this and the concept of standard scores in mind, we may rewrite (4.1) or (4.2) as

$$h = \frac{1}{\sqrt{2\pi}}\, e^{-z^2/2} \qquad (4.5)$$

as the equation for the unit normal curve (unit area, unit standard deviation; and mean of 0). Note that this is a general equation in which $z$ as a *relative deviate* may be either a standard score or, as we shall see, the deviation of a value (not necessarily a score) taken relative to an appropriate standard deviation.

The value of $1/\sqrt{2\pi}$ is about .39894, and therefore at $z = 0$ (i.e., at the mean) $h$ will equal .39894, which is the maximum $h$ for the normal curve of unit area and unit standard deviation. The ordinates for other values of $z$ will be less. For instance, at $\pm 1z$, $h = .24197$, and at $\pm 2z$, $h = .05399$.

The percentage area under any part of the curve can be determined by methods of the calculus. The area under the curve between any two values, $z_1$ and $z_2$, is obtained as the value of the integral

$$A = \int_{z_1}^{z_2} h\, dz \qquad (4.6)$$

Perhaps this expression will be more meaningful to the student who has not studied integral calculus if the given area is regarded as composed of a large number of strips, each having a tiny base $dz$ and a height of $h$. For each such strip the area will be nearly $h\, dz$, and the integral sign in formula (4.6) simply means the "sum of" the areas of these tiny strips.

The student of the calculus will also note that the first derivative of either equation (4.1) or (4.5) set equal to zero and solved will yield a maximum for the curve when $x$ or $z$ equals zero, thus proving more rigorously that the mean and mode coincide. If the second derivative is set equal to zero and solved for $x$ or $z$, it will be found that the points of inflection of the curve are located where $x$ is $\pm\sigma$ or $z$ is $\pm 1$.

**Normal curve table.** Because of the widespread use of the normal curve, tables of proportionate frequencies and ordinates for various $z$ values are available. The student need not be able to integrate equation

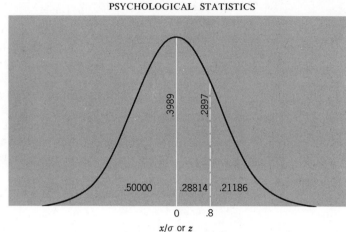

**Fig. 4.2.** Normal curve.

(4.6) in order to understand a table of the normal curve functions. Table A of the Appendix contains four columns, the first of which is $z$ values. The second column gives the area of the curve from the mean out to the corresponding $z$ value, this area being the same whether $z$ is positive or negative; a given $z$ divides the curve into two parts, and the third column gives the area of the smaller part. The area of the larger part can be obtained by adding .5 to the entries in column 2. If we wish to determine the proportionate area between plus and minus a given $z$, we should double the values in column 2. The fourth column gives the $h$ or ordinate for each of the $z$ values. For purposes of reference, the meanings of the several entries in Table A are illustrated in Fig. 4.2, in which an ordinate (dotted) has been erected at a $z$ value of $+.8$. The area from the mean to $+.8$ is found from column 2 as .28814; the area below this point is .78814, and that above is .21186, of the total area. Note that .78814 plus .21186 equals unity and that .78814 is .50000 plus .28814. The height of the curve at $z = .8$ is found from column 4 as .2897, whereas the maximum height of .3989 is at the mean.

It is frequently useful to know the relationship between the various measures of dispersion for a normal distribution. It can be shown that the following hold true:

$$Q = \quad .8453AD = \quad .6745S$$
$$AD = 1.1829Q \quad = \quad .7979S$$
$$S = 1.4826Q \quad = 1.2533AD$$

It is also useful to know that for an $N$ of 50 the $S$ will be about one-fifth the range, that for an $N$ of 200 the $S$ will be about one-sixth the range, and that for an $N$ of 1000 the $S$ will be about one-seventh the range.

The tabled values for the normal curve are often used in connection with problems similar to the following. If a distribution of the heights of men is normal with a mean of 68.0 inches and a standard deviation of 2.5, what percentage of men are more than 6 feet tall? We find $z$ as the difference between 72 and 68, divided by $S$, or $z = 1.6$; then from Table A we find the percentage of cases that fall above this $z$ value to be 5.48. Suppose that the mean IQ of 10-year-old boys is 100 and the standard deviation 16. What percentage have IQs between 90 and 110? What percentage of 10-year-old boys would be classified as "gifted" (IQ above 140)?

In practice, the answers to the foregoing questions would be approximate because $M$ and $S$ would be used in lieu of the population values, and because obtained distributions will not be exactly normal in form.

The student will have noted that the answers to problems similar to the foregoing are possible by virtue of the fact that the areas and ordinates of Table A are for the standard score form of the normal curve with total area set equal to unity. By formula (4.4), we can pass from raw scores to standard scores and vice versa, and knowing $N$, we can readily convert proportionate areas to frequencies or frequencies to proportions. Thus the table can be used with any normal distribution regardless of the original measurement units.

**Standard scores.** Perhaps it should be pointed out at this place that transforming scores, when distributions are normal or approximately so, to standard scores leads to new sets of scores which are comparable. For example, inches and pounds are not comparable units. If a man is 71 inches in height and weighs 170 pounds, it is impossible to say whether he is taller than he is heavy, but when the 71 inches is transformed to a $z$ of .9 and the 170 pounds to a $z$ of 1.3, we are able to say that, relative to his position in the two distributions, he is heavier than he is tall. Likewise, the raw scores on two psychological tests will seldom be comparable; changing to standard scores permits comparison, so that it can be decided whether a boy's performance on one test is better or worse than his performance on another. This assumes, of course, a close approximation to normality, and that the means and standard deviations used in the transformations are based on the same or highly similar groups.

Standard scores, as defined by formula (4.4), will involve both positive and negative values and decimal scores. Since these are awkward to use, a further transformation is frequently made in such a way as to yield a distribution with a preassigned mean and standard deviation, instead of the 0 and 1 that hold for the standard scores defined by formula (4.4). If we wish a distribution with a mean of 50 and a standard deviation of 10, we simply multiply each $z$ by 10 and add 50. Multiplying each $z$ by 20 and adding 100 would yield a mean of 100 and a $S$ of 20. Either of these transformations will get rid of negative values and permit a sufficient number of score values without the use of decimals. In general, if we wish to transform a set of scores having a mean, $M$, and a standard deviation, $S$, to new values to be called $Z$s, with mean equal to any value $K$ and $S$ equal to $S'$, all we need to do is to apply the relationship

$$Z = z(S') + K, \quad \text{or} \quad Z = \left(\frac{X - M}{S}\right)(S') + K$$

which becomes

$$Z = \frac{S'}{S}(X) - \frac{M}{S}(S') + K \tag{4.7}$$

The last form is the easier to use in practice, particularly with a calculating machine. Note that the last two terms will combine numerically and therefore can be placed in the lower dial as a positive or negative number; then the numerical value of $S'/S$ can be set in the keyboard as a constant to be multiplied in turn upon the varying values of $X$. If the machine has a continuous upper dial, the best procedure is to multiply by the highest $X$ first, and then, without clearing the dials, to subtract once for each successively lower value of $X$. Care is needed in aligning decimals, a check on which can be obtained by multiplying by the $X$ nearest $M$. This should lead to a value, in the lower dial, that is near $K$. With this setup, we can readily run off a table that gives the values of $Z$ for varying values of $X$.

The comparability of two sets of standard scores, either as $z$s or as $Z$s with the same mean ($K$) and same $S'$, does not hold for skewed distributions unless the two distributions show the same degree and direction of skewness. This is unlikely to be the case in practice. There is a scheme for use with skewed distributions which not only leads to comparable units but which also normalizes the distributions, i.e., changes the distributions from skewed to normal. This procedure is known as $T$ scaling, and the resulting scores are known as $T$ scores. They are usually so calculated as to yield a mean of 50 and a $S'$ of 10, but other values for these constants are possible. The detailed procedure may be found in McCall's *Measurement*,* which also includes a table for expediting the transformation. Suffice it to say here that $T$ scaling basically involves determining the proportion (or percentage) of cases exceeding a given value plus half those on that value, and then entering such proportions in a table of the

---

* McCall, W. A., *Measurement*, New York: Macmillan, 1939, pp. 505–508.

normal curve function to find the corresponding $z$ values. Standard scores based on a normal distribution of original scores and $T$ scores based on any shape distribution are comparable, provided they have been determined so as to yield the same mean and standard deviation. They differ only in the way in which they are computed, the standard score being a linear transformation which leaves the shape of the distribution unchanged, whereas $T$ scaling changes the distribution to the normal form. If we begin with an exactly normal distribution and convert the scores to both $z$s and $T$s, there will be a linear correspondence between the two sets of transformed scores. If their means and $S'$ values are set equal, the $Z$s and $T$s will be equal to each other.

It will be recalled that the use of percentiles is another way of expressing scores on different tests so as to have comparability. The student should give sufficient thought to percentiles and standard scores to see how they are interrelated when the original scores are normal in distribution. Hint: The tabled functions (Table A) of the normal curve may help. The student might also demonstrate to his own satisfaction that the difference between the 50th and 60th percentile points is *not* apt to be equal to the difference between the 80th and 90th percentile points.

**Kinds of distributions.** In anticipation of topics to be discussed, it might be well to mention some possible ways of regarding frequency distributions. We can have an *observed*, or *sample*, distribution of scores for a group of $N$ individuals; we can imagine a *population* distribution of scores for either a finite or for an infinite $N$; and we can conceive of a distribution curve defined by a *mathematical* equation (or function). Because of chance factors (as yet undefined herein) we do not expect an observed sample distribution to be exactly like the distribution of the population from which the sample is drawn or like a defined mathematical distribution.

Since we are seldom able to measure all members of a population, we can only assume that population scores follow some defined mathematical distribution. The form of mathematical curve assumed is usually decided upon by a consideration of the shape of an observed sample distribution. As will be seen later, the reasonableness of the assumption can be checked statistically.

It is possible, however, to show mathematically that under prescribed conditions given measures will follow a defined distribution curve exactly. We shall refer to such a distribution as *theoretical* or *expected*. Strictly speaking a mathematical distribution curve holds only for a continuous variable. If we had the distribution for a discrete variable, such as number of children per family, we would never expect that increasing $N$ would produce a curve—the variable takes on only *point* values 0, 1, 2, etc.;

hence we cannot allow the interval size (see p. 31) to approach zero, which is necessary for a smooth curve.

As implied previously, there are distribution curves which are not normal. We shall introduce other curves (or functions) when needed. Thus far, the normal curve has been discussed as a *frequency* curve, and the area interpretation has been in terms of the number of individuals or percentage of cases falling between certain score limits. This same curve is often spoken of as the normal *probability* curve, and as such it is regarded as a theoretical curve. We shall see, moreover, that there are theoretical curves other than the normal curve which may be regarded as probability curves.

# Chapter 5

# PROBABILITY AND HYPOTHESIS TESTING

Statistical inference and the testing of hypotheses involve the concept of chance, or probability. A simple example will serve to illustrate the probabilistic nature of hypothesis testing. Suppose a chap claims that he can distinguish between Camels and Lucky Strikes. To test his claim we could blindfold him and present him with either a Camel or a Lucky Strike (the brand to be presented is determined by tossing a coin). If on this one trial he correctly names the brand, we would not be inclined to accept his claim since he would have a 50–50 chance of being correct on a sheer guessing basis. So we give him a second trial (again, and for any subsequent trials, we toss a coin to determine which brand to present to him). If he were again successful we might give some credence to his claim but someone might ask whether making two correct discriminations could happen on the basis of chance. We shall presently see that the chances are 1 in 4 of getting two correct, i.e., success on two trials could easily occur on the basis of chance.

But suppose he is correct on three trials, then on the fourth trial, and also on the fifth; or perhaps he is correct on ten trials, or perhaps on 9 of 10 trials? Regardless of the number of trials and the number of successes we certainly should have some information about chance success, or the probability of correctly naming the brands on the basis of chance guessing, before we reach a decision regarding the claimed ability to distinguish between the two brands of cigarettes. This and similar decision problems involve notions of probability, to which we now turn.

**Probability.** If we had a box containing 70 white and 30 black balls, well mixed, and were to draw 1 ball at random, the chance of the drawn ball's being black is said to be 30 out of 100, and the chance of its being white would be .70. This can be interpreted to mean that, if we made 1000 random draws, each time replacing the drawn ball and remixing the contents of the box, the percentage of black balls drawn would be about 30, and of white draws about 70. If we roll a die, the probability of obtaining a 4 is $\frac{1}{6}$; i.e., a large number of rolls would yield a 4 about $\frac{1}{6}$ of the time. If one tosses a symmetrical coin, it is usually said that there is a 50–50 chance of its landing "heads up," or the probability of a head is $\frac{1}{2}$. This is another way of saying that in the long run the proportion of times that the coin lands as a head will be the same as the proportion of times it lands as a tail.

These very simple examples illustrate a *definition of probability:* if an event can happen in $A$ ways and fail in $B$ ways, all possible ways being equally likely, the probability of its occurring is $A/(A + B)$ and of its failing is $B/(A + B)$. That is, a probability figure is the ratio of the number of favorable events to the total number of events, and it is therefore necessary that we be able to enumerate events in order to arrive at a probability figure.

If we draw a card from a pack, the probability of obtaining a spade is $\frac{1}{4}$, and the probability of drawing a club is also $\frac{1}{4}$, but the probability of drawing *either* a spade *or* a club is $\frac{1}{4}$ plus $\frac{1}{4}$, or $\frac{1}{2}$. If we roll a die, the probability of obtaining *either* a 4 *or* a 5 is $\frac{1}{6}$ plus $\frac{1}{6}$, or $\frac{1}{3}$. These two situations illustrate the *addition theorem* of probability: the probability that *either* one event *or* another event will happen is the sum of the probabilities of their occurrences as single events. (The events must be mutually exclusive; i.e., if one occurs, the other cannot.)

If we roll a pair of dice, the probability of a 2 on the first *and* a 5 on the second is $\frac{1}{6}$ times $\frac{1}{6}$, or $\frac{1}{36}$. If we toss 2 coins, the probability that the first will land a head *and* the second a head is $\frac{1}{2}$ times $\frac{1}{2}$, or $\frac{1}{4}$, which is, of course, the probability that both will land as heads. Notice that the result obtained with the second die or coin is independent of the outcome of the first die or coin. These two examples illustrate the *multiplication theorem:* the probability of two (or more) independent events' occurring simultaneously or in succession (one *and* the other) is the product of their separate probabilities.

As just indicated, if we toss 2 coins, the probability that the first will land a head *and* also the second a head will be $\frac{1}{2}$ times $\frac{1}{2}$, or $\frac{1}{4}$, which is the probability that both will fall as heads. The probability that the first will land a head and the second a tail will also be $\frac{1}{2}$ times $\frac{1}{2}$, or $\frac{1}{4}$. But 1 head and 1 tail can be obtained in a manner mutually exclusive to the above; i.e., the first can land as a tail and the second as a head, and this combination or event has a probability of $\frac{1}{4}$, whence the probability of obtaining 1 head and 1 tail will be $\frac{1}{4}$ plus $\frac{1}{4}$, or $\frac{1}{2}$. This same result can be arrived at by listing all the possible combinations and taking the ratio of the number of favorable to the total number of possible combinations. The possible combinations are *HH, HT, TH, TT*, from which we see that 2 out of the 4 possible events are favorable for the occurrence of 1 head and 1 tail. We also note that 1 out of 4 is favorable to 2 heads.

Suppose we were to toss 3 coins; we would have the following possible combinations:

| Coin 1 | H | H | H | H | T | T | T | T |
|--------|---|---|---|---|---|---|---|---|
| Coin 2 | H | H | T | T | H | H | T | T |
| Coin 3 | H | T | H | T | H | T | H | T |

The total number of possible "events" is 8, 1 of which is favorable to 3 heads, 3 to 2 heads, 3 to 1 head, and 1 to no heads, thus giving the respective probabilities of $\frac{1}{8}$, $\frac{3}{8}$, $\frac{3}{8}$, and $\frac{1}{8}$. If we were to toss 4 coins, we would have the following probabilities:

| 4 heads | $\frac{1}{16}$ | 1 head | $\frac{4}{16}$ |
|---------|---------------|--------|---------------|
| 3 heads | $\frac{4}{16}$ | 0 head | $\frac{1}{16}$ |
| 2 heads | $\frac{6}{16}$ | | |

The student should satisfy himself that these are the correct figures by writing down all the combinations possible and counting those favorable to any particular number of heads.

## BINOMIAL DISTRIBUTION

The process of determining possible combinations becomes quite laborious for, say, 10 coins, but the several probabilities can be obtained by the coefficients in the expansion of the binomial $(a + b)^n$. Thus for $n = 2$ (i.e., 2 coins) we have $a^2 + 2ab + b^2$, or 1, 2, 1; for $n = 3$, $a^3 + 3a^2b + 3ab^2 + b^3$, or 1, 3, 3, 1; for $n = 4$ the coefficients are 1, 4, 6, 4, 1. In each case the sum of the coefficients, $2^n$, will be the total possible combinations, and the coefficients taken as ratios with the common denominator, $2^n$, will represent the probabilities for $n$, $n - 1$, $n - 2, \cdots, 0$ heads.

The student may recall that the general expansion of $(a + b)^n$ is

$$a^n + na^{n-1}b + \frac{n(n - 1)}{1 \times 2} a^{n-2}b^2 + \frac{n(n - 1)(n - 2)}{1 \times 2 \times 3} a^{n-3}b^3 + \cdots$$

This expansion will contain $(n + 1)$ terms and will terminate in $b^n$. For $n = 10$, we have the following coefficients: 1, 10, 45, 120, 210, 252, 210,

120, 45, 10, 1, which sum to 1024, or 2 to the tenth power. Thus the probability that all 10 coins will fall as heads is 1/1024; 9 heads, 10/1024; etc. If we plot these values as a frequency polygon—these coefficients are frequencies in the sense that they represent the expected number of times for 10 heads, 9 heads, etc., out of a total of 1024 tosses—we will have a bell-shaped graph which will resemble somewhat the normal curve.

Another and more useful way, for our purpose, of considering the binomial expansion is to use $p$ and $q$, in the place of $a$ and $b$, with $p$ defined as the probability of success on a single element and $q$ as the probability of failure, or $q = 1 - p$. Thus we would have $(p + q)^n$. Suppose $n = 2$; the expression would be $p^2 + 2pq + q^2$. If $p = \frac{1}{2}$, as in the coin situation, this would give $(\frac{1}{2})^2 + 2(\frac{1}{2})(\frac{1}{2}) + (\frac{1}{2})^2$, or $\frac{1}{4}, \frac{2}{4}$, and $\frac{1}{4}$ as the probabilities for securing 2 heads, 1 head, and 0 head respectively. Each term is itself a probability fraction; the numerators are 1, 2, and 1 as before. For $n = 10$, we would have $(\frac{1}{2})^{10}$ or 1/1024, $10(\frac{1}{2})^9(\frac{1}{2})$ or 10/1024, $45(\frac{1}{2})^8(\frac{1}{2})^2$ or 45/1024, etc., as the probabilities for obtaining 10 heads, 9 heads, 8 heads, etc.

The chief advantage of using the $p$ and $q$ notation is that we can readily see what happens when $p$ is not equal to $\frac{1}{2}$. Consider the expectation when we roll a pair of dice with "success" defined as the rolling of "snake eyes." We would have $(p + q)^2 = (\frac{1}{6} + \frac{5}{6})^2 = \frac{1}{36} + 2(\frac{5}{36}) + \frac{25}{36}$ as indicating the probability of obtaining 2 one-spots, 1 one-spot, and 0 one-spot. If 3 dice were rolled, we would have $\frac{1}{216} + 3(\frac{5}{216}) + 3(\frac{25}{216}) + \frac{125}{216}$ or $\frac{1}{216}, \frac{15}{216}, \frac{75}{216}$, and $\frac{125}{216}$ as the respective probabilities for 3, 2, 1, and 0 one-spots. The important thing for the student to note is that these probabilities are definitely skewed—not all probability distributions are of the symmetrical type. The student can, as a tedious exercise, work out the probabilities for 4, 5, 6, 7, and 8 dice, and therefrom learn that the shape of the distribution changes from marked skewness to less and less skewness as the number of dice is increased. It can be easily shown that, if $p = \frac{5}{6}$ and $q = \frac{1}{6}$, the skewness will be in the opposite direction. Another proposition which the student can demonstrate to himself is that, for a fixed $n$, the skewness increases as $p$ is taken farther from $\frac{1}{2}$ in either direction— extremely small or extremely large $p$s (near unity) lead to very marked skewness.

The binomial expansion provides the probabilities of the theoretically expected frequencies for given $n$s, $p$s, and $q$s. Such theoretical distributions can be described as to central value, variation, skewness, and kurtosis. The numerical values for these measures may be obtained by direct computation from the distributions built up by the binomial expansion, or these measures may be obtained by simple formulas, which can be derived by simple algebra, without having the actual distributions available. The formulas are:

$$\mu = np$$

$$\sigma = \sqrt{npq}$$

$$\gamma_1 = \frac{q-p}{\sqrt{npq}} \quad \text{(skewness)}$$

$$\gamma_2 = \frac{1-6pq}{npq} \quad \text{(kurtosis)}$$

Since these formulas, which are for theoretical distributions, specify parameters (not values based on a sample), Greek instead of Latin letters are used as symbols.

It should be noted that $n$ is the number of elements, not the number of cases. The formula for skewness permits several deductions. When $p = \frac{1}{2}$, $q$ also equals $\frac{1}{2}$, and hence the skewness is zero; the degree of skewness for a fixed $n$ depends upon the deviation of $p$ from $\frac{1}{2}$, i.e., the smaller or the larger the probability of success for each element, the more skewed the distribution. Note also that, since $n$ is in the denominator, the larger the number ($n$) of elements, the smaller the skewness for fixed values of $p$ and $q$.

The above formulas describe the theoretically expected distribution for given $n$s, $p$s, and $q$s. As will be seen later, any empirical distribution obtained by tossing 10 coins or rolling 3 dice will yield values which, for reasons to be discussed, will only approximate these values.

It is of interest to consider plotting the binomial distribution as a histogram—the height of the successive bars will indicate the several expected frequencies, each of which is the numerator for a probability fraction. Now, if we work out the expected frequencies for number of heads when 20 coins are tossed, and if in drawing the histogram we scale the ordinate so as to have the over-all height about the same as that for the 10-coin situation and also squeeze the base-line scale (ranging from 0 to 20) into about the same over-all distance as for 10 coins, the vertical bars will be narrower, and the resulting picture will look more like a normal histogram than that obtained for 10 coins. If we repeat the process with $n$ larger and larger, each time scaling our axes to about the same size as used for 10 coins and for 20 coins, the several bars of the histograms will become narrower and narrower, and with $n$ sufficiently large the bars will seem to merge and the contour of the graph will tend to appear indistinguishable from a normal curve.

The normal curve is for a continuous variable on the $x$ axis, whereas the binomial distribution involves a discrete variable, or point series. For example, it is impossible to have any values between, say, 22 and 23 heads. As *n* is taken larger and larger, and the total base line is kept fixed, the obtained values or possible points become more and more closely spaced so that the point series approaches, or at least takes on the appearance of, continuity. As *n* approaches infinity, the binomial distribution approaches the normal distribution as a limit.

**Approximation of probabilities.** The foregoing suggests the possibility of using the normal curve as a basis for approximating the probabilities obtainable by the binomial expansion. In order to see how this might be done we shall consider the binomial distribution for $n = 16$ for the coin tossing situation, as shown in Table 5.1. Suppose we wish to ascertain the

**Table 5.1. Binomial distribution for 16 coins**

| Number of Heads | Expected Frequencies | Number of Heads | Expected Frequencies |
|---|---|---|---|
| 16 | 1 | 7 | 11,440 |
| 15 | 16 | 6 | 8,008 |
| 14 | 120 | 5 | 4,368 |
| 13 | 560 | 4 | 1,820 |
| 12 | 1,820 | 3 | 560 |
| 11 | 4,368 | 2 | 120 |
| 10 | 8,008 | 1 | 16 |
| 9 | 11,440 | 0 | 1 |
| 8 | 12,870 | | |
| | | | 65,536 |

probability of getting at least 12 heads. This would be the sum of the separate probabilities of tossing 12, 13, 14, 15, and 16 heads. These probabilities would be the respective "expected frequencies" each divided by 65,536; hence the sum of the probabilities would be obtained by summing the numerators: 1, 16, 120, 560, and 1820, then dividing this sum, 2517, by 65,536. Thus the probability of securing at least 12 heads (12 or more) would be 2517/65,536, or a decimal equivalent of .03841 (to 5 places).

Now let us attempt to find the same probability by using the normal curve approximation. First we note that for the distribution in Table 5.1 the mean will be $np = 16(.5)$ and the $\sigma$ will be $\sqrt{npq} = \sqrt{16(.5)(.5)} = 2$. It will help us understand the method of approximation if we superimpose on the histogram of the frequencies in Table 5.1 a normal curve having a mean of 8 and a $\sigma$ of 2 (see Fig. 5.1). If we regard the area of each bar as

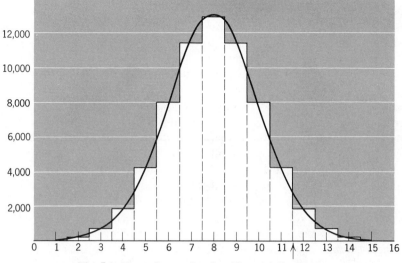

**Fig. 5.1.** Normal curve fitted to binomial distribution.

representing an expected frequency, we see that the sum of the areas for the bars based on 12, 13, 14, 15, and 16 heads divided by the total area of all the bars (= 65,536) will give the probability value of .03841 reported previously. To approximate this by the normal curve we need to consider the area under the curve for that part of the curve which spans the bars with base-line values of 12, 13, 14, 15, and 16. Obviously we need the area under the curve beyond an $X$ value of 11.5, a value which does not make much sense in terms of number of heads but which does make sense when it is recalled that we are here treating a point (discrete) variable as though it were a continuous variable, normally distributed. Hence we have $X - \mu = 11.5 - 8 = 3.5 = x$, and $x/\sigma = 3.5/2 = 1.75$. Turning to Table A we find that the proportionate area under a normal curve beyond a $z$ of 1.75 is .04006. This is our approximation to the exact probability value of .03841; the error in this approximation is of the order of .002. In general, when $n$ is fairly large the failure to shift .5 (e.g., from 12 to 11.5 as done here) leads to a negligible error. This shift of .5 is referred to as *correction for continuity*.

We can, of course, use the normal curve to approximate any of the exact probabilities obtainable from Table 5.1 (or from the binomial with $n$ other than 16). For example, the exact probability of obtaining 10 or 11 or 12 heads is $(8008 + 4368 + 1820)/65,536$, or .21661. The normal curve approximation, calculated as the proportionate area under the curve from 9.5 to 12.5, is .21441.

It is fortunate for us that for $n$ larger and larger the normal curve approximation becomes better and better since for $n$ large the computation of exact probabilities by the binomial method becomes very arduous.

Notice that, in approximating the probability, we have utilized an *area under a curve*; i.e., we have said that the area between two $X$ values taken relative to a total area may be interpreted as a probability figure. This is not inconsistent with our original definition of probability involving number (frequency) of events favorable relative to a total number of events (total frequency). Since, as previously indicated, the total area under a frequency curve for a continuous variable (or function) can be regarded as the total frequency, and the area for a particular segment can be regarded as the frequency with which values (or scores) fall in the given segment, it follows that the ratio of the segmental to the total frequency may be spoken of as a probability—the probability that a score falls between the two $X$ values defining the segment. When we are dealing with a distribution of the normal type, the probability associated with a given segment is found by converting the two $X$ values, which define an interval, into $z$ values and then determining the area from Table A. The obtained proportionate area represents the probability expressed as a decimal fraction.

It should be obvious, when we consider the unit normal curve, that we can readily specify the proportionate area between any two $z$ values, say $z_1$ and $z_2$, and interpret the proportion as the probability of obtaining $z$ values between the given $z_1$ and $z_2$. By reference to tables more extensive than Table A, it can be found that the area between a $z$ of $-1.96$ and a $z$ of $+1.96$ is very nearly .95; hence it would be said that .95 represents the probability of obtaining $z$ values between these two points. Furthermore, it can be said that .05 represents the probability that a $z$, drawn at random from a normally distributed supply of $z$s, will be numerically larger than 1.96. Similarly it can be said that the probability of drawing a $z$ between $\pm 2.576$ is very nearly .99, whereas the probability for a $z$ falling outside these limits is .01.

The foregoing interpretation of proportionate areas under the normal curve as probabilities is, in a sense, the basis for sometimes calling this curve the *normal probability curve*. It has been noted that for $p$ not equal to $q$, the point binomial leads to skewed probability distributions. For continuous functions it is also possible to have distributions, other than the normal, which permit probability statements on the basis of proportionate areas. Later we shall consider the use of three nonnormal probability distributions.

## HYPOTHESIS TESTING

We may now return to a consideration of the blindfold test of the claimed ability to distinguish between two cigarette brands. By using the binomial expansion we can readily specify the probability of being correct (by chance) $n$ times out of $n$ trials. The answer is simply $1/2^n$; if there were 10 trials the probability of 10 correct choices (by chance guessing—no real discriminatory ability) would be $1/1024$, or about .001; the probability of being correct 16 out of 16 trials would be $1/65,536$, or about .000015. If our self-proclaimed expert did succeed in 10 of 10 trials we would, because of the small probability of 10 successes by chance, concede that he really possessed the ability to discriminate between the two brands.

But suppose he was successful on 9 trials of a 10-trial series? We could readily specify the probability of 9 successes by chance (it would be $10/1024$) but for reasons which will become apparent later, it is better to ascertain the probability of as many as 9 successes in 10 trials (at least 9, or 9 or more, successes). This probability will be the probability of exactly 9 successes plus the probability of exactly 10 successes, or $10/1024 + 1/1024 = 11/1024 =$ about .01, which is sufficiently small that we might decide that his performance was based on ability rather than on chance. Note that such a record would occur by chance about 1 time in 100, so we couldn't be sure that he really had the ability.

Next, let us suppose that he was correct on 8 of the 10 trials. The probability of at least 8 successes occurring on a chance basis would be $45/1024 + 10/1024 + 1/1024 = 56/1024 =$ about .05. Would we now conclude that he had the claimed ability? If we did so conclude we wouldn't be as sure of our inference as when there were 9 successes, and far less sure than when there were 10 successes. In other words, the smaller the probability of attaining an obtained number of successes by chance the surer we would be of our conclusion. If he were successful on 7 trials (probability $= P = .17$ for 7 or more successes) we would no doubt hesitate before conceding that his performance was based on ability to discriminate, since 7 successes can too easily occur on the basis of chance alone.

We are thus led to the question: What level of probability should be adopted as a criterion for deciding whether an observed performance is based on ability rather than chance? We are not yet ready to attempt an answer to this, but it might be remarked here that in choosing a level of probability it is necessary to consider the risk of being wrong in concluding that the fellow can discriminate vs. the risk of attributing his performance to chance when in reality he does have some ability.

Whether a person can discriminate between two brands of cigarettes is a simple illustration of the problem of statistical inference, or the testing of hypotheses. For purpose of inference we set the hypothesis that our friend cannot discriminate between brands. This readily permits us to calculate the probability ($P$) of as many successes by chance as he attains on a series of trials; if $P$ is sufficiently small we reject the hypothesis of no ability, and in so doing we are saying that his number of successes is *statistically significant*, that is, nonchance. The *level of significance* associated with rejection of the hypothesis is represented by a probability—if we agree to reject the hypothesis only when the probability of chance success is as low as .01, we will have adopted the $P = .01$ level of significance. If we are willing to be less sure and require $P$ to be as low as .05 we will be working at the .05 level of significance. Whether we adopt the .01 or the .05 level is somewhat arbitrary—for this chapter let us quite arbitrarily choose $P = .01$ as our working level of significance. After considering the more detailed discussion of this issue later in the chapter, the reader may prefer to adopt the .05 or some other level for judging significance.

The binomial expansion (and normal curve approximation thereto) may be used in a wide variety of situations as a means of testing hypotheses. A general requisite is that we be able to specify the probability of success (or something analogous to success) for a single element (coin, die, trial, etc.). In other words, we need to specify $p$ (and $q$) so as to use $(p + q)^n$ or we need to calculate the mean and $\sigma$ in order to utilize the normal curve approximation when $n$ is not small.

Consider the problem of public opinion polling. In polling studies we are usually interested in whether or not a population of potential voters is split 50–50 on an issue. Accordingly we set the hypothesis that there is a 50–50 split in the population. This hypothesis is to be accepted or rejected on the basis of information yielded by a sample of $N$ persons, who are asked to respond "yes" (agree) or "no" (disagree) to a statement of the given issue. Suppose for sake of simplicity we take $N = 64$ and that 42 of them give a yes response. Is this result consistent with the hypothesis of a 50–50 split?

To answer this we note that so far as the opinion poller is concerned there is, by hypothesis, a 50–50 chance that any individual in the sample will say yes (this despite the fact that the individual so far as *he* is concerned is not giving a chance response). Thus the probability of a yes response for a single individual is $1/2$; that is, $p = .5$ and $q = .5$ (since $q$ is always $1 - p$). Now our sample of 64 is analogous to a trial toss of 64 coins, so we consider the binomial distribution with $n = N = 64$. The mean $= Np = 32$, and the $\sigma = \sqrt{Npq} = 4$. The number of yes responses, 42, deviates 10 from the mean. (Our normal curve approximation would be slightly better if we used $41.5 - 32 = 9.5$ as our deviate—correction for continuity.) Thus we have $z = 10/4 = 2.50$. Turning to Table A we find that the probability of obtaining this large a deviation in an *upward* direction from 32 is about .006, but in testing our hypothesis of 50–50 split we need also to include the probability of obtaining as large a deviation in the opposite direction; hence we double .006 and have $P = .012$ as the probability of as large a deviation irrespective of direction. Since this $P$ is very near our arbitrarily (and temporarily) agreed upon $P = .01$ level for judging significance, we reject the hypothesis of an equal split in the population being sampled, and this rejection implies that a majority of the population would endorse the given statement.

In passing, it should be noted that had the number of yes responses been, say, 35 we would accept the hypothesis of an equal split. But this acceptance would not prove the hypothesis since 35 could easily be a chance deviation from any of a number of splits, 55–45, 54–46, etc. We will have more to say about this later.

Opinion poll results are usually expressed in percentage form, that is, as proportions multiplied by 100. Thus the hypothesis of a 50–50 split in the population implies .50 or 50 per cent yeses, and our result of 42 yeses for a sample of 64 leads to .656 or 65.6 per cent yeses. Accordingly it would appear that in testing the significance of the deviation of 42 from 32 we are also testing the deviation of .656 from .50 (in proportion units) or 65.6 from 50 (in percentage units).

Actually, what we did above was to take

$$z = \frac{x}{\sigma} = \frac{42 - 32}{\sqrt{64(.5)(.5)}} = 2.50$$

In converting to proportions we divided both numerator terms by 64 (or $N$), and if we also divide the denominator by 64 (or $N$) we will not change the value of $x/\sigma$. Thus we have

$$\frac{x}{\sigma} = \frac{42/64 - 32/64}{\sqrt{64(.5)(.5)/(64)^2}} = \frac{.656 - .50}{.062} = 2.52$$

which differs from 2.50 only because of rounding errors. This implies that dividing by $N$ somehow preserves the $x/\sigma$ nature of the result. The numerator, or $x$, is a deviation, the deviation of an observed sample proportion from a hypothetical proportion. We might, therefore, deduce that the denominator is a $\sigma$, but $\sigma$ of what?

Let $f$ = number of yeses (frequency of yeses); $f$ can vary from zero to $N$, with $\mu_f = Np$ and $\sigma_f = \sqrt{Npq}$. If we divide every possible $f$ by $N$ we have proportions. The mean in proportion units will be $\mu_f/N = Np/N$, or simply $p$, and by a principle hinted at on page 25 the standard deviation in proportion units will be $\sigma_f/N = \sqrt{Npq}/N = \sqrt{(pq)/N}$. This last term is precisely what we had previously as the denominator, hence as a $\sigma$ it is the standard deviation of a distribution of proportions; we may symbolize this as $\sigma_p$.

In summary, we have $np$ and $\sqrt{npq}$ as the mean and $\sigma$ for a chance distribution of successes (on $n$ coins or $n$ dice or $n$ trials, etc.). We have $Np$ and $\sqrt{Npq}$ as the mean and $\sigma$ of a chance distribution of number of yes responses for $N$ individuals. We have $p$ and $\sqrt{pq/N}$ as the mean and $\sigma$ of a chance distribution of proportion of yeses based on samples of $N$ individuals. In the coin tossing and analogous situations, each toss or trial leads to a countable number of successes, and the distribution of the number of successes for successive trials follows the binomial. For the polling situation, each sample of $N$ cases leads to a calculable proportion of yeses, and the distribution of proportions for successive samples (of same size) also tends to follow the binomial. Such a distribution is referred to as the *random* (chance) *sampling* distribution of proportions.

It is customary to refer to $\sigma_p$ as the *standard error* of a proportion. The term "error" is used here because, in effect, we are specifying the variability due to chance (sampling) error. Actually, the sampling distribution of proportions is a theoretical distribution—we usually have just one sample proportion (or a few at most). Statistical theory provides us with information concerning the central value, variability, and shape of the distribution to be expected if we did have a very large number of sample proportions.

The scheme outlined previously for testing hypotheses is not, of course, restricted to the cigarette blindfold test and the polling situation. In the first place the $p$ for the binomial need not be $1/2$—our setup might involve a $p$ of $1/3$ (e.g., identifying 1 of 3), nor are we confined to the hypothesis of 50–50 split when polling (e.g., we might be interested in whether there is a 2 to 1 split). In the second place, we need not limit ourselves to number of successes or number of yeses. The fundamental requirement is that we be able to categorize observations (or individuals) into two classes (a dichotomy) such as pass or fail, agree or disagree, like or dislike, present or absent, cured or not cured, etc.

When a hypothesis involving a proportion is tested, the general procedure is to express the observed proportion, $p_{ob}$, as a deviation from $p_h$, the proportion expected on the basis of a statistical hypothesis, then to divide this deviation by

$$\sigma_p = \sqrt{p_h q_h/N} \qquad (5.1)$$

This gives a $z$, sometimes called a critical ratio ($CR$), which for $N$ not too small and $p_h$ not too extreme will follow the unit normal curve, the table of which permits us to ascertain the probability of a deviation as great as that observed. Note that the proviso that $p_h$ cannot be extreme follows from the fact that the binomial distribution is skewed when $p$ is extreme, say when $p$ is greater than .90 or less than .10 (see the formula for skewness on

p. 45). Since the skewness is also a function of $n$, it follows that any rule that we might adopt to prevent unjustifiable use of the normal curve approximation will be a function of $N$ and $p_h$. In general when both $Np_h$ and $Nq_h$ exceed 5 we can safely use the normal curve; if either product is between 5 and 10 we should deduct $.5/N$ from the *numerical* value of the deviation of $p_{ob}$ from $p_h$. This is another way of incorporating the correction for continuity (p. 47).

Formula (5.1) for $\sigma_p$ has been written with $p_h$ as a value specified by the hypothesis to be tested. As such the formula measures the chance variation in proportions when the hypothesis is true. Actually, saying, "if there is a 50–50 split in opinion," is the same as saying "if the proportion of yeses is .50 in the population." If we let $p_{pop}$ stand for population proportion then the variation of sample proportions is given by substituting $p_{pop}$ (and $q_{pop}$) in (5.1). When we have an obtained proportion, $p_{ob}$, and do not know $p_{pop}$ (usually the case) and have no hypothesis in mind, we use $p_{ob}$ as an estimate of $p_{pop}$, and

$$S_p = \sqrt{p_{ob}q_{ob}/N} \qquad (5.2)$$

as an approximation of the standard error of an observed proportion.

At this point the student may be somewhat confused by the use of $p$, first as the probability of, say, success on a single element and then as a proportion. Note, however, that if we were told that .30 (a proportion) of a given group have brown eyes, we could say that the probability that a randomly selected person has brown eyes is .30. Furthermore, when we say that the probability of rolling a snake eye is 1/6 or .1667, we mean that the proportion of snake eyes for a large number of rolls will tend to be .1667.

**Some sampling theory.** To facilitate later discussion we shall now introduce some notions of sampling theory. We will confine our attention to what is known as simple *random sampling*. The conditions for random sampling are that each individual (person, plant, animal, observation, etc.) in a defined population (universe, or supply) shall have an equal chance of being included in the sample, and that the drawing of one individual shall in no way affect the drawing of another (that is, the drawings must be independent of each other). The first condition is not easily met in practice. The aim is, of course, to obtain a sample which will be, within limits of random or chance errors, representative of the population from which it is drawn.

When dealing with attributes, or the classification of individuals into two (or more) categories, for which the proportion in a given category is a useful descriptive measure, we can conceive of a population proportion, $p_{pop}$, and a proportion, $p_{ob}$, obtained on a random sample of $N$ cases. Now if we could draw successive samples of $N$, determine $p_{ob}$ for each sample, and then make a distribution of the several $p_{ob}$ values, we would expect this distribution to follow the normal curve for $N$ not small and $p_{pop}$ not extreme. This follows from our discussion of the binomial distribution and normal curve approximation thereto, the only difference being that we were then speaking of a chance distribution about some hypothetical proportion, $p_h$. If $p_h$ happened to equal $p_{pop}$ we would be dealing with precisely the same distribution of sample values. If, for example, the hypothesis of a 50–50 split is true we would expect the distribution of successive sample proportions to center at .5 and have $\sigma_p = \sqrt{p_h q_h/N}$ $= \sqrt{(.5)(.5)/N}$; if the population proportion, $p_{pop}$, is .5 we would expect the successive sample proportions to have a mean of .5 and $\sigma_p = \sqrt{p_{pop}q_{pop}/N} = \sqrt{(.5)(.5)/N}$.

## DIFFERENCES BETWEEN PROPORTIONS

The testing of hypotheses need not be confined to a single proportion. This is fortunate because in research involving attributes we are more apt to have two proportions, and since each is subject to chance (sampling) error, it follows that the difference between them will also be subject to chance error. To test a hypothesis regarding the difference between two proportions it will be necessary that we have information concerning the theoretical random (chance) sampling distribution of the differences between proportions. We will need to distinguish two different types of situations: (1) proportions based on two samples drawn independently from two populations and (2) proportions for responses or observations obtained under two different conditions on just one sample. For either situation we set up a statistical hypothesis known as a *null hypothesis*. This hypothesis, which states that there is no difference between the population proportions, will be rejected if the obtained difference reaches some prescribed level of significance but will be accepted otherwise. Stated differently, if the observed difference could readily arise on a chance basis we accept the null hypothesis; if the probability of its occurrence by chance is small we reject the null hypothesis. Note that our statistical hypothesis of no difference may be, and often is, diametrically opposed to the research hypothesis being checked by the data. That is, on the basis of theory or prior observations we may expect a difference, yet for statistical reasons we set the null hypothesis. If the obtained difference is statistically significant in the expected direction we regard the data as tending to support the research hypothesis.

**Nonindependent proportions.** We shall consider first the situation in which the two proportions being compared are not based on independent groups but on just one group (or on two related groups). Suppose we are interested in whether a movie leads to a change of opinion, i.e., to an increase in the proportion favorable to some issue. We select a random sample from some defined population, get a yes (favorable) or no (unfavorable) response from each individual, show them the movie, then again get a yes or no response from each. Our next step is that of tabulation and, since we are concerned with possible changes in opinion, we will need to arrange our tabulation so as to show how many changed from no to yes, how many from yes to no, and how many "stood pat." This can be done by placing tally marks in a 2 by 2, or fourfold, table such as that depicted in Table 5.2. For an individual who gave a yes response the first time and

**Table 5.2. Tabulation plan for handling proportions based on the same individuals**

| | Frequencies 2nd | | | | Proportions 2nd | | |
|---|---|---|---|---|---|---|---|
| | No | Yes | | | No | Yes | |
| 1st Yes | $A$ | $B$ | $A + B$ | 1st Yes | $a$ | $b$ | $p_1$ |
| 1st No | $C$ | $D$ | $C + D$ | 1st No | $c$ | $d$ | $q_1$ |
| | $A + C$ | $B + D$ | $N$ | | $q_2$ | $p_2$ | $1.0$ |

a yes response the second time, a tally would go in the upper right-hand cell; for a yes at first followed by a no, a tally would go in the upper left quadrant; and so on. Let $A$, $B$, $C$, and $D$ represent the respective frequencies for yes-no, yes-yes, no-no, and no-yes responses. Then $A + B$ is the total number of yeses at first and $B + D$ is the total number of yeses the second time. If each of these totals is divided by $N$, we will have the proportions of yeses, $p_1$ and $p_2$, respectively, for the first (or pre-) and the second (or post-) set of responses. (Note: the right-hand part of Table 5.2 is obtained by dividing the 8 frequencies in the left-hand part by $N$.)

Before proceeding to develop a scheme for testing the statistical significance of the difference between the proportions, $p_1$ and $p_2$, let us note that $p_1$ and $p_2$ can differ only in case the frequency $A$ differs from the frequency $D$, since $p_1 = (A + B)/N$ and $p_2 = (B + D)/N$ have $B$ in common. Our null hypothesis is that the movie produces no change, i.e., that if the movie could be shown to the entire defined population, the proportion of yeses before and after would be exactly the same. This does not mean that an individual cannot change, but it does mean that the number of changes from yes to no balances off the number of changes from no to yes.  Thus we come to the proposition that on the basis of the null hypothesis we would expect those individuals who gave a changed response to split 50–50 as to direction of change.  Stated differently, we would expect $\frac{1}{2}$ of the $A + D$ individuals (the changers) to change from yes to no and $\frac{1}{2}$ of them to change from no to yes.

Since this is precisely analogous to tossing $A + D$ coins, we would expect that when $A + D$ persons change, the chance distribution of no to yes changes would follow, under null hypothesis conditions, the binomial distribution with mean of $(A + D)/2$ and $\sigma = \sqrt{(A + D)(.5)(.5)}$; that is, with $n = A + D$ and $p = \frac{1}{2}$.  Note that for $A + D$ fixed the number of yes to no changes is complementary to the number of no to yes changes, just as when coins are tossed the number of tails is complementary to the number of heads—we need not count both.  Thus a test of the significance of the deviation of either $D$ or $A$ from $(A + D)/2$ tells us whether $D$ differs significantly from $A$.

For $A + D$ small, say 10 or less, we may use the actual binomial expansion to evaluate the change, but for $A + D$ large we will need to resort to the normal curve approximation.  The latter is readily accomplished by expressing $D$ as a deviation from $(A + D)/2$ and dividing by $\sigma$, or by $\sqrt{(A + D)(.5)(.5)}$, which gives a critical ratio,

$$z = \frac{x}{\sigma} = \frac{D - (A + D)/2}{\sqrt{(A + D)(.5)(.5)}} = \frac{.5D - .5A}{.5\sqrt{A + D}} = \frac{D - A}{\sqrt{A + D}} \qquad (5.3)$$

as a value with which to enter Table A to find the probability of as large a deviation as that obtained.  If this $z$ is 2.58 (or larger) the $P = .01$ level of significance will have been reached (we are here dealing with the probability of as large a deviation irrespective of direction).  When $A + D$ is not large, say 11 to 20, our approximation will be appreciably improved by deducting 1 from the absolute value of $D - A$; this is the correction for continuity again.

If we wished to carry our computations through on the basis of proportions, we could express $D$ as a proportion of $A + D$ (similar to what we did on p. 51) but, as we shall see, it is more appropriate to introduce the sample size, $N$, into the picture.  Dividing both numerator and denominator of (5.3) by $N$ will not change the value of the fraction, that is,

$$z = \frac{x}{\sigma} = \frac{D/N - A/N}{\sqrt{(A + D)/N^2}}$$

If we let $a = A/N$ and $d = D/N$, this may be written as

$$z = \frac{x}{\sigma} = \frac{d - a}{\sqrt{(a + d)/N}} \qquad (5.4)$$

This form for $x/\sigma$ will make more sense if we again consider Table 5.2, particularly the right-hand part. Note that since $a + b = p_1$ and $b + d = p_2$, it follows that $d - a = p_2 - p_1$ and accordingly a test of the significance of $D$ as a deviation from $(A + D)/2$ is also a test of the significance of the difference between the proportions of yeses obtained on the two occasions.

To incorporate the correction for continuity, deduct $1/N$ from the absolute value of $d - a$.

The denominator of the right-hand side of (5.4) must be a standard deviation. Of what? Actually it is the standard deviation of the theoretical sampling distribution of differences between proportions, each difference being based on one sample of size $N$. Such a standard deviation, as we have noted previously, is referred to as a standard error. Thus we have

$$\sigma_{D_{p(r)}} = \sqrt{(a + d)/N} \qquad (5.5)$$

as the standard error of the difference between correlated proportions. The subscript $r$ has been added to indicate that this formula holds for related or correlated proportions. The relationship, or correlation, concept needs a brief word of explanation. If, by chance sampling, $p_1$ were lower than the population value, we would expect $p_2$ also to be somewhat low; if $p_1$ were by chance high, we would expect $p_2$ to be somewhat high; if $p_1$ were near the population value (near average), we would expect $p_2$ to be near average. This varying together is referred to as a co-relationship or correlation. Stated differently, we would not expect the two proportions to vary independently of each other for successive samples.

The proportions need not be based on the same individuals to be correlated. For example, if we were interested in sex differences in opinion we might randomly choose families and then ascertain the proportion of yeses among the husbands and also among the wives; for successive samplings the two proportions might be correlated because of a possible tendency for husbands and wives to agree on the given issue. As a second example, consider the setup involving the pairing of individuals for the purpose of having comparable experimental and control groups. The fact of pairing signifies that the two groups have not been drawn independently in the sampling sense; hence there might be a tendency for the proportions based on the two groups to be more or less alike. (About pairing we will have more to say in Chapter 6.)

Another instance for which formulas (5.3), (5.4), and (5.5) are applicable is the problem of judging the significance of the difference between proportions of yeses for two different questions asked of the same sample of $N$ cases. Since the responses to the two questions might tend to vary together there could be a correlation between the proportions on successive samplings.

In each of the foregoing situations we have pairs of responses, and our tabulation must follow the scheme set forth in Table 5.2; i.e., our tabulation will lead to the frequency of yes-no, yes-yes, no-no, and no-yes responses.

Formulas (5.3), (5.4), and (5.5) are usable in other situations. When judging whether or not two test items differ significantly in difficulty we ordinarily have pass-fail data for both items on the same sample of $N$ cases. Our tabulation leads to the frequencies for pass-fail, pass-pass, fail-fail, and fail-pass. The kind of response is irrelevant—it need only be such that a dichotomy is involved for each item or question.

These formulas may be safely used for any size sample provided $A + D$ is 10 or more (and the correction for continuity is included when $A + D$ is 10 to 20). If $A + D$ is less than 10, the binomial expansion provides an easily computed test of significance leading to an exact probability for as great a difference between the proportions as that observed. The $P$ so obtained needs to be doubled to get the probability for as great a difference irrespective of direction; otherwise it is the probability for as large a difference in one direction only. About this we shall have more to say later under the heading, "One-tailed vs. two-tailed tests," pp. 64–65.

**Independent proportions.** It is not easy to build up a general formula for evaluating the difference between two proportions based on two independent samples. We can, however, learn something about formula construction and, incidentally, illustrate a general statistical theorem by considering a special case involving differences between independent proportions.

We have already seen how the binomial expansion, $(p + q)^n$, can be used as a basis for ascertaining theoretical, or expected, frequencies for various possible outcomes (events). Let us now see whether we can set up expected frequencies for the joint occurrence of events. Suppose persons $J$ and $K$ decide to while away some time at coin tossing. Each uses $n = 5$ coins, for which the binomial yields expected frequencies of 1, 5, 10, 10, 5, 1 for 5, 4, 3, 2, 1, 0 heads, with mean $= np = 2.5$ and $\sigma^2 = npq = \frac{5}{4}$. But instead of making just 32 tosses, each makes 1024 tosses, for which the expected frequencies would be 32 times the 1, 5, 10, 10, 5, 1, or 32, 160, 320, 320, 160, 32.

$J$ and $K$ decide to make simultaneous tosses in order to learn something about joint outcomes, that is, to see how often both get 5 heads or how often $J$ gets 4 heads while $K$ gets 3 heads, and so on. Now a little thought will indicate that the total number of possible joint outcomes will be 6 times 6, or 36. To keep a record of their results, $J$ and $K$ would be wise to lay out a 6 by 6 table with 0 to 5 (heads) along the bottom and also along the left-hand side. When a particular combination occurs, say, 2 heads by $J$ and 4 by $K$, a tally mark is entered in the cell to the right of 2 and above 4 (enter with $J$'s along the ordinate and $K$'s along the abscissa).

Can we anticipate the frequencies in the 36 cells of the table? This we cannot do, but we can specify the theoretically expected frequencies in either of two ways. The first method involves use of the multiplication theorem of probability. The probability of $J$ obtaining 5 heads is 1/32; the probability of $K$ obtaining 5 heads is also 1/32. The product of these two is 1/1024, which permits us to enter a 1 in the upper-right cell as the expected number of times (out of 1024 simultaneous tosses) that each gets 5 heads. The probability of the joint outcome, $J$ 2 heads and $K$ 4 heads, is 10/32 times 5/32, or 50/1024, which permits us to enter 50 as the expected frequency in the cell defined by 2 along the left and 4 along the bottom.

Table 5.3. Expected frequencies for joint outcomes when $J$ and $K$ each make 1024 simultaneous tosses of 5 coins

| | $E_k$ | 32 | 160 | 320 | 320 | 160 | 32 | $E_j$ |
|---|---|---|---|---|---|---|---|---|
| | 5H | 1 | 5 | 10 | 10 | 5 | 1 | 32 |
| | 4H | 5 | 25 | 50 | 50 | 25 | 5 | 160 |
| $J$ | 3H | 10 | 50 | 100 | 100 | 50 | 10 | 320 |
| | 2H | 10 | 50 | 100 | 100 | 50 | 10 | 320 |
| | 1H | 5 | 25 | 50 | 50 | 25 | 5 | 160 |
| | 0H | 1 | 5 | 10 | 10 | 5 | 1 | 32 |
| | | 0H | 1H | 2H | 3H | 4H | 5H | 1024 |
| | | | | $K$ | | | | |

Each of the other 34 cells can be similarly filled in by the multiplication theorem. The second method is simpler. For the 32 times we expect $J$ to get 5 heads, we would expect $K$'s results to follow the binomial, hence we can immediately write down 1, 5, 10, 10, 5, 1 in the top row of the 6 by 6 table. For the 160 times we expect $J$ to obtain 4 heads we would again expect $K$'s outcomes to follow the binomial but, since 160 is five times 32, we would need to multiply the 1, 5, 10, 10, 5, 1 by 5, giving 5, 25, 50, 50, 25, 5 as entries in the second row in the 6 by 6 table. By exactly the same line of reasoning the other rows can easily be filled in, with results as shown in Table 5.3.

When a particular cell frequency in Table 5.3 is divided by 1024 we have a probability for a joint occurrence. Another way of interpreting a particular cell frequency is to regard it as a mean value in the sense that if $J$ and $K$ performed a very, very large number of series of 1024 tosses we would expect the average of the obtained frequencies for that cell to correspond to the given theoretically expected frequency. That is, any expected frequency is to be regarded as the mean over an infinitely large number of trials.

But we built up Table 5.3 for the ultimate purpose of saying something about the difference between independent proportions. Suppose $J$ and $K$ decide to make two additional tabulations for each pair of simultaneous tosses: the sum of their separate outcomes, that is, the number of heads for all 10 coins; and also the difference in number of heads, expressed arbitrarily as $J$'s count minus $K$'s count. Thus for tabulating the sum of their results they would need "intervals" $10H, 9H, \cdots 1H, 0H$, whereas for the difference they would need $+5, +4 \cdots 0 \cdots -4, -5$. Again, let us attempt to determine the expected results.

It is easy to write down the expected frequencies for the various outcomes as sums—these would simply come from the binomial $(p + q)^{10}$. We can, however, write them from Table 5.3. A sum of 10 (heads) can occur only when both $J$ and $K$ obtain 5 heads, for which the expectation is 1 out of 1024. A sum of 9 can occur either when $J$ gets 5 and $K$ gets 4 or when $J$ gets 4 and $K$ gets 5. Since the expectation for each of these is 5, the expectation for 9 as a sum becomes 10. A sum of 8 results from 5 and 3, 4 and 4, or 3 and 5 for $J$ and $K$ respectively, and these joint outcomes have expectations of 10, 25, and 10, which add to 45. Note now that diagonal adding, upper-left to lower-right in Table 5.3, will lead to 1, 10, 45, 120, 210, 252, 210, 120, 45, 10, 1 as expected frequencies for the possible outcomes when $J$ and $K$ sum the results for each of their simultaneous tosses.

As to the difference in "scores," when $J$ gets 5 heads and $K$ none we have a difference of $+5$ for which the expectation is 1 (out of 1024). A difference of $+4$ can arise when $J$ gets 5 and $K$ gets 1 or when $J$ gets 4 and $K$ gets none; summing the two expectations, $5 + 5 = 10$ as the expected number of times for a difference of $+4$. A difference of $+3$ can occur in three ways with expectations of 10, 25, and 10, which add to 45 as the expected frequency for a difference of $+3$. Note that we are again summing diagonally in Table 5.3, this time from lower-left to upper-right.

The results both for sums and for differences, given in Table 5.4, are worth scrutinization. The two distributions are identical except for their location parameters, the mean being 5 for one and 0 for the other. Obviously, the variances are equal. The fact that the differences have a mean of 0 might have been anticipated, since every time $J$ and $K$ toss their 5 coins, each is, in effect, making a trial—a trial which represents a sample. But each is sampling from the same universe, the universe of events when 5 coins are tossed. (It is presumed that the coins are unbiased.) $J$ and $K$'s "universes" have the same mean ($np = 2.5$); hence in the long run it would be expected that chance will operate in such a way that the average of obtained differences will be zero.

Chance will also operate to produce variability in the differences, the standard deviation of which can be specified. We have seen that the variance of the difference is equal to the variance of the sum. The variance of the sum is nothing more than the variance of the distribution of heads when 10 coins are tossed an infinite number of times, hence the variance of the difference is also simply $npq = 10(.5)(.5)$. Note that $10(.5)(.5) = 5(.5)(.5) + 5(.5)(.5)$. In general, when $n_t = n_j + n_k$ we can say that the variance of the sum will be the sum of the separate variances, i.e., $n_t pq = n_j pq + n_k pq$.

Table 5.4. Expected frequencies ($E_f$) for sums ($\Sigma_h$) and differences ($D_h$) for 1024 simultaneous tosses of two sets of 5 coins, and differences in proportions ($D_p$)

| Sums | | Differences | | | |
|---|---|---|---|---|---|
| | | For heads | | For proportions | |
| $\Sigma_h$ | $E_f$ | $D_h$ | $E_f$ | $D_p$ | $E_f$ |
| 10 | 1 | +5 | 1 | +1.0 | 1 |
| 9 | 10 | 4 | 10 | .8 | 10 |
| 8 | 45 | 3 | 45 | .6 | 45 |
| 7 | 120 | 2 | 120 | .4 | 120 |
| 6 | 210 | 1 | 210 | .2 | 210 |
| 5 | 252 | 0 | 252 | .0 | 252 |
| 4 | 210 | −1 | 210 | −.2 | 210 |
| 3 | 120 | −2 | 120 | −.4 | 120 |
| 2 | 45 | −3 | 45 | −.6 | 45 |
| 1 | 10 | −4 | 10 | −.8 | 10 |
| 0 | 1 | −5 | 1 | −1.0 | 1 |

At this point, it should be obvious to the student that the variance of the sum of heads obtained on an infinite number of simultaneous tosses for any values of $n_j$ and $n_k$, not necessarily equal, will be given by summing the separate variances. It is not obvious that this also holds generally for the variance of the differences. Later we will have an algebraic proof, showing that the variance of a sum (or difference) is always equal to the sum of the separate variances when the events (scores) being summed are independent.

In Table 5.4 we have a chance expected, or random, sampling distribution of differences in number of heads, with $\mu = 0$ and $\sigma = \sqrt{10(.5)(.5)}$. Suppose that $J$ and $K$ changed their "scoring" system from number of heads per toss to the proportion of heads per toss by simply dividing the former by $n = 5$. Thus, they would have a scale running as .0, .2, .4, .6, .8, 1.0 along the ordinate and abscissa of Table 5.3. The differences between these proportions as scores would run $+1.0$, $+.8$, $+.6$, $+.4$, $+.2$, .0, $-.2$, $-.4$, $-.6$, $-.8$, $-1.0$, as shown in the not yet discussed right-hand part of Table 5.4. Note that in changing the scale from number of heads per toss to proportion of heads per toss, both $J$ and $K$ divided the former by $n = 5$. Note further that the scale for differences in proportions ($D_p$) in Table 5.4 can be obtained by dividing the $D_h$ scale values (center of the table) by $n = 5$. This change of scale leaves $\mu = 0$ unchanged; however, the standard deviation is changed: $\sigma_{D_p} = \frac{1}{5}\sigma_{D_h}$. More generally, if $J$ and $K$ each toss $n$ coins (or roll $n$ dice) an infinite number of times, the variance of the random sampling distribution of the differences, in proportion units, for their simultaneous tosses (or rolls) will be given by

$$\sigma^2{}_{D_p} = \frac{1}{n^2}\sigma^2{}_{D_h} = \frac{1}{n^2}(npq + npq) = \frac{pq}{n} + \frac{pq}{n}$$

The foregoing rather lengthy development shows one way of arriving at a formula for the variance of the sampling distribution of the differences between independent proportions under the specified conditions, but these conditions ($n_j = n_k = n$, and known $p$) are seldom, if ever, encountered in research work. In practice we will have two proportions, $p_1$ and $p_2$, based on $N_1$ and $N_2$ cases. Both $p_1$ and $p_2$ will be subject to sampling variation, hence their difference will also be influenced by sampling error. We will not know the two population proportions necessary for specifying exactly the standard errors for $p_1$ and $p_2$ and for their difference. We must, therefore, resort to estimation. For this purpose we will assume the null hypothesis to be true; if true, the proportions for the populations will be the same. The best available estimate for this unknown common population proportion will be obtained by pooling the two samples, i.e., by taking $p_c$, the proportion for the two samples combined, as the estimate. Then with $q_c = 1 - p_c$, we take the following as our estimate of the standard error of the difference between two independent proportions:

$$S_{D_{p(i)}} = \sqrt{\frac{p_c q_c}{N_1} + \frac{p_c q_c}{N_2}} = \sqrt{p_c q_c(1/N_1 + 1/N_2)} \qquad (5.6)$$

The value of $p_c$ is readily obtained by combining the two frequencies of yeses (or whatever the given category is) and dividing by $N_c = N_1 + N_2$, and as usual $q_c = 1 - p_c$. An observed difference divided by $S_{D_{p(i)}}$ will give a $z$ interpretable as a unit normal curve deviate provided the $N$s are not too small and $p_c$ is not too extreme. The rule-of-thumb is that $p_c$ or $q_c$

(whichever is smaller) times $N_1$ or $N_2$ (whichever is smaller) shall exceed 5. When this product is between 5 and 10, a correction for continuity should be incorporated. This may be done by reducing the numerical (absolute) value of the difference, $p_1 - p_2$, by the quantity $\frac{1}{2}\left(\frac{1}{N_1} + \frac{1}{N_2}\right)$.

## SOME GENERAL CONSIDERATIONS

Before going further we should stop long enough to delineate the general problem of hypothesis testing, discuss the question of one-tailed vs. two-tailed tests, and consider the problem of what level of significance to adopt.

**Which hypothesis?** In general, successive samplings will yield a sampling distribution of frequencies or of proportions or of differences between statistical measures or certain ratios (such as $z$ or other ratios, to be discussed later). Hypotheses, whether statistical or research, are usually concerned either with differences or with deviations. By research hypothesis we mean the hypothesis set up on the basis of theory or prior observation or on logical grounds. Such a hypothesis usually involves a prediction regarding the outcome of an experiment. By statistical hypothesis we usually mean a null hypothesis set up for the purpose of evaluating the research hypothesis.

When we are considering possible differences the null hypothesis, frequently symbolized as $H_0$, is pitted against an alternate hypothesis, $H_1$. Now $H_0$ specifies that, for example, $p_{pop(1)} = p_{pop(2)}$ or that two population values do not differ, whereas $H_1$ might specify that $p_{pop(1)} > p_{pop(2)}$ or that $p_{pop(1)} < p_{pop(2)}$ or that $p_{pop(1)} \neq p_{pop(2)}$. Which of these alternatives is appropriate depends on the research hypothesis to be tested by experiment or what question is to be answered by experiment. An experiment is carried out which yields sample values, $p_1$ and $p_2$, and the difference between $p_1$ and $p_2$ is used to test $H_0$ against $H_1$; that is, on the basis of the obtained difference we are to make a decision as to whether $H_0$ or $H_1$ is true.

If $H_0$ is true we can specify the probability of obtaining by chance a difference as great as $p_1 - p_2$ or as great as $p_2 - p_1$ or as great as the numerical (irrespective of sign) difference, $p_1 - p_2$. Let $\alpha$ represent a chosen level of significance—any level such as $P = .10$ or $P = .05$ or $P = .01$ or $P = .001$. We reject $H_0$, the null hypothesis, if the probability of the obtained result is as small as the chosen $\alpha$, and this rejection implies the acceptance of $H_1$. If $\alpha$ is not reached we accept $H_0$, but this acceptance merely says that $H_0$ could be true—any of a whole series of differences near zero could also be true. This acceptance-rejection business involves risks, to be discussed under "Choice of level of significance."

**One-tailed vs. two-tailed tests.** The three possible alternatives listed previously for $H_1$ have to do with hypotheses admissible on the basis of either the research hypothesis or the question for which we seek an answer by way of an experiment. In general, if $H_1$ states that $p_{pop(1)}$ does not equal $p_{pop(2)}$, a two-tailed test is in order; if $H_1$ specifies which population value is the larger, a one-tailed test is used. The issue as to whether we should use a one-tailed test or a two-tailed test depends on whether the scientific hypothesis being tested (or at times the practical decision to be made) demands that we be concerned with chance deviations in just one direction or in both directions. For situations in which we wonder whether a performance is better than chance, as in blindfold cigarette discrimination, we are concerned only with results in one direction, since any performance in which the subject is successful on less than .50 of the trials leads us, without further statistical ado, to accept the hypothesis that he cannot discriminate better than chance. Thus a one-tailed test is appropriate. But for situations in which we wish to decide whether a population is split 50–50 on some question, we need to consider chance sampling deviations in both directions; hence we should use a two-tailed test.

Next consider the problem of testing the significance of the difference between two proportions. If, for example, we have the proportion of yeses to some question for a sample of Republicans and for a sample of Democrats as a basis for deciding whether Republicans and Democrats differ on the given issue, we would need to use a two-tailed test—we reject the hypothesis of no difference in case the obtained difference, irrespective of direction, has a probability of occurrence which is as small as $\alpha$, the chosen level of significance. A one-tailed test would be utilized for judging significance in an experiment in which, for example, we were trying a new drug to see if it is better as a preventative than some commonly used drug. The decision to adopt the new drug is made only if the new drug leads to a greater proportion of immunities—results in only one direction are crucial to the decision to change drugs. But if we were trying out two drugs with the idea of adopting the one which is most promising we would use a two-tailed test since significance in either direction is the basis for decision.

It is sometimes argued that whenever the outcome of an experiment is predicted on the basis of theory or previous observation, a one-tailed test is appropriate since some benefit should accrue to the researcher who has predicted the direction of the results as opposed to the investigator who, though obtaining similar results, has not predicted the outcome. The benefit comes about in that the $z$ for, say, the $P = .01$ level of significance need reach only 2.33 for a one-tailed as compared with 2.58 for a two-tailed test. For the $P = .05$ level the respective values are 1.64 and 1.96. In other words a difference, to be significant, does not have to be as large for a one-tailed as for a two-tailed test. Since the situation involving prediction is equivalent to taking $H_1$ as the hypothesis that the difference between two population values is in a specified direction, it is not only defensible to use a one-tailed test but actually better in the sense that if there is a real difference in the predicted direction it will be more apt to be detected by a one-tailed than by a two-tailed test. However, a few words of caution are in order.

First, the prediction should be made prior to the collection of data, that is, independently of the data to be used in testing the hypothesis. Second, we must be on guard against habit—instances can be cited where an investigator after making a series of one-tailed tests failed to shift to a two-tailed test when he should have. Third, in case the results are significant in the direction opposite to the prediction, the investigator must, in effect, have a red face because the outcome is not consistent with either of the admissible hypotheses: no difference (as set forth by the null hypothesis) or a difference in the predicted direction (as set forth by the research hypothesis being tested). It is one thing to have results which simply fail to support a hypothesis, and quite a different thing to have an outcome which is diametrically opposed to the hypothesis.

**Choice of level of significance.** How large should $z$ be before the investigator claims significance? Asked differently, How does he choose $\alpha$, the value of $P$ to be required for judging significance? There is no one answer to this question. For a long time psychologists insisted on a $z$ of 3.00 (equivalent to $P = .003$ level for two-tailed test) as a rule-of-thumb value for judging significance. There might be occasions when one would desire the assurance represented by a $P$ of .003, but it should be noted that the acceptance of the null hypothesis whenever $z$ does not reach 3.00 may lead too frequently to another type of erroneous conclusion. To understand this, we must consider what it means when an observed difference does not lead to the rejection of the null hypothesis. Acceptance of the null hypothesis does not prove that no difference exists. For example, a difference of 1 per cent, in number of yeses for two samples, which yields a $z$ of .8 does not prove that there is no difference in the two universe values—it merely indicates that the real difference could easily be zero. However, the obtained difference of 1 could be a chance departure from a real difference of .5 or 1.2 or 1.8 or any of a whole series of values near 1. In other words, the null hypothesis is one which can be rejected but can never be proved; therefore to accept it too often because we insist on a high level of significance for rejection means that we are apt to overlook real differences. This, plus the fact that we do not ordinarily need the assurance represented by a significance level of .003, would suggest that a $z$ of 3.00 is too high.

At the other extreme, a few are willing to accept as significant a difference which is 1.5 times its standard error. Since $P = .13$ (two-tailed) for a $z$ of 1.5, it is readily seen that such persons would all too frequently have their publics believing that chance differences are real. A less lax level, which is now generally accepted by psychologists, is represented by a $P$ of .05. This also may be a rather low level of significance for announcing something as "fact." Those writers who advocate the .05 level for research workers in psychology cite R. A. Fisher, an eminent statistician, as their authority, *but* they fail to point out that Fisher's applications were to experimental situations in agriculture and biology where there is far better control of sampling than is ordinarily the case in psychology.

If the findings of a study are to be used as the basis either for theory and further hypotheses or for social action, it does not seem unreasonable to require a higher level of significance than the .05 level. The answer as to what level, in terms of probability, should be adopted in order to call a finding statistically significant is not uninvolved. There is the balancing of risks: that of accepting the null hypothesis when to do so may mean the overlooking of a real difference against that of rejecting the null hypothesis which may lead to the acceptance of a chance difference as real. There is the question of the likelihood of independent verification, and, finally, there is the whim of personal preference: some individuals are more eager than others to announce a "significant" finding; others are more cautious. It follows that no hard and fast rule can be given; a finding may be interpreted in terms of the probability of its occurrence by chance and then it may be noted whether the $P$ is near the significance level adopted *prior* to the experiment because it seemed appropriate when all factors were weighed.

The reader will have noted from the foregoing that the testing of hypotheses involves the possibility of two types of erroneous conclusions. These are usually referred to as type I and type II errors, which we shall now more specifically define. Consider again the null hypothesis that no difference exists between two population values. If we reject this hypothesis when in fact it is true, we will have committed a type I error. If we accept the hypothesis when in fact it is false, we will have made a type II error. Possible outcomes of our conclusion are shown in Table 5.5.

The factors in choosing a level of significance might be further clarified by a somewhat different approach. Notice that when we adopt $P = \alpha$ as our level of significance we are definitely specifying the probability of committing a type I error; it is simply $\alpha$. By taking $\alpha$ smaller and smaller we can reduce the risk of making a type I error. But what happens to the probability of making a type II error as we thus reduce the risk of a type I error? Before answering this, we need to show how for a given $\alpha$ one can

**Table 5.5. Correct and incorrect statistical conclusions**

True Situation

|  |  | No difference | Real difference |
|---|---|---|---|
| Conclusion | Real difference | Type I error ($\alpha$) | Correct ($\beta$) |
|  | No difference | Correct ($1 - \alpha$) | Type II error ($1 - \beta$) |

proceed to calculate the probability of making the type II error, herein symbolized by $1 - \beta$ with $\beta$ defined as the probability of correctly rejecting the null hypothesis when a true difference exists.*

Suppose the aim of a research project is to ascertain whether drugs $A$ and $B$ differ in effectiveness in providing immunity to a disease. The drugs are administered to two independently drawn samples from which we can derive observed proportions of immunity. Let us presume that the proportions and sample sizes are such that the standard error of the difference is .02. The exposition will be somewhat simplified if we change to percentage units, which is readily accomplished by shifting decimals for the proportions and also for the standard error. The latter becomes 2 in percentage units. Since the decision will be to adopt the better drug, we will need a two-tailed test for judging significance. Suppose we adopt the .05 level of significance, thus fixing the probability of the type I error as $\alpha = .05$.

By definition of the type II error, such an error can be committed only when the two drugs differ in effectiveness. But we do not, of course, know how much they differ—a bit of ignorance which handicaps us in calculating *the* probability of a type II error. We can, however, specify the error for any supposed real difference in the effect of the two drugs.

We illustrate the calculation of $\beta$ (and $1 - \beta$) by presuming that the true difference is 4 percentage points in favor of drug $A$. That is, populationwise, $\%A - \%B = 4$. Now the mechanism for testing the null hypothesis is to divide an obtained difference between the sample percentages by the standard error of the difference. With a two-tailed, .05 level test, we know that $D/\sigma_D$ must equal or exceed 1.96. Since $\sigma_D$ is specified as 2, it follows from $D/2 = 1.96$ that $D$ must equal or exceed 3.92 in order for us to reject the null hypothesis. Stated differently, under

---

* The reader should know that some textbooks use $\beta$ as the symbol for the probability of a type II error. We are here following the notation used by Neyman and Pearson, the first to delineate the two types of error. In following this notation, this book finds itself in company with several, but not all, texts authored by mathematical statisticians.

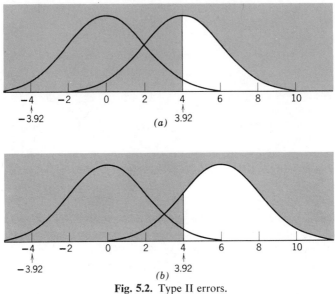

**Fig. 5.2.** Type II errors.

the null condition we would have a random sampling distribution of differences centering at zero, having a standard deviation of 2, with .025 of the area to the right of 3.92 and .025 of the area to the left of −3.92. Sample differences falling beyond (to the right and to the left of, respectively) these two points would be in what are termed *critical regions* for rejecting the null hypothesis at the given level of significance. These regions are in the tails beyond the base-line points depicted by arrows in the two curves centered at 0 in Fig. 5.2.

Next consider what would be expected to happen when the true difference is 4 points in favor of drug *A*. For convenience we are calling this a positive difference (*B*'s percentage subtracted from *A*'s). If the true difference were 4, the random sampling distribution of differences would center at 4, with standard deviation = $\sigma_D = 2$. How many of these differences would one expect to fall in the critical region(s) for rejecting the null hypothesis as established by the random sampling distribution under the null condition? Stated differently, how many of the differences for the curve centered at 4 in Fig. 5.2a would one expect to fall in the region beyond +3.92? Each time a difference falls in this region we correctly reject the null hypothesis; each time a difference fails to reach +3.92, we falsely accept the null hypothesis. These times, expressed as proportions, will yield $\beta$ and $1 - \beta$, respectively. The calculations are easy. The mean for the right-hand curve is 4. So we take $(3.92 - 4)/2 = -.04$ as the $z$ value of 3.92 in the distribution centered on 4. From the normal curve table we learn that the area from 3.92 to 4 is .02, which when added to .50 gives .52 as $\beta$, the probability of correctly rejecting the null hypothesis. Then $1 - \beta = .48$ gives the relative number of times that sample differences, under the condition of a true difference of 4, would lead us to accept falsely the null hypothesis. Thus we have a probability of .48 of committing the type II error. The unshaded part of the distribution centered at 4 represents $\beta$, the shaded part represents the typeII error.

Can we regard .48 as *the* probability of a type II error? Let us consider what we would expect to happen if the true difference were $+6$. The critical points which determine the regions of rejection remain unchanged. We have the same null-hypothesis curve, along with a distribution curve centered at 6 (Fig. 5.2*b*). For this curve, the area to the right of $+3.92$ represents $\beta$. This area is obviously larger than holds for the curve centered at 4. Again we get $\beta$ by calculating $z = (3.92 - 6)/2 = -1.04$. From Table A we get the area to the left of a $z$ value of $-1.04$ as .15, which is the probability of a type II error when the true difference is $+6$. The corresponding value of $\beta$ is .85. By now you should deduce that $\beta$, the probability of detecting a difference, varies with the true difference and that *the* probability of committing the type II error is not simply one value—it also varies with the magnitude of the true difference.

Are $\beta$ and $1 - \beta$ functions of $\alpha$, the chosen level of significance? Again, consider the drug situation, but this time let us say that our clinical trial is to answer the question as to whether *new* drug $A$ is better than the currently used drug $B$. This calls for a one-tailed test. Again we presume a $\sigma_D = 2$ percentage points.

Figure 5.3 shows a series of sampling distribution curves, all with $\sigma = 2$, but with locations differing according to supposed true, or population, differences of 0, 4, and 8. The top part (*a*) is for $\alpha = .10$, the middle (*b*) for $\alpha = .05$, and the bottom (*c*) for $\alpha = .01$. In each part an ordinate has been erected at the difference required for significance at the given $\alpha$ level of significance. These required differences spring from the fact that for a one-tailed test the $z$ values that cut off .10, .05, and .01 of a normal curve are 1.28, 1.64, and 2.33 respectively, and since $\sigma$ is 2, the respective required differences in percentages would be 2.56, 3.28, and 4.66. Sample differences falling beyond these values would, of course, be in the *critical regions* for rejecting the null hypothesis at the three respective $\alpha$ values. For example, values beyond 4.66 would be in the critical region when the $P = .01$ level of significance is adopted.

From these several sampling distribution curves and with the help of a table of the normal curve functions, we can specify the probability of committing a type II error for a specified (supposed) true difference. If we keep in mind that the probability of a type I error is $\alpha$ ($= .10$, .05, or

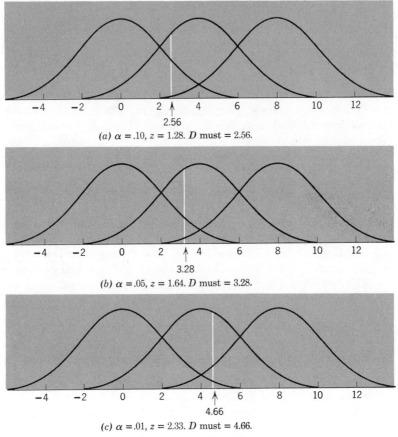

(a) $\alpha = .10$, $z = 1.28$. $D$ must = 2.56.

(b) $\alpha = .05$, $z = 1.64$. $D$ must = 3.28.

(c) $\alpha = .01$, $z = 2.33$. $D$ must = 4.66.

**Fig. 5.3.** Type I and type II errors.

.01), and that we can make a type I error only when the true difference is zero, we see that the proportionate areas beyond 2.56, 3.28, and 4.66 for the three curves centering at zero represent the probabilities of making a type I error for the respective $\alpha$ values. For all sample values in the regions to the left of 2.56, 3.28, and 4.66 we would correctly accept the null hypothesis when in reality it is true. The probabilities for correct acceptance are given by $1 - \alpha$, or .90, .95, and .99 respectively.

Let us now consider the supposition that the true difference is 4. If 4 is the true difference, any obtained difference falling in the region to the right of 2.56, 3.28, and 4.66 will, for the respective levels of significance, lead to the *correct* decision that a true difference exists. The probabilities for these correct inferences are obtained by expressing 2.56, 3.28, and 4.66 as deviations from 4 (the supposed true value being considered), taking each deviation relative to the standard error of the difference ($= 2$), and thus obtaining $z$ values of $(2.56 - 4)/2 = -.72$, $(3.28 - 4)/2 = -.36$, and $(4.66 - 4)/2 = .33$. Looking these values up in a table of the normal curve we get probabilities, $\beta$s, for correctly rejecting the null hypothesis, of .76, .64, and .37, for the respective specified levels of significance, *when* the true difference is 4 percentage points. Note that all sample values falling in the region to the left of 2.56, 3.28, and 4.66 (for the curves centering at 4) will lead to the false acceptance of the null hypothesis. The probabilities of making type II errors will correspond to the proportionate areas, for the curves centering at 4, to the left of these three points (when we have the one-tail test as considered here). These probabilities will, of course, be given to us by $1 - \beta$. Thus we have .24, .36, and .63 as the probabilities of making a type II error, when the true difference is 4 and for the .10, .05, and .01 levels of significance. Note that taking $\alpha$ smaller and smaller increases the probability of making a type II error.

For a true difference of 8, we can by a similar line of reasoning obtain the probability of correctly rejecting the null hypothesis and the probability of falsely accepting the null hypothesis, when using any one of the specified values of $\alpha$. These probabilities will involve the areas, under the curves centering at 8, to the right of 2.56, 3.28, and 4.66 (for the $\beta$s) and to the left of these same points (for the type II errors). The student can readily verify that areas to the right of 2.56, 3.28, and 4.66 are approximately .997, .99, and .95 respectively. Subtracting each of these from unity will yield the probabilities, .003, .01, and .05, of falsely accepting the null hypothesis or committing a type II error when the true difference is 8 and for $\alpha$s of .10, .05, and .01. Again, the smaller we take $\alpha$ the larger the probability of making a type II error.

The probabilities given in the last two paragraphs, along with similar figures for other supposed true differences, have been assembled in Table 5.6. A careful study of this table reveals the general rule that the smaller the value of $\alpha$ the smaller the probability ($\beta$) of correctly rejecting the null hypothesis and the larger the probability ($1 - \beta$) of committing a type II error. Thus when we reduce the probability of making a type I error by choosing $\alpha$ small, we do so at the risk of more often making a type II error. Note also that regardless of $\alpha$, the probability of making a type II error decreases as the true differences deviate farther and farther from zero. This is another way of saying that the larger the true difference the more apt we are to detect it by experiment, and conversely the smaller the difference the less likely we are to discover it.

Does the probability of making a type II error depend on factors other than $\alpha$ and the magnitude of the true difference? Let us reconsider the $\alpha = .05$, two-tailed situation illustrated by Fig. 5.2. Suppose that sample

**Table 5.6. Probability ($\beta$) of correctly rejecting the null hypothesis and probability $(1 - \beta)$ of type II error associated with three levels of significance ($\alpha$s of .10, .05, .01) when certain true differences are supposed to exist**

| | $\beta$ | | | $1 - \beta$ | | |
|---|---|---|---|---|---|---|
| $\alpha \rightarrow$ | .10 | .05 | .01 | .10 | .05 | .01 |
| True difference | | | | | | |
| 1 | .22 | .13 | .03 | .78 | .87 | .97 |
| 2 | .39 | .26 | .09 | .61 | .74 | .91 |
| 3 | .59 | .44 | .20 | .41 | .56 | .80 |
| 4 | .76 | .64 | .37 | .24 | .36 | .63 |
| 5 | .89 | .79 | .57 | .11 | .21 | .43 |
| 6 | .96 | .91 | .75 | .04 | .09 | .25 |
| 7 | .99 | .97 | .88 | .01 | .03 | .12 |
| 8 | .997 | .99 | .95 | .003 | .01 | .05 |
| 9 | >.999 | .997 | .975 | <.001 | .003 | .025 |
| 10 | >.999 | >.999 | .996 | <.001 | <.001 | .004 |

sizes are increased fourfold, i.e., we take 4 times each $N$ instead of the $N$s (unspecified) that led to 2 as the standard error of the difference. The effect of quadrupling the sample sizes will be to halve the standard error of the difference. Thus we have $\sigma_D = 1$, which for a two-tailed, .05 level test will lead to plus and minus 1.96(1), or $\pm 1.96$ on the percentage scale, as the points that define the critical regions for rejection. With a true difference of $+4$, the area in the sampling distribution centering at 4, with $\sigma_D = 1$, falling beyond $+1.96$ is found by using $z = (1.96 - 4)/1 = -2.04$. This leads to $\beta = .979$, hence .021 is the probability of a type II error. With a true difference of $+6$, we have $z = (1.96 - 6)/1 = -4.04$, which leads to a $\beta$ of near .99997, and .00003 as $1 - \beta$. Thus increasing $N$s fourfold has decreased the probability of making a type II error from .48 to .027 and from .15 to .00003 for true differences of 4 and 6.

That $1 - \beta$ is reduced when other true differences exist is seen from an examination of Table 5.7. Strictly speaking, Table 5.7 should, since based on the two-tailed test, also include values for true differences in the opposite direction. However, the values of $\beta$ and $1 - \beta$ will be symmetrical, i.e., the same for a true difference of, say, $-4$ as for a true difference of $+4$.

**Power of a statistical test.** The value of $\beta$ for any true difference is referred to as the *power* of the statistical test for detecting that true difference. When we plot $\beta$ against supposed true differences (see Fig.

**Table 5.7. Probability (β) of correctly rejecting the null hypothesis and probability (1 − β) of type II error, two-tailed, .05 level test with sample sizes of $N$ and $4N$, leading to $\sigma_D$ of 2 and 1 respectively**

| True difference | $\beta$ | | $1 - \beta$ | |
|:---:|:---:|:---:|:---:|:---:|
| | $N$ | $4N$ | $N$ | $4N$ |
| 1 | .08 | .17 | .92 | .83 |
| 2 | .17 | .52 | .83 | .48 |
| 3 | .32 | .85 | .68 | .15 |
| 4 | .52 | .979 | .48 | .021 |
| 5 | .71 | .999 | .29 | .001 |
| 6 | .85 | | .15 | |
| 7 | .94 | | .06 | |
| 8 | .979 | | .021 | |
| 9 | .995 | | .005 | |
| 10 | .999 | | .001 | |

5.4) we have a picture of the *power function* of the test. Note that power increases with the magnitude of the true difference. Note further that the power of a one-tailed test is higher than that of a two-tailed test for given $N$s and same $\alpha$. Note also that the power of the two-tailed test is much increased by quadrupling sample size(s). These curves would be higher for $\alpha = .10$ and lower for $\alpha = .01$. (See Table 5.6 for comparison of power, $\beta$, for three levels of significance.)

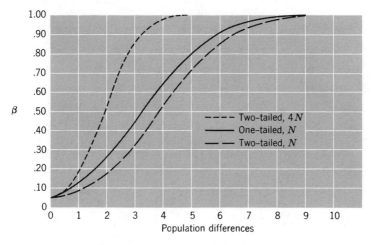

**Fig. 5.4.** Power curves, $\alpha = .05$, $\sigma_D = 2$ for $N$.

Since $1 - \beta$ gives the probability for a type II error, it follows that any increase in power decreases the probability of erroneously accepting the null hypothesis. We may increase power and decrease the risk of a type II error three different ways: (a) by taking $\alpha$ larger, say .10 instead of .05, with consequent increase in risk of the type I error; (b) by using a one-tail test, when appropriate, instead of a two-tailed test; (c) by increasing sample size(s) and thereby decreasing the standard error. Later we will discuss the possibility of sometimes reducing standard errors by suitable experimental design. We shall also see that a significance test for a difference in location parameters (i.e., central values) is more powerful when based on means instead of on medians, because the means have smaller sampling errors for given sample sizes.

The foregoing delineation concerning the type II error involves a seldom mentioned paradox. $\beta$ was defined as the probability of detecting an existing difference, or rejecting the null hypothesis when a real difference exists. Hence $\beta$ apparently is the probability of a correct conclusion and is usually so designated in tabular forms similar to Table 5.5. Let us consider what happens if for the situation involving a two-tailed test, $\alpha = .05$ and $\sigma_D = 2$, we suppose that the true difference is only $+0.1$. If so, the sampling distribution curve would center at $+0.1$, with a proportionate area of .0281 in the region of rejection to the right of $+3.92$ and an area of .0222 in the region of rejection to the left of $-3.92$. Thus the probability of rejecting the null hypothesis becomes the sum of these two areas; i.e., $\beta = .0503$, but is this the probability of *correctly* rejecting the null hypothesis in the sense of leading to a correct conclusion (or decision)? The answer is "No," as can be seen from the fact that .0222 of the times we would falsely conclude that the difference is negative, whereas it is positive 0.1. The third type of error is less apt to occur when the true difference is not small. For instance, if the true difference is $+1.0$ (or one-half the standard error), the $\beta$ value of .08 has a correct component of about .07 and a third type error part of about .01. The probability of the type III error is only .002 when the true difference is $+2.0$, or once the standard error. This sort of thing cannot happen in the one-tailed test situation simply because one does not consider differences in the opposite direction as admissible. A strict adherence to the one-sided hypothesis as an alternative to the null hypothesis does not lead to a test of significance when the result fails to be in the specified direction.

The reader may have thought correctly that the two power curves in Fig. 5.4 for the two-tailed tests should have included components for negative true differences. Such components will be symmetrical to those for positive differences and will swing upward as one proceeds leftward from zero, or no difference. What the reader may not have surmised is that the usual depiction of the power curve for a one-tailed test includes a segment for true values opposite to (not included in) the admissible values subsumed under the directional hypothesis as an alternate to the null hypothesis. This segment descends as the true values deviate from zero, becoming asymptotic to the base line. We have omitted this segment because the power to detect a difference in direction opposite to nonnull, one-sided admissible values is not relevant for the hypothesis being tested by a one-tailed test.

One further point on the power curves. We have already noted that when the true difference is $+0.1$, the value of $\beta$ based on the two-tailed test is .0503. Note how close this is to .05, the value of $\alpha$. The one-tailed test yields a $\beta$ of .0559 for a true difference of $+0.1$. The nearer the true difference is to zero, the closer $\beta$ is to $\alpha$ for both the one- and the two-tailed tests. What happens when the true difference is exactly zero? The answer is not what you would deduce from the curves in Fig. 5.4 or from similar curves in other textbooks. By definition, $\beta$ is the probability of detecting a specified true difference; if the true difference is zero, there is nothing to detect and therefore $\beta$ is not definable. It is not .05. Similarly, the probability of making the type II error approaches $1 - \alpha$ as the true difference approaches zero, but at exactly zero the probability of a type II error is not $1 - \alpha$, simply because one cannot falsely accept the null hypothesis when no true difference exists.

**Another strategy.** Although contemporary mathematical statisticians usually consider hypothesis testing in terms of a definite reject-accept decision according to whether the chosen level of significance is or is not reached, there is another possibility. We might follow the rule of rejecting the null hypothesis when $P$ is less than .01 (say), accepting it when $P$ is greater than .10, and reserving judgment when $P$ is between .10 and .01. This, in effect, introduces a region of indecision, or calls for a postponement of decision until the experiment is repeated or more data are collected. Another possibility, when a decision is not required for some practical reason, is simply to report that a difference is significant at the .09 or the .04 or the .002 or whatever level is reached, and then let the reader evaluate the finding according to his own preferred level of significance (which he is apt to do anyway unless he is too naive).

There are a couple of other points regarding significance. First, a statistically significant difference doesn't necessarily mean a difference either of practical significance or of scientific import. Sometimes a "what of it" is not an impertinence. Second, the habit of merely checking to see whether a result reaches a chosen level of significance should not lead us to overlook the possibility of claiming, when appropriate, that a much higher level of significance was attained than the preresearch chosen level.

## FURTHER NOTE ON JOINT PROBABILITIES

While indicating a method of attack on the problem of testing the difference between independent proportions (p. 59), we learned that starting with two symmetrical binomial distributions we can produce a table of joint probabilities (Table 5.3). What happens when the binomial distributions are skewed, i.e., when $p$ is not equal to $1/2$?

Let us say that (persons) $J$ and $K$ become bored with their coin tossing and turn to rolling dice. Each has a pair of dice, and "success" is defined as the getting of snake eyes. By the binomial, $(p + q)^n = (1/6 + 5/6)^2$, the probabilities for 2, 1, and 0 snake eyes are $1/36$, $10/36$, and $25/36$, respectively. With these in mind we can quickly, by use of the multiplication theorem, calculate the probabilities for joint outcomes for simultaneous rolls. The results are assembled in Table 5.8, which is analogous

**Table 5.8. Expected frequencies for joint outcomes when $J$ and $K$ each roll 2 dice simultaneously 1296 times**

| $E_k$ | 900 | 360 | 36 | $E_j$ |
|-------|-----|-----|-----|-------|
| 2 | 25 | 10 | 1 | 36 |
| 1 | 250 | 100 | 10 | 360 |
| 0 | 625 | 250 | 25 | 900 |
|   | 0 | 1 | 2 | 1296 |

to Table 5.3. The "scores" on the left and bottom margins are 2, 1, and 0 as number of successes (snake eyes). Notice that the marginal ($E_j$ and $E_k$) values result from a "blowup" of the binomial based on 36 tosses for a total of 1296 tosses. Each of the three rows (and columns) contains binomial distributions for 36, 360, and 900 rolls. Each cell entry provides the numerator for a joint probability figure, all with a common denominator of 1296.

Basically, as was also the case in Table 5.3, $J$'s separate probabilities are the same for any and all of $K$'s possible outcomes. This is as it should be since it is difficult to see how $J$'s outcomes can in any way depend on $K$'s outcomes. We have an a priori basis for saying that their outcomes are independent. With this in mind, it is interesting that probability theorists say that independence holds when it can be shown that the *conditional probability* (the probability of event $E$, given that event $F$ has occurred) is the same as the probability that event $E$ will occur irrespective of event $F$.

Later in the text we will encounter a situation in which events are not independent, hence the conditional probabilities will not be the same as marginal (unconditional) probabilities.

Next, we proceed to utilize Table 5.8 to establish the chance distribution for sum "scores" and for difference "scores" for simultaneous rolls. The sum values, obviously, can range from 0 to 4 successes, while the difference scores ($J$'s minus $K$'s) will range from $-2$ through 0 to $+2$. Again, as happened when we used Table 5.3 to generate Table 5.4, we can sum the joint expectancies in Table 5.8 diagonally from upper-left to lower-right to get expected frequencies for the sum scores, and from lower-left to upper-right to get the expectancies for the difference scores. The results are set forth in Table 5.9. Unlike those in Table 5.4, the two $E_f$ columns

Table 5.9. Expected frequencies for sums ($\Sigma$) and differences ($D$) in number of snake eyes for 1296 simultaneous rolls of 2 pairs of dice

| Sums | | Differences | |
|---|---|---|---|
| $\Sigma$ | $E_f$ | $D$ | $E_f$ |
| 4 | 1 | 2 | 25 |
| 3 | 20 | 1 | 260 |
| 2 | 150 | 0 | 726 |
| 1 | 500 | $-1$ | 260 |
| 0 | 625 | $-2$ | 25 |
| | 1296 | | 1296 |

are not identical. Those for the sums are, of course, nothing more than the expectancies for snake eyes when 4 dice are rolled, obtainable from the binomial $(p + q)^4$ with $p = 1/6$. The variance of the distribution of sum scores is given by $npq = 4(1/6)(5/6)$, which can be partitioned in an additive fashion as $4(1/6)(5/6) = 2(1/6)(5/6) + 2(1/6)(5/6)$. This again illustrates a theorem concerning variances: the variance of sum scores equals the sum of the variances (of the scores being added).

What may have come as a surprise to some readers is the fact that the distribution of difference scores arising from two skewed distributions turns out to be symmetrical. But a hasty generalization that this is always true is not tenable. (The skeptic should figure out what would be expected to happen if $K$ changes his scoring scheme: $J$ continues with success as the getting of snake eyes, whereas $K$ redefines his success as the getting of any outcome other than a snake eye. The $p$ for $J$ remains $1/6$ but that for $K$ becomes $5/6$, which reverses the skewness of his distribution from positive to negative. The distribution of differences in number of successes for simultaneous rolls will be skewed. Actually, the $E_f$ for the difference scores will be exactly the same as the $E_f$ for sums in Table 5.9, and the $E_f$ for the new sums will be the same as those for the differences in Table 5.9.)

One should not be surprised that the mean of the difference scores in Table 5.9 is zero. What of the variance? If you compute it directly from the distribution you will learn that it is $4(1/6)(5/6)$, which again can be partitioned into two added variances. The variance of the difference between two independent scores is the sum of the separate variances. Now this last variance for differences, $4(1/6)(5/6)$, which is identical with the variance of the sums, would seem to be the variance of a binomial distribution. But do the frequencies $(E_f)$ of differences in Table 5.9 follow a binomial distribution? Recall that all basic symmetrical binomial distributions begin and end with 1, with adjacent values of $n$ as frequencies (see, e.g., Table 5.1). Can the distribution of differences in Table 5.9 be regarded as a blown-up binomial? If so, we can reduce it back to basic frequencies by a process of division. Dividing by 25 will obviously yield the appropriate end values of 1 and 1, but when all 5 frequencies are divided by 25 we obtain 1, 10.4, 29.04, 10.4, and 1. The fact that we have some fractional values leads us to suspect that this is not a binomial distribution, and the fact that the only basic symmetrical binomial with 5 terms is 1, 4, 6, 4, and 1 confirms the suspicion. We conclude that differences between binomial variates can be specified a priori as to mean and variance but the shape of the distribution is not easily foreseen. Because of the obvious connection between scores as number of successes and scores as proportions, a corollary is that the differences between independent proportions may not follow exactly a binomial distribution, as required for use of the normal curve approximation of probabilities for differences. These considerations should help the reader understand the rule-of-thumb restriction given after formula (5.6) for its use.

## SUMMARY

In this chapter we have given a brief account of the concept of probability and have sketched procedures for applying probability notions in the testing of hypotheses involving frequencies and proportions (or percentages). We have noted the conditions for which it is safe to use a $z$ and the normal curve to approximate probabilities. If these conditions do not hold (when samples are small or proportions are extreme), we can obtain $P$ exactly by way of the actual binomial expansion for situations involving one proportion and for two correlated proportions. For proportions based on independent samples, exact $P$s may be ascertained by another, and more complicated, method to be presented later (p. 272).

The discussion of this chapter is only an introduction to the theory of statistical inference, or the use of probability in the testing of hypotheses. We have, however, developed the general principles. The extension of the theory to hypotheses involving continuous variables for relatively simple situations will be given in Chapters 6 and 7, with methods for more complex situations being postponed to later chapters (14–19). In Chapter 13 we shall discuss more extensive procedures for handling hypotheses regarding frequencies and proportions.

# Chapter 6

# INFERENCE: CONTINUOUS

# VARIABLES

As will be recalled, a frequency distribution for measurements on a continuous variable is describable with respect to central value, variability, skewness, and kurtosis; hence hypotheses involving continuous variables will be concerned with at least one of the descriptive measures of these four features of a frequency distribution. To test a given hypothesis, we need information regarding the sampling behavior of the descriptive measure being used (or of some ratio containing the measure).

In Chapter 5 we were able to make certain easy deductions. We saw, at the intuitive level, that the sampling distribution of proportions and of differences between proportions tends, under specified conditions, to be normal in distribution with specifiable standard deviation, and that we could set up a deviation, $x$, such that the ratio $x/\sigma$ tends to follow the unit normal curve. Unfortunately, the behavior of random sampling distributions of the measures that describe frequency distributions cannot so readily be determined. Accordingly, we will need to lean on the deductions made by the mathematical statistician, who has the task of ascertaining mathematically the characteristics of random sampling distributions when certain conditions and assumptions hold. We can learn how to use his results as a basis for testing hypotheses without necessarily understanding his mathematical derivations.

Since hypotheses involving means arise frequently in practice and since inferences based on means serve to illustrate further the general theory of statistical inference, we shall give considerable detail on sampling errors connected with means. We shall present first an easily duplicated demonstration of the chance variation of means, and then a discussion of some theory and its use as a basis for hypothesis testing. This chapter will be restricted to the large sample situation, with requisite sample size specified at appropriate times.

## EMPIRICAL DEMONSTRATION

The operation of chance sampling errors for means and standard deviations can be illustrated by tossing, say, 7 coins 50 times and tabulating the number of heads per toss. The obtained frequencies will usually vary somewhat from those expected, which would be proportional to 1, 7, 21, 35, 35, 21, 7, 1 (as obtained by the binomial expansion). When the mean number of heads for 50 tosses is computed, it is not likely to be exactly 3.5 ($np$, the mean of the expected distribution), and the discrepancy from 3.5 can be attributed to chance. Likewise, 100 tosses will show departures from the expected frequencies, and consequently the mean based on 100 tosses will differ more or less from 3.5. Furthermore, and for the same reason, the standard deviation of the obtained distribution of heads will likely differ from 1.323 ($\sqrt{npq}$, the $\sigma$ of the expected frequencies). As an exercise the student can demonstrate the foregoing statements by actually tossing coins. Indeed it will be quite instructive if each class member tosses 7 coins 50 times, each time tallying the number of heads that turn up. This will lead to a frequency distribution running (possibly) from 0 to 7 heads, with an $N$ of 50. Then a second series of 50 tosses should be made, thus providing a second distribution. The two frequency distributions can be combined, so each student will have three distributions, two with $N$s of 50 and one with an $N$ of 100. Note that chance is so operating as to produce a distribution somewhat similar to the expected, but at the same time is operating in such a manner as to lead to discrepancies between observed and expected frequencies.

Each student should compute the mean and the standard deviation for each of the three distributions. Note how far these values depart from the expected mean of 3.5 and the expected standard deviation of 1.323. Then the several means and standard deviations secured by the class members should be brought together. In order better to understand what happens when each of several persons tosses 7 coins 50 times, i.e., takes a sample of 50 tosses, a frequency distribution of the $M$s, also of the $S$s, based on 50 tosses should be made. Likewise a separate distribution should be made for the $M$s based on 100 tosses; also, the $S$s. A study of these distributions should provide answers to such questions as: Their central tendencies are near what values? What is the extent of dispersion for these distributions of $M$s and $S$s? Is there any difference in the

dispersion for the distribution of means based on 50 tosses and that based
on 100 tosses? How would you account for this difference? In general,
what is the shape of these distributions of $M$s and $S$s?

Table 6.1 shows the distributions of the means obtained by several of
the author's classes. Though these are not models for number of intervals,
they are nevertheless sufficient as a basis for answering the foregoing
questions. Note that both distributions appear to be normal, that both
center very near the mean of the theoretical distribution (3.5), and that the

Table 6.1. Distribution of 600 means based on 50 tosses and
300 means based on 100 tosses of 7 coins

|  | 50 Tosses | 100 Tosses |
| --- | --- | --- |
| 4.00–4.09 | 3 | |
| 3.90–3.99 | 14 | |
| 3.80–3.89 | 35 | 4 |
| 3.70–3.79 | 50 | 23 |
| 3.60–3.69 | 98 | 58 |
| 3.50–3.59 | 119 | 78 |
| 3.40–3.49 | 120 | 85 |
| 3.30–3.39 | 85 | 32 |
| 3.20–3.29 | 52 | 17 |
| 3.10–3.19 | 21 | 3 |
| 3.00–3.09 | 2 | |
| 2.90–2.99 | 1 | |
| Number of means | 600 | 300 |
| Mean of means | 3.516 | 3.513 |
| $S^*$ of distribution of means | .190 | .135 |
| Expected $S$ | .187 | .132 |

\* Corrected for grouping.

variability for means based on 100 tosses is less than that based on 50
tosses. It would thus seem that means based on 100 tosses are somewhat
more stable or less variable than those based on 50 tosses. Does this
suggest that a larger number of tosses, i.e., a larger sample, would tend to
iron out the chance factors that operate to produce discrepancies between
the observed distribution of number of heads and the expected distribution
calculated by the binomial expansion? Do you think that means based on
500 tosses would show less dispersion than means based on 100 tosses?

According to the mathematical statisticians, the standard deviation of
the distribution of means is expected to be equal to 1.323 (expected $\sigma$ of
the distribution of number of heads) divided by the square root of the

sample size. Note at the bottom of Table 6.1 that the $S$s of the distributions of means, .190 and .135, are very near the expected values of .187 and .132 obtained from $1.323/\sqrt{50}$ and $1.323/\sqrt{100}$, respectively.

Summarizing the results of the foregoing empirical work, we see that the means for successive samples tend to distribute themselves normally about the expected or universe mean, $\mu$, with a spread or standard deviation which is very near the value predicted by mathematical theory. The student should keep these empirical distributions and deductions therefrom in mind as we now proceed to a more detailed consideration of what the mathematical statistician says will happen when successive samples of a given size are drawn from a defined universe or population or supply.

## MORE SAMPLING THEORY

The discussion here holds for what is known as simple *random sampling*. As specified in Chapter 5, the conditions for simple random sampling are that the sample should be drawn in such a way that each individual (person, plant, animal, etc.) in the defined universe shall have an equal chance of being included in the sample, and that the drawing of one individual shall in no way affect the drawing of another. The aim is, of course, to obtain a sample which will, within limits of random or chance errors, be representative of the universe from which it was drawn.

Let

$N$ = the number of cases, or size of sample.
$M$ = the mean of any sample (known, i.e., computed).
$S$ = the standard deviation of any sample (known, i.e., computed).
$\mu$ = the mean of the defined population (unknown).
$\sigma$ = the standard deviation of the defined population (unknown).

The $\mu$ and $\sigma$ are for the distribution of scores or measurements for all the individuals in the defined universe. It is not assumed that this universe distribution is exactly normal; it may be skewed slightly. Strictly speaking, the number, $N_{pop}$, of cases in the universe should be infinitely large, but failure to meet this requirement is not serious. As will be seen later, the adjustment necessary when a sample of $N$ cases is drawn from a limited (finite) universe of $N_{pop}$ cases is of the order of $N/N_{pop}$; if it is known that $N_{pop}$ is very large relative to $N$, the formulations about to be presented will be sufficiently accurate for all practical purposes.

Now suppose we draw a sample of $N$ cases, compute the mean and standard deviation, then draw another sample of the same size and compute its mean and standard deviation, and so on until a large number of samples, say 10,000, have been drawn. We will then have 10,000 means and 10,000 standard deviations, each based on $N$ cases. When we make a distribution of the 10,000 means and of the 10,000 standard deviations, we have random sampling distributions. From the point of view of mathematical rigor, the number of successive samples should be much larger than 10,000, certainly far larger than the 600, or 300, successive samples of Table 6.1, in which we have only the beginning of two random sampling distributions.

By rather complex mathematical methods it can be shown that, *if* successive samples of constant size, $N$, are drawn randomly from a normally distributed universe or population with mean equal to $\mu$ and standard deviation equal to $\sigma$, the successive sample means will be normally distributed about $\mu$, and the standard deviation of this sampling distribution will be $\sigma/\sqrt{N}$. The random sampling distribution of the successive standard deviations will center at $\sigma$ (there is a small bias here which need not concern us at this time). For $N$ large (100 or more) this distribution of $S$s will be approximately normal with standard deviation equal to $\sigma/\sqrt{2N}$. These mathematical findings have often been checked empirically. Table 6.1 provides a limited check on the sampling theory regarding the mean.

We are now in position to consider a term used in Chapter 5. In general, the *standard error* of a statistical measure is the standard deviation of the sampling distribution for the given measure. The square of the standard error is called the *sampling variance*. For the practical statistician, the sampling distribution is hypothetical, and hence its standard deviation must be determined by a different formula from that used for computation from an actual distribution. The value given by $\sigma/\sqrt{N}$ is called the standard error of the mean and may be designated as $\sigma_M$. Each sample mean can be expressed in relative deviate form as $(M - \mu)/\sigma_M$, and these relative deviates will form a normal distribution with mean of zero and standard deviation of unity. By reference to Table A we can readily specify the chances of obtaining a sample mean yielding a deviation as great as that for a given $M$, provided the value of $\mu$ is known. But in practical work $\mu$ is the unknown about which we desire to make an inference on the basis of just one sample.

Before resolving this practical problem, we must call attention to the fact that the universe standard deviation, $\sigma$, needed to obtain $\sigma_M$ is also an unknown. A single sample will yield a standard deviation, $S$, which, being a sample value, will of course deviate more or less from $\sigma$. In order that an inference about $\mu$ may be made from a single sample, $\sigma_M$ is estimated by using $S/\sqrt{N}$; i.e., the unknown $\sigma$ is replaced by the sample $S$ as an estimate. Instead of the true value for the standard error of the mean as given by $\sigma/\sqrt{N}$, we have an approximate value, $S/\sqrt{N}$. Let $S_M$, defined as $S/\sqrt{N}$, stand for the approximate standard error.

The ignorance concerning $\sigma$, and the consequent approximate value for the standard error of a given mean, lead to a reconsideration of the sampling distribution of means expressed as relative deviates. As already pointed out, the means from successive samples will be distributed normally, and the relative deviates, $(M - \mu)/\sigma_M$, will likewise be distributed normally since $\sigma_M = \sigma/\sqrt{N}$ is a constant. When (as is nearly always the case) we have $S$ instead of $\sigma$ and wish to make an inference about a universe mean, we need to know something of the sampling behavior of successive sample means expressed as relative deviates from $\mu$ where $S_M$ is not a constant but varies from sample to sample because the several sample standard deviations vary. Thus the relative deviate of the first sample mean will be $(M_1 - \mu)$ divided by $S_1/\sqrt{N}$; for the second sample, $(M_2 - \mu)$ divided by $S_2/\sqrt{N}$; and so on. The distribution of these relative deviates will *not* approximate normality unless $N$ is fairly large. Thus the use of an estimate of $\sigma$ in determining $\sigma_M$ imposes the restriction that $N$ shall not be too small. If $N$ is not less than 30, we can safely use the normal curve as the basis for drawing an inference or testing a hypothesis regarding $\mu$. This chapter's discussion of sampling is therefore not applicable unless $N$ is greater than 30. The refinements necessary for $N$s less than 30 will be given in Chapter 7.

## HYPOTHESES REGARDING A SINGLE MEASURE

Whether the foregoing theory is used as a basis for making an inference about a population value or for testing some hypothesis depends on the practical problem faced by the investigator. We shall now consider hypothesis testing, and later we shall discuss a type of inference which is useful both when we do and do not have a research hypothesis in mind.

**Single mean.** The procedure for testing a hypothesis about a population mean on the basis of a sample mean (and $S$) for $N$ cases is very similar to that for testing a hypothesis when we have a sample proportion (discussed earlier, pp. 50–52). We let $M_h$ stand for a hypothesized value of $\mu$. Our sample mean, $M$, taken as a deviation from $M_h$, is expressed in the form of a $z$, that is, as $(M - M_h)/S_M$. The theory tells us that if $M_h$ is true (i.e., corresponds to $\mu$), successive sample $M$s will be distributed normally about $M_h$ with standard deviation $= S_M$ (approximately). In testing the given hypothesis we are merely raising the question as to whether it is reasonable to believe that our observed sample mean, $M$, belongs to a sampling distribution centering at $M_h$. Put differently, does $M$ deviate significantly from $M_h$? To answer this we need to know the probability of as large a deviation on the basis of chance sampling errors, and to get this probability we need only enter Table A with $(M - M_h)/S_M$ as a $z$. If we have decided to adopt the $P = .01$ level for judging significance, we reject the hypothesis when $(M - M_h)/S_M$ reaches 2.58 (for a two-tailed test).

Actually, there are relatively few occasions in psychological research for which either scientific theory or prior observation provides us with a hypothesis concerning the mean of a population on some variable. An exception is the mean of changes, to be discussed shortly.

As an example of a situation for which the testing of a hypothesis about a mean is appropriate we cite the IQ tests. For reasons which we shall not discuss here, a properly constructed test should yield 100 as the average IQ for the population of children for any given age level. Consider Form L of the 1937 Revision of the Stanford-Binet Scale. For age 7, a sample of 202 gives a mean of 101.78 and an $S$ of 16.18. The value of $S_M$ becomes $16.18/\sqrt{202} = 1.14$. From these figures we have $(M - M_h)/S_M = (101.78 - 100)/1.14 = 1.56$ as a $z$. Turning to Table A we find that the $P$ for as large a deviation (irrespective of direction—a two-tailed test is needed here) from 100 is .12. Since this probability is not as small as our arbitrarily chosen $P = .01$ level of significance, we accept the hypothesis that the 1937 Stanford-Binet meets the requirement of yielding an average of 100 at age 7. That the scale was not entirely satisfactory in this regard is evident when we consider the $M$ of 104.28 and $S$ of 16.42 for a sample of 204 nine year olds. We have $S_M = 1.15$, which leads to a $z$ of $(104.28 - 100)/1.15 = 3.72$. Since the probability of as large a deviation is about .0002, we reject the hypothesis that the scale would yield a population mean of 100 at age 9.

**Significance of mean change.** A frequently encountered problem is that of evaluating changes in order to say whether some provided experience or change in conditions leads to a shift in performance.

Let

$X_1 =$ score prior to experience (or under one condition).

$X_2 =$ score after the experience (or under second condition).

$D = X_2 - X_1 =$ change score.

Or we might take $D = X_1 - X_2$ if losses instead of gains are of interest, but regardless of which way we define the $D$ score, the subtraction is made in the same direction for all $N$ cases and negative signs are kept. A sample of $N$ individuals will give us $N$ changes, or $N$ $Ds$. We can either make or conceive of a distribution of the $Ds$. This distribution will have a mean, $M_D$, and a standard deviation, $S_D$, whence we can get the standard error of the mean difference: $S_{M_D} = S_D/\sqrt{N}$. In other words, a mean change is treated just like any other mean. Regardless of any hunch or prediction about the effect of the experience (or the effect of the change in conditions), the null hypothesis is set that there is no effect. This is equivalent to saying that, if we had $X_1$ and $X_2$ scores on the defined population, the value of $\mu_D$ would be zero. If this hypothesis is true and if we were to take successive samples of size $N$, we would expect that the sample means would be distributed normally about zero with $S = S_{M_D}$. To test the null hypothesis we simply take our obtained $M_D$ as a deviation from the null value of zero and divide by $S_{M_D}$. That is, $(M_D - 0)/S_{M_D} = M_D/S_{M_D}$. This as a $z$ is then used as an entry into Table A in order to specify the probability of as large a mean difference as our sample $M_D$ arising solely on the basis of chance sampling. Whether we reject or accept the hypothesis of no effect depends on whether $P$ does or does not reach the chosen level of significance. We could use a one-tailed test here if the research hypothesis predicted the direction of the change, but if we had no a priori hypothesis as to the direction of change we would need to use the two-tailed test.

A word should be inserted about the required computations since there is some danger of confusion when we are confronted with the calculation of $M$ and $S$ for scores (changes) which are both positive and negative, and sometimes zero. The gross score formula for the mean (3.2) and that for the standard deviation (3.7) are applicable provided we take $\Sigma D$ (equivalent to $\Sigma X$) as the algebraic sum. The equivalent of $\Sigma X^2$, that is, $\Sigma D^2$, raises no problem since the squaring process automatically eliminates negative signs. There are two reasons why we should make a frequency distribution of the $D$s. First, the theory assumes that the $D$s approximate a normal distribution; if a distribution is made we have at least a rough check on this assumption (there are statistical methods for checking this assumption; see p. 89 and also p. 267). Second, if $N$ is sizable, computation from a frequency distribution is more economical of time than use of the gross score formulas. In laying out the intervals, we must provide a place for tabulating zero $D$s. This can conveniently be accomplished by the following illustrative scheme which includes only the four intervals near zero: 2–3, 0–1, −1–2, −3–4 (for $i = 2$); 3–5, 0–2, −1–3, −4–6 (for $i = 3$); 4–7, 0–3, −1–4, −5–8 (for $i = 4$); etc. Note that the last given intervals in each set are for negative $D$s. $AO$ taken as the midpoint of the bottom interval will be a negative number, and must be treated as such when entered into formula (3.3).

**Other single measures.** The general theory of statistical inference may be extended to testing hypotheses concerning any descriptive measure, provided information is available (from the mathematical statistician) concerning the characteristics of the random sampling distribution of the measure. When the sampling distribution is normal in form with known or estimable variability, we may proceed to test hypotheses by setting up a $z$, or $x/S$, or $x/\sigma$. For this purpose we need formulas for the standard errors of different measures. The formulas about to be presented are based on the assumption that the score distribution is normal or approximately so.

As previously noted, for $N$ greater than 30 we may safely use

$$S_M = \frac{S}{\sqrt{N}} \tag{6.1}$$

as the standard error of the mean. For $N$ greater than 100 it is safe to take

$$S_{mdn} = \frac{1.253S}{\sqrt{N}} \tag{6.2}$$

as the standard error of the median. A comparison of the standard error of the mean with that of the median indicates that the mean fluctuates less than the median; i.e., the mean is a more stable measure of central value than the median. In order to reduce the standard error of the median to the same magnitude as that of the mean it is necessary to take 57 per cent more cases, i.e., increase $N$ by 57 per cent. It follows from this that the use of the median for distributions which are reasonably normal in form is equivalent to throwing away a large proportion of the cases.

The sampling errors involved in measures of dispersion are

$$S_S = \frac{S}{\sqrt{2N}} = \frac{.707S}{\sqrt{N}} = .707S_M \tag{6.3}$$

$$S_{AD} = \frac{.756(AD)}{\sqrt{N}}$$

$$S_Q = \frac{1.166(Q)}{\sqrt{N}}$$

From these error formulas it will be seen that, considering the error relative to the magnitude of the measures of dispersion, $S$ is the most stable measure of variation. Provided $N$ is 100 or more, the sampling distributions for these measures of dispersion are such that their standard errors can be utilized in exactly the same way as the standard error of the mean.

The standard errors for measures of skewness and kurtosis, as defined on p. 26, are

$$\sigma_{g_1} = \sqrt{\frac{6}{N}} \tag{6.4}$$

$$\sigma_{g_2} = \sqrt{\frac{24}{N}} = 2\sqrt{\frac{6}{N}} = 2\sigma_{g_1} \tag{6.5}$$

These two formulas are based on the assumption that the sample has been drawn from a normally distributed population, and therefore they can be legitimately used in testing the assumption of normality. It will be recalled that, for normal distributions, both $g_1$ and $g_2$ are equal to zero, but for a sample they may not be zero; however, sample values should not show a greater deviation from zero than can be reasonably attributed to chance. If a sample yields a $g_1$ value which is more than, say, 2.58 times its sampling error, we would suspect that the sample was not drawn from a symmetrically distributed supply. Likewise, if $g_2$ deviates more than 2.58 times its standard error, we would question whether it is reasonable to believe that the population or supply is distributed with normal kurtosis. A two-tailed test is appropriate here, and consequently choosing 2.58 is equivalent to adopting the .01 level of significance.

## HYPOTHESES ABOUT DIFFERENCES

One of the foremost problems in practical statistics is the comparison of group trends. We may wonder whether one college group is superior to another, whether practice on a task improves performance, whether rats learn more rapidly when food or when water is the incentive, whether reaction time is faster to sound than to light, whether the sexes show a difference in variational tendency, whether one learning method is better than another, etc. In order to answer questions like the above, it is necessary to make observations on samples from two groups or on the same group under two different experimental conditions, and then to compute appropriate statistical measures for the variable on which we wish to make the comparison.

Thus, typically, we have two samples of $N_1$ and $N_2$ cases or two sets of scores on just $N$ cases under two different conditions, with means $M_1$ and $M_2$ and standard deviations $S_1$ and $S_2$, where the subscripts refer to the two sets of scores. As we have learned, each mean is subject to sampling fluctuations; therefore the difference between the means will also be subject to sampling fluctuations. Even though $\mu_1 = \mu_2$ there may be a difference between sample means because of chance sampling errors. To test an obtained difference for significance we will need a measure of the sampling error of differences, i.e., the standard error of the difference between two means. Knowing this standard error we can set up the null hypothesis that there is no difference between the two population means and then reject or accept this hypothesis according to whether the obtained difference does or does not reach an appropriate level of significance.

Here, as in the case of the difference between proportions, we must distinguish between the situation where our two means are based on independent as opposed to nonindependent (correlated) scores.

**Difference between correlated means.** Let us again consider the method outlined previously for testing the significance of a mean change. As implied there, the $X_1$ and $X_2$ scores could stand for performance for $N$ individuals under two different conditions. A little simple algebra at this point will lead to some interesting results. As before, we let

$$D = X_2 - X_1$$

By definition the mean of the distribution of these $N$ difference scores will be

$$M_D = \frac{\Sigma D}{N} = \frac{\Sigma(X_2 - X_1)}{N}$$

$$= \frac{\Sigma X_2}{N} - \frac{\Sigma X_1}{N}$$

hence

$$M_D = M_2 - M_1 = D_M$$

by which we see that the *mean of the differences is equal to the difference between the means.* This will, of course, be true for every sample. It follows therefore that when we test the significance of $M_D$ as a deviation from zero we are also testing the significance of $D_M$ as a deviation from zero. In other words, we are testing the significance of the difference between two means based on the same $N$ cases.

When testing $M_D$, we calculated $S_D$, thence $S_{M_D}$. Let us consider a bit further the standard deviation of the distribution of differences, $S_D$. We first express the $D$s as deviations from their own mean, i.e., $d = D - M_D$. Since $D = X_2 - X_1$ and $M_D = M_2 - M_1$, we have

$$d = (X_2 - X_1) - (M_2 - M_1)$$

which, when the parentheses are removed and the terms shifted, becomes

$$d = X_2 - M_2 - X_1 + M_1$$

or

$$d = (X_2 - M_2) - (X_1 - M_1)$$

Both these new parentheses terms define deviation units of the type $x = X - M$, so that $d = x_2 - x_1$. The standard deviation squared, or variance, of the differences can be expressed by substituting $d$ for $x$ in formula (3.4); thus

$$S^2{}_D = \frac{\Sigma d^2}{N}$$

If we replace $d$ by its equivalent, we have

$$S^2{}_D = \frac{\Sigma(x_2 - x_1)^2}{N} = \frac{\Sigma x^2{}_2}{N} + \frac{\Sigma x^2{}_1}{N} - \frac{2\Sigma x_2 x_1}{N}$$

The first two of the three terms on the right are obviously the variances for the second and first sets of scores. The last term, involving the sum of the cross products of $x_2$ and the $x_1$ with which it is paired, has to do with the degree of correlation between, or similarity of, the scores that belong to the same individual. The reader is asked to take on faith, without further explanation here, the fact that the last term becomes $2r_{12}S_1S_2$, in which $r$ is a measure of correlation. Hence we can write

$$S^2{}_D = S^2{}_2 + S^2{}_1 - 2r_{12}S_1S_2 \tag{6.6}$$

or

$$S_D = \sqrt{S^2{}_2 + S^2{}_1 - 2r_{12}S_1S_2}$$

Since the standard error of any mean is given by dividing the standard deviation by the square root of $N$, we secure the standard error of the mean difference by dividing $S_D$ by $\sqrt{N}$, i.e.,

$$S_{M_D} = \frac{S_D}{\sqrt{N}} = \frac{\sqrt{S^2{}_1 + S^2{}_2 - 2r_{12}S_1S_2}}{\sqrt{N}}$$

$$= \sqrt{\frac{S^2{}_1}{N} + \frac{S^2{}_2}{N} - \frac{2r_{12}S_1S_2}{N}}$$

The first two terms under the last radical are the sampling variances of the two means, and since $2r_{12}S_1S_2/N$ can be written as

$$2r_{12}\frac{S_1S_2}{\sqrt{N}\sqrt{N}}$$

we have finally that

$$S_{M_D} = \frac{S_D}{\sqrt{N}} = \sqrt{S^2{}_{M_1} + S^2{}_{M_2} - 2r_{12}S_{M_1}S_{M_2}}$$

Since each $M_D = D_M$, it follows that $S_{M_D} = S_{D_M}$, or that the *standard error of the mean difference is equal to the standard error of the difference between the two means*. Thus we have two ways for evaluating a difference between nonindependent means. We can compute $M_D$, $S_D$; thence

$$S_{M_D} = \frac{S_D}{\sqrt{N}} \tag{6.7}$$

or we can compute $M_1$, $M_2$, $S_1$, $S_2$, and $r_{12}$, and then obtain

$$S_{M_D} = \sqrt{S^2{}_{M_1} + S^2{}_{M_2} - 2r_{12}S_{M_1}S_{M_2}} = S_{D_M} \qquad (6.8)$$

Formula (6.8) is usually referred to as the standard error of the difference between correlated means, hence the symbol $S_{D_M}$.

But by working with the difference between paired scores, we can obtain the standard error of the mean difference ($=$ difference between means) without computing $r$. Even after we have learned how to compute $r$, it matters not whether we compute the standard error of the difference between means of related scores by formula (6.8), or whether we compute its equivalent, the standard error of the mean of the differences.

Strictly speaking, the $r_{12}$ in (6.8) should be written as $r_{M_1 M_2}$ so as to indicate that it is a measure of the extent to which successive pairs of means vary together, but it can be shown that the correlation between means is the same as $r_{12}$, the correlation between the scores entering into the means.

Since $M_D = D_M$ and $S_{M_D} = S_{D_M}$, it should be obvious that when testing the null hypothesis we have

$$\frac{x}{S} = \frac{M_D}{S_{M_D}} = \frac{D_M}{S_{D_M}} = z$$

That is, the procedure for testing the null hypothesis that $M_D$ is zero for a population is equivalent to testing the null hypothesis that $\mu_1 = \mu_2$ where the subscripts 1 and 2 indicate that we are considering two populations of *scores*, one for each condition.

Formulas (6.7) and (6.8) are appropriate in a number of situations in which an $X_1$ score is somehow paired with an $X_2$ score. Some of the possibilities are the following:

*a.* $X_1$ as first trial—practice—$X_2$ as later trial; same person.
*b.* $X_1$ as initial—experience—$X_2$ as final; same person.
*c.* $X_1$ as pretest—experience—$X_2$ as posttest; same person.
*d.* $X_1$ under experimental conditions vs. $X_2$ under normal (or control); same person.
*e.* $X_1$ in one experimental condition vs. $X_2$ in another; same person.
*f.* $X_1$ as experimental vs. $X_2$ as control; twin or litter pair.
*g.* $X_1$ as experimental vs. $X_2$ as control; unrelated persons, but matched by pairing on pertinent variables. Ditto, for two experimental conditions.

For situation (*g*), which is commonly employed in experimental work, we can think of having drawn $N$ individuals at random for one group, then forming the second group by selecting individuals who can be paired with the members of the first group on the basis of variables which need to be controlled. Thus any found difference between $M_1$ and $M_2$ will not be attributable to differences between the two groups with respect to the variables used in forming the pairs, since the pairing tends to make the groups equivalent on the pairing variables. This same pairing procedure, and also twin or litter pairs, can be used for situation *e*. Furthermore, as we shall see below, the $X_1$ and $X_2$ scores can themselves stand for changes: $X_1$ the change from pretest to posttest under an experimental condition and $X_2$ the change under another experimental condition or under control conditions.

The statistical advantages of having scores which are somehow related will be discussed later under the caption "Reduction of sampling errors."

**Difference between independent means.** When we have means for two samples which have been drawn independently, there will be no way of pairing scores except on a chance basis and chance pairing will tend to produce a zero correlation. In fact, if we took all possible pairs the correlation would be exactly zero. Thus the correlation term in (6.8) vanishes, so that the standard error of the difference between means based on independent samples becomes

$$S_{D_M} = \sqrt{S^2{}_{M_1} + S^2{}_{M_2}} = \sqrt{\frac{S^2{}_1}{N_1} + \frac{S^2{}_2}{N_2}} \qquad (6.9)$$

This formula is not restricted to samples of the same size; i.e., $N_1$ need not equal $N_2$. The right-hand form of (6.9) has an obvious computational advantage.

The $S_{D_M}$ obtainable by formula (6.9) may be used in exactly the same manner as the standard error of the difference by formulas (6.7) and (6.8). Again, we set the null hypothesis that $\mu_1 = \mu_2$ or that the difference between the population means is zero. If it is zero, the sampling distribution of $D_M$ resulting from successive replications will center at zero with standard deviation $= S_{D_M}$. If $D_M/S_{D_M}$ (or $z$) is sufficiently large, the null hypothesis is rejected; if not, it is accepted. In other words the general procedure for testing hypotheses about differences is precisely the same for means (and other statistical measures) as that outlined in Chapter 5. The student would do well to review the discussion dealing with admissible hypotheses, one-tailed vs. two-tailed tests, choice of level of significance, and the two types of error one risks in testing hypotheses.

**Differences between other descriptive measures.** The general theory of hypothesis testing is applicable for descriptive measures other than proportions or means. The general pattern for the standard error of the difference between any two statistical measures, say $C_1$ and $C_2$ is

$$S_{D_C} = \sqrt{S^2{}_{C_1} + S^2{}_{C_2} - 2r_{C_1 C_2} S_{C_1} S_{C_2}}$$

That is, we need to know the standard error for both $C_1$ and $C_2$ and a measure of the correlation between $C_1$ and $C_2$ in case of nonindependence (the $r$ term drops out for independently drawn samples). This correlation, which is a measure of the extent to which $C_1$ and $C_2$ vary together when successive samples are drawn, is known to be $r_{12}$ when the $C$s are means and to be $r^2_{12}$ when the $C$s are standard deviations, with $r_{12}$ being the correlation between the scores entering into the means and standard deviations. Accordingly, the standard error of the difference between two $S$s based on the same individuals or on scores related consanguineously or related by pairing on pertinent variables is given by

$$S_{D_S} = \sqrt{S^2_{S_1} + S^2_{S_2} - 2r^2_{12}S_{S_1}S_{S_2}} \qquad (6.10)$$

and for $S$s based on independent samples

$$S_{D_S} = \sqrt{S^2_{S_1} + S^2_{S_2}} = .707S_{D_M} \qquad (6.11)$$

These formulas are valid for large $N$s (100 or more), and to test the null hypothesis we simply take $D_S/S_{D_S}$ as a unit normal $z$. (For $N$s small, see Chapter 14.)

The difference between medians based on correlated scores cannot be tested because the needed $r$ is unknown, but for independent samples we have

$$S_{D_{mdn}} = \sqrt{S^2_{mdn_1} + S^2_{mdn_2}}$$

Expressions for $S_{D_{AD}}$ and for $S_{D_Q}$ can be similarly written for the case of independent samples.

Any student who is worried because formula (5.5) for the standard error of the difference between correlated proportions does not include an $r$ term may rest assured that the correlation has been allowed for even though not visibly so. Formula (5.5) is analogous to formula (6.7), which we have seen is equivalent to the longer formula (6.8) in which there is an $r$.

## REDUCTION OF SAMPLING ERRORS

One of the aims of scientific method is to attain as great precision in results as is practicable. In statistical work this can be accomplished by increasing the accuracy or dependability of the scores or individual measurements or responses and by decreasing the chance sampling errors of the various descriptive measures. One way to reduce sampling errors is to employ either the stratified or the area method of sampling, both of which are too complicated for us to discuss here. If the random sampling method is being used in projects which aim to study the difference between groups (or populations), the obvious, and only, way for decreasing the standard error of the difference is to increase $N$ for either or for both samples. Most field investigations are of this type.

In contrast, the experimentalist can define his population with reference to two laboratory or experimental situations, i.e., a population of individuals under situation $A$ and a population of individuals under situation $B$. His sample individuals for the two situations may be the same individuals, first under the $A$ and then under the $B$ condition. In general, the use of the same individuals, if feasible in view of possible practice or fatigue effects, will usually involve a fairly high degree of correlation, the net effect of which is to reduce the standard error of the difference considerably; i.e., it is sometimes possible to reduce sampling error simply by using the same individuals as the "two" samples. Thus, if we wish to study the effect of two different degrees of humidity on mental output or efficiency, it will be a more economical and better controlled experiment if we make observations on the same individuals under the two conditions $A$ and $B$, rather than on $N_1$ individuals under condition $A$ and $N_2$ individuals under condition $B$.

If it is not feasible to use the same individuals in the two experimental situations, we can make up two groups by pairing or matching individuals on the basis of one or more characteristics. Such a procedure leads to more nearly comparable groups for our experiment than can be obtained by choosing individuals at random and, by using either formula (6.7) or (6.8) instead of (6.9), we can make allowance for the fact that the individuals for the two samples have not been chosen independently. The use of individuals who have been paired is considered good experimental technique—it cannot be said that a found difference between means for the variable being studied may be due to a lack of comparability of the two groups with respect to the matching variables. The use of paired individuals has a statistical as well as experimental advantage in that the sampling error of the difference between means is thereby reduced without the necessity of increasing the number of cases. If pairing produces an $r$ of .75, the reduction in $S_{D_M}$ is equivalent to that achieved by quadrupling the number of cases when the random method of forming groups is employed. After the student has learned about correlation he will better appreciate the fact that the gain in pairing depends on the extent to which the variables used in pairing are correlated with the variable being studied.

It is thus seen that, for some types of investigations, greater precision can be obtained by judicious planning. If we had unlimited resources, we could always attain any desired degree of precision by simply taking sufficiently large samples.

Frequently the question is raised as to how many cases should be secured for a given study. The answer might be in terms of the number needed to reach a given degree of accuracy, but this in turn would raise the question of what degree of precision is needed, and this in turn depends on how small a difference we wish to detect. When group comparisons are made and when the $N$s are relatively small, the null hypothesis is apt to be accepted too often for the simple reason that a real difference has to be sizable before it is demonstrable by small samples. On the other hand, if a real difference is so small that its statistical demonstration requires thousands of cases, we may question whether it has practical or scientific importance.

## COMPARISON OF CHANGES

Although the comparison of changes involves nothing new in the way of statistical theory, such comparisons are somewhat more complicated than the tests of significance so far discussed. The researcher may be interested in either of two questions. First, he may wish to evaluate the effect of only one experimental condition or, second, he may wish to contrast the changes produced under two (or more) different experimental conditions.

For the first of these, a sample is selected, measurements are made prior to (pretest) and subsequent to (posttest) the provided condition, but, since changes from a first to a second measure might occur because of practice effect or because of some other experience beyond the control of the investigator, it is necessary to set up a control group the members of which are measured and then remeasured, at chronological times corresponding as closely as possible to those of the pretest and posttest of the experimental group. It is presumed that all uncontrollable effects will be operating similarly on both groups so that any difference in change for the two groups will have resulted from whatever was done to the members of the experimental group. The statistical problem is that of evaluating the change shown by the experimental group compared with that shown by the controls.

For the second type of question the investigator starts with two experimental groups, one of which is subjected to one experimental condition and the other to a second experimental condition, both groups having been measured prior to the experience (pretest), and then again after the experience (posttest). Since the question is concerned with contrasting gains (or losses) associated with the two conditions, a control group is not needed. Presumably, uncontrollable factors are alike for the two groups. The statistical analysis consists of testing for significance the difference between the changes shown by the two groups.

Whether we are dealing with a problem calling for an experimental and a control group or for two experimental groups, the two groups may be drawn at random or formed on the basis of the pairing of individuals on pertinent variables. If the groups are set up on the basis of pairing, we need to allow for that fact when determining the required standard error of the difference between changes.

Parenthetically, it may be said that the setup which involves an experimental and a control group (or two experimental groups) for studying shifts has led to a great deal of confusion regarding the proper statistical handling of the data. We have a total of four means, for the pretest and the posttest for each of the two groups. By using a combination of subscripts, 1 and 2 for the pretest and posttest, and $E$ and $C$ to represent the two groups, we can specify the means as $M_{E1}$, $M_{E2}$, $M_{C1}$, and $M_{C2}$. Not all the possible differences between these four will have meaning. Those that have meaning may be set forth as:

$D_E = M_{E1} - M_{E2}$, the change shown by the experimental group.
$D_C = M_{C1} - M_{C2}$, the change shown by the control group.
$D_1 = M_{E1} - M_{C1}$, the pretest difference between the groups.
$D_2 = M_{E2} - M_{C2}$, the posttest difference between the groups.

Which of these four meaningful differences should we test for significance? Obviously, it is insufficient to test only $D_E$ because we cannot be sure that the shift shown, even though nonchance, is really due to the interpolated experience. In fact, the reason for having the control group is to enable us to evaluate the shift which takes place as a result of causes other than the experimentally provided experience. Now it might be thought that if $D_E$ is significant while $D_C$ is less, or not at all, significant, an effect has been demonstrated. This type of comparison, however, does not provide a check on the net change. Some have argued that if $D_2$ is significant while $D_1$ is not, it may be safely concluded that the interpolated experience has had an effect. This comparison also fails to test the net change. We should test the significance of the difference between the two changes, i.e., $D_D = D_E - D_C$, in order to gauge properly the net shift. Although, as regards absolute magnitude, $D_E - D_C$ will always equal $D_2 - D_1$, it is easier to evaluate the former difference.

To get the standard error of $D_D(= D_E - D_C)$ when the groups have been independently drawn we need the sampling variance of $D_E$ and $D_C$ so as to substitute in

$$S_{D_D} = \sqrt{S^2{}_{D_E} + S^2{}_{D_C}} \tag{6.12}$$

Now since $D_E = M_{E1} - M_{E2}$ is the difference between two means based on the same persons, we could get the standard error of $D_E$ by using formula (6.8), but since the difference between correlated means is equal to the mean difference, $M_{D_E}$, we can use formula (6.7) to get the required

$S^2_{D_E}$. This same situation holds for the control group, so (6.7) would also be used to get $S^2_{D_C}$.

If the experimental and control groups have been formed by pairing, our standard error of the difference between changes will require an $r$ term to enable us to take advantage of the fact that we have a better controlled experiment. The required $r$ is the correlation between the changes shown by the members of the pairs; to compute it we need to consider the paired changes. We can, however, get the standard error of the difference by way of the algebraic difference between the changes shown by the members of the pairs, without computing an $r$.

Let $X_1$ and $X_2$ stand for pretest and posttest scores and let the members of the $J$th pair be designated as $J$ and $J'$, with $J$ assigned to the experimental, and $J'$ to the control, group. Each individual will have a change score which is nothing more than his pretest score minus his posttest score. Thus the change score for the members of the $J$th pair will be

$$C_j = D_j = X_{1j} - X_{2j} \quad \text{and} \quad C_{j'} = D_{j'} = X_{1j'} - X_{2j'}$$

Hence the difference between the changes (or differences) shown by the members of any pair will be

$$D = (C_j - C_{j'}) = (D_j - D_{j'})$$
$$= (X_{1j} - X_{2j}) - (X_{1j'} - X_{2j'})$$

For $N$ pairs we will have $N$ $Ds$. These $Ds$ are tedious to compute since one must preserve the same direction for each subtraction and keep track of signs. The process can be made somewhat simpler by removing the parentheses, thus

$$D = X_{1j} - X_{2j} - X_{1j'} + X_{2j'}$$

Simply add $X_{1j}$ and $X_{2j'}$ and then substract the sum of $X_{2j}$ and $X_{1j'}$ with the sign for $D$ depending on whether the first or the second of these two sums is the larger.

Once the $N$ $Ds$ have been determined, we can get $M_D$, $S_D$, and thence $S_{M_D}$ by formula (6.7). This $M_D$ will equal $D_E - D_C$, or $(M_{E1} - M_{E2}) - (M_{C1} - M_{C2})$, and this $S_{M_D}$ will be exactly the same as

$$S_{D_D} = \sqrt{S^2_{D_E} + S^2_{D_C} - 2r_{D_E D_C} S_{D_E} S_{D_C}}$$

After the student has learned how to compute $r$, he may prefer to use this longer formula for $S_{D_D}$ (equivalent to $S_{M_D}$) rather than go through the tedium of differencing differences. Regardless of how the standard error of the difference is obtained, the null hypothesis is tested by calculating a $z$, as the net difference between the two changes divided by its standard error. The foregoing procedures are also applicable when we are dealing with two experimental conditions. We need only to use appropriate subscripts in place of $E$ and $C$.

## INFERENCE: ESTIMATION

So far we have discussed statistical inference from the point of view of hypothesis testing, but there are occasions when we may wish to use information from a sample as a basis for estimating population values. There are two general types of estimation: point and interval. We shall discuss the first briefly in order to introduce some concepts which the student might encounter, and the second because of its practical implications.

**Point estimation.** We may regard a sample statistic as an estimator for the corresponding population value (parameter). How "good" an estimator it is depends on whether or not it is unbiased and consistent, and on its relative efficiency.

An estimator is said to be *unbiased* if the average of a large number of sample estimates tends to equal the parameter being estimated. The mean is unbiased because the mean of sample means will approach nearer and nearer $\mu$ as we take more and more samples, but $S^2$ defined as $\Sigma x^2/N$ is biased in that the mean of sample variances tends to be smaller than the population variance. An unbiased estimate is given by $s^2 = \Sigma x^2/(N-1)$, but for subtle mathematical reasons $s$, or $\sqrt{\Sigma x^2/(N-1)}$, involves a negligible bias as an estimator of the population standard deviation. Note that the bias is small when $N$ is large.

An estimator is said to be *consistent* if it approaches nearer and nearer the population value as sample size is increased indefinitely. All the measures so far discussed satisfy this criterion.

The *efficiency* of an estimator is a function of its sampling error. Thus, in terms of efficiency the sample mean is far better than the median as an estimator of the central value of a population of normally distributed scores even though both are unbiased and consistent estimators.

**Interval estimation: confidence interval.** Interval estimation, which takes into account the sampling error of an estimator, provides limits, or an interval, for the population value, and at a prescribed level of confidence. Given a sample mean and its standard error, one could set up a whole series of "trial" hypothesis values for the population mean. All trial hypothesis values well above and below the sample mean could be rejected at a high level (small $P$) of significance, but rejection would become more and more risky as we approached nearer and nearer the sample mean, and for a whole series of values near the sample mean all trial hypotheses would be acceptable. This implies that at some point above the sample mean and at some point below the sample mean we change from rejection to acceptance of the trial values. If we have adopted, say, the $P = .05$ level, the change will obviously be at $M \pm 1.96S_M$. In rejecting trial values outside these limits and accepting values within these limits, we are in effect inferring that the population value is in an interval defined by these limits.

It would seem that there should be some way of expressing our degree of confidence that the population mean lies between the limits $M \pm 1.96S_M$, since, as we have seen, we can be somewhat sure that the sample mean is not a chance deviation from a population mean outside the limits so determined. Note that, given a population mean and sigma, we can legitimately speak of the probability of a sample mean falling in a specified region, but given a sample mean we cannot speak of the probability of the population mean being in a certain region (or interval) for the simple and compelling reason that $\mu$, being definitely just one value, has no distribution. We can in no way enumerate events so as to conceive of a probability fraction since just one event (value) is possible.

In order to arrive at a statement which expresses our degree of confidence, we note that, if we draw a second sample, we would be apt to have a different set of limits for the simple reason that the second sample mean may differ from the first. If we take additional samples of the same size, we would have a distribution of sample means, hence a sort of distribution of sets or pairs of limits, since each sample mean would provide a set. Our discussion can be greatly simplified by taking sets of limits given by $M \pm 2S_M$ (as approximating the $M \pm 1.96S_M$ values). For simplicity of exposition, let us assume that we are drawing successive samples from a population having a mean of 10, and that the $\sigma$ and $N$ are such that $\sigma_M$ can be taken as 2. Then $M \pm 2\sigma_M$ will be $M \pm 2(2)$, or $M \pm 4$. It will also facilitate our exposition if we think of the random sampling distribution of means in terms of intervals of $\frac{1}{2}\sigma$ distances on the base line with the approximate percentage area for the several intervals, as shown in the top curve of Fig. 6.1.

Now each possible sample mean will lead to a lower limit of $M - 4$ and an upper limit of $M + 4$. If we consider the 19 per cent of sample means expected between 9 and 10, we see at once that these 19 will lead to intervals with lower limits between 5 and 6 and upper limits between 13 and 14. That is, the sample means falling between 9 and 10 will generate that part of the lower limit ($LL$) curve of Fig. 6.1 between 5 and 6 and that part of the upper limit ($UL$) curve between 13 and 14. Likewise the 15 per cent of sample means falling between 8 and 9 will lead to the 4 to 5 part of the $LL$ curve and to the 12 to 13 part of the $UL$ curve. Similarly, as can be seen by careful study (a requirement for most students if understanding is

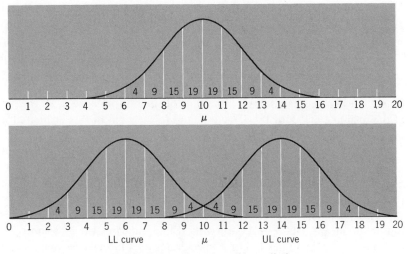

**Fig. 6.1.** Generation of confidence limits.

to be achieved) of the three curves of Fig. 6.1, every left-hand segment of the top curve generates a left-hand segment for each of the bottom curves. Stated differently, the left half of the top curve leads to a distribution of intervals with lower limits less than 6 and upper limits of less than 14. In exactly the same fashion it can be seen that the right half of the top curve leads to the right half of the LL curve and also the right half of the UL curve. Thus we have a sampling distribution of intervals (sets of limits) as found by taking $M \pm 4$ (or $M \pm 2\sigma_M$). Our next task is to ask how many of these various intervals actually include 10, or the population mean. Reference to Fig. 6.1 will verify that, out of 100 tries, we would expect to get:

 4 times an interval with LL of 2 to 3 and UL of 10 to 11
 9 times an interval with LL of 3 to 4 and UL of 11 to 12
15 times an interval with LL of 4 to 5 and UL of 12 to 13
19 times an interval with LL of 5 to 6 and UL of 13 to 14
19 times an interval with LL of 6 to 7 and UL of 14 to 15
15 times an interval with LL of 7 to 8 and UL of 15 to 16
 9 times an interval with LL of 8 to 9 and UL of 16 to 17
 4 times an interval with LL of 9 to 10 and UL of 17 to 18

Notice that for every set of limits in the foregoing groups the population mean *is* in the range or interval defined by the upper and lower limits of the set. When we sum these expected frequencies, we see that 94 per cent of the sets of limits lead to intervals within which the population mean lies. If we had not rounded to the nearest per cent, these would sum to 95.45 per cent. This implies that 4.55 per cent of the time the intervals so defined would not include the population value. This can be verified by noting that sample means of less than 6 (top curve) lead to *upper* limits of *less* than 10, and do so 2.27 per cent of the time, whereas sample means of more than 14 produce *lower* limits of *more* than 10 about 2.27 per cent of the time. These percentages are for the tails of the bottom curves, to the left of the ordinate at 10 for the *UL* curve and to the right of this ordinate for the *LL* curve.

There is a somewhat different pictorial device that may help the student grasp the confidence interval concept. Consider Fig. 6.2, for which it is again supposed that a population mean, $\mu$, is 10 and that $\sigma_M = 2$. Let the range of possible sample means run from 0 to 20. The figure includes expected proportionate frequencies, $E_{pf}$, for the random sampling distribution of means. The heavy horizontal line indicates the location of $\mu$ and the two broken lines hold for specific means that will become crucial to the meaning of a confidence coefficient. Each of the paired double-arrowed vertical lines originates from a sample mean falling just inside the limits for a specific interval. There are 10 such pairs shown.

Consider the first pair of verticals, which spring from sample means between 9 and 10. Obviously, each overlaps or includes the population value. All other possible sample means between 9 and 10 will also lead to limits that overlap the population mean. Out of 100,000 trials, we would

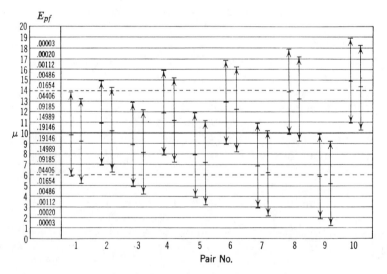

**Fig. 6.2.** Generation of confidence limits.

expect to have 19,146 sample means falling between 9 and 10, thus yielding 19,146 correct deductions. Consider the eighth pair and note that each overlaps the population value and that sample values, between 13 and 14, which produce intervals including the population mean will occur 4,406 times in 100,000 tries. Now note that sample values falling below 6 and above 14, as illustrated by the ninth and tenth pairs, will result in limits that do not contain $\mu$, hence engender a false inference. The expected proportionate number of such inferences, quickly calculated by simply summing the $E_{pf}$ above 14 and below 6, is .04550. The number of correct inferences, obtained by summing the $E_{pf}$ between 6 and 14, is expected to be .95452. (The failure of the two expectancies to add to 1.00000 is a result of roundings.)

In summary, if we were to make in our lifetime 100 inferences concerning population means on the basis of sample values by each time taking the limits as $M \pm 2\sigma_M$, the limits so established would include the population value about 95 per cent of the tries. That is, in the long run we would be correct about 95 per cent of the time in concluding that the population value is within the intervals so determined, and about 5 per cent of the time we would be in error. If we used $M \pm 1.96\sigma_M$ for setting limits, we would be correct 95 per cent, and in error 5 per cent, of the time. When we take $M \pm 1.96\sigma_M$ as a confidence interval, the degree of faith in such limits is represented by a $P$ of .95; i.e., the *level of confidence* for such an inference is represented by a probability-type figure of .95. If we wish to be surer of our inferences, we might choose the .99 level of confidence, which in practice can be attained by taking $M \pm 2.58\sigma_M$ as limits.

The limits set by the confidence interval method are so very similar to *fiducial limits*, and the level of confidence, sometimes referred to as the *confidence coefficient*, is so much like *fiducial probability* that the beginning student can well let the mathematical statistician worry about the theoretical difference between what seems to be two ways of doing the same thing.

The preceding illustration of the meaning of interval estimation was based on a presumed known $\sigma$; in practice we will have a sample estimate, $S$, hence $S_M$ as a basis for calculating limits. Since $S_M$ will vary from sample to sample (because of varying $S$s), the width of the interval will vary from sample to sample and, therefore, it might be inferred that using $M \pm 1.96S_M$ would not lead to intervals that overlap $\mu$ 95 per cent of the time. But since the width of the interval will sometimes be too short and sometimes too long, there is a balancing effect for $N$ not too small.

Confidence intervals can be set up for statistical measures other than the mean, but if the random sampling distribution of a given measure is nonnormal, the method will not be the simple stunt of taking $C \pm 1.96S_C$ or $C \pm 2.58S_C$ where $C$ stands for any statistical measure. It should be obvious that, since the standard errors for all statistical measures are a function of $N$, it is possible by increasing the sample size to narrow the confidence interval without any loss in the degree of confidence with which we accept the limits.

Since Chapter 5 was devoted to the development of the general principles of statistical inference, the critical reader may ask why the concept of confidence limits was not discussed there, in connection with proportions (or percentages), as a part of inference. This was not done because of two complicating factors, one of which is the frequent lack of normality for the sampling distribution of proportions. This factor could have been side-stepped by specifying situations involving either proportions in the middle range or large $N$ for high (or low) proportions. This is what was done for inference as hypothesis testing. The second complication, which arises from the fact that the standard error of a proportion varies with the value of the population proportion, is not too disturbing for proportions near .50 but is such that its effect is not eliminated for extreme proportions by requiring $N$ to be sizable.

In order to understand this second factor, let us rewrite the previous (p. 99) introduction to the concept of interval estimation. Given that a sample proportion of, say, .70 is based on $N = 50$. If we take a trial value of .60 for the population proportion and ask how often a sample value of .70 or more will arise, we need to use $\sqrt{(.60)(.40)/N}$, or .069, as the standard error. If we take a trial value of .80 and ask how often sample values will fall below .70, we have to use $\sqrt{(.80)(.20)/50}$, or .057, as the (only) appropriate standard error. Note that since the two standard errors differ, we would not reject a trial value deviating .10 above .70 at the same level of significance as that for a trial value of .10 below .70. Stated differently, if we proceeded with a series of "trial" hypotheses, upward and downward toward the sample value of .70, with .025 as our $\alpha$ value for a one-tail level of significance, we would not, because of the differing sizes of the standard errors, arrive at limits equidistant from .70. This systematic directional difference in standard errors does not hold for the means of continuous variables.

If we attempted to redo Fig. 6.2 for the proportion (percentage) situation with, say, 10 as the population percentage, and did so with full knowledge that in practice a confidence interval must be calculated from information based on the sample at hand, we would have to say that when we have a sample percentage of 7 the interval will be shorter than when the sample percentage is 13. This is true because the estimated standard errors are functions of $7 \times 93$ and $13 \times 87$, respectively. If $N$ were 163, the standard errors would be 2.0 and 2.6 percentage points. Since there are no simple methods, generally applicable, for ascertaining confidence limits for a population proportion, we merely note that $p \pm 1.96S_p$ and $p \pm 2.58S_p$ will yield reasonably satisfactory limits when $p$ is not extreme and/or $N$ is large.

**Confidence interval for a difference.** There are times when it is desirable not only to know whether a difference is significant but also to specify limits for the population difference. Such specification does not presume that a significant difference has been found. Even when a difference fails to reach significance, the specification of confidence limits gives some idea of the possible difference between population values, and such information may help answer the nonstatistical question of whether the population difference is apt to be large enough to be of practical or scientific importance. This procedure may be helpful in evaluating the consequences of accepting the null hypothesis when in reality the hypothesis is false.

Furthermore, the setting up of a confidence interval may be particularly helpful when we have obtained a difference which is highly significant. Consider the case of a difference of 4.78 inches in mean height between men and their sisters. Because of large $N$s and the presence of brother-sister correlation, the standard error of the difference is very small; its value is about .07. When we compute $D/S_D$ we have a $z$ of 68. This would, if we could evalute it, yield a probability, for as large a difference by chance, which would be so microscopically small that we could not comprehend it. However, when we set confidence limits at, say, the .99 level, we have $4.78 \pm 2.58(.07)$, or 4.60 and 4.96, as limits for the population difference. This permits a down-to-earth way for evaluating the obtained difference.

**Level of confidence vs. level of significance.** The term "level of confidence" should not, as it frequently is, be misused in place of "level of significance." The first term pertains to interval estimation, the other to hypothesis testing.


## QUESTION OF ASSUMPTIONS

It may be well to consider briefly the assumptions underlying the procedures so far discussed for making statistical inferences, since assumptions restrict the applicability of a method.

**Independence of sampling units.** It is assumed that the conditions of random sampling hold, but the frequency with which the requirement of independence is violated by researchers suggests that a warning is needed. The violation usually comes about when multiple measurements or observations are made on each of the individuals in a sample and each measurement (or response) is treated as a sample value, thereby inflating $N$ $n$-fold times when $n$ repeated measurements (or responses) are available for each person. The lack of independence comes about in that, for instance, if the sample of individuals happened to include one high scoring person there would automatically be $n$ high scores. The effect of such an inflation of $N$ is an illegitimate reduction in standard errors.

**Infinite vs. finite universe.** If we are sampling from a finite universe, particularly a universe with a rather small number of cases, it seems reasonable to think that as the sample size becomes large relative to the number of cases in the universe, the sample mean, for example, will tend to fluctuate less from the universe mean than it does when we are drawing from an infinite population. This suggests that the standard error formulas need to be modified for the finite population situation. The required modifications are available for only a few statistical measures. If we let $N$ represent the sample size and $N_{pop}$ the size of the finite universe, the standard errors for the mean and for a proportion are approximately as follows:

$$S_M = \frac{S}{\sqrt{N}} \sqrt{1 - N/N_{pop}} \quad \text{and} \quad S_p = \sqrt{pq/N} \sqrt{1 - N/N_{pop}}$$

In a given research it is sometimes difficult to decide whether the universe being sampled is finite or infinite in size, and, if finite, it is not always easy to determine the value of $N_{pop}$. It might be argued that psychologists never study an infinite universe. It can readily be seen, however, that the corrective factor in the sampling error formulas becomes negligible when $N_{pop}$ is large. Thus, if $N_{pop}$ is known to be large relative to $N$, it matters little whether the given universe is wrongly conceived as being infinite. For example, when $N$ is .01 of $N_{pop}$, the corrective term leads to a reduction in the sampling error of about .005 of the value obtained by the ordinary formulas.

These formulas for the finite universe situation are frequently useful when we wish to compare a subgroup with a total group which contains the subgroup. Such a comparison is sometimes erroneously made by taking $\sqrt{S^2_t/N_t + S^2_s/N_s}$ as the standard error of the difference between the subgroup mean, $M_s$, and the total mean, $M_t$. This makes no allowance for the fact that the two means are not based on independent groups. An appropriate procedure is to regard $M_s$ as based on a sample drawn from a finite universe of $N_t$ cases with mean and standard deviation of $M_t$ (as $\mu$) and $\sigma_t$; then with the standard error of $M_s$ taken as

$$\sigma_{M_s} = \frac{\sigma_t}{\sqrt{N_s}} \sqrt{1 - N_s/N_t}$$

we can test the significance of the deviation of $M_s$ from $M_t$ by using the ratio $(M_s - M_t)/\sigma_{M_s}$, which is interpretable as a $z$. This ratio will give a very close approximation to the $z$ which would be obtained if we were to compare the subgroup with the remainder (the total cases less the subgroup) as two independent groups, using the usual formula for standard error of the difference. The foregoing scheme would also be applicable in case proportions instead of means were the descriptive measures used as a basis for comparison.

**Skewed distributions.** The standard error formulas given in this chapter assume normal or nearly normal score distributions for the population being sampled. Skewness is the most frequently encountered evidence for nonnormality, and accordingly it is of interest to consider the effect of skewness on the sampling distribution of the mean, the measure most apt to be involved in testing hypotheses. The relationship between the degree of skewness, $g_1$, for a variable and the amount of skewness for the sampling distribution of means is $g_M = g_1/\sqrt{N}$. Thus the skewness in the distribution of means rapidly disappears as $N$ is taken larger and larger. For example, if $g_1$ is .77 (see Fig. 3.1, p. 28) and $N$ is 35, the skewness for the sampling distribution of means will be only .13 (see Fig. 3.1 again). Accordingly, the procedures in this chapter may be safely used with moderately skewed distributions when $N$ is large and with markedly skewed distributions when $N$ is very large. Some methods for handling nonnormal data will be discussed in Chapter 19.

## A FURTHER WORD ON PROPORTIONS

The student will have noted that the general principles of statistical inference set forth in Chapter 5 have been utilized and extended in the present chapter. There are many points of obvious similarity in the two chapters, but there is an additional parallelism which is not obvious. For an attribute involving a dichotomy such as yes-no, like-dislike, pass-fail, etc., we may arbitrarily assign a score of 1 to one category and a score of 0 to the other. That is, $X = 0$ or 1.

Let $f_0$ and $f_1$ stand for the frequency of, say, no and yes responses respectively in a sample of $N$ cases. Thus we have a miniature frequency distribution, with the two categories being analogous to two intervals. Let us consider the mean and standard deviation of this miniature frequency distribution, both in terms of gross score formulas. Notice that in Table 6.2 we have a score column, $X$, a frequency column, $f$, and an $fX$ and an $fX^2$ column (analogous to $fd$ and $fd^2$, with $d = X$). It will be seen that $\Sigma X = f_1$; hence the mean of the distribution is $M = \Sigma X/N = f_1/N = p$, where $p$ is the proportion of yeses. Hence a proportion may be regarded as a mean. It will also be seen that $\Sigma X^2 = f_1$; hence when we utilize formula (3.7)

**Table 6.2. Scheme for mean and standard deviation of a dichotomous variable**

| Response | $X$ | $f$ | $fX$ | $fX^2$ |
|----------|-----|-----|------|--------|
| Yes | 1 | $f_1$ | $f_1(1)$ | $f_1(1)^2$ |
| No | 0 | $f_0$ | $f_0(0)$ | $f_0(0)^2$ |
| Sums | | $N$ | $f_1(1)$ | $f_1(1)$ |
| | | | $= \Sigma X$ | $= \Sigma X^2$ |
| | | | $= f_1$ | $= f_1$ |

to write the variance of the distribution we have

$$S^2 = \frac{1}{N^2}[N\Sigma X^2 - (\Sigma X)^2]$$

$$= \frac{1}{N^2}[Nf_1 - (f_1)^2]$$

$$= \left[\frac{Nf_1}{N^2} - \frac{f_1^2}{N^2}\right]$$

$$= (p - p^2) = p(1 - p) = pq$$

Hence $S = \sqrt{pq}$ as the standard deviation of the dichotomous distribution. (Any connection with the binomial?)

In this chapter we have given $S_M = S/\sqrt{N}$ as the standard error of a mean. If this holds for the dichotomous distribution we would have $S_M = \sqrt{pq}/\sqrt{N} = \sqrt{pq/N}$. But this is the same as $S_p$ given by formula (5.2). This is as it should be since $p = M$ for the dichotomous distribution.

Furthermore, formula (5.5) for the standard error of the difference between correlated proportions has its analogue in the development on pp. 90–92 for the difference between correlated means, and formula (5.6) involves a pattern similar to that of formula (6.9).

## NOTE ON THE PROBABLE ERROR

An antiquated procedure is the use of the probable error, $pe$, instead of the standard error in connection with sampling. The $pe$ of the mean is $.6745S_M$, and therefore we would expect 50 per cent of successive sample means to fall between $\mu \pm pe_M$. Similarly, the $pe$ for any other statistical measure is .6745 times its standard error. The student who attempts to survey the research literature on a given topic is apt to encounter $pes$ and he therefore must know the relationship of the $pe$ to the standard error.

## NOTE ON NOTATION

We have used the Greek letter $\mu$ as the symbol for population mean and the corresponding Latin letter $M$ for a sample mean. Another frequently used symbol for a sample mean is $\bar{X}$ (read $X$ bar); later in this text we will use the bar to indicate a sample mean. The student needs to know both $M$ and $\bar{X}$ as symbols. We have used $\sigma$ as a symbol for the standard deviation of a population and also for the standard deviation of a theoretical distribution, such as the binomial or (the definition formula of) the normal curve; the Latin equivalents, $S$ and $s$, stand for sample standard deviation, one biased the other unbiased. As shall be seen, we need in the sequel both $S$ and $s$. Consistency in notation would call for $p$ (or $P$) as a sample proportion and the corresponding Greek $\pi$ as a population value, but the letter $\pi$ was long ago taken by mathematicians as the symbol for something else, so to avoid confusion we used $p_{pop}$ instead of $\pi$. Later we will use $r$ and $r_{pop}$ as symbols for sample and population correlation coefficients because $\rho$ (rho) has, as we shall see, been used to signify a particular kind of correlation coefficient.

# Chapter 7

## SMALL SAMPLE OR
## $t$ TECHNIQUE

Although the general principles of statistical inference are the same for both large and small samples, the techniques differ. We shall confine our attention in this chapter to the technique for dealing with a single mean and with the difference between two means. Chapter 14 will deal with inferences concerning variabilities.

It will be recalled that the sampling distribution of the mean is normal when the trait distribution is normal. This holds regardless of sample size. The sampling distribution of means centers at the population mean with a true standard deviation $\sigma_M = \sigma/\sqrt{N}$, which sigma we termed the true standard error of the mean. Recall also that the relative deviates, $(M - \mu)/\sigma_M$, follow the unit normal curve. When successive samples are drawn and a $S_M$ is computed for each sample by using the sample $S$ instead of $\sigma$ (an unknown), the ratios of given $(M - \mu)$s to their $S_M$ values so computed will be distributed normally for very large $N$s and approximately so for $N$s of moderate size, but for $N$s as small as 30 the approximation is none too good. The value 30 is arbitrarily chosen—the approximation to normality becomes progressively worse as we go from large to small $N$s rather than becoming abruptly worse in the vicinity of $N = 30$.

We have already mentioned the fact that $S^2 = \Sigma x^2/N$ suffers from bias, whereas $s^2 = \Sigma x^2/(N - 1)$ is an unbiased estimator of the population variance. Since the bias in $S$ increases with a decrease in $N$, it is important to use the unbiased estimator when $N$ is small. We will accordingly use $s_M = s/\sqrt{N}$, in place of $S_M = S/\sqrt{N}$, as a nonnegligible improvement in the estimate of the standard error of a mean based on a small sample.

110

Even so, the successive sample ratios, $(M - \mu)/s_M$, with $s_M$ computed from each sample, will not follow the unit normal curve because the sampling distribution of $s$ (also $S$) is skewed for $N$ small; hence the distribution of successive values of $s_M$ will be skewed. That is, the successive sample values of $(M - \mu)/s_M$ will involve a variable numerator which is normally distributed and a variable denominator which has a skewed distribution. The distribution of the resulting ratios will be symmetrical about zero but will be leptokurtic. That is, it is characteristic of the sampling distribution of $(M - \mu)/s_M$ that the tails of the curve beyond a ratio of about 1.7 tend to be higher than the tails of the normal curve. Thus, there will be relatively more large ratios.

**The $t$ distribution.** It can be shown that such ratios, involving a normally distributed deviate divided by an unbiased estimate of its sampling error, will follow the so-called $t$ distribution, defined by

$$ h = \frac{\Gamma\left(\dfrac{n+1}{2}\right)}{\Gamma\left(\dfrac{n}{2}\right)\sqrt{n\pi}}\left(1 + \frac{t^2}{n}\right)^{-(n+1)/2} $$

in which $\Gamma$ indicates the gamma function as defined in texts in advanced calculus. Although this equation will be beyond the mathematical comprehension of most students, it should be noted that $h$ is the height of a curve, that since $t$ is squared the distribution is symmetrical, and that the equation contains an $n$ as yet undefined. This $n$ has to do with the number of degrees of freedom, a concept which is to be discussed. Suffice it to say just now that $n$ will be a function of sample size (or sizes) and accordingly that there will be not just one but many distributions of $t$, one for each possible value of $n$.

Figure 7.1 shows the curve of $t$, when $n = 7$ and when $n = 3$, as compared to the normal curve. For $n$ larger and larger, the curve of $t$ approaches that of the normal distribution. Table E of the Appendix gives the values of $t$, for $ns$ of 1 to 30, which will be exceeded by chance a specified proportion of times. Thus for $n = 30$ we see from Table E that the $P = .05$ point is at a $t$ of 2.04 as compared to a normal deviate of 1.96. For $n = 10$, the point corresponding to the .05 level is $t = 2.23$. The .01 level is at $t = 2.75$ for $n = 30$, and at 3.17 for $n = 10$, as compared with 2.58 for the normal curve.

**Degrees of freedom.** The $n$ of the equation for $t$, and in the $t$ table, is the number of degrees of freedom ($df$) involved in the estimate of the population variance. The $df$ depends on how many of the $x$s in $\Sigma x^2$, or $\Sigma(X - M)^2$, are "free to vary." Suppose two scores, 3 and 5. Their

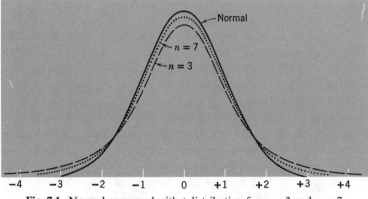

**Fig. 7.1.** Normal compared with $t$ distribution for $n = 3$ and $n = 7$.

mean is 4, and the sum of squares (of deviations) is $(3 - 4)^2 + (5 - 4)^2$
$= 2$. Recall (from p. 29) that we have already shown that $\Sigma x = \Sigma(X - M)$
$= 0$, always. Therefore, as soon as one of two deviations is known, the
other $x$ is determinable. Thus, if $x_1$ is $-1$, the other deviation, $x_2$, must
satisfy the equation $-1 + x_2 = 0$. One deviation and hence its square can
be thought of as dependent on the other deviation, which has some independence, and therefore 1 degree of freedom. Suppose that we have three
scores, 3, 4, and $X$, which yield a mean of 4. The deviations must satisfy
the requisite that they sum to zero; i.e., $(3 - 4) + (4 - 4) + (X - 4)$
$= 0$. Thus one of the three deviations is fixed by the other two, i.e., is not
independent of their values, because the three deviations must sum to zero.

It may be more enlightening to start with symbols for scores. Suppose
that $X_1$, $X_2$, $X_3$, and $X_4$ represent four scores, and it is reported that their
mean equals 40. How many of the four deviations can we assign at will?
Stated in deviation units, we have $(X_1 - 40) + (X_2 - 40) + (X_3 - 40)$
$+ (X_4 - 40)$ as a sum which must equal zero. It is readily apparent that
only three deviations can "vary freely"—the fourth is fixed by the numerical values of the other three. Hence $df = 4 - 1$; i.e., 1 degree of freedom
in the deviations or their squares is lost because of the one restriction
imposed. Actually, this restriction comes about because we are taking
deviations about one constant, the mean, computed from the set of scores
at hand. The $df$ for a sum of squares (of deviations) about a mean is
always $N - 1$ when $N$ scores are used to compute the mean. In general,
the $df$ for the sum of squares is equal to the number of squares minus
the number of restrictions imposed by constants computed from the data.

Note that the unbiased estimate of the population variance, $s^2$
$= \Sigma x^2/(N - 1)$ involves dividing by $df$, the number of degrees of freedom.
This is a general rule.

**Computation of $s^2$ or $s$.** For $N$ small the mean and $s^2$ or $s$ are readily computed from gross score formulas. Thus $M = \Sigma X/N$. To compute $s^2$ or $s$ we need $\Sigma x^2$ in terms of gross scores. This was given earlier as

$$\Sigma x^2 = \frac{1}{N} [N\Sigma X^2 - (\Sigma X)^2] \tag{3.6}$$

Division of this by $N - 1$ yields $s^2$, the square root of which is the required $s$. An easily derived relationship between $s^2$ and $S^2$ is $s^2 = \dfrac{N}{N-1} S^2$. Although we do not need a frequency distribution for purpose of computations, a distribution should be made anyway so as to permit at least a rough check on the assumption that the scores have been drawn from a normally distributed population of scores.

**$t$ for a single mean.** We can test the significance of $M$ as a deviation from any hypothesized value for the mean, $M_h$, by taking $t = (M - M_h)/s_M$ as an entry in Table E, with $n = df = N - 1$, to see whether the obtained $t$ reaches the $t$ value required for certain levels of significance. If the $t$ does not reach the value required for the chosen level of significance, the deviation would be attributed to chance and the hypothesis accepted.

If we wish to specify the confidence limits for the unknown population mean and to do so with a level of confidence indicated by $P = .99$, we first note from the table of $t$ how large $t$ must be, for the given $df$, to correspond to the .01 probability level. Then $M$ plus and minus the $t$, so found, times $s_M$ will give the desired limits. For example, suppose nine cases yield a mean of 80 and a sum of squares of 1152. Dividing the sum of squares by $df$, or 8, we get $s^2 = 144$, $s = 12$ as an estimate of $\sigma$ and $s_M = 12/\sqrt{9} = 4$. For 8 $df$ we find from Table E that $t = 3.355$ for the .01 level. Then $80 \pm (3.355)(4)$ gives 66.58 and 93.42 as the .99 confidence limits for the population mean. If we used the large sample method of Chapter 6 we would have $S^2 = 1152/9$, giving $S$ as 11.31, from which we would get $S_M = 11.31/\sqrt{9} = 3.77$. Since for the normal distribution a relative deviate of 2.575 corresponds to the .01 level, we have $80 \pm (2.575)(3.77)$ or 70.29 and 89.71 as the .99 confidence limits for the universe mean. These values for the confidence interval differ appreciably from those obtained previously when proper allowance was made for the smallness of the sample.

**Difference between correlated means.** It will be recalled that when we have two means based on the same individuals or on paired cases, the test of significance of the difference must make allowance for the fact that the two sets of scores are not random with respect to each other. In Chapter 6 we saw that this could be done by including the $r$ term in the standard error of the difference, as in formula (6.8), or by working directly with the differences between paired scores. It was shown that $M_D = D_M$ and that $S_{M_D} = S_{D_M}$. When we have small samples, it is easier to work with $M_D$, an estimate of the $\sigma$ of the distribution of differences between paired scores, and thence $s_{M_D}$. To get the best estimate of the sampling error of $M_D$, we need the sum of squares of the deviations of the pair differences from the mean difference, i.e., $\Sigma(D - M_D)^2$, which when divided by the proper $df$, or $N - 1$, where $N$ is the number of differences or the number of paired scores, gives the best estimate of the variance of the universe distribution of differences. Let $s^2_D$ stand for this estimate. Then

$$s_{M_D} = \frac{s_D}{\sqrt{N}} \tag{7.1}$$

The computation is straightforward. Each of the $D$s is the difference between two scores, the subtraction being made in the same direction for all, and the sum of squares, $\Sigma(D - M_D)^2$, is obtained by formula (3.6) with the $X$s replaced by $D$s; that is $\Sigma(D - M_D)^2 = \frac{1}{N}[N\Sigma D^2 - (\Sigma D)^2]$.

The $D$s are summed algebraically, and their squares are summed. After $s_{M_D}$ has been calculated, we get $t$ as $M_D/s_{M_D}$. The hypothesis to be tested is that the universe value of $M_D$ is zero; the table of $t$ is entered with the obtained $t$ and with $df = N - 1$ in order to see whether it reaches a prescribed level of significance. Note that the $df$ is 1 less than the number of $D$s, not 1 less than the total number of scores (see "Further note" on $df$s, p. 116).

The assumption of normality pertains to the $D$s; hence, again, even though a frequency distribution is not needed for computational purposes, it should be made so as to provide a rough check on the assumption. A confidence interval for $M_D$ (and consequently $D_M$) can be set up in precisely the same manner as indicated previously for a single mean.

**Difference between independent means.** Given: two groups of $N_1$ and $N_2$ cases, and that we wish to test the significance of the difference, $D_M = M_1 - M_2$. By the procedure of Chapter 6 for large $N$s, we would make the necessary calculations for determining $D_M/S_{D_M}$ or $z$. As an aid to transition in thought from $z$ to $t$, let us first write the expression for $z$, thus,

$$z = \frac{D_M}{S_{D_M}} = \frac{M_1 - M_2}{\sqrt{S^2_{M_1} + S^2_{M_2}}} = \frac{M_1 - M_2}{\sqrt{\dfrac{S^2_1}{N_1} + \dfrac{S^2_2}{N_2}}}$$

which involves the two sample variances. Now, for the small sample situation, we need $t = D_M/s_{D_M}$ where $s_{D_M}$ is to be the best possible estimate of the standard error of the difference. To get this we apparently need the best possible estimates of the two variances of the two populations from which the samples have been drawn. But here we encounter an assumption underlying *t* for this situation: the two populations must have the same variance. Hence, we need just one estimate, an estimate of the variance common to the two populations. Calling this estimate $s^2$, by analogy with the $z$ technique, we need

$$t = \frac{D_M}{s_{D_M}} = \frac{M_1 - M_2}{\sqrt{\dfrac{s^2}{N_1} + \dfrac{s^2}{N_2}}}$$

The best estimate, $s^2$, of the common population variance is obtained by computing the sum of squares separately for the two samples, then combining these sums, and dividing by the proper *df*, or

$$s^2 = \frac{\Sigma(X - M_1)^2 + \Sigma(X - M_2)^2}{N_1 + N_2 - 2} \tag{7.2}$$

The two separate sums are computed by formula (3.6). Note that 2 degrees of freedom are lost because the sum of squares is about two means, which leads to two restrictions. Substitution of the obtained $s^2$ in the foregoing expression leads to a *t*, which is looked up in Table E with *df*, or *n*, equal to $N_1 + N_2 - 2$ in order to see whether it reaches a chosen level of significance.

There is one point in the method of determining the $s^2$, needed for testing the significance of the difference between means, which may have puzzled the student. The setting of the null hypothesis, in combination with the assumption of equal population variances, implies that the two samples have been drawn from a single universe or from two universes which have the same mean and equal variances, for the given and measured trait. It might accordingly be assumed that the best estimate of the population variance would be obtained by taking the sum of squares about the combined mean rather than about the separate means. The former would give the better estimate of the variance if it were actually known that the two universe means were the same (or that only one universe was involved), but there is always the possibility that the two universe means really differ. If they do differ, the taking of the sum of squares about the combined mean would, in general, yield too large an $s^2$ for the simple reason that the real difference between groups would be contributing to the variability of the two groups combined. (The student who has difficulty seeing this point should imagine what would happen to the variance of scores when two groups markedly different in means were combined.) It follows, therefore, that in the long run the best value for $s^2$ will be provided by summing the sums of squares about the two means.

The procedure for setting a confidence interval when we have independent means is no different from that for correlated means. Simply take $D_M \pm t_\alpha s_{D_M}$ where $t_\alpha$ is the $t$, for the given $df$, required for significance at the $P = \alpha$ level. This will give limits for the $P = 1 - \alpha$ level of confidence. Suppose we wish the .99 confidence interval; this requires an $\alpha$ of .01, or as sometimes written, $t_\alpha = t_{01}$ where $t_{01}$ is found under the $P = .01$ column, opposite the $df$.

**Further note on degrees of freedom.** Suppose two independent groups with $N_1 = N_2 = N$, and also two groups of scores based on $N$ cases (or $N$ paired persons). For the former the $df$ is $N_1 + N_2 - 2 = 2N - 2$, whereas for the latter the $df$ is $N - 1$ even though in the paired situation the total number of persons is $2N$. This may be (and has been) confusing to some; it seems as though the obviously better plan (matching) leads to a loss in $df$ compared to the setup involving independent groups. It is sometimes argued that the $df$ would perhaps be larger if we worked not with the difference scores but with the two sets of scores in terms of the sums of squares of deviations for each set and the sum of cross products since, as can be seen from p. 91,

$$\Sigma(D - M_D)^2 = \Sigma x^2_1 + \Sigma x^2_2 - 2\Sigma x_1 x_2$$

The $df$ for the left-hand sum of squares is obviously $N - 1$, and since the right-hand side of the equation is merely an algebraic variant of the left-hand side, it does not seem reasonable to believe that the $df$s will differ for the two sides. Note that if we consider $\Sigma x^2_1$ as having $N - 1$ degrees of freedom, we cannot have any more degrees of freedom for the other sums on the right side because the $x_2$ values are not independent of the $x_1$ values; they (the $x_2$ scores) are not "free to vary."

**Comparison of changes.** In Chapter 6 (p. 96) we discussed the procedures for testing the differences between changes shown by two groups. For the situation involving paired persons, a $D$ for the difference between changes for the members of a pair was defined (p. 98), and the test of significance involved computing, for $D$s so defined, an $M_D$, $S_D$, and thence $S_{M_D}$. For the small sample, or $t$, technique we need $s_D$ and $s_{M_D}$, just as given previously for correlated means. The $df$ is 1 less than the number of pairs. For the setup involving the changes for independent groups, we would need an $s_{D_D}$ instead of the $S_{D_D}$ of (6.12). The required $s_{D_D}$ is given by

$$s_{D_D} = \sqrt{\frac{s^2_D}{N_E} + \frac{s^2_D}{N_C}}$$

in which

$$s^2{}_D = \frac{\Sigma(D - M_{D_E})^2 + \Sigma(D - M_{D_C})^2}{N_E + N_C - 2}$$

with the subscripts $E$ and $C$ referring to experimental and control groups. Thus, the procedure for testing hypotheses involving changes for two groups is precisely the same as that for testing the difference between two independent means, discussed previously—$X$ is replaced by $D$, a difference score.

**One-tailed versus two-tailed test.** Our discussion of the $t$ technique so far has been in terms of the $t$ value needed for a two-tailed test at a given level of significance. If the hypothesis to be tested or the decision to be made logically warrants a one-tailed test, the $t$ required for significance at the .01 level would be found under the .02 column of Table E, and for the .05 level the .10 column would be used. Those who do not wish to be restricted to the $P$ levels given in Table E will find for $df$s up to 20 the $P$ associated with any $t$ in Table XLV of Peters and Van Voorhis' *Statistical procedures and their mathematical bases*. This table gives one-tailed values, which need, of course, to be doubled for two-tailed tests.

**Question of assumptions.** When we use the tabled values of the $t$ distribution as a basis for judging significance or for setting confidence limits, we are in effect presuming that some quantity, usually a ratio such as $(M - M_h)/s_M$ or $M_D/s_{M_D}$ or $(M_1 - M_2)/s_{D_M}$, will in the sampling sense follow the $t$ distribution. The mathematical proof thereof is based on certain assumptions: normality for the population of $X$ scores and of $D$ scores for the first two ratios, and normality of $X$s for both populations with common, or equal, variances for the third ratio. Whether or not these assumptions hold will usually be unknown.

It might be thought that the assumption of normality underlying the use of $t$ could be tested on the basis of the sample (or samples) at hand either by testing the departure of $g_1$ (skewness) and $g_2$ (kurtosis) from zero (or by a chi square technique, discussed in Chapter 13), but these methods of testing for normality are not sensitive enough to lead us to reject, on the basis of a small sample, the hypothesis of normality unless the departure therefrom is very marked. Likewise, the as yet undiscussed test (see Chapter 14) for a possible difference between variances is too insensitive when used with small samples to lead to rejection of the hypothesis of equal variances unless the difference between the two universe variances is sizable; hence it is difficult to be sure that the assumption of equality of variances is tenable when two groups are being compared by the $t$ technique. The foregoing statements are, of course, based on the proposition that by statistical methods it can be proved, at a desired level of significance, that a sample distribution did *not* arise from a normally distributed universe or that two universe values are different, but such methods will not prove normality nor prove that two universe values are identical.

Since it is difficult to be sure that the assumptions will hold for a given batch of data, the question may be raised as to the effect of violations of the assumptions. Will too many or too few calculated *t*s reach the tabled value for the .05 or the .01 levels of significance? Or stated differently, does the chosen level of significance actually represent the probability of making the type I error? Over the years there have accumulated both mathematical deductions and empirical evidence indicating that the *t* test is "robust" under violation of assumptions; that is, calculated *t*s tend to follow closely the *t* distribution. There are exceptions to this rule, as is shown by the recent empirical study by Boneau.*

Boneau, with the indispensable help of an electronic computer, calculated 1000 *t*s for the difference between independent means for each of 20 different combinations of conditions with regard to *N*s, shapes of distributions in the "universes," and equality or inequality of universe variances. The percentage of the *t*s reaching the .05 and the .01 levels is indicative of the disruption produced by specified violations of assumptions.

First, differences in variances ($\sigma$s of 1 and 2, or one population variance four times that of the other; both distributions normal) for *N*s very small, 5 and 5, produced about 1 per cent too many *t*s at the .05 and the .01 levels, but for *N*s of 15 and 15 the discrepancies were only one tenth of a per cent. With samples of size 5 from the universe having the smaller variance and 15 from the universe having the larger variance, too few reached the .05 and the .01 levels—the .05 level being reached only .01 of the times and the .01 level only .001 of the trials. But when 15 cases and 5 cases were drawn, respectively, from the universes having the smaller and larger variances, far too many calculated *t*s reached "significance"—16 per cent at the 5 per cent level and 6 per cent at the 1 per cent level. The moral is clear: if we suspect that the variances may be unequal, we should make the two sample sizes equal or nearly so. Presumably, the disruption of the *t* test will depend on the relative magnitude of the two universe variances—the larger the variance difference, the greater the disruption. In psychological research, when sample sizes are large enough to permit any firm statement about a difference between $\sigma$s, it is rarely possible to conclude that the $\sigma$ for one population is twice that of the other, the ratio of the $\sigma$s in the Boneau study.

Second, when sampling from platykurtic (actually rectangular shaped) distributions Boneau found negligible effects, but when sampling from

* C. A. Boneau. The effects of violations of assumptions underlying the *t* test. *Psychol. Bull.*, 1960, **57**, 49–64.

markedly skewed (J-shaped; $g_1 = 2.0$) distributions, $N$s equal, he found that 3 and 4 per cent of the *t*s reached the 5 per cent level and too few reached the 1 per cent level. It is comforting to know that such extreme skewness, rarely encountered in practice, will not lead to too many significant *t*s.

Third, although the foregoing results hold for both one- and two-tail tests, Boneau found that when one sample was from a J-shaped distribution and the other from either a normal or a rectangular distribution, the distributions of the resulting calculated *t*s were skewed: a doubling of the risk of falsely concluding that the mean of the J-distribution is lower than that for the rectangular, also normal, distribution; conversely, "significant" differences in the opposite direction occurred only half as often as expected from the theoretical *t* curve. These results should give pause to the advocates of one-tail tests; they also have obvious implications for two-tail tests even though the number of *t*s, irrespective of direction, exceeded only slightly the expected number.

Suppose that in one study the difference between two means for two small samples leads to a *t* which falls at the .01 level and that in another study two large samples yield means, for another trait, which are also significantly different at the .01 level. Can we place as much reliance on the first difference as on the second? The answer is yes, provided the two studies have been carried out with the same degree of care as regards controls and adequate sampling techniques, *and* provided it is safe to presume that the fundamental assumptions underlying *t* are tenable. Thus our confidence in a result based on small samples is a function not only of the probability level of significance attained but also of our faith that assumptions have been met. Since, as we have suggested, the conditions of trait normality and equality of variances are exceedingly difficult to demonstrate when the only information available is based on the small samples at hand, we are forced to conclude that, in general, we cannot place as much reliance on the results from small samples as on those from large samples.

Although this last statement can, in light of Boneau's results, be qualified, we still have the question of the place of small samples in psychological research, and about this there will be a diversity of opinion. We do not propose to settle the issue or even debate it; instead, we shall mention a few points which we feel are pertinent. There are, of course, types of research for which it is impossible or practically impossible to secure more than a few cases, either because of their scarcity or because of prohibitive costs. For such situations it is fortunate that the small sample or *t* technique, which permits some allowance for the smallness of the sample or samples, is available. Quite frequently small samples may be useful in a preliminary study carried out solely for the purpose of guiding the experimenter. If given hypotheses seem to be verified, the next step should be to secure more cases for further verification rather than to rush into print with positive conclusions.

It seems to the writer that those who publish statistical results based on a small number of cases should, unless they are positively sure that the basic assumptions underlying $t$ have been met (and this assurance can seldom be attained), adopt a more stringent level of significance than they would adopt if they had large samples. Admittedly, a more stringent criterion of significance means that the null hypothesis may be less frequently rejected and consequently that a real difference may be overlooked. At this point some readers may need to be reminded that the best way to avoid committing type II errors is to avoid the use of small samples: the greater the number of cases the greater the likelihood of detecting a difference.

An illustration of the fact that small samples are not conducive to rejection of the null hypothesis unless the difference between universe values is sizable may be in order. Let us suppose that the means for the heights of two populations are 64.5 and 68.0 and that the universe standard deviations are both equal to 2.7. An investigator who does not know these facts draws a random sample of eight cases from each universe; and in order to help him a little (and also simplify this discussion), we tell him that each $\sigma = 2.7$. The standard error of the difference between means becomes $2.7\sqrt{\frac{1}{8} + \frac{1}{8}}$ or 1.35. If the investigator accepts the .01 level of significance, it is immediately apparent that an obtained difference would have to be at least (2.58)(1.35), or 3.48, for him to reject the null hypothesis. (Why are we justified in using the normal deviate, 2.58, with such small samples?) A little consideration of the fact that the sampling distribution of differences between means will center at 3.5 indicates that the chances are nearly 50–50 that the investigator will be accepting the null hypothesis even though the real difference is more than a standard deviation in magnitude.

There are times when an investigator may be so anxious to accept the null hypothesis that he will seize upon a very high level of significance in order to better his chances for accepting the hypothesis of no difference. Another way for increasing the odds in favor of accepting the null hypothesis is to use exceedingly small samples. Now those who desire to claim that no difference exists must face the simple fact that such a proposition can never be proved on a sampling basis. The most convincing way to demonstrate that a difference is of no practical or scientific importance is to use large samples and the confidence interval method for specifying limits for the population difference.

**Apportioning cases to groups.** Suppose you have facilities for utilizing, say, $N_t$ subjects for an experiment. These cases are to be assigned to two groups (experimental and control or two experimental groups), and the cost per case for collecting data is nearly the same for the two groups. Does it make any difference whether or not you divide the $N_t$ subjects equally between the two groups? The formula for the estimated standard error of the difference between means can be rewritten as

$$s_{D_M} = \sqrt{\frac{s^2}{N_1} + \frac{s^2}{N_2}} = s\sqrt{\frac{1}{N_1} + \frac{1}{N_2}}$$

Suppose $N_t$ is 40. If we assign 5 and 35 subjects for the groups, the value under the radical sign becomes $1/5 + 1/35 = .20 + .0286 = .2286$; if we assign 10 and 30, the value is $1/10 + 1/30 = .10 + .0333 = .1333$; if we assign 20 and 20, we have $1/20 + 1/20 = .05 + .05 = .10$. The three square roots are .48, .37, and .32, from which we quickly see that the optimal split is 50–50. Not only will a 50–50 split result in the smallest standard error for the difference, it will also free one of worry about the dire effect of unequal population variances when the $N$s are not equal (p. 118).

# Chapter 8

## CORRELATION: INTRODUCTION AND COMPUTATION

One of the chief tasks of a science is the analysis of the interrelations of the variables with which it deals. In the physical sciences, and frequently in the biological sciences, the interrelations can be determined by noting how much of a change in one variable is associated with change in another. The physicist studying the relationship between temperature and pressure exerted by a gas can vary the former at will so as to determine the pressure at different temperatures. In the social sciences, and sometimes in the biological sciences, the variables studied are apt to be characteristics of individuals (plant or animal); thus to study relationships the experimenter is compelled to make measurements on several individuals. For example, if two variables such as height and weight are under consideration, the measured height and weight of $N$ individuals will provide $N$ pairs of observations from which it can be determined whether the two vary together. In either case it is important to determine the form (mathematical) of the relationship and the accuracy with which it is possible to make predictions.

Many relationships are expressible in terms of the simplest of all mathematical forms, $Y = A + BX$, in which $X$ and $Y$ represent variables and $A$ and $B$ are constants determinable from the observations. The accuracy of prediction can be determined, and it is convenient that we have some general measure of this accuracy. One such measure which can be computed and which will yield information as to the degree of accuracy and the *degree of relationship* is the correlation coefficient, designated $r$. This measure of co-relation, as we shall soon see, not only tells us the degree of relationship, but will also, in conjunction with the two means and standard deviations, permit us to write the linear equation for predicting $Y$ from $X$ or $X$ from $Y$.

Our present discussion will be concerned with the determination of relationship between such typical variables as height, weight, strength, age, intelligence, social status, attitudes—i.e., with those variables which show variation from individual to individual. The question of the relationship between variables of this type can be stated quite simply: Is there a tendency for the individual who ranks high (or low) on one characteristic to be high (or low) on another also? It should be noted that at times a relationship may involve just one variable: Are heights of sons related to the heights of their fathers? Are the IQs of adults related to their childhood IQs?

## THE SCATTER DIAGRAM

The first task is that of tabulation. If we have observations on the height and weight of a large number of individuals, using cross-sectional or coordinate paper, we can lay off on the $y$ axis convenient tabulating intervals for, say, height and on the $x$ axis intervals for weight. The rules for choosing intervals stated on p. 6 should be followed here. Tabulation then consists first of finding on the $y$ axis the interval in which an individual's height falls and locating the interval on the $x$ axis for his weight. A tally or dot is then placed in the *cell* formed by the intersection of these two intervals. The result of such a two-way or cross tabulation is referred to as a *scatter diagram* or correlation table. It will contain as many tallies as there are pairs of observations. The tallies in each row, or horizontal array, can be counted and recorded, separately by rows, to the right of the diagram. This procedure will, of course, yield the frequency distribution for all individuals with respect to the variable on the $y$ axis. A similar count, and recording at the top, of tallies for each column, or vertical array, will yield the distribution for the other variable. The sum of the frequencies for either of these marginal distributions should equal $N$, or the number of pairs of observations.

Figures 8.1 and 8.2 are illustrative scatter diagrams, but not models so far as number of grouping intervals is concerned. In practice, from 12 to 20 intervals should be used in order to reduce the grouping error to a negligible amount. It is to be understood that the intervals in these charts are 40–44, 30–39, 50–59, etc. The student should study these diagrams so as to grasp some of the mechanical details involved in their construction. It should be noted that the number and size of the intervals for the two variables need not be the same, and that the zero points on the scales of measurement need not appear or even be indicated on the axes.

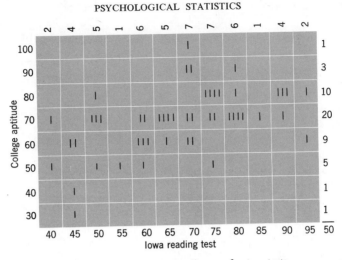

Fig. 8.1. Correlation scatter diagram for two tests.

It can readily be seen that these two diagrams represent different degrees of relationship. A precise method for measuring or describing degree of relationship or association or correlation will be discussed in detail in the pages to follow. We shall begin with a symbolic definition of a basic correlation coefficient, indicate its computation, and then discuss its meaning, interpretation, assumptions, and finally its limitations. Certain

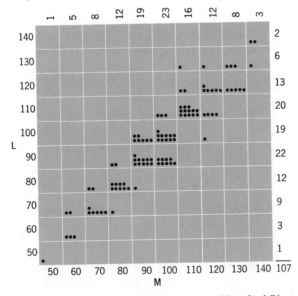

Fig. 8.2. Correlation scatter for two forms of Stanford-Binet.

elementary mathematical derivations will be either indicated or given whenever it is thought that their inclusion will be useful in clarifying a point or clinching an assumption.

The Pearson *product moment correlation coefficient* is defined by

$$r = \frac{\Sigma xy}{N S_x S_y} \tag{8.1}$$

in which $x$ and $y$ represent deviation measures from the respective means of the two variables, i.e., $x = X - M_x$ and $y = Y - M_y$, the $S$s in the denominator are the standard deviations of the two distributions, and $N$ is the number of individuals measured. With reference to a scatter diagram, $M_x$ and $S_x$ hold for the marginal distribution at the top, whereas $M_y$ and $S_y$ hold for the distribution to the right. The numerator term, $\Sigma xy$, implies that the product of each individual's $x$ and $y$ is determined, and that all such products are summed algebraically. There will, of course, be $N$ products in this sum, some of which will be positive, some negative, and perhaps some zero.

Definition formula (8.1) is seldom used for computation. For $N$ small a usable computational equivalent is

$$r = \frac{N\Sigma XY - \Sigma X \Sigma Y}{\sqrt{N\Sigma X^2 - (\Sigma X)^2} \sqrt{N\Sigma Y^2 - (\Sigma Y)^2}} \tag{8.2}$$

which involves four familiar sums, and the sum of the products of the paired raw scores. This formula is unwieldy for large $N$ and/or scores which are numerically large. For reasons which will become apparent later, the careful researcher will always make a scatter diagram, and once this has been done it is economical to compute $r$ in terms of step-interval deviations from arbitrary origins. An appropriate formula is

$$r = \frac{N\Sigma d_x d_y - \Sigma d_x \Sigma d_y}{\sqrt{N\Sigma d^2_x - (\Sigma d_x)^2} \sqrt{N\Sigma d^2_y - (\Sigma d_y)^2}} \tag{8.3}$$

in which $d_x$ is defined as an individual's score deviation, in step intervals, from an arbitrary origin on the $X$ scale, and $d_y$ is defined similarly for the $Y$ scale. The student will note the similarity of the radical terms to formula (3.5) for computing $S$. Formula (8.3) calls for two sums, two sums of squares, and a sum of cross products, all in terms of step or interval deviations from arbitrary origins. The arbitrary origins may be taken at the center or at the bottom of each distribution. The former will involve handling smaller figures but will have the disadvantage of introducing negative numbers. The latter scheme is better if a calculating machine is available.

## CALCULATION OF $r$

The computation of $r$ will be illustrated for both hand and machine calculating methods. The hand calculation scheme here used may not be quite as economical as other available schemes, but the particular setup has the advantage that it forms an economical basis for machine computation, and the author presumes that practically all those who are apt to compute more than a few $r$s will have access to a calculating machine of the Monroe or Marchant or Friden type. Once the steps involved in the hand calculation form are grasped, it becomes easy to transfer them to machine work. The writer has never found the commercial correlation charts helpful. All that is necessary is a sheet of cross-section paper ruled four lines to the inch, on which we can readily lay out the axes, in intervals, for tabulating or tallying. When the scatter diagram has been made and the tally (or dot) marks have been summed across and up to get the marginal frequencies (as shown in Figs. 8.1 and 8.2), the $d$ values, taken from an arbitrary origin at the bottom-most interval for each variable, can be written, preferably with colored lead, alongside the marginal frequencies (see Table 8.1). The columns of $fd$ and $fd^2$ values along each margin can be obtained by multiplying in exactly the same manner as was previously done for calculating the standard deviation. The sums of these columns provide four of the five sums needed for $r$.

In order to obtain $\Sigma d_x d_y$, each individual's $d_x$ must be multiplied by his $d_y$, and all such products then summed. In the 140 interval on the $y$ axis we find one individual whose score on the $X$ variable falls in the 50 interval on the $x$ axis. In terms of step deviations his $d_y$ value is 8 and his $d_x$ value is 5, and therefore 5 times 8, or 40, represents his $d_x d_y$ product. Another individual with the same $d_y$ value has a $d_x$ value of 6, whence 6 times 8 is his contribution to $\Sigma d_x d_y$. The third individual in the 140 interval has a $d_x$ value of 7, whence 7 times 8 is his product. These three individuals contribute $5 \times 8 + 6 \times 8 + 7 \times 8$, or 144, to the sum of products. The $d_y$ value of 8 is a common factor to these three products, whence $8(5 + 6 + 7)$ or $8 \times 18$ yields 144. This suggests a scheme, for computing the $d_x d_y$ sum, which involves first summing the $d_x$ values for a particular $Y$ interval or array and then multiplying this sum by the $d_y$ value. Thus the $d_x$ values of the individuals in the 130 interval sum to 34, and in the 120 interval to 34, and so on down to the 60 interval, which yields 2 as the sum of the $d_x$ values. The determination of these $d_x$ sums is greatly facilitated by the use of a runner on which the $d_x$ values $0, 1, 2, 3, \cdots$, have been labeled to correspond exactly with the deviations in step intervals alongside the marginal distribution at the top of the diagram. Since each of these $d_x$ sums is to be multiplied by a $d_y$ value and then all the products summed,

### Table 8.1.* Computation of $r$

| | 25 | 30 | 35 | 40 | 45 | 50 | 55 | 60 | $f_y$ | $d_y$ | $fd_y$ | $fd^2_y$ | $d_x$ sums | $d_y d_x$ |
|---|---|---|---|---|---|---|---|---|---|---|---|---|---|---|
| $fd^2_x$ | 0 | 5 | 40 | 99 | 208 | 225 | 288 | 147 | 1012 | $\Sigma d^2_x$ | | | | |
| $fd_x$ | 0 | 5 | 20 | 33 | 52 | 45 | 48 | 21 | 224 | $\Sigma d_x$ | | | | |
| $d_x$ | 0 | 1 | 2 | 3 | 4 | 5 | 6 | 7 | | | | | | |
| $f_x$ | 2 | 5 | 10 | 11 | 13 | 9 | 8 | 3 | | | | | | |
| 140 | | | | | | 1 | 1 | 1 | 3 | 8 | 24 | 192 | 18 | 144 |
| 130 | | | | | 1 | 1 | 3 | 1 | 6 | 7 | 42 | 294 | 34 | 238 |
| 120 | | | | 1 | 2 | 2 | 1 | 1 | 7 | 6 | 42 | 252 | 34 | 204 |
| 110 | | | 1 | 2 | 3 | 2 | 2 | | 10 | 5 | 50 | 250 | 42 | 210 |
| Y 100 | | | 2 | 3 | 4 | 2 | 1 | | 12 | 4 | 48 | 192 | 45 | 180 |
| 90 | | 1 | 2 | 3 | 2 | 1 | | | 9 | 3 | 27 | 81 | 27 | 81 |
| 80 | | 2 | 3 | 2 | 1 | | | | 8 | 2 | 16 | 32 | 18 | 36 |
| 70 | 1 | 2 | 1 | | | | | | 4 | 1 | 4 | 4 | 4 | 4 |
| 60 | 1 | | 1 | | | | | | 2 | 0 | 0 | 0 | 2 | 0 |
| | 25 | 30 | 35 | 40 | 45 | 50 | 55 | 60 | 61 | | 253 | 1297 | 224 | 1097 |
| | | | X | | | | | | N | | $\Sigma d_y$ | $\Sigma d^2_y$ | Ck | $\Sigma d_x d_y$ |

$$r = \frac{(61)(1097) - (224)(253)}{\sqrt{(61)(1012) - (224)^2} \sqrt{(61)(1297) - (253)^2}} = .776$$

\* Space limitations account for the use of too few intervals in this table. A complete labeling of intervals would be 25–29, etc., and 60–69, etc.

it is convenient first to record the $d_x$ sums to the right as a separate column and then to multiply each $d_x$ sum by the corresponding $d_y$ value, thus leading to the last column of figures. Before these final multiplications are made, the column of $d_x$ sums should be added to see whether it agrees with the $\Sigma d_x$ already computed from the marginal distribution of $X$ scores. Thus an internal check is provided for the column of $d_x$ sums; all other computations should be done twice in order to insure accuracy.

When a calculator is available, the work sheet need not include the $fd$ and $fd^2$ columns, since the sums of these two columns can readily be obtained by the method discussed on pp. 22–23. This means that the column of $d_x$ sums can be placed alongside the $d_y$ values; then each $d_x$ sum can be multiplied by the juxtaposed $d_y$ value, with the products allowed to accumulate in the dial as the needed $\Sigma d_x d_y$. Thus the right-hand column figures need not appear on the work sheet.

The substitution of the five sums into formula (8.3) is straightforward. The denominator factors are evaluated as explained on p. 23, and the numerator is obtained by punching $\Sigma d_x d_y$ into the keyboard and multiplying by $N$; then, with the product left in the lower dial, $\Sigma d_x$ is subtracted $\Sigma d_y$ times. If needed, the two means can be obtained by substituting $\Sigma d_x$ and $\Sigma d_y$ into (3.3), and the two standard deviations by multiplying the proper radical by the interval size and dividing by $N$ [equivalent to substituting the sum and sum of squares into (3.5)].

# *Chapter 9*

# CORRELATION: INTERPRETATIONS AND ASSUMPTIONS

Intelligent use of the correlation coefficient and critical understanding of its use by others are impossible without knowledge of its properties. It is not sufficient that we be able merely to recognize $r$ as a measure of relationship. It is a peculiar kind of measure which permits certain interpretations provided certain assumptions are tenable and provided we consider possible disturbing factors. Since the interpretations of $r$ are so closely related to assumptions, no attempt will be made to present a separate discussion of these two aspects. The factors which affect $r$, and which are therefore limitations additional to assumptions, will be discussed in Chapter 10.

## STUDY OF SCATTERGRAM

We shall begin by making a somewhat detailed study of certain properties of a typical scatter diagram. The columns and rows of the diagram have already been referred to as vertical and horizontal *arrays*, the intersection of two arrays has been called a *cell*, and the meaning of the marginal distributions has been given. If the scatter diagram depicted in Table 9.1 is examined, it will be noted that each vertical (and also each horizontal) array contains a frequency distribution, and that the marginal totals really represent the number of cases in these array distributions. These array distributions are very much like any other typical distribution: bell-shaped with a clustering or scattering about a central value. The mean

**Table 9.1.  Correlation table for height of fathers $(X)$ and height of sons $(Y)$**

|  | 2 | 6 | 12 | 19 | 27 | 26 | 26 | 26 | 20 | 15 | 8 | 5 | $M_a$ |
|---|---|---|---|---|---|---|---|---|---|---|---|---|---|
| 75 |  |  |  |  |  |  |  |  |  | 1 |  |  | 1  71.0 |
| 74 |  |  |  |  |  |  |  |  |  | 1 |  | 1 | 2  72.0 |
| 73 |  |  |  |  |  | 1 | 1 |  |  | 1 | 1 | 1 | 5  70.6 |
| 72 |  |  |  |  | 1 | 1 | 2 | 2 | 2 | 1 | 1 |  | 10  70.0 |
| 71 |  |  |  | 1 | 2 | 2 | 2 | 3 | 4 | 2 | 2 | 1 | 19  69.1 |
| 70 |  |  | 1 | 1 | 4 | 2 | 4 | 4 | 4 | 3 | 1 | 1 | 25  68.5 |
| 69 |  | 1 | 1 | 3 | 4 | 3 | 5 | 6 | 4 | 2 | 1 |  | 30  67.8 |
| 68 |  | 1 | 2 | 2 | 5 | 6 | 5 | 6 | 3 | 2 | 2 |  | 34  67.7 |
| 67 | 1 | 1 | 3 | 4 | 5 | 5 | 4 | 2 | 3 | 1 |  |  | 29  66.7 |
| 66 |  | 1 | 2 | 2 | 2 | 4 | 3 | 1 |  |  |  |  | 15  66.3 |
| 65 |  | 1 | 2 | 3 | 2 | 2 | 1 | 1 |  |  |  |  | 12  65.8 |
| 64 | 1 | 1 | 1 | 2 | 2 | 1 |  |  |  |  |  |  | 8  64.7 |
| 63 |  |  |  | 1 | 1 |  |  |  |  |  |  |  | 2  65.5 |
|  | 62 | 63 | 64 | 65 | 66 | 67 | 68 | 69 | 70 | 71 | 72 | 73 |  |
| $M_a$ | 65.5 | 66.5 | 66.8 | 66.8 | 67.6 | 67.8 | 68.6 | 69.1 | 69.5 | 70.6 | 70.3 | 72.0 |  |

$$r = \quad .56 \qquad\qquad N = 192$$
$$M_x = 67.69 \qquad\qquad S_x = \quad 2.49$$
$$M_y = 68.44 \qquad\qquad S_y = \quad 2.33$$

and standard deviation again become useful descriptive terms.  Thus, in Table 9.1, the mean height of sons whose fathers were 64 inches tall is found to be 66.8 inches.  This is simply the mean of the twelve cases which fall in this particular array.  Similarly for all the vertical arrays we have the means as recorded along the bottom of Table 9.1.  The means of the horizontal array distribution have been recorded to the right of the scatter diagram.  For example, the mean height of the 10 fathers whose sons were 72 inches tall is 70.0 inches.

If the means of the vertical arrays are plotted (see crosses in Fig. 9.1) two things will be noticed: the means are progressively greater as we pass

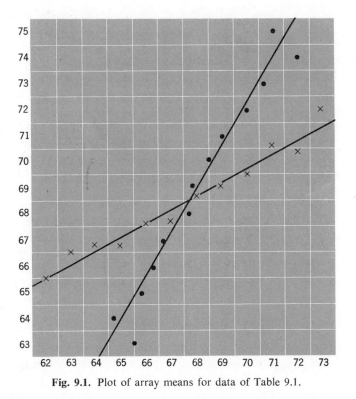

**Fig. 9.1.** Plot of array means for data of Table 9.1.

from short to tall fathers, and they fall approximately on a straight line. It will be noted (see dots in Fig. 9.1) that the means for the horizontal arrays also approximate a line and show progression. Now, with reference to the means of the vertical arrays, each represents the mean height of sons of fathers of a particular height and therefore may be used as a basis for predicting the height, if unknown, of a man if we have been told the height of his father. Thus, if the father is 66 inches tall, the best estimate of his son's height is 67.6, the observed mean height of men whose fathers are 66 inches in height.

Obviously such an estimation would be subject to considerable error, since we have also the observable fact that the heights of sons of fathers 66 inches tall show a large amount of variation about the array average. This variation tells us something about the possible magnitude of the error involved in using 67.6, the array mean, as our estimated value. The unknown height, of which we take an array mean as an estimate, may actually fall anywhere within rather wide limits on either side of the array mean. These limits can be described in terms of the standard deviation of the array distribution. The standard deviation for the distribution of heights of sons whose fathers were 66 inches in height is about 2.1. Now, if we take 67.6 as the best estimate, we can say that, if we were to predict the height of 100 sons (fathers 66 inches), about 68 per cent of the time the error would be within the limits 67.6 ± 2.1, 95 per cent within 67.6 ± 4.2, and nearly always within the limits 67.6 ± 6.3. Likewise, when the $S$s for the several arrays have been computed, a statement of the limits of the error in predicting any son's height from his father's height can be made. Such a procedure will yield as many measures of error as there are vertical arrays. We shall soon see that a convenient assumption can be made which will usually allow us to use a single indication of the error of estimate.

Let us return again to the line of the means. Two such lines have been drawn in Fig. 9.1; one line "fits" the means of the vertical, the other the means of the horizontal, arrays. Let us for the present confine our attention to the means of the vertical arrays. They do not lie exactly on the drawn line; some are above, some below. If they fell exactly on the line, a prediction based on an array mean would be precisely the same as a prediction obtained by noting the $Y$ value of the line where it cuts the middle of the array. Furthermore, if the means were exactly on a straight line, we might write the equation for this line in the form $Y = BX + A$, where $A$ equals the $y$ intercept (value of $Y$ where the line crosses the $y$ axis) and $B$ equals the slope of the line (the inclination of the line to the $x$ axis). With $A$ and $B$ known, the value of $Y$ for a particular $X$ can be readily estimated.

But, since the means do not lie exactly on a straight line, the foregoing reasoning would not seem offhand to yield us anything of practical value. From many points of view, however, it is desirable that we determine the equation of the straight line which best "fits" the means, i.e., the equation of a line which passes near all the means. Then we can use this equation instead of the array means in making predictions. The justification for this procedure depends on the validity or tenability of an assumption: we assume that the failure of the means to fall exactly on a straight line is due to chance fluctuations in the means. Each array mean is based on a sample and consequently deviates more or less from the true or population value of the mean for the array. This is equivalent to saying that, if all the array means were based on a much larger number of cases, we could assume that they would approximate more exactly a straight line. This is an assumption which can always be made provided the array means for a particular scatter do not show marked deviations from linear form. (Adequate checks in terms of probability, to be described later, can be utilized to ascertain whether the fluctuations from linearity are larger than is reasonable on the basis of chance.)

## THE BEST-FIT LINE

We can now consider one of the advantages of using a line instead of the several array means as a basis for prediction. The location of the line is dependent on all the means, or rather on all the cases. It therefore seems reasonable to believe that the line would be more stable from the sampling point of view than would the array means, each of which is based on a rather small number of cases.

If we accept the assumption of *linearity* of array means, our problem is that of determining $A$ and $B$ so that we can write the equation of the line of means. We need the equations of two lines: $Y = BX + A$ for the means of the vertical arrays and $X = B'Y + A'$ for the horizontal array means. We shall consider the determination of the constants $A$ and $B$ for the first equation, but before doing so something must be said concerning what is meant by a "best-fit" line. The constant $A$ gives the $y$ intercept, i.e, tells us where the line cuts the $y$ axis. Suppose we think of several possible lines having the same slope (the same $B$) as the line in Fig. 9.1 which passes near the crosses. Obviously, if we considered a line passing near the top or bottom of the scatter diagram, it would be a "worse fit" than that drawn in Fig. 9.1. Likewise, if we think of pivoting the line about some point, thereby altering its slope, it can be readily seen that rotating it to a vertical or horizontal position would give a worse fit. It should now be clear that the assigning of some values to $A$ and $B$ will lead to a worse fit than that obtained by certain other values, or conversely that some values will yield a much better fit than others.

One criterion accepted as a basis for a best-fit line is that the sum of the squares of the deviations from the line shall be as small as possible. With respect to determining the best-fit line to the means of the vertical arrays, this criterion or definition of fit implies that the values of $A$ and $B$ are to be such that the sum of the squared deviations of the observed heights of sons—deviations in an up and down or vertical direction—about the line will be a minimum. Stated in symbols, let $Y' = BX + A$, where $Y'$ (read $Y$ prime) is the value estimated from a given $X$, and let $Y$ be the observed value. Then $(Y - Y')^2$ represents the squared deviation of any $Y$ from the line or estimated value. The problem is so to choose $A$ and $B$ as to make $\Sigma(Y - Y')^2$ as small as possible. It is more convenient to deal with both the equation, $y' = bx + a$, and the sum, $\Sigma(y - y')^2$, in deviation units, with $y'$ and $y$ as deviations from $M_y$ and $x = X - M_x$. This is merely the translation of the axes which makes the origin or reference point coincide with $M_x$ and $M_y$. The student should visualize the meaning of this shift of axes. Note that the pattern of tallies is not changed by this simple transformation. Do you think that the slope $B$ will equal the slope

$b$? Will $A = a$? Let us keep the first question in abeyance and examine now the second question. Both $A$ and $a$ represent the $y$ intercepts of the desired prediction line. If it is not immediately obvious to the student that $A$ may not equal $a$, he should imagine that in Table 9.1 and Fig. 9.1 the axes have been moved so that the origin is at the center of the scatter diagram, and then ask himself where the line through the means of the vertical arrays would cut the new $y$ axis. (Incidentally, it should be noted that the value of $A$ cannot be read directly from Fig. 9.1 for the simple reason that the reference frame as drawn does not include the origin. The real $y$ and $x$ axes of the original measures would be, respectively, to the left of, and lower than, the indicated axes.)

It is of interest to speculate concerning the value of $a$ in the equation $y' = bx + a$. Common sense would suggest that, if an individual were average on $X$, the best guess would be that he would be average on $Y$. That is, if $X = M_x$, we would expect $Y'$ to equal $M_y$. But, if an individual's $X$ measure fell at $M_x$, his deviation, or $x$ value, would be 0, and the estimated value of $Y$ as being equal to $M_y$ would in terms of deviation scores become 0. This would imply that the prediction line would pass through the origin of the deviation score reference axes, and consequently that the $y$ intercept would be zero; hence $a = 0$. For the purpose of simplifying the determination of the best value for $b$, we ask the reader to accept, on the basis of the foregoing reasoning, that $a = 0$ for the best-fitting line. If we carried both $a$ and $b$ along in the following development, $a$ would in fact turn out to be zero.

This permits us to write $y' = bx$ as the equation for estimating $y$, in deviation units, from $x$, or deviation values of $X$. Our task becomes that of determining the value of $b$ which will make $\Sigma(y - y')^2$ a minimum. Incidentally, it should be obvious that the discrepancy of any particular $y$ value from the desired line has the same numerical value as the deviation of its corresponding original $Y$ value from the line, and that $\Sigma(y - y')^2 = \Sigma(Y - Y')^2$. When we have determined the optimal value for $b$ in $y' = bx$, we can readily pass back to the original reference frames, the gross score axes, by substituting for $y'$ the value $Y' - M_y$, and for $x$, $X - M_x$. With $a$ fixed as zero, i.e., with the $y$ intercept equal to zero, we can think of the line as passing through the origin (deviation axes); i.e., its up and down location is fixed. Obviously, many lines could be drawn through the origin, and they would differ only as to slope, i.e., as to $b$. Of all possible lines which may be drawn through the origin, some will be closer than others to the observations (tallies) *in toto*. The student might imagine several lines any of which would seem to constitute a good fit. As he takes lines with either greater or lesser slope than those of apparently good fit, the fits will become worse; and of those which seem to fit, some will actually be better than others. The student might think it would only be necessary to draw what seems by inspection to be the best-fitting line, and then obtain its slope by actually measuring the angle which it makes with the horizontal (with needed adjustment to allow for the measurement units). The trouble with this procedure is that individuals would tend to disagree regarding which of several lines was really best; also, the measurement of angles would be none too exact. What we need is an objective procedure, a method that will yield the value of $b$ which leads to the best possible fit in the sense of reducing the sum of the squares of the discrepancies to a minimum.

We set up the function

$$f = \frac{\Sigma(y - y')^2}{N} = \frac{\Sigma(y - bx)^2}{N}$$

in which we have $N$ deviations of the form $y - y'$ or $y - bx$ (since $y' = bx$). These deviations when squared, summed, and divided by $N$ give us a quantity or function which is to be minimized by the proper choice of $b$. The value to be assigned to $b$ can best be ascertained by the calculus.* This is done by taking the derivative of the function with respect to $b$ setting this derivative equal to zero, and then solving for $b$. Thus

$$\frac{df}{db} = \frac{-2\Sigma x(y - bx)}{N}$$

which, set equal to zero and divided by $-2$, gives

$$\frac{\Sigma x(y - bx)}{N} = 0$$

or

$$\frac{\Sigma xy - b\Sigma x^2}{N} = 0$$

then

$$\frac{\Sigma xy}{N} - b\frac{\Sigma x^2}{N} = 0$$

The first or cross-product term involves the correlation coefficient as defined by formula (8.1), from which definition formula we see that $\Sigma xy/N = rS_xS_y$; and since $\Sigma x^2/N = S^2_x$, we have

$$rS_xS_y - bS^2_x = 0$$

* The student who has not studied the calculus will either take the first part of the following derivation on faith or, if skeptical, will dig into a calculus text to satisfy himself that no magic is involved here.

or

$$rS_y - bS_x = 0$$

which gives

$$b = r\frac{S_y}{S_x}$$

as the optimal value for $b$. We therefore have

$$y' = r\frac{S_y}{S_x}x \qquad (9.1)$$

as the equation for the best-fit line. This equation is in terms of deviation measures, and by proper substitution we get

$$Y' - M_y = r\frac{S_y}{S_x}(X - M_x)$$

or

$$Y' = r\frac{S_y}{S_x}X + \left(M_y - r\frac{S_y}{S_x}M_x\right) \qquad (9.2)$$

as the equation in terms of the original or gross scores. This is the form which we would use in predicting $Y$ from $X$. Note that $B = b = r(S_y/S_x)$ is the slope of this line and that the constant $A$ is equal to the parentheses term.

By similar reasoning the equation of the best-fit line to the means† of the horizontal arrays is found to be

$$x' = r\frac{S_x}{S_y}y \qquad (9.3)$$

which becomes

$$X' = r\frac{S_x}{S_y}Y + \left(M_x - r\frac{S_x}{S_y}M_y\right) \qquad (9.4)$$

When both equations are written in the $B$ and $A$ notation, we may attach subscripts to differentiate between the $B$s and between the $A$s:

$$Y' = B_{yx}X + A_{yx}$$
$$X' = B_{xy}Y + A_{xy}$$

**Regression.** Equations (9.1) and (9.3) in deviation score form and (9.2) and (9.4) in gross score form are known as *regression equations*, and the constants denoting slope are known as *regression coefficients*. It is assumed that prediction will be as accurate by means of a regression

† More strictly speaking, we are fitting a line to means weighted according to their respective $N$s; i.e., we are fitting a line to the observations.

equation as by way of array means, and it can readily be seen that by using a regression equation we can predict from intermediate values, e.g., $64\frac{1}{2}$. This is of especial advantage with grouped data: the array mean is associated only with the midpoint value of the grouping interval, whereas the regression line is not so limited since it is continuous.

**Rate of change.** The results of the foregoing derivation make it clear that the correlation coefficient, along with the two means and the two standard deviations, enables us to write the equation by which either variable can be predicted from the other. The regression coefficients indicate the *rate of change*—unit of change in one variable per unit of change in the other—and in case the two standard deviations are equal, *r* itself indicates the rate of change. Thus we have *one* of the possible interpretations of the correlation coefficient.

For the correlation table in Table 9.1 we get, by proper substitution, the following as the regression equations:

$$Y' = .52X + 33.24 \text{ (to estimate son's height)}$$
$$X' = .60Y + 26.63 \text{ (to estimate father's height)}$$

The student should study Fig. 9.1 sufficiently to convince himself that .52 is the slope of the line passing near the crosses, and that .60 represents the slope (with reference to the vertical) of the line through the dots. The student should also satisfy himself that the constants 33.24 and 26.63 really represent the points at which the two lines intercept the $y$ and $x$ axes. Finally he should show that, if a father's height is at the mean of all fathers, the mean of the heights of all the sons is the best estimate of his son's height.

## ACCURACY OF PREDICTION

The next problem to which we turn is concerned with the accuracy of prediction by means of a regression equation. It has already been indicated that, when the mean of an array is used in prediction, the error of estimate is a function of the spread within that array. By introducing an assumption it becomes possible to substitute *one* measure of error in place of the several, numerically different, array standard deviations. An examination of the array distributions in Table 9.1 reveals that the vertical arrays differ from each other very little in dispersion (likewise, the horizontal arrays). If we were to compute the standard deviations for the vertical arrays, we would find differences, for this diagram, of such size as could readily be attributed to chance or sampling fluctuations; i.e., we assume that, if we had a much larger $N$, the array dispersions would be very nearly equal. Ordinarily this assumption of *homoscedasticity* can be met, and one measure of dispersion can be used for all the vertical arrays (and another for all the horizontal arrays).

**Error of estimate.** One such measure might be an average of the array $S$s, but to determine this we would need first to compute all the $S$s, a somewhat laborious job. Since we are to use the regression line, instead of array means, as a basis for prediction, we really need something corresponding to the $S$ about this line. Such a value can be obtained by noting that $y - y'$ (or $Y - Y'$) represents the discrepancy between estimated and observed values and that $\Sigma(y - y')^2/N$ is the mean of the squared deviations, the root of which will be the standard deviation of the discrepancies between estimated and observed values. This will be taken as the one standard deviation to replace the several standard deviations as our measure of the error of prediction. This particular standard deviation, defined as the square root of $\Sigma(y - y')^2/N$, is called the *standard error of estimate*. It may be determined in two ways. First we can take a roundabout way which involves these steps: the prediction of each $Y$ by use of equation (9.2), or each $y$ by use of (9.1); the calculation of the discrepancies $(Y - Y')$ or $(y - y')$; squaring, summing, dividing by $N$, and taking the square root. A quicker method for determining the standard error of estimate is readily derived algebraically.

Let $S_{y \cdot x}$ stand for the standard error of $Y$ as estimated from $X$; then by definition,

$$S^2_{y \cdot x} = \frac{\Sigma(Y - Y')^2}{N} = \frac{\Sigma(y - y')^2}{N}$$

but

$$y' = r \frac{S_y}{S_x} x$$

by formula (9.1) whence

$$S^2_{y \cdot x} = \frac{1}{N} \Sigma \left( y - r \frac{S_y}{S_x} x \right)^2$$

$$= \frac{1}{N} \Sigma \left( y^2 - 2r \frac{S_y}{S_x} xy + r^2 \frac{S^2_y}{S^2_x} x^2 \right)$$

$$= \frac{\Sigma y^2}{N} - 2r \frac{S_y}{S_x} \left( \frac{\Sigma xy}{N} \right) + r^2 \frac{S^2_y}{S^2_x} \left( \frac{\Sigma x^2}{N} \right)$$

$$= S^2_y - 2r \frac{S_y}{S_x} r S_x S_y + r^2 \frac{S^2_y}{S^2_x} S^2_x$$

$$= S^2_y - r^2 S^2_y$$

then

$$S_{y \cdot x} = S_y \sqrt{1 - r^2} \tag{9.5}$$

By a similar line of reasoning it can be shown that

$$S_{x \cdot y} = S_x \sqrt{1 - r^2} \tag{9.6}$$

which gives the standard error of $X$ as estimated from $Y$.

Thus the correlation coefficient not only enters into the prediction equations (9.1) to (9.4), but also permits us to gauge the accuracy of prediction. It should be noted in passing that we can write the equation of a best-fit line without first determining $r$ and that the error of prediction can also be ascertained without recourse to $r$. Such a method for determining the error of estimate has already been indicated; the square root of $\Sigma(Y - Y')^2/N$, in which $Y - Y'$ represents the computed discrepancy between observed and predicted values. This need not involve $r$ unless the prediction equation is written in terms of $r$, as was done in (9.2). The equation $Y' = A + BX$ can be written in the form

$$Y' = \frac{\Sigma X^2 \Sigma Y - \Sigma X \Sigma X Y}{N \Sigma X^2 - (\Sigma X)^2} + \frac{N \Sigma X Y - \Sigma X \Sigma Y}{N \Sigma X^2 - (\Sigma X)^2}(X) \tag{9.7}$$

in which $X$ and $Y$ stand for gross or original measures. Formula (9.7) for the best-fitting line (least squares solution) does not involve means, $S$s, or the correlation coefficient. If, as is frequently the case, we are interested in obtaining the equation for $Y$ only, it will be noticed that it is unnecessary to compute the sum of the $Y$ squares, which is not, however, a tremendous saving of time. Perhaps the quickest way for determining the equation is by direct substitution in (9.7), but the determination of the error of estimate (sometimes called the closeness of fit of the line) is certainly facilitated by calculating $r$ and $S_y$ and substituting in (9.5).

The standard error of estimate is to be interpreted as a standard deviation, and in so doing we are tacitly assuming that the array distributions are not only equal in dispersion but also normal. For the correlation diagram in Table 9.1, we have $S_{y \cdot x} = 1.9$, which is to be considered the standard deviation of the $Y$ values about the regression line, $Y' = .52X + 33.24$. By use of this equation we would predict that the height of the son of a man 70 inches tall ($X = 70$) would be 69.6, and the error of estimate, 1.9, would be interpreted by saying that, if we made many such predictions, 68.26 times out of a hundred the actual height of sons of 70-inch fathers would be within the limits $69.6 \pm 1.9$, and nearly always within the limits $69.6 \pm 3(1.9)$.

This is a *second method* for interpreting the correlation coefficient: in terms of the accuracy of prediction or closeness of fit of regression lines. If no correlation exists, the errors of estimate are $S_{y \cdot x} = S_y$ and $S_{x \cdot y} = S_x$. In this connection it can be seen from formulas (9.2) and (9.4) that, when

$r = 0$, the estimated $Y$, $Y'$, becomes $M_y$, and $X'$ becomes $M_x$. For example, if it has been established that the correlation between toe length and IQ is zero, we would always take 100 (the mean) as our best guess for an individual's IQ regardless of toe length. The error of estimate would of course be the standard deviation of the distribution of IQs, and it would be said that toe length is useless in predicting IQ. The scatter diagram for IQ as $Y$ and toe length as $X$ would exhibit the following characteristics: first, the regression line $Y' = A + BX$ would be horizontal, i.e., $B$ would equal zero, and the means of the arrays would fluctuate about the value $M_y$, or $A$ would equal $M_y$; and, second, all the array distributions would have dispersions approximately equal to $S_y$. What would be the best guess as to the other regression line and the standard deviations of the horizontal arrays?

Now suppose the correlation between the variables were perfect ($r = +1$ or $-1$). The tallies in the scatter diagram would lie in a line, there would be no spreading about this line, the two regression lines would coincide, and no error would be involved in estimating $X$ from $Y$ or $Y$ from $X$. That $S_{y \cdot x}$ and $S_{x \cdot y}$ would both be zero in case of perfect correlation is quite evident when we consider formulas (9.5) and (9.6).

At this point the student should note the difference between positive and negative correlation. In the case of a positive $r$, a high score goes with high and low with low, whereas, for a negative $r$, high goes with low and low with high. With reference to the scatter diagram, a negative $r$ typically involves a swarm of tallies stretching from the upper-left to the lower-right corner, whereas for a positive $r$ the trend is from lower left to upper right (this assumes that the axes have been laid off in the conventional fashion). With reference to the regression equations, a negative $r$ yields negative regression coefficients or negative slope for the lines. The student should be warned that an apparently negative $r$ may in reality be positive. Thus, if one variable is a test or performance scored in terms of time (or errors) and the other variable is scored in terms of amount done, the scatter diagram might show large time scores as going with small amounts of work done, i.e., high with low, which might be wrongly taken to indicate negative rather than positive correlation. Instead of asking whether high goes with high and low with low, it is safer to ask whether best goes with best. This rule, however, is difficult to apply when we are dealing with the interrelation of personality traits, especially those which do not readily permit of a statement as to which is the desirable end of the trait scale. The sign of the correlation coefficient in such cases always needs a qualifying statement which explicitly tells the direction of the relationship between the variables. Obviously, as far as accuracy of prediction is concerned, the error is the same for a negative and positive $r$ of the same magnitude.

**Alienation.** To return to the interpretation of the correlation coefficient by way of the standard error of estimate, we see that the factor in formulas (9.5) and (9.6) which involves $r$ is $\sqrt{1 - r^2}$. It is the value of this which, when multiplied by the proper $S$, leads to the error of estimate. The expression $\sqrt{1 - r^2}$ is called the *coefficient of alienation*. If $r$ is zero, its value is 1 and the error of estimate is the $S$ for the variable being estimated. Table 9.2 gives the value of the coefficient of alienation for varying values of $r$. The student will do well to fix in mind the trend in this table. It will be noted that, compared to a correlation of zero, an $r$ of .60 reduces the error of estimate by 20 per cent, whereas an $r$ of .30 reduces it by about 5 per cent; that $r$ must be as high as .866 before the error of estimate is

Table 9.2. **Values of the coefficient of alienation**

| $r$ | $\sqrt{1 - r^2}$ | $r$ | $\sqrt{1 - r^2}$ |
|---|---|---|---|
| .00 | 1.000 | .60 | .800 |
| .10 | .995 | .70 | .714 |
| .20 | .980 | .80 | .600 |
| .30 | .954 | .866 | .500 |
| .40 | .917 | .90 | .436 |
| .50 | .866 | .95 | .312 |

reduced by one-half; and that the difference in reduction between an $r$ of .70 and an $r$ of .90 is approximately the same as that between .20 and .70. This interpretation of $r$ is most useful and at the same time most disturbing, since the errors of estimate for $r$s in the vicinity of .40 to .70, values usually found and utilized in predicting success from test results, are discouragingly large.

A somewhat different way of grasping the meaning of $r$, as it is applied to accuracy of prediction, is to square both sides of formula (9.5) and then solve explicitly for $r$. This leads to

$$r^2 = 1 - \frac{S^2_{y \cdot x}}{S^2_y} \qquad (9.8)$$

from which it is readily seen that the correlation coefficient depends on the accuracy of prediction *relative* to the total variance of the variable being predicted.

It might be well at this time to bring together a few remarks concerning the assumptions involved in using and interpreting a correlation coefficient in terms of either rate of change or accuracy of prediction. When an $r$ is reported, and no evidence to the contrary is given, we have a right to expect that the assumptions of linearity of regression and homoscedasticity have been met. The interpretation of $r$ as rate of change definitely assumes linearity, and the interpretation in terms of the error of estimate definitely assumes both linearity and homoscedasticity. In certain special cases where the investigator is interested only in a one-way prediction, say $Y$ from $X$, and there is no likelihood of ever reversing to predict $X$ from $Y$, it will suffice if the regression of $Y$ on $X$, i.e., for predicting $Y$ from $X$, be linear and the $Y$ or vertical array distributions be homoscedastic. The use of the correlation coefficient in predicting performance from age may be cited as an instance in which there is no need to worry about the possible nonlinear regression of age on score or the lack of homoscedasticity about this regression line.

Although there are adequate checks for linearity and homoscedasticity, a careful scrutinization of the scatter diagram is usually sufficient to warn us of violent departures from these assumptions. Formula (8.2) and other nonplotting schemes for computing $r$ give no inkling as to whether these assumptions are being violated and therefore cannot command the confidence of the careful investigator. The purpose of a research project might very well be the study of the relationship between two variables, but an end result in terms of a correlation coefficient, with no attention given to the form of the relationship, is inadequate.

## VARIANCE AND CORRELATION

A *third method* of interpreting $r$ is in terms of variance. Before discussing this interpretation, we must introduce an important theorem concerning the variance of a sum (or difference). Suppose that variable $W$ is made up of two parts $U$ and $V$ such that $W = U + V$. For example, the score on an arithmetic test might consist of two parts: score in addition and score in multiplication. Obviously, $w = u + v$, and therefore the variance of the $W$ variable is

$$S^2_w = \frac{\Sigma w^2}{N}$$

$$= \frac{1}{N}\Sigma(u + v)^2$$

$$= \frac{1}{N}(\Sigma u^2 + \Sigma v^2 + 2\Sigma uv)$$

$$= S^2_u + S^2_v + 2r_{uv}S_uS_v \qquad (9.9)$$

and in case $U$ and $V$ are independent, we have

$$S^2_w = S^2_u + S^2_v \qquad (9.10)$$

If we are dealing with the difference, $W = U - V$, we have

$$S^2_w = S^2_u + S^2_v - 2r_{uv}S_uS_v \tag{9.11}$$

and for $U$ and $V$ independent, we have

$$S^2_w = S^2_u + S^2_v$$

which is identical with (9.10). In words, *the variance of a sum (or difference) of two independent variables is equal to the sum of their separate variances.* Variances are additive, whereas standard deviations are not. It can be shown that, when $U$ and $V$ are distributed normally, their sum or difference will also yield a normal distribution.

Now, with regard to the third method for interpreting $r$, let us note that in deviation units an observed $y$ can be thought of as made up of two independent parts, the part which can be predicted from $x$, namely $y'$, and the residual or unpredictable part, $(y - y')$. Before going further we must demonstrate that $y'$ and $(y - y')$ are really independent. The numerator for the correlation between $y'$ and $(y - y')$ can be expressed as $\Sigma y'(y - y')$. But, since $y' = r\dfrac{S_y}{S_x}x$ and $(y - y') = y - r\dfrac{S_y}{S_x}x$, we have

$$\Sigma y'(y - y') = \Sigma r\frac{S_y}{S_x}x\left(y - r\frac{S_y}{S_x}x\right)$$
$$= r\frac{S_y}{S_x}\Sigma xy - r^2\frac{S^2_y}{S^2_x}\Sigma x^2$$
$$= r\frac{S_y}{S_x}NrS_xS_y - r^2\frac{S^2_y}{S^2_x}NS^2_x$$

which is seen to be zero; hence $y'$ and $(y - y')$ are uncorrelated.

We have $y = y' + (y - y')$; whence, by the foregoing variance theorem,

$$S^2_y = S^2_{y'} + S^2_{y\cdot x} \tag{9.12}$$

in which $S^2_{y\cdot x}$ is the variance of the residuals, $(y - y')$. If we divide both sides of this equation by $S^2_y$, we get

$$1 = \frac{S^2_{y'}}{S^2_y} + \frac{S^2_{y\cdot x}}{S^2_y} \tag{9.13}$$

from which we see that, since the two ratios add to unity, either one can be interpreted as a proportion (or a percentage by shifting the decimal point). Thus the ratio of $S^2_{y'}$ to $S^2_y$ is the proportion of the variance in $Y$ which can be predicted from $X$, and the ratio of $S^2_{y\cdot x}$ to $S^2_y$ represents the proportion of the variation (variance) of $Y$ which is left over or remains or cannot be predicted from $X$. A little reflection as to the meaning of this residual variance should convince the student that we are here dealing with the same variance which results if we square formula (9.5), thus

$$S^2_{y \cdot x} = S^2_y(1 - r^2)$$

which means that

$$\frac{S^2_{y \cdot x}}{S^2_y} = 1 - r^2$$

When we substitute this value into (9.13), we have

$$1 = \frac{S^2_{y'}}{S^2_y} + 1 - r^2$$

from which it is readily seen that the ratio

$$\frac{S^2_{y'}}{S^2_y} = r^2 \qquad (9.14)$$

That is, the square of the correlation coefficient gives the proportion of the total variance of $Y$ which is predictable from $X$, or $r^2$ measures the proportion of the $Y$ variance which can be attributed to variation in $X$. The proportion of the variance of $Y$ which is due to variables other than $X$ is given by $1 - r^2$. By shifting decimals, we can think of $r^2$ as indicating a percentage, the percentage of variance which has been explained, and $1 - r^2$ as the percentage of variance due to other causes. It will be noted that $r^2$, not $r$, can be so interpreted. This is true because variances are additive, whereas standard deviations are not. It should be emphasized that $r^2$ as a proportion has to do with variation expressed technically as variance.

It is of some interest to examine the meaning of $S^2_{y'}$. It is the square of the standard deviation of the estimated values, and, with reference to the scatter diagram, $S_{y'}$ corresponds approximately to what we would obtain if we were to compute the standard deviation about $M_y$ of the vertical array means, each weighted according to the number of cases in its array. As an excercise, the student can prove $r^2 = S^2_{y'}/S^2_y$ by determining directly, rather than by formula (9.5), that $S^2_{y'} = r^2 S^2_y$. (HINT: use the deviation score form of the regression equation.)

This third method of interpreting a correlation coefficient assumes linearity of the regression line involved in predicting $Y$, or the dependent variable, from $X$ as the independent variable; i.e., the regression of $Y$ on $X$ must be linear. If $X$ were considered as the dependent variable, the interpretation that $r^2$ indicates the proportion of the variance of $X$ explained by $Y$ would assume linearity for the regression of $X$ on $Y$. The assumption of linearity becomes explicit if it is proved directly that $S^2_{y'}$ $= r^2 S^2_y$, and it was implied when we used $S^2_{y \cdot x}$ in that this residual variance was taken about a straight line. This interpretation does not assume homoscedasticity, nor does it assume normality either for the marginal or for the array distributions.

The investigator who is interested in analyzing variation and its possible causes will prefer the interpretation of the correlation coefficient in terms of variance. The problem is frequently one in which an attempt is made to explain variation in one trait in terms of variation of another which is conceived of as being more basic. The use of $r^2$ as the percentage of the variance of a trait which is predictable by, or attributable to, variation in a second variable becomes a valuable tool in the analysis of variation. Of course we must use caution in assuming causation of one variable by another. Logic, not statistical method, must be invoked to determine whether a causal relationship exists, and the statistical interpretation modified accordingly. Variation in $X$ might cause variation in $Y$, or vice versa, or variation in both $X$ and $Y$ might be due to the influence of some other variable or variables.

To illustrate the interpretation of $r^2$ as a percentage, let us suppose we have the performance of a group of school children on a substitution test. Considerable variation in scores will be present, and we may rightfully ask whether a portion of this variation is due to age differences. We can determine the correlation between age and performance. Suppose $r = .60$; this can be interpreted by saying that 36 per cent of the *variance* in performance is due to age differences, and 64 per cent is due to other causes. Likewise, the variance in crop yield due to variation in rainfall can be determined; or the variance in the height of a group of men may be analyzed into two or more parts, one of which might be the portion due to variation in the heights of their fathers.

## CORRELATION AND COMMON ELEMENTS

A *fourth* possible interpretation of the correlation coefficient assumes that each of the two variables can be thought of as a summation of a number of equally potent, equally likely, independent elements, which can be either present or absent. Then the degree of correlation is a function of the number of elements common to the two variables. The general formula is

$$r_{xy} = \frac{n_c}{\sqrt{n_x + n_c} \sqrt{n_y + n_c}} \tag{9.15}$$

in which $n_x$ equals the number of elements unique to $X$, $n_y$ the number unique to $Y$, and $n_c$ the number common to both variables. If the number of elements in $X$ equals the number in $Y$, $r$ gives the proportion of elements common to $X$ and $Y$; if $X$ is determined only by elements common to $Y$, whereas $Y$ has additional elements, $r^2$ gives the proportion of elements entering into $Y$ which determine $X$. There is little, if any, factual basis for believing that the assumptions stated are tenable so far as psychological variables are concerned, and therefore the interpretation of the correlation coefficient in terms of common elements may be viewed with scepticism.

## NORMAL CORRELATION

A *fifth* interpretation of $r$ is more mathematical but of little practical value. We have already seen how a frequency distribution and its polygon can be thought of as smooth, conforming perhaps to the equation of the normal curve. A correlation table is a frequency distribution, a picture or graph of which requires a third dimension. If we were to replace each tally in a scatter diagram by a thin block, there would result something analogous to the histogram except that it would be three dimensional—the heights of the stacks of blocks would indicate the frequencies for the various cells. Now suppose that this mound of blocks is by some method smoothed to a surface, and we consider the total volume under the surface (between the surface and the $XY$ plane) as representing $N$. Then the number of cases falling between two given $X$ values and simultaneously between two given $Y$ values will be approximately the volume of that portion of the mound which has as its base the rectangle or square formed by the intersections of the two $X$ and two $Y$ values. If the regression lines are linear, if the array distributions are normal and homoscedastic, and if the marginal distributions are normal, the resulting surface is termed the *normal correlation surface*, and the equation of the surface can be written as

$$h = \frac{N}{2\pi\sigma_x\sigma_y\sqrt{1 - r^2}} e^{-\frac{1}{2(1-r^2)}\left(\frac{x^2}{\sigma^2_x} + \frac{y^2}{\sigma^2_y} - \frac{2rxy}{\sigma_x\sigma_y}\right)} \qquad (9.16)$$

A number of important properties of the normal correlation surface can be deduced from this equation and its integral. For instance, the standard error of estimate can be derived from formula (9.16), and it can also be shown that the contour lines which represent different altitudes on the mound, i.e., different frequencies, will be concentric ellipses, and that if $r = 0$, the contour lines will become concentric circles. If the equation is written with $N$ equal to unity, by double integration the probability of an individual's falling between two particular $Y$ values and between two $X$

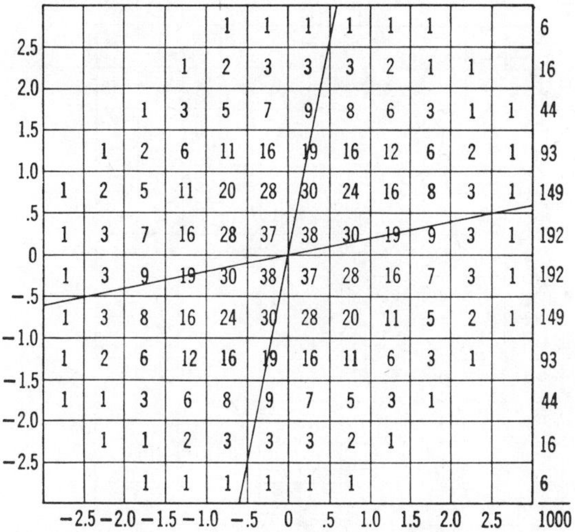

| | | | | | | | | | | | | 6 |
|---|---|---|---|---|---|---|---|---|---|---|---|---|
| | | | 1 | 1 | 1 | 1 | 1 | 1 | | | | 6 |
| | | 1 | 2 | 3 | 3 | 3 | 2 | 1 | 1 | | | 16 |
| | 1 | 3 | 5 | 7 | 9 | 8 | 6 | 3 | 1 | 1 | | 44 |
| 1 | 2 | 6 | 11 | 16 | 19 | 16 | 12 | 6 | 2 | 1 | | 93 |
| 1 | 2 | 5 | 11 | 20 | 28 | 30 | 24 | 16 | 8 | 3 | 1 | 149 |
| 1 | 3 | 7 | 16 | 28 | 37 | 38 | 30 | 19 | 9 | 3 | 1 | 192 |
| 1 | 3 | 9 | 19 | 30 | 38 | 37 | 28 | 16 | 7 | 3 | 1 | 192 |
| 1 | 3 | 8 | 16 | 24 | 30 | 28 | 20 | 11 | 5 | 2 | 1 | 149 |
| 1 | 2 | 6 | 12 | 16 | 19 | 16 | 11 | 6 | 3 | 1 | | 93 |
| 1 | 1 | 3 | 6 | 8 | 9 | 7 | 5 | 3 | 1 | | | 44 |
| 1 | 1 | 2 | 3 | 3 | 3 | 2 | 1 | | | | | 16 |
| 1 | 1 | 1 | 1 | 1 | 1 | | | | | | | 6 |

y-axis labels (top to bottom): 2.5, 2.0, 1.5, 1.0, .5, 0, −.5, −1.0, −1.5, −2.0, −2.5

x-axis labels: −2.5 −2.0 −1.5 −1.0 −.5 0 .5 1.0 1.5 2.0 2.5 1000

**Fig. 9.2.** Normal bivariate, $r = .20$.

values can be determined. Tables are available which can be utilized for this purpose.‡

By use of the tables of the function defined by (9.16) it is possible to generate sets of expected, or theoretical, frequencies for scattergrams which depict normal bivariate correlations of certain magnitudes (.00, .05, .10 · · · .90, .95). A study of Figs. 9.2 to 9.4 will help the reader appreciate more precisely the meaning of degrees of relationship represented by $r$s of .20, .50, and .80. It should be pointed out that calculating the expected frequencies so as to have $N = 1000$ for each scatter involved rounding, the effect of which can lead one to question certain array distributions as being nonnormal. Note, for example, that in Fig. 9.3 the distribution in the bottom array is rectangular in shape and in the adjacent array it is triangular. The rectangular distribution resulted from rounding frequencies such as those shown in the last right-hand array. It should also be mentioned that none of the $r$s computable from these scatters will correspond exactly to the (given) $r$s which underlie the theoretical determined frequencies. There are three reasons for this: rounding, grouping into intervals, and lumping together frequencies at the ends of the distributions.

It will be observed that the regression lines "scissor" nearer and nearer together as one proceeds from the lowest to the highest $r$. It will also be noticed that the variations in the arrays about the regression lines are

‡ Pearson, Karl, *Tables for statisticians and biometricians, part II*, Cambridge: Cambridge University Press, 1931. See Tables 8 and 9.

**Fig. 9.3.** Normal bivariate grid (frequencies per cell; rows run from $z_y = +2.5$ at top to $z_y = -2.5$ at bottom, columns are $z_x$; right‑hand column is the row total $n$).

| $z_x \rightarrow$ | -2.5 | -2.0 | -1.5 | -1.0 | -.5 | 0 | .5 | 1.0 | 1.5 | 2.0 | 2.5 | $n$ |
|---|---|---|---|---|---|---|---|---|---|---|---|---|
|  |  |  |  |  |  | 1 | 1 | 1 | 1 | 1 | .7 | 6 |
|  |  |  |  |  | 1 | 2 | 3 | 4 | 3 | 2 | .9 | 16 |
|  |  |  | 1 | 2 | 5 | 7 | 10 | 9 | 6 | 3 | 1.3 | 44 |
|  |  | 1 | 2 | 6 | 13 | 20 | 21 | 16 | 9 | 4 | 1.3 | 93 |
|  |  | 2 | 6 | 15 | 26 | 34 | 31 | 21 | 10 | 3 | 1.1 | 149 |
|  | 1 | 5 | 13 | 26 | 40 | 43 | 34 | 20 | 7 | 2 | .6 | 192 |
|  | 1 | 2 | 7 | 20 | 34 | 43 | 40 | 26 | 13 | 5 | 1 | 192 |
|  | 1 | 3 | 10 | 21 | 31 | 34 | 26 | 15 | 6 | 2 | .1 | 149 |
|  | 1 | 4 | 9 | 16 | 21 | 20 | 13 | 6 | 2 | 1 |  | 93 |
|  | 1 | 3 | 6 | 9 | 10 | 7 | 5 | 2 | 1 |  |  | 44 |
|  | 1 | 2 | 3 | 4 | 3 | 2 | 1 |  |  |  |  | 16 |
|  | 1 | 1 | 1 | 1 | 1 | 1 |  |  |  |  |  | 6 |
| | -2.5 | -2.0 | -1.5 | -1.0 | -.5 | 0 | .5 | 1.0 | 1.5 | 2.0 | 2.5 | 1000 |

| 33 | 38 | 57 | 68 | 77 | 84 | 90 | 95 | 97 | 98 | 100 | % above $z_y = -1.0$ |
|---|---|---|---|---|---|---|---|---|---|---|---|
| 50 | 62 | 77 | 85 | 91 | 95 | 97 | 99 | 99 | 100 | | % above $z_y = -1.5$ |

**Fig. 9.3.** Normal bivariate, $r = .50$.

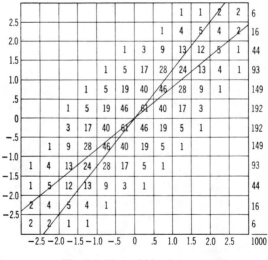

**Fig. 9.4.** Normal bivariate, $r = .80$.

disturbingly large, even for $r$s of .50 and .80. The reader would be well advised to consider appropriate values of alienation coefficients (Table 9.2) while studying Figs. 9.2 to 9.4.

There is one reason why the student should give Fig. 9.2 careful study. As we shall see later, an $r$ of .20 based on $N = 100$ would be judged statistically significant (as a deviation from zero). Very frequently one finds much ado, in the research literature in psychology, about an $r$ of .20 (even lower for large $N$) as being "significant." This takes the form of saying that the $r$ supports a theory; or if no theory is being subjected to test, the $r$ sets off a near welter of speculation as to why it was "significant"; and along either of these sometimes windy avenues there is apt to be a statement that would have the unsuspecting reader believing that high scores go with high scores and low with low to an extent far greater than a study of the (nonavailable) scatter would warrant. This is not to suggest that integrity should be questioned, but rather that enthusiasm engendered by a statistically "significant" finding can in reality be concerned with a rather trivial degree of relationship.

## CORRELATION AND EXPECTANCY

A *sixth* way for interpreting an $r$ is useful in applied work when the degree of correlation has to do with a test's predictive validity for a criterion. For example, if the criterion were grade point average expressed in standard scores it could be that $-1.0$ represents the $C$ average that is crucial for surviving the freshman year. To what extent will the expectancy of survival depend on score level on the test? Since an $r$ of .50 is typical of aptitude test validity coefficients we will use the (theoretical) frequencies in Fig. 9.3 to illustrate the calculation of expectancies. Take, for example, the third from the left vertical array. Simply sum the frequencies above $z_y = -1.0$ and divide by the total frequency, 44, in the array and you have a proportion, .57, which may be interpreted as the expectancy of survival for those with a test score, $z_x$, in the interval $-2.0$ to $-1.5$. A similar operation will provide the expectancies for the other test score intervals. The resulting proportions can be regarded as giving (empirical) probabilities for survival; or they can be converted to percentages for "successes." The two bottom rows of Fig. 9.3 contain expectancies, as percentages, for two critical levels on $z_y$, one with an over-all success rate of 84.1 per cent (above $z_y = -1.0$), the other with an over-all rate of 93.4 per cent (i.e., above $z_y = -1.5$).

Such arrays of expectancies form the bases for graphic presentation. In Fig. 9.5 will be found "curves" for expectancies, as dependent on test score level, when the cut-off for defining "success" is a standard score position of $-1.0$. The three curves are for expectancies calculated from Figs. 9.2, 9.3, and 9.4, representing $r$s of .20, .50, and .80. Despite the fact that the error of estimate attached to prediction for an individual is

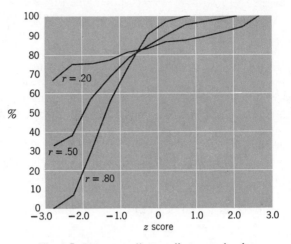

**Fig. 9.5.** Percentage "success" at score levels.

large for $r = .50$, the increase in "success" with test level is impressive in the actuarial (group) sense. The greater worth of an $r$ of .80 and lesser worth of an $r$ of .20 is obvious.

There is a second type of expectancy which can be used to illustrate the value of utilizing a test in personnel selection. To perform the needed calculations one must have: (a) the scattergram for criterion (performance) versus test score for a sample of on-the-job individuals, none of whom was hired on the basis of test scores, and no applicants rejected because of scores; (b) the percentage deemed to produce at a satisfactory rate, or a critical point on the criterion; and (c) the selection ratio, or the proportion of applicants needed for filling the jobs. How can we express the gain that would result if the test were adopted as a basis for selection from among future applicants? Suppose the scattergram is that shown in Fig. 9.3, that the over-all success rate is .841, and that the supply of applicants is such that the selection ratio is .50. This selection ratio would mean that selection on the test would involve a cutting score at the median. In order to ascertain an expectancy for satisfactory production, we would simply say that if the test had been used for selection, we would be dealing only with cases above $z_x = 0$ on the test. To determine the percentage of success for employees so selected we can erect, on the scattergram, a vertical at $z_x = 0$ and draw a horizontal line at $z_y = -1.0$, or the $z_y$ value that is critical for success. These two lines divide the scatter into four parts, or a fourfold table for the frequencies.

Of the 500 cases above $z_x = 0$, 469 are above $z_y = -1.0$. The ratio $469/500 = .938$ provides the expectancy of success that would have resulted from utilizing the test as a basis for selection under the specified conditions. Stated differently, selection based on the test would have increased the success proportion from .841 to .938, and, conversely, it would have reduced the failure expectancy from .159 to .062. If (unlikely in practice) we had a test yielding a correlation of .80, the figures would be an increase from .841 to .986 for success expectancy and a decrease from .159 to .014 for failure expectancy. In contrast, for a test correlated only .20 with productivity, the increase would be from .841 to .878, and the decrease from .159 to .122, respectively, for success and failure expectancies. The reader who is interested in this sort of thing will wish to turn to H. C. Taylor and J. T. Russell in *J. Appl. Psych.*, 1939, **23**, 565–578.

## CORRELATION, INDEPENDENCE, AND PROBABILITIES

Obviously, when there is correlation between two variables it cannot be said that the variates are varying independently. Does zero correlation, as measured by the product moment $r$, mean independence? The answer is no in that the slope of the best fitting line can be zero even when there is a close curvilinear relationship (U-shaped, for example). Independence of variates means zero correlation; zero $r$ does not necessarily mean independence.

In our discussion of probability we encountered the word "independence." Table 5.3 was constructed on the basis of independence of events, the events being "scores" from pairs of tosses of coins. If you re-examine that table, you will see that it exhibits zero correlation. The frequencies in the table provided the basis for the probabilities of joint outcomes of independent events. Can we have a joint probability for events that are not independent? Consider Fig. 9.3 again. The cutting of the two axes at the means will permit us to set up a fourfold table of frequencies. The frequency in the upper-right quadrant will be 335, from which we may say that the probability is .335 that a pair of $X$ and $Y$ scores will be above the mean. Note that if we were considering only $X$, the probability of an $X$ falling above the mean of $X$ would be .50; ditto for a $Y$ score. If we applied the multiplication theorem, we would have $.50 \times .50 = .25$ as the probability of an $X$ and a $Y$, as a pair, being above their means. Now the .335 is just as meaningful as the .25; obviously, application of the multiplication theorem can lead to false probabilities.

On p. 76 mention was made of conditional probabilities. Once again we turn to Fig. 9.3 from which we see, for example, that *given* that a person's $z_x$ is in the interval 1.5–2.0, the probability is $(9 + 6 + 3 + 1)/44 = 19/44 = .43$ that his $z_y$ is above $+1.0$. Now the unconditional probability for drawing a person with a $z_y$ above $+1.0$ is .16 (obtained from the marginal distribution, which is the distribution for $z_y$ irrespective of $z_x$). The fact that the conditional ("given that") and the unconditional proba- bilities differ simply indicates that the events (scores) are not independent. The distribution on the right-hand margin of Fig. 9.3 and the distribution in any vertical array permit probability statements about $z_y$ scores falling in a given interval, but obviously the probabilities derived therefrom differ. This is not true for Table 5.3.

## LIMITS FOR $r$

Attention is called to the fact that definition formula (8.1) becomes $r = \Sigma z_x z_y / N$, when written in terms of standard scores for both variables. This indicates specifically that the correlation coefficient is a statistical average, the average of the cross products of standard scores. Suppose that we ask what happens when the correlation is perfect in the sense that each individual's $z_x$ score equals his $z_y$ score. If this is true, the sum $\Sigma z_x z_y$ would be the same as $\Sigma z^2$, which when divided by $N$ gives 1.00. Thus the upper limit for $r$ is $+1.00$. Now suppose a perfect inverse relationship, such that an individual's $z_x$ and $z_y$ are the same except for sign, one being positive whereas the other is negative. If this holds true for all the cases, the sum $\Sigma z_x z_y$ can be written as $\Sigma z(-z)$ or $-\Sigma z^2$, which when divided by $N$ gives $-1.00$ as the limit for perfect negative correlation.

As exercises, the student should show that multiplying or dividing either $X$ or $Y$ or both by a constant, or $X$ by one constant and $Y$ by another, will not change $r$, and that adding or subtracting a constant does not affect the value of $r$.

## SUMMARY

The six suggested methods for interpreting the correlation coefficient may be briefly summarized here.

1. $r$ is associated with the rate at which one variable changes with another. This assumes that the regression line so interpreted is linear.

2. $r$ tells us how accurately we can predict by a regression equation. The standard error of estimate permits one to infer the possible magnitude of the prediction error, whereas the coefficient of alienation indicates the reduction in error over that error which would exist if there were no correlation. This interpretation assumes that the regression line used in predicting is linear and that variation about this line is normal and homoscedastic.

3. $r^2$ gives the proportion of variance in $Y$ predictable from, or attribut- able to, variation in $X$. This assumes linearity for the regression of $Y$ on $X$ and requires caution in assuming the direction of cause and effect.

The student should attempt to visualize the meaning of these three principal methods of interpreting correlation. In particular, he should note the meaning of $S_y$, $S_{y'}$, and $S_{y \cdot x}$ (or their counterparts with the subscripts $y$ and $x$ interchanged). The first, $S_y$, holds for the marginal distribution of all $Y$s; $S_{y'}$ pertains to the variability of all $Y$ values as predicted from $X$; the third, $S_{y \cdot x}$ is a measure of the variation about the regression line for predicting $Y$ from $X$.

For none of these three interpretations of $r$ do we have to assume normal distribution on the margins. However, it is possible and likely that nonlinearity, lack of homoscedasticity, and nonnormality of arrays will tend to be associated with skewness in one or both the marginal distributions.

4. $r$ or $r^2$ can be interpreted in terms of the proportion of elements common to two variables provided we are willing to make rather hazardous and unrealistic assumptions as regards the nature of the variables.

5. $r$ can be interpreted mathematically in terms of the equation for the normal correlation surface. This assumes that both regressions are linear, that homoscedasticity and normality hold for both the horizontal and vertical array distributions, and that both marginal distributions are normal in form.

6. $r$ can be interpreted in two different ways in terms of "success" expectancies when dealing with the predictive validity of tests. Such expectancies were illustrated by normal bivariate scattergrams, but in practice the restrictive assumptions of normal bivariate correlation can be avoided by calculations based on a scattergram of observed scores.

The nature of the investigation will usually dictate or suggest the appropriate interpretation to be placed on a correlation.

# Chapter 10

# FACTORS WHICH AFFECT THE CORRELATION COEFFICIENT

Before we interpret or draw conclusions from a particular correlation coefficient, it is necessary that we ask ourselves what factors might have affected its magnitude. The size of an obtained $r$ depends upon several specific conditions, and, even though it is not always essential that corrections be applied, the investigator must forever be on the lookout for correlations which deviate from their "true" value because of the operation of disturbers. This chapter is devoted to a discussion of the more common factors which influence $r$.

It is assumed that errors in computation have not been permitted—that all arithmetical work has been checked. It is also assumed that sufficient intervals have been used so as to make unnecessary the application of Sheppard's correction for grouping; if more than twelve intervals have been used, the slight increase in $r$ which results from correcting the standard deviations will be negligible. Certain textbooks have advocated a correction to $r$ for smallness of the sample, which correction reduces $r$ by a negligible amount. In view of the magnitude of the effects of other factors on $r$, these two possible corrections seem trifling.

## SELECTION

One of the first questions which must be faced is: Do the cases on which $r$ is based represent a random sampling of some defined population, or have selective factors so operated as to increase or decrease $r$? The literature of psychology is not free from correlation coefficients which are decidedly different from values that would have been obtained had the

154

sampling been random.  This is not to say that any investigator has willfully selected his cases so as to produce correlation, but rather to say that unwitting errors are frequently present in spite of an effort to avoid selective factors.

## SAMPLING ERRORS

Even though we may feel reasonably sure of the randomness of the sample on which an $r$ is based, it is still necessary to consider the obtained $r$ in terms of variable errors due to sampling.  Any $r$ based on $N$ pairs of observations will differ more or less from the universe, or population, value which is here conceived of as the value of the correlation coefficient which we would obtain if we had an infinitely large sample.  Many of the older texts gave $(1 - r^2)/\sqrt{N}$ as the standard error of $r$, but failed to point out a serious limitation as regards interpretation: that this is an approximation and that $r$s for successive samples are not distributed normally unless $N$ is large and/or the universe value is near zero.

Before further discussion it should be said that some measure of the sampling fluctuation of the correlation coefficient is highly desirable for any of three reasons.  (1) We may wish to say whether an obtained $r$ can be taken as representing a real, nonchance, correlation, i.e., whether it deviates sufficiently far from zero so that we cannot regard it as a chance fluctuation from no relationship; (2) we may wonder whether a given $r$ deviates significantly from some a priori or expected value; or (3) we may raise the question of whether two obtained $r$s are significantly different from each other.  The answers to these questions must be in terms of probability, and the probability figure which we accept as indicating significance determines the confidence with which we regard any such conclusions as we set forth.

If $N$ is greater than 100, and if we are interested in saying whether or not an $r$ (of .50 or less, usually) is significantly different from zero, we can determine its standard error by

$$\sigma_r = \frac{1}{\sqrt{N}} \tag{10.1}$$

and then divide the obtained $r$ by this standard error in order to secure a $z$ value with which to enter the normal probability table.  If $r/\sigma_r$ is greater than 2.58, we can conclude with a fairly high degree of sureness that the true or universe value of $r$ is likely to be greater than zero.

For $N$ less than 100, it is necessary to follow a different procedure.  It can be shown that, if the correlation coefficient is computed for successive samples drawn from a population for which the correlation is zero, the successive values of

$$t = r \frac{\sqrt{N-2}}{\sqrt{1-r^2}} = \frac{r}{\sqrt{(1-r^2)/(N-2)}} \qquad (10.2)$$

will follow the $t$ distribution with $df = N - 2$. If a sample $t$ reaches the .01 level of significance, one would conclude that it is not a chance deviation from zero, or that some correlation exists between the two variables involved.

From the foregoing expression, it would appear that the $t$ for testing the significance of correlation is nothing more than an $r/s_r$, with $s_r = \sqrt{(1-r^2)/(N-2)}$ as an estimate of the sampling error of $r$. However, there are subtle mathematical reasons why such an interpretation is not permissible.

The student may wonder why the $df$ is taken as $N - 2$. Actually, when we test the significance of an $r$, we are testing the significance of regression. If $r$ is zero, the regression is zero in the sense that the regression coefficient or slope of the regression line is zero. Now a linear regression line involves two constants, its slope and its intercept; hence 2 degrees of freedom are lost in fitting the line. Suppose $N = 2$, and that the two $X$ scores differ; likewise, the two $Y$ scores. Imagine these pairs of scores plotted in a scatter diagram, and a regression line fitted or a correlation coefficient computed. The regression line would go through both plotted points; therefore for the sample of two cases the prediction would be perfect and $r$ would be unity. The student may, as an exercise, prove algebraically that when $N = 2$ *and* when there is variation in both $X$ and $Y$, the correlation, must be $+1$ or $-1$. In other words, with $N = 2$ there is no freedom for sampling variation in the numerical value of $r$.

The $t$ test of $r$ assumes normality and homoscedasticity either for the vertical array or for the horizontal array distributions. Nothing is assumed about the total $X$ and $Y$ distributions. There is evidence, as with the $t$ test for means, that sizable violations of the assumptions are tolerable, but there is always comfort in knowing that the assumptions are fairly well met.

It might be remarked at this place that if the sum of squares of the deviations about a regression line were divided by $N - 2$, the $df$, we would have an unbiased estimate of the error of estimate variance. This added precision seems unnecessary in the practical situation where prediction (regression) equations are usually based on sizable $N$s.

Formulas for the standard error of $r$, when $r_{pop}$ is large, are misleading, because for high values of $r_{pop}$ the distribution of successive sample values is markedly skewed. This skewness becomes noticeable when $r_{pop}$ reaches .40 or .50 and increases rapidly as $r_{pop}$ nears unity. The skewness is also a function of $N$. Because of this skewness the standard error of $r$ loses its meaning; it cannot be expected to yield a trustworthy answer as to whether an obtained $r$ deviates significantly from some a priori value, nor can the significance of the difference between two $r$s be determined by substituting in the ordinary formula for the standard error of a difference.

**The $r$ to $z$ transformation.** R. A. Fisher developed a very useful and accurate technique for handling sampling errors for high values of $r$. This procedure is also applicable for low $r$s and can be used when $N$ is large or small. He found a transformation

$$z = \tfrac{1}{2} \log_e (1 + r) - \tfrac{1}{2} \log_e (1 - r) \tag{10.3}$$

or

$$z = 1.1513 \log_{10} \frac{1 + r}{1 - r} \tag{10.4}$$

which has two distinct advantages: (1) the distribution of $z$ for successive samples is independent of the universe value, i.e., for a given $N$ the sampling distribution will have the same dispersion for all values of $r_{pop}$; (2) the distribution of $z$ for successive samples is so nearly normal that it can be treated as such with very little loss of accuracy. The standard error of $z$ is

$$\sigma_z = \frac{1}{\sqrt{N - 3}} \tag{10.5}$$

*Note on notation:* Since the standard error of $z$ is a theoretical value, dependent solely on $N$, and hence does not involve estimation from the sample, it is symbolized as $\sigma_z$ rather than as $S_z$ or $s_z$.

If we wish to state the .99 confidence limits for $r_{pop}$ we transform the obtained $r$ to $z$ by formula (10.4) or by Table B of the Appendix, determine $\sigma_z$, find $z + 2.58\sigma_z$ and $z - 2.58\sigma_z$, and then transform these two $z$ values back to $r$s by using Table C. As an example and in contrast to the less exact procedure of taking $r \pm 2.58S_r$, where $S_r = (1 - r^2)/\sqrt{N}$, let us suppose an $r$ of .90 based on an $N$ of 50. The standard error of $r$ by the usual formula is .027; whence .90 $\pm$ (2.58)(.027) yields the values .830 and .970 as confidence limits for the universe value. Now, if we utilize the $z$ transformation, we find $z = 1.47$, and $\sigma_z = .146$, whence 1.47 $\pm$ (2.58)(.146) gives 1.093 and 1.847. These two values are then transformed back to the two $r$ values, .798 and .951, which it will be noted differ from the confidence limits for $r_{pop}$ as determined by using the classical $S_r$.

*Note:* Since the foregoing $z$ is not a relative deviate, it should not be referred to as a standard score.

**Difference between $r$s.** If we wish to determine the significance of the difference between two $r$s, both are transformed into $z$s, and the standard error of the difference between the two $z$s is obtained by

$$\sigma_{z_1 - z_2} = \sqrt{\frac{1}{N_1 - 3} + \frac{1}{N_2 - 3}} \qquad (10.6)$$

and then the ratio of the difference to its standard error is treated in the usual manner. If the $z$s are significantly different, we conclude that the two $r$s are significantly different.

Suppose we have the correlation between $X_1$ and $X_2$ and also between $X_1$ and $X_3$, with both $r$s based on the same sample of $N$ cases, and we wish to decide whether there is a significant difference between $r_{12}$ and $r_{13}$. The foregoing method is not applicable because we need to allow for the fact that, for successive samplings, $r_{12}$ and $r_{13}$ are not independently distributed, but correlated. The standard error of the difference must include a subtractive $r$ term involving the correlation between the correlation coefficients. The methods for estimating this needed correlation are none too satisfactory, but there is a test which is interpretable by way of the $t$ table for $N$ small and by way of the normal table for $N$ large. It has been shown that

$$t = \frac{(r_{12} - r_{13})\sqrt{(N-3)(1 + r_{23})}}{\sqrt{2(1 - r^2_{12} - r^2_{13} - r^2_{23} + 2r_{12}r_{13}r_{23})}} \qquad (10.7)$$

follows the $t$ distribution with $N - 3$ degrees of freedom when the null hypothesis of no difference is true. If $t$ is significant, we conclude that one variable correlates higher than the other with $X_1$.

**Averaging correlations.** When we have two (or more) sample values for the correlation between two variables we may wish to average the $r$s (1) in case it is known that the samples have been drawn from the same population or (2) in case it can be assumed (because the $r$s are not significantly different from each other) that the samples have been drawn from equally correlated populations. An appropriate procedure is to convert each $r$ to $z$, then take a weighted (each $z$ by the inverse of its sampling variance) average of the $z$s. Thus, for three sample values this weighted average is given by

$$z_{av} = \frac{(N_1 - 3)z_1 + (N_2 - 3)z_2 + (N_3 - 3)z_3}{(N_1 - 3) + (N_2 - 3) + (N_3 - 3)}$$

This $z_{av}$ can be transformed back to an $r$, and any significance test concerning such an average $r$ would be made on $z_{av}$ which has a standard error of

$$1/\sqrt{(N_1 - 3) + (N_2 - 3) + (N_3 - 3)}$$

**Sampling errors of regression coefficients.** Those who are ascertaining relationships involving two variables, one of which can definitely be characterized as independent $(X)$ and the other as dependent $(Y)$, or one as an antecedent and the other as a consequent variable, may prefer to specify the relationship in terms of the regression constants, $B$ and $A$. Both of these are, of course, estimates of unknown population values and are therefore subject to sampling errors. Ordinarily $A$ is of little interest whereas $B$ specifies the rate of change and may be used for testing hypotheses. Does the sample slope vary significantly from some hypothetical value, $B_h$? Or does the slope differ significantly from zero? Occasionally, we may have two regression coefficients (same variables) and wish to test the significance of the difference between the two slopes, $B_1$ and $B_2$.

To test whether or not a single slope, $B$, differs significantly from zero or some other a priori value, we need an estimate of its standard error. The classical standard error of $B_{yx}$ was taken as

$$S_{B_{yx}} = \frac{S_{y \cdot x}}{S_x \sqrt{N}} = \frac{S_y \sqrt{1 - r^2}}{S_x \sqrt{N}}$$

Note that the formula involves a ratio which is a function of the $Y$ measurement units relative to the $X$ measurement units. This is reasonable since the slope is also a ratio of $Y$ to $X$ units. (The standard error of any statistic must be in the same units as the statistic.) The upstairs part of the formula is the familiar standard error of estimate. It is reasonable that the sampling stability of $B_{yx}$ should depend on the variability within the arrays because, in effect, we are fitting the regression line to the array means (weighted by their $N$s), and the stability of these means is a function of the array variances.

Greater precision in testing hypotheses regarding $B_{yx}$ will be achieved by having an unbiased estimate of its sampling error. It will facilitate exposition to note that for any variable, $V$, the variance is given by $S_v^2 = \Sigma v^2 / N$. If we have $S_v^2$ and wish to recover the sum of the squares of the deviations, $\Sigma v^2$, we can use $\Sigma v^2 = N S_v^2$. To secure an unbiased estimate we would have $s_v^2 = \Sigma v^2 / (N - 1) = N S_v^2 / (N - 1)$. Parenthetically, it might be noted that since $S_y / S_x = s_y / s_x$, the slope would be unchanged by introducing unbiased estimates of the two standard deviations.

For the sampling variance of $B_{yx}$ we need

$$s_{B_{yx}}^2 = \frac{s_{y \cdot x}^2}{N S_x^2} \tag{10.8}$$

in which for reasons not yet specified we have allowed a mixture of biased and unbiased estimates. To get $s_{y \cdot x}^2$ we would need to divide $\Sigma(Y - Y')^2$

by its $df$, $N - 2$. But $S^2_{y \cdot x} = \Sigma(Y - Y')^2/N$, hence $\Sigma(Y - Y')^2 = NS^2_{y \cdot x} = NS^2_y(1 - r^2)$. Therefore

$$s^2_{y \cdot x} = \frac{\Sigma(Y - Y')^2}{N - 2} = \frac{NS^2_y(1 - r^2)}{N - 2}$$

which leads to

$$s^2_{B_{yx}} = \frac{s^2_{y \cdot x}}{NS^2_x} = \frac{\dfrac{NS^2_y(1 - r^2)}{N - 2}}{NS^2_x} = \frac{\dfrac{S^2_y(1 - r^2)}{N - 2}}{S^2_x}$$

the square root of which gives

$$s_{B_{yx}} = \frac{S_y\sqrt{1 - r^2}}{S_x\sqrt{N - 2}} \tag{10.9}$$

To test the null hypothesis that the slope for the population is zero, we have

$$t = \frac{B_{yx}}{s_{B_{yx}}} = \frac{r(S_y/S_x)}{S_y\sqrt{1 - r^2}/S_x\sqrt{N - 2}}$$

with $N - 2$ degrees of freedom. Note that since $S_y/S_x$ cancel, it follows that if we used unbiased estimates, $s_y$ and $s_x$, they would also cancel. With the $S$s cancelled we have the ratio

$$t = \frac{r}{\sqrt{(1 - r^2)/(N - 2)}}$$

which is the $t$ test for the significance of $r$. When testing the null hypothesis of zero slope we are doing nothing more than testing the null hypothesis of zero correlation, a fact that certainly could have been anticipated since the only way that $B_{yx} = r(S_y/S_x)$ can be zero is for $r$ to be zero. The mathematical purist can say that $B$ could be zero when $S_y$ is zero, but when $S_y$ is zero the slope and $r$ become indeterminates: you cannot have a relationship or a slope in the absence of variation for either variable.

For those who have an aversion to $r$ and standard deviations and who persist in the erroneous belief that a test of $B_{yx}$ differs from a test of $r$, it should be noted that $r$ and the $S$s can be avoided by expressing $B_{yx}$ as in equation (9.7) and taking the following as an expression for the unbiased estimate of the sampling error of $B_{yx}$:

$$s_{B_{yx}} = \frac{\sqrt{\dfrac{\Sigma(Y - Y')^2}{N - 2}}}{\sqrt{\dfrac{N\Sigma X^2 - (\Sigma X)^2}{N}}}$$

in which

$$\Sigma(Y - Y')^2 = \Sigma Y^2 - A_{yx}\Sigma Y - B_{yx}\Sigma XY \qquad (10.10)$$

with $A_{yx}$ and $B_{yx}$ as in equation (9.7).

**Difference between regression coefficients.** Let $B_1$ and $B_2$ be the values for $B_{yx}$ for two independent samples of $N_1$ and $N_2$ persons. The $t$ test for the difference between the two $B$s is analogous to that for testing the difference between independent means. We need $s_{D_b}$ which, it might be guessed, will follow the usual pattern:

$$s_{D_b} = \sqrt{s^2_{b_1} + s^2_{b_2}}$$

for which we need the best unbiased estimates of the two variances under the radical. Instead of using the two residual variances separately, a combined estimate is obtained by combining the sums of squares of deviations about the respective regression lines and then dividing by the combined $df$, or $N_1 + N_2 - 4$. The best estimate of the (assumed to be) common residual variance may be expressed in symbols as

$$_cs^2_{y \cdot x} = \frac{\Sigma(Y - Y')^2_1 + \Sigma(Y - Y')^2_2}{N_1 + N_2 - 4}$$

with each numerator term calculable by equation (10.10) written separately for the two groups. If the $r$s and $S$s have been computed, we may use the exact equivalent

$$_cs^2_{y \cdot x} = \frac{N_1 S^2_{y_1}(1 - r^2_1) + N_2 S^2_{y_2}(1 - r^2_2)}{N_1 + N_2 - 4}$$

Then by utilizing (10.8), we have

$$s_{D_b} = \sqrt{\frac{_cs^2_{y \cdot x}}{N_1 S^2_{x_1}} + \frac{_cs^2_{y \cdot x}}{N_2 S^2_{x_2}}}$$

and $t = (B_1 - B_2)/s_{D_b}$ as a ratio that follows the $t$ distribution with $N_1 + N_2 - 4$ degrees of freedom.

It is of interest to note that whereas the $t$ tests for the two null hypotheses, $r_{pop}$ zero and $B_{pop}$ zero, are identical, the test for the difference between $B$s is not the same as that for the difference between the two respective $r$s. That this should be so may be understood by considering two $r$s and two $B$s based on two samples one of which has a larger range of $X$s (larger $S_x$) than the other. The two slopes could very well be nearly identical, yet one $r$ be considerably higher than the other. This constitutes one argument for "regression" instead of "correlational" analysis: for the $X$ and $Y$ as independent-dependent variable situation, the slope is not a function of curtailment on $X$, or the range of $X$ scores.

## RANGE OR SPREAD OF TALENT

The magnitude of the correlation coefficient varies with the degree of heterogeneity (with respect to the traits being correlated) of the sample. If we are drawing a sample from a group which is restricted in range with regard to either or both variables, the correlation will be relatively low. Thus the restricted range of intelligence is one factor which leads to lower correlation between intelligence and grades for college students than that usually found for high school groups. If the range with respect to one variable has been curtailed, and we know the standard deviation for an uncurtailed distribution, it is possible to adjust the correlation for the difference in range, provided we can be sure of the tenability of two assumptions: that the regressions are linear and that the arrays are homoscedastic for the scatter based on the uncurtailed distribution. If the curtailment is on variable $X$, and we let

$$sd_x = S \text{ for curtailed distribution}$$
$$SD_x = S \text{ for uncurtailed distribution}$$
$$r_{xy} = \text{correlation for curtailed range}$$
$$R_{xy} = \text{correlation for uncurtailed range}$$

the relationship by which we would predict $R_{xy}$ from $sd_x$, $SD_x$, and $r_{xy}$ is given by

$$R_{xy} = \frac{r_{xy}(SD_x/sd_x)}{\sqrt{1 - r^2_{xy} + r^2_{xy}(SD_x/sd_x)^2}} \qquad (10.11)$$

Obviously, if we have $R$ instead of $r$, the value of $r$ for a restricted range can be estimated by formula (10.11). All we need to do is interchange $SD_x$ and $sd_x$, $R$ and $r$, and then substitute to find $r$. The estimation of $r$ need not be made in ignorance of whether the assumptions of linearity and homoscedasticity can be met; an examination of the accessible scatter for the uncurtailed range will reveal the facts.

Formula (10.11) indicates definitely that the magnitude of the correlation coefficient is a function of the degree of heterogeneity with respect to one of the traits being correlated. A better appreciation of the extent of this influence can be had by examining Table 10.1 which gives, for varying values of $R_{xy}$ along the top and different $sd_x/SD_x$ ratios along the left, the corresponding values of $r_{xy}$. It can be shown that double selection, i.e., curtailment on both variables, tends to depress the correlation coefficient. Since the formulas for "correcting" for double curtailment are not too satisfactory, none is given here.

One important rule emerges from the foregoing: standard deviations should always be reported along with correlation coefficients, and some indication should be given as to variation typically found for the variables.

Table 10.1. Values for $r_{xy}$ for $R_{xy}$s of .30, .40, $\cdots$ .80 with $sd_x/SD_x$ values of .90, .80, $\cdots$ .50

| | $R_{xy}$ | | | | | |
|---|---|---|---|---|---|---|
| $sd_x/SD_x$ | .30 | .40 | .50 | .60 | .70 | .80 |
| .90 | .272 | .366 | .461 | .559 | .662 | .768 |
| .80 | .244 | .330 | .419 | .514 | .617 | .730 |
| .70 | .215 | .292 | .375 | .465 | .566 | .682 |
| .60 | .185 | .253 | .327 | .410 | .507 | .625 |
| .50 | .155 | .213 | .277 | .351 | .440 | .555 |

## EFFECTS OF UNRELIABILITY

Before considering the effect of unreliability, or errors of measurement, on the correlation between two variables, it is necessary that we digress to explain briefly what is meant by reliability. If we were assigned the task of determining the height of an individual by the use of a tape measure, we might be satisfied with one measurement, but unfortunately a single determination might not be entirely free from error. To overcome this, two or more measures are averaged on the assumption that the chance or *variable* errors will more or less cancel out. If we compute the standard deviation of the distribution of several measurements (of the same thing), a summary figure indicating the possible magnitude of the variable errors will be obtained. This $S$ neither pertains to nor measures the magnitude of a possible *constant* error, i.e., an error which affects all the measurements in the same direction. We are here concerned only with the magnitude of variable errors, or inaccuracies in measurement which are of a chance nature.

**Reliability.** If we had the problem of determining the error in the measurement of height, we could make several measurements on one person and compute a measure of accuracy, or we might make just two measures on each of several persons and take some function of the difference between the two measurements for all $N$ individuals as our gauge of accuracy. Either scheme leads to an estimate of the size of the variable errors that may be involved.

In psychological measurement, it is not always feasible or possible to obtain more than two measures on an individual for a given trait; hence it is necessary to use the second-mentioned scheme for determining the accuracy of measurement. The mean or median *absolute* error may suffice, but, as in physical measurement, we sometimes need to know the extent of the variable errors in relation to the magnitude of the thing being measured, i.e., the *relative* or percentage error. Psychologists have found it useful to interpret variable errors, not with regard to the magnitude (a nearly meaningless word in psychological tests) of the measures, but relative to the variability of the trait for a specific group of individuals. The correlation between two determinations is, as we shall soon see, one method of expressing the accuracy of measurement relative to the trait dispersion. Such a correlation is termed the *reliability coefficient*.

Suppose

$X$ = an obtained score or measure for an individual

$X_t$ = his true score

$e$ = a variable error, positive or negative

Then we can consider that

$$X = X_t + e$$

or in deviation units

$$x = x_t + e$$

The variance of the obtained scores will be

$$S^2_x = S^2_t + S^2_e \qquad (10.12)$$

provided we can assume $x_t$ and $e$ uncorrelated. This assumption seems reasonable since the variable error, $e$, is supposed to be a chance affair, as often positive as negative, and therefore its magnitude and direction should not be related to anything else. Equation (10.12) can be stated in words: the variance of the distribution of scores can be broken up into two portions, the variance of the true scores and the variance due to errors of measurement.

Suppose that for a given trait we have two measurements, each of which is in error but not necessarily to the same extent or in the same direction. Symbolically,

$$x_1 = x_t + e_1$$

$$x_2 = x_t + e_2$$

in which the $e$s represent the errors which go with the two obtained scores. The reliability coefficient is defined as the correlation between two comparable measures of the same thing, i.e., the correlation between $x_1$ and $x_2$. (We need an $x_1$ and $x_2$ for each measured individual.) Thus we have the reliability coefficient,

$$r_{xx} = r_{x_1 x_2} = \frac{\Sigma x_1 x_2}{NS_1 S_2} = \frac{\Sigma (x_t + e_1)(x_t + e_2)}{NS_1 S_2}$$

$$= \frac{\Sigma x^2_t + \Sigma x_t e_2 + \Sigma x_t e_1 + \Sigma e_1 e_2}{NS_1 S_2}$$

Dividing by $N$ gives

$$r_{xx} = \frac{S^2_t + r_{te_2}S_tS_{e_2} + r_{te_1}S_tS_{e_1} + r_{e_1e_2}S_{e_1}S_{e_2}}{S_1S_2} \qquad (10.13)$$

If we assume all three $r$s in the numerator equal to zero, we have

$$r_{xx} = \frac{S^2_t}{S_1S_2}$$

It is assumed that we are correlating *comparable* measures of the *same* thing or trait—comparable in the sense that $S_{e_1} = S_{e_2}$, and $S_1 = S_2$. (The same trait is implied in that $x_1$ and $x_2$ are measures of $x_t$.) Whence we have

$$r_{xx} = \frac{S^2_t}{S^2_x} \qquad (10.14)$$

where $S_x = S_1 = S_2$. The reliability coefficient can be interpreted as a proportion, since from formula (10.12) we have

$$\frac{S^2_t}{S^2_x} + \frac{S^2_e}{S^2_x} = 1$$

i.e., the reliability coefficient represents the proportion of the variance of the obtained scores which is due to the variance of the true scores. It follows that $1 - r_{xx}$ gives the proportion of the variance which is due to errors of measurement.

Obviously, the reliability coefficient can, by substitution from formula (10.14) into the foregoing expression, also be written as

$$r_{xx} = 1 - \frac{S^2_e}{S^2_x} \qquad (10.15)$$

which indicates clearly that the reliability coefficient is a function of the magnitude of the variable error *relative* to the variability of the trait in question. It also follows from formula (10.15) that the error of measurement can be stated in terms of the reliability coefficient and $S_x$; thus,

$$S_e = S_x\sqrt{1 - r_{xx}} \qquad (10.16)$$

That $S_e$ is to be interpreted as the *standard error of measurement* may be clarified if we note that, when $x$ ($= x_1$ or $x_2$) is taken as evidence of the true score, $x - x_t$ becomes the error, and the standard deviation of such errors will be $S_e$, as can be shown by easy algebra. If it were possible to secure a large number of measures on an individual, we would expect these measures to distribute themselves normally about the true score with a standard deviation corresponding to $S_e$. Thus, if the result of one testing yields an IQ of 80, and if $S_e = 3$, we can conclude with high confidence that the individual's true position, on the scale of measured (obtained) IQs, is somewhere between 71 and 89 (80 $\pm$ 3$S_e$), and with fair confidence that it is somewhere between 74 and 86.

**Determination of reliability.** The foregoing argument regarding the interpretation of the reliability coefficient either as an indicator of relative accuracy or in terms of $S_e$ rests on the supposition that we have obtained the reliability coefficient as the result of correlating *comparable* measures of the *same* thing and that the *variable errors are uncorrelated* with themselves and with the true scores. The practical determination of the reliability coefficient involves more, therefore, than the mere correlating of two sets of measurements. The conditions under which the two sets of scores are obtained must be scrutinized for possible violation of the requisite assumptions. Some of the difficulties involved in ascertaining the reliability of a psychological measurement are suggested in the following paragraphs.

First let us note that the chance variable error, $e$, can be broken up into many smaller components, at least logically, although not necessarily experimentally. Thus we might set

$$e = e_a + e_b + e_c + e_d + e_f + \cdots$$

in which

$e_a$ = error in the instrument or test
$e_b$ = error due to extraneous physical disturbance
$e_c$ = error due to physiological condition of individual
$e_d$ = error in scoring or in reading instrument
$e_f$ = error due to day-to-day fluctuations

Other sources of variable error might be added, or some of those listed might be broken up into more minute parts. It is not assumed that these several sources contribute an equal amount to the variance of $e$, nor is it assumed that these several components are entirely independent of each other. For instance, daily fluctuations might be influenced by physiological condition.

The assumption of uncorrelated errors implies that $e_1$ is not correlated with $e_2$. Of course the two scores for an individual might by chance contain a variable error of the same magnitude and sign; we are here interested, however, in whether an error which is chance for one score might tend in general to affect the second score in the same manner. For example, an upset stomach might lead to a reduced performance score, and if the second test was administered the same day, this same chance factor would affect the second performance score in the same direction. Thus in examining any proposed scheme for determining the reliability of a test, we must inquire as to whether any of the sources of error can affect the two measurements on an individual in the same direction. If it seems reasonable to suspect that errors are correlated, it follows that the obtained reliability coefficient will be spuriously high since the presence of correlated errors will not allow formula (10.13) to be reduced to (10.14).

The presence of $e_a$ as a source of error may be appreciated if we regard the items as providing a sampling of the performance (or responses) of the individual. Consider the simple problem of measuring vocabulary level. We can easily conceive of the individual's true score as the proportion, $p_t$, of words in Webster's Unabridged Dictionary that he can satisfactorily define. Instead of the time-consuming tedium of asking him to define each word, we might resort to sampling. We could get a fairly good sample of, say, $n = 30$ words by taking the fourth word from the top of the right-hand column of every page ending in the numeral 55. The standard error of his score would be given by $\sqrt{p_t q_t / 30}$, or more generally by $\sqrt{p_t q_t / n}$. Even though $p_t$ is unknown (estimable as $p_{ob}$), it is readily seen that the larger the number of items the smaller the error, and vice versa. Thus to reduce $e_a$, as instrumental error, the length of the test is increased. This general principle holds even though our vocabulary illustration is not quite analogous to what goes on in test construction because a "universe" of items is rarely, if ever, available for sampling and because test constructors tend to improve on randomness by *selecting* items that possess certain desirable characteristics.

Let us consider a few of the "accepted" schemes for ascertaining reliability in order to see whether they are "acceptable" in light of the assumptions requisite to a sound reliability coefficient. These assumptions may be recapitulated in the form of three questions. Do the two tests or determinations represent measures of the same thing? Are the two series of measures comparable (comparable tests or instruments)? Is it possible or likely that the errors of measurement are correlated; i.e., can the error on the first test be correlated with the error on the second, or can the error on either be correlated with the true measure?

For the ordinary mental, personality, or achievement test, reliability is usually ascertained by correlating supposedly equivalent (comparable?) forms, by correlating split halves (odd vs. even items or first half vs. second half of test), or by correlating test-retest scores. The test-retest method is of limited value in that there may be a memory carry-over from test to retest, in which case the retest will measure the same trait as the original test plus memory effects. In order to overcome this memory transfer, the retest may be administered some months after the first test, but this permits of a possible change in the trait or ability as a result of maturation or experience.

Split-half reliability involves the correlating of two halves and applying the Brown-Spearman formula to determine the reliability of scores based on the whole test. This formula is easily derived. Let $X_1$ and $X_2$ stand for the respective halves. Now $r_{12}$ would be the reliability for scores based on either half, but in practice we always use total scores, defined as $X_a = X_1 + X_2$. The reliability of $X_a$ can be thought of in terms of the correlation between $X_a$ and an imaginary set of comparable scores, $X_b = X_3 + X_4$, where $X_3$ and $X_4$ are scores on the two respective halves of a nonexistent form of the test. Given information about $X_1$ and $X_2$, we seek an expression for $r_{ab}$. In deviation units, $x_a = x_1 + x_2$ and $x_b = x_3 + x_4$; hence we may write

$$r_{ab} = \frac{\Sigma x_a x_b}{NS_a S_b} = \frac{\Sigma (x_1 + x_2)(x_3 + x_4)}{NS_a S_b}$$
$$= \frac{\Sigma x_1 x_3 + \Sigma x_1 x_4 + \Sigma x_2 x_3 + \Sigma x_2 x_4}{NS_a S_b}$$

Dividing through by $N$ and utilizing formula (8.1), and with formula (9.9) as a basis for specifying $S_a$ and $S_b$, we have

$$r_{ab} = \frac{r_{13}S_1 S_3 + r_{14}S_1 S_4 + r_{23}S_2 S_3 + r_{24}S_2 S_4}{\sqrt{S_1^2 + S_2^2 + 2r_{12}S_1 S_2}\, \sqrt{S_3^2 + S_4^2 + 2r_{34}S_3 S_4}}$$

Now it is assumed that the $X_1$ and $X_2$ scores are comparable (equivalent sets, with $S_1 = S_2$), and we simply say that our imaginary scores, $X_3$ and $X_4$, are comparable with each other and also with $X_1$ and $X_2$; hence all four $S$s have the same value, and therefore cancel out, leaving

$$r_{ab} = \frac{r_{13} + r_{14} + r_{23} + r_{24}}{\sqrt{2 + 2r_{12}}\, \sqrt{2 + 2r_{34}}}$$

Comparable or equivalent sets of scores will correlate equally with each other, that is, the five unknown $r$s in this expression will all equal $r_{12}$, our known value. Therefore we have

$$r_{xx} = r_{ab} = \frac{4r_{12}}{\sqrt{2 + 2r_{12}}\, \sqrt{2 + 2r_{12}}} = \frac{2r_{12}}{1 + r_{12}} \qquad (10.17)$$

as the reliability of scores based on the whole test.

The only assumption underlying formula (10.17) is that the two halves being correlated are comparable (equivalent or parallel). If the test items have been arranged according to difficulty, a first-half vs. second-half reliability will not satisfy the notion of comparable measures. Ordinarily the odd-even item technique will satisfy the criteria of comparability and sameness of trait. Neither of the split-half methods will satisfy the assumption of uncorrelated errors. Since both measures are determined at the same sitting, any chance fluctuations due to physiological conditions or to chance factors in the test situation will influence the two scores of an individual in the same direction. It is to be expected, therefore, that the correlation of halves will in general lead to a reliability coefficient which is too high, giving us an exaggerated notion of the accuracy with which we can place an individual on the trait continuum.

By far the best method for determining the reliability of a test is to have two forms which have been made equivalent and comparable by careful selection and balancing of items. No item in one form should be so nearly identical with an item in the other form as to permit a direct memory transfer. Two forms, equivalent yet not identical, can be administered within, say, 2 weeks' time—a procedure which properly includes in the estimate of variable error the daily fluctuations due to either physiological or psychological conditions and variations due to chance factors in the physical situation in which the tests are given. With so short an interval between testings, the trait being measured will have changed only a negligible amount as a result of maturation or ordinary environmental influences.

The form versus form method for calculating reliability may reflect two major sources of unreliability: instrumental error and trait instability. If it is claimed that $X$ indicates a person's position on a scale and if we wish to know something of the precision of the score, we should so determine $S_e$ as to include both major sources of error. High precision for a score earned today is a necessary but not sufficient condition for score stability; if the day-to-day variation happens to be large we can not have a very dependable score—its lack of dependability associated with the accident of measurement on a particular day should be incorporated in $S_e$.

When we attempt to obtain the reliability of a learning score or of any performance which is influenced by practice, we encounter difficulties which are baffling to the researcher who rigorously adheres to the fundamental requisites of the reliability coefficient. The chief difficulty is the obvious fact that the "thing" being measured changes as a result of each measurement or trial. Test-retest, or first half vs. second half (of trials), or today's trials vs. tomorrow's will not represent measures of the same function, nor will any scheme analogous to equivalent forms avoid this difficulty, since "forms" which are comparable will permit transfer. The use of scores on odd vs. even trials will have the advantage of balancing somewhat the influence of practice, especially if several trials are given; but the possibility that a chance error affects odds and evens alike is present, in that a slip in the experimental procedure or a temporary discouragement on the part of the testee or the adoption by the subject of a poor approach to the problem will have a similar effect on both scores. If trials were spaced, say, a day apart, the factors just mentioned might not greatly disturb the reliability determination. In general, it can be said that the odd-even trial method will yield a reliability coefficient which is higher than the "true" reliability.

The same shortcomings are present in the aforementioned methods when they are employed in determining the reliability of animal (or human) maze-learning scores. Other techniques, peculiar to the maze situation, have been proposed. Performances on the odd and even blinds, somewhat similar to odd and even items, have been correlated for the purpose of reliability, but since blinds differ considerably as regards difficulty, we cannot be sure that the two halves are comparable. We can also question the comparability of the first half and second half of the maze, since in general the last part tends to be learned more quickly than the first. Attempts to ascertain the reliability of one maze by correlating performance on it with that on another maze involve several difficulties. In the first place, there seems to be a general positive transfer (perhaps a general adaptation to the maze situation) from a first to a second maze; secondly, the second maze must be similar to the first in order to satisfy the requisite of comparable measures of the same ability, but if this similarity approaches identity the second maze becomes a retest; and thirdly, a close degree of similarity will lead to possible interference effects which may act differentially from animal to animal.

The foregoing brief discussion of the requisites for and difficulties in arriving at a meaningful reliability coefficient should make obvious the necessity for examining critically any proposed method of determining the reliability of a psychological measurement. The interpretation of the reliability coefficient in terms of the standard error of measurement definitely assumes homoscedasticity, which is another way of saying that the reliability coefficient is valid only when the error of measurement is of the same order of magnitude for the entire range of scores. That this may not always hold true is evident from findings with the 1937 Stanford Revision of the Binet Test.

It should be noted that the magnitude of the reliability coefficient is influenced by the trait homogeneity of the sample on which it is based. Let $sd$ represent the standard deviation for the restricted range, $SD$ the standard deviation for the unrestricted range, $r_{xx}$ the reliability for the restricted, and $R_{xx}$ the reliability for the unrestricted. If we may assume that $S_e$ for the smaller range equals $S_e$ for the larger range, we may write

$$(sd)^2(1 - r_{xx}) = (SD)^2(1 - R_{xx}) \qquad (10.18)$$

as a formula from which we can infer $r_{xx}$ from $R_{xx}$, and vice versa. The more homogeneous the group, the lower the reliability coefficient.

**Attenuation.** Now we return to the question which led to this lengthy detour: How does unreliability affect the correlation between variables? Let

$$x = x_t + e$$
$$y = y_t + d$$

where $e$ and $d$ represent the variable errors in the two scores, $x$ and $y$. Then

$$r_{xy} = \frac{\Sigma(x_t + e)(y_t + d)}{NS_xS_y}$$
$$= \frac{\Sigma x_t y_t + \Sigma x_t d + \Sigma y_t e + \Sigma ed}{NS_xS_y}$$

If we assume that $d$ is uncorrelated with $x_t$, that $e$ is uncorrelated with $y_t$, and that $e$ and $d$ are uncorrelated, we have

$$r_{xy} = \frac{\Sigma x_t y_t}{NS_xS_y} = \frac{r_{x_t y_t} S_{x_t} S_{y_t}}{S_x S_y}$$

Since, in general, $S_t = S_x\sqrt{r_{xx}}$ by formula (10.14), we have

$$r_{xy} = \frac{r_{tt} S_x\sqrt{r_{xx}}\, S_y\sqrt{r_{yy}}}{S_x S_y} \qquad (r_{tt} = r \text{ between true scores})$$

$$r_{xy} = r_{tt}\sqrt{r_{xx}}\,\sqrt{r_{yy}} \qquad\qquad (10.19)$$

which, since the reliability coefficients are less than unity, shows clearly that the correlation between obtained scores will be less than the correlation between true scores; i.e., errors of measurement tend to reduce or attenuate the correlation between traits.

We can rearrange formula (10.19) as

$$r_{tt} = \frac{r_{xy}}{\sqrt{r_{xx}}\,\sqrt{r_{yy}}} \qquad\qquad (10.20)$$

by which we can estimate what the correlation would be if perfect, errorless, measures were available. This is known as *correction for attenuation.* Correlation coefficients corrected for attenuation are of theoretical importance in the analysis of relationships in that allowance can be made for variable errors of measurement, but such corrected *r*s are of little practical value since they cannot be used in prediction equations. The prediction of one variable from another and the accompanying error of estimate must necessarily be based on obtained, or fallible, rather than true scores.

Since the correlation between variables is a function of the reliability of their measurement, we may examine the limits imposed on $r$ as a result of fallible scores. By reference to formula (10.19), we observe that, if the correlation between true scores is unity and if the reliability for one variable is perfect, the obtained correlation between the two cannot exceed the square root of the reliability coefficient for the other variable. If the correlation between the true scores is perfect, and if each variable is subject to errors of measurement, the obtained correlation cannot exceed the product of the square roots of the two reliability coefficients. Obviously, if the reliabilities are the same, the obtained correlation cannot be greater than the reliability coefficient.

In addition to the assumptions which were made specifically in deriving the formula for correcting for attenuation, it is also necessary to meet all the assumptions required for a sound reliability coefficient. Since obtained correlations and also reliability coefficients are functions of the homogeneity, with respect to the two traits, of the sample on which they are based, it follows that the reliability coefficients used in correcting an obtained $r$ should be based on the same sample as $r$ or on a sample which is of comparable homogeneity. Corrected $r$s greatly in excess of unity have been reported. Such absurd results lead us to ask whether the assumptions have been met, but this question should be raised concerning any corrected $r$, even though it does not exceed unity, since the assumptions may have been violated. It has been said that a corrected $r$ can legitimately exceed unity by as much as 2 or 3 times its sampling error. Formulas for the standard error of a corrected $r$ are available, but nothing is known concerning the nature of the distribution of corrected $r$s for successive samples. Presumably this distribution would be markedly skewed for high values; hence the use of an ordinary standard error technique to determine whether a corrected $r$ exceeds unity (or any other magnitude) by more than can reasonably be expected on the basis of sampling is an unsound procedure.

**Measurement error and comparison of means.** Will the presence of errors of measurement affect the large sample, $z$, and the small sample, $t$, tests for the difference between means? It seems reasonable to presume that $D_M = M_1 - M_2$ would not be systematically affected by measurement errors, because positive and negative errors would tend to balance so that each $M_x$ would tend to equal the $M_{x_t}$ for the sample. The value of $S_{D_M}$ or $s_{D_M}$ for independent means will be increased by errors of measurement because score variance is increased by such errors [see formula (10.12)]. For correlated means, $S_{M_D}$ and $s_{M_D}$ will be increased by measurement errors because the variance of the difference scores is increased by unreliability. Thus for either independent or correlated means, the $z$ and $t$

are systematically reduced by errors of measurement. Moral: the use of unreliable measures is not conducive to the finding of significant differences.

**Slope and measurement errors.** The slope for the regression of $Y$ on $X$, $B_{yx} = r(S_y/S_x)$ can be written in terms of the correlation between true scores and the $S$s for true scores by utilizing (10.20) and (10.14). Thus

$$B_{y_t x_t} = \frac{r_{xy}}{\sqrt{r_{xx}}\sqrt{r_{yy}}} \cdot \frac{S_y\sqrt{r_{yy}}}{S_x\sqrt{r_{xx}}} = \frac{r_{xy}S_y}{r_{xx}S_x}$$

from which we see that the slope in terms of true scores is larger than the the slope based on obtained scores; that is, the slope is reduced by errors in $X$. Note that the errors in $Y$ do not affect the slope; this is reasonable because for a fixed value of $X$ (or for an interval on $X$), the average value of $Y$ will involve a balancing of the chance errors in the $Y$s, that is, the means of the vertical arrays will not be systematically affected by the errors in the $Y$s. Therefore the slope of the line "fitting" the array means will not be systematically affected by the $Y$ errors. For those primarily interested in slopes, it would seem unimportant to have high reliability for $Y$, but it must be noted that the sampling variance of $B_{yx}$ will be increased by errors in $Y$ via an increase in the residual variance.

**Reliability of difference scores.** There are three situations for which we may wish to say whether two scores differ more than expected on the basis of errors of measurement. First, two persons each with a score on a given test. Second, a change score based on an "initial" and a "final" score for a person. Third, the difference score for a person on two different tests.

For the first situation, the standard error of the difference is given by

$$\sqrt{S^2_e + S^2_e} = S_e\sqrt{2}$$

A difference would need to be approximately $2S_e\sqrt{2}$ to be significant at the .05 level.

For the second situation, let us be a bit unconventional in terminology by letting $Y$ stand for an initial score and $X$ stand for a second score, on the same variable, taken after an experience or time interval that might produce a change. (Ordinarily, we let $X_1$ and $X_2$ or $X_i$ and $X_f$ stand for such scores; the $Y$ and $X$ notation will have certain conveniences in the sequel.) Thus the change score is given by $D = X - Y$ or, in deviation units, as $d = x - y$. Although $Y$ and $X$ are based on the same instrument or test, the second score, or $X$, might be regarded as measuring the same thing as $Y$ plus the experimentally produced effect variable over individuals.

Let the error in $D$ be represented by

$$e_d = d - d_t$$
$$= (x - y) - (x_t - y_t)$$
$$= (x - x_t) - (y - y_t)$$

or

$$e_d = e_x - e_y$$

Assuming that the errors in $X$ and $Y$ are uncorrelated, we have by the variance theorem,

$$S^2_{e_d} = S^2_{e_x} + S^2_{e_y} = S^2_x(1 - r_{xx}) + S^2_y(1 - r_{yy}) \qquad (10.21)$$

the square root of which will yield the standard error (of measurement) for the change score. In order to determine the reliability coefficient for change (difference) scores, we may utilize (10.15) with a shift to subscripts appropriate to the present problem:

$$r_{dd} = 1 - \frac{S^2_{e_d}}{S^2_d}$$

in which $S^2_d$ is the variance of difference (change) scores and is given (see p. 91) by

$$S^2_d = S^2_x + S^2_y - 2r_{xy}S_xS_y$$

Substituting, we get

$$r_{dd} = 1 - \frac{S^2_x(1 - r_{xx}) + S^2_y(1 - r_{yy})}{S^2_x + S^2_y - 2r_{xy}S_xS_y} \qquad (10.22)$$

An estimate of $r_{dd}$ based solely on the $N$ cases in an investigation involving change scores poses a problem: How ascertain the needed $r_{xx}$ and $r_{yy}$ values? Apparently, the most feasible procedure would be to use the odd-even, Spearman-Brown, method for calculating each value. We might, however, secure a fair approximation of $r_{dd}$ when the scores are based on a standardized test or a scale of known reliability provided it seems safe to assume that the error of measurement variance is the same for the $N$ cases at hand as that for the group on which the known reliability coefficient was calculated and provided it can be assumed that the error of measurement variances for the first and second set of scores on our $N$ cases are equal. The first assumption is tenable unless our $N$ cases are highly atypical compared with those in the group yielding the known reliability coefficient. (Note that we need to assume the equivalence of two $S_e$ values for this first assumption, not the equivalence of two reliability coefficients. Why?) The second assumption would be questionable if the imposed experimental condition led to drastic changes (unlikely in most investigations). The approximation of $r_{dd}$ would be obtained by replacing each of the numerator terms by the already available $S^2_e$ of the test. Since the odd-even method is less fraught with assumptions, its use is preferable.

It is of interest to simplify (10.22) to

$$r_{dd} = \frac{r_{xx}S^2_x + r_{yy}S^2_y - 2r_{xy}S_xS_y}{S^2_x + S^2_y - 2r_{xy}S_xS_y} \tag{10.23}$$

and then note what happens when $S_x = S_y$ (i.e., when no change in variation occurs). The $S$s cancel, giving

$$r_{dd} = \frac{r_{xx} + r_{yy} - 2r_{xy}}{2 - 2r_{xy}} \tag{10.24}$$

Under these conditions $r_{yy}$ could very well equal $r_{xx}$, so that we would have

$$r_{dd} = \frac{r_{xx} - r_{xy}}{1 - r_{xy}} \tag{10.25}$$

which makes it quite apparent that as $r_{xy}$ approaches $r_{xx}$ the value of $r_{dd}$ approaches zero, a proposition that also tends to hold for $r_{dd}$ by way of (10.23) and its exact equivalent, (10.22). This means that for change scores to be reliable the experimentally produced effect must lead to a shift in the ordering of individuals (regardless of the presence or absence of an over-all mean change). With an $r_{xx}$ of, say, .90 and $r_{xy}$ of, say, .80, the value of $r_{dd}$ by (10.25) is only .50.

Although the foregoing was couched in terms of experimentally produced changes, it is obvious that the same deductions hold for long-time changes in longitudinal studies. For either case the reliability of change scores may be surprisingly low despite high reliability for initial and final scores. The most serious consequence occurs when changes on one variable are being correlated either with changes on another variable or with scores on some other variable: the $r$s will be attenuated, sometimes so much as to make it difficult to obtain statistically significant $r$s.

The (usual) unreliability of change scores poses a paradox when the question of the significance of the mean, or over-all, change is considered. Suppose $S_x = S_y$ and suppose the mean change is appreciable, say half of $S$ in magnitude, and suppose further that $r_{xy}$ is very nearly equal to $r_{xx}$ with a consequent $r_{dd}$ of near zero. How could a mean change based on changes so lacking in reliability possess statistical significance? In answering this, it should be noted that two things happen as $r_{xy}$ approaches $r_{xx}$ as its limit: not only do the change scores become more unreliable but also the standard error of the difference (= standard error of the mean change) is progressively reduced by the increasing $r_{xy}$ in the standard error of the difference between initial and final means. Such an occurrence —very high $r_{xy}$ and a substantial mean change—is highly academic because produced changes are usually different from person to person rather than nearly constant over persons as implied by a very high $r_{xy}$.

The third situation for which we may need and in fact should have information regarding the reliability of difference scores occurs when for each of several persons we have a difference between scores on two *different* tests, or variables. For such difference scores to have meaning, the two sets of scores should be in comparable units such as standard scores or $T$ scores, with equal means and $S$s. Under these conditions the reliability of difference scores will be given by (10.24) with the subscripts referring to the two different variables, or tests. This time it is high correlation between the two variables that tends to lead to unreliable difference scores even though $r_{xx}$ and $r_{yy}$ are satisfactorily high. Again, the unreliability will limit the degree of correlation of such difference scores with other differences or with other variables. The instability of difference scores would seem to limit their possible usefulness in diagnostic work or in guidance programs. But it should be noted that even though difference scores may not provide a very reliable basis for differentiating among individuals, a difference for a particular individual may be dependable provided it is sufficiently large, say $1.96S_{e_d}$, for which $S_{e_d}$ would be obtained as the square root of the middle or right-hand part of (10.21), with the components in standard score form. Needless to say, even large differences may not have diagnostic or guidance significance—empirical study is needed to demonstrate their value.

**Measurement errors and a regression phenomenon.** Suppose we have the scatterplot for the scores on two comparable forms of a test for $N$ persons. With a form versus form correlation, or reliability coefficient, of, say, .85 the regression line for the second form score on the first score (or the line one might use to predict $X_2$ from $X_1$) will have a slope of .85 (assumes $S_1 = S_2$), which means, of course, that those initially below average and those initially above average have "regressed" toward the mean on the second testing. That is, there would seem to be a tendency for the initially low to gain and the initially high to lose, which implies a negative correlation between initial score and gain.

Let's take an algebraic look at the situation.

Let $\qquad\qquad X_i =$ initial score

$\qquad\qquad\qquad X_f =$ final score

$\qquad\qquad\qquad X_g = X_f - X_i = x_f - x_i = x_g =$ gain

Then

$$r_{ig} = \frac{\Sigma x_i x_g}{NS_i S_g} = \frac{\Sigma x_i (x_f - x_i)}{NS_i S_g} = \frac{\Sigma x_i x_f - \Sigma x^2_i}{NS_i S_g}$$

$$r_{ig} = \frac{r_{if} S_i S_f - S^2_i}{S_i S_g} = \frac{r_{if} S_f - S_i}{\sqrt{S^2_i + S^2_f - 2 r_{if} S_i S_f}} \qquad (10.26)$$

as a general formula for the correlation of gain with initial score. No gain
scores ever need to be calculated to obtain this correlation. For the special
case where initial and final are simply based on two comparable forms
with a short interval between the two testings, the $S$s will be so nearly
equal that they cancel out, leaving

$$r_{ig} = \frac{r_{if} - 1}{\sqrt{2 - 2 r_{if}}} \qquad (10.27)$$

from which it is quite evident that the lower the value of $r_{if}$ (that is, the
lower the form versus form reliability) the greater the negative correlation
between gain and initial score. If $r_{xx} = .85$, the value of $r_{ig}$ becomes $-.27$,
sufficiently high to be statistically significant at the .01 level if $N$ is in excess
of 90. But this "significant" $r$ has been produced by nothing more than
errors of measurement—no real gain for the initially low or real loss for
the initially high.

More generally, it is seen from (10.26) that in follow-up studies involving
test (initial) and retest (final) scores there will be a negative correlation
between gain and initial score unless $S_f$ has increased appreciably over $S_i$.
This negative correlation is produced in part by the attenuating effect of
unreliability on $r_{if}$ and in part by whatever factors contribute to real
differential changes from initial to final. Before concluding that gain and
initial status are really correlated, we need to get rid of that part caused by
errors of measurement; that is, we need a correction for the regression
that occurs solely on the basis of unreliability. This can be achieved
provided we have an $r_{xx}$ that holds for the group at hand [say an odd-even
Brown-Spearman estimate or the best (in sampling sense) available reliability coefficient adjusted for difference in variances for the two groups, the
adjustment by way of formula (10.18)]. The calculated $r_{xx}$ would represent
an estimate of $r_{12}$, the correlation between the first and second test under
the condition of no changes taking place other than those attributable to
measurement errors. It would be presumed that with no real changes
occurring, $S_2$ would tend to equal $S_1$, hence the prediction equation in
deviation units would be $x'_2 = r_{12} x_1 = r_{xx} x_1$, or in raw scores,

$$X'_2 = r_{xx} X_1 + M_2 - r_{xx} M_1$$

with the unknown $M_2$ taken as equal to $M_1$. Then the gain would be taken as the actual $X_f$ minus the predicted $X_f$ given by the foregoing $X'_2$. We are taking $X'_f$ as given by $X'_2$ as our best regression estimate of what the final score would be in case there were no change-producing conditions intervening between the testings. Actually $X'_f$ may be regarded as an initial score adjusted so as to eliminate that part of the regression of final on initial score that is produced by measurement errors. A little algebra will help at this point.

In terms of deviation units we have $x_i$, $x_f$, $x'_f = r_{xx}x_i$, and $g = x_f - x'_f = x_f - r_{xx}x_i$, hence the correlation between the adjusted initial scores, $r_{xx}x_i$, and gains from the adjusted initial scores becomes

$$
\begin{aligned}
{}_a r_{ig} &= \frac{\Sigma x'_f g}{NS_{x'_f}S_g} = \frac{\Sigma r_{xx}x_i(x_f - r_{xx}x_i)}{NS_{x'_f}S_g} \\
&= \frac{r_{xx}\Sigma x_i x_f - r^2_{xx}\Sigma x^2_i}{NS_{x'_f}S_g} \\
&= \frac{r_{xx}r_{if}S_iS_f - r^2_{xx}S^2_i}{r_{xx}S_i\sqrt{S^2_f + r^2_{xx}S^2_i - 2r_{if}r_{xx}S_iS_f}}
\end{aligned}
$$

$$
{}_a r_{ig} = \frac{r_{if}S_f - r_{xx}S_i}{\sqrt{S^2_f + r^2_{xx}S^2_i - 2r_{if}r_{xx}S_iS_f}} \tag{10.28}
$$

which, as a computational formula, gives the desired adjusted correlation without the tedium of first calculating $N$ $X'_f$ values and $N$ differences, and then a correlation coefficient. Note from (10.28) that when $S_f = S_i$ and when $r_{if} = r_{xx}$, the adjusted correlation between gain and initial becomes zero. Contrast this with the artifactual $r_{ig}$ given by (10.26) for the same conditions.

Another approach to the elimination of the effect of errors of measurement on the correlation between initial and gain is to work with so-called *regressed* scores, a concept which we will need in another connection. Consider the problem of estimating a true score, $x_t$, from an obtained score, $x$. We can write a prediction equation as

$$
x'_t = r_{x_t x}\frac{S_t}{S_x}x \tag{10.29}
$$

Now

$$
r_{x_t x} = \frac{\Sigma x_t x}{NS_tS_x} = \frac{\Sigma x_t(x_t + e)}{NS_tS_x} = \frac{\Sigma x^2_t + \Sigma x_t e}{NS_tS_x}
$$

As earlier, it is assumed that $x_t$ and $e$ are uncorrelated, so we have

$$
r_{x_t x} = \frac{S^2_t}{S_tS_x} = \frac{S_t}{S_x} \quad \text{which from (10.14)} = \sqrt{r_{xx}}
$$

Also from (10.14) we have $S_t = S_x\sqrt{r_{xx}}$
Substituting in (10.29),

$$x'_t = \sqrt{r_{xx}}\,\frac{S_x\sqrt{r_{xx}}}{S_x}\,x = r_{xx}x \qquad (10.30)$$

Note that the estimate, $x'_t$, is identical to $x_2$ estimated from $x_1$ or to $x_f$ estimated from $x_i$, both $x'_2$ and $x'_f$ being estimates under the condition of no changes except those due to errors of measurement. We see now that in saying that $x'_2$ or $x'_f$ may be regarded as a sort of adjusted value for the first or initial score, we are in effect so adjusting the initial score as to yield an estimate of the true score. The $x'_t$ values are referred to as *regressed scores.* Obviously, the utilization of regressed initial scores as a basis for determining whether or not gains are correlated with initial (regressed) scores will lead to (10.28). Conceptually, it may seem preferable to think in terms of $x'_t$ as an adjusted initial score that makes allowance for the error of measurement part of the regression of final on initial standing.

**A matched group fallacy.** Occasionally, in the comparison of group changes, either long-term or experimentally produced, we may be dealing with samples from two populations that are known to differ appreciably; that is, random sampling will yield groups that will differ in initial score level. In order to have groups with comparable initial standing, cases are paired, a procedure which is ordinarily desirable but which for this given situation introduces a difficulty in that the matching will involve pairing some persons from the top half of one population with persons from the bottom half of the other population. Upon subsequent testing and without any interpolated change-producing experience, one group will show gain whereas the other group will show loss, but this difference in change represents nothing more than the regression of scores in each group toward the respective mean of the larger group from which it was drawn. And of course, with change-producing conditions this type of regression due to errors of measurement will contaminate (either increase or decrease) any real difference in change between the two groups.

A second type of situation in which matching may be disrupted by measurement errors occurs when pairing is used to obtain two groups that are comparable on some variable, $Y$, which should be controlled in making a comparison on variable $X$. If the matching on $Y$ involves pairing a person from the upper-half of one supply group with a person from the lower-half of the other supply group, the two matched samples, so set up, will tend to have equal means for $Y$, but will nevertheless differ on the $Y$ variable because of the errors in the $Y$ scores used in pairing. An immediate retest on $Y$ will show that those from the top part of one "population" (supply) will average lower on $Y$, whereas those from the bottom portion of the other supply will average higher on $Y$, than originally.

For either of these situations—control on initial score or on another variable—the failure to obtain really comparable groups will occur even though not all pairs involve top-half versus bottom-half status. The larger the percentage of such pairs, the greater the disruption. For both situations, the trouble can be avoided by pairing on the basis of regressed scores. Separate regression equations of the form $x'_t = r_{xx}x_i$ (or $y'_t = r_{yy}y$) will be needed for each group from which the (matched) samples are to be formed, with the value of $r_{xx}$ (or $r_{yy}$) taken as the reliability most appropriate for the particular group. The calculation of the regressed score is facilitated by casting the regression equation into raw score form: $X'_t = r_{xx}X + M_x - r_{xx}M_x$.

## INDEX CORRELATION

A possible source of error in correlational work may be introduced when two indexes having a common variable denominator are correlated, such as $X/Z$ and $Y/Z$. Before considering this special case, it might be well to turn our attention to more general formulas for indexes. These formulas involve the coefficient of variation, namely, $v = S/M$, and their use leads to serious error when the $v$s are large—$v^3$ and higher-power terms having been dropped in the derivations.

Let $I = X_1/X_2$; then it can be shown that the mean and standard deviation of such an index or ratio will be approximately

$$M_I = \frac{M_1}{M_2}(1 - r_{12}v_1v_2 + v^2_2) \qquad (10.31)$$

$$S_I = \frac{M_1}{M_2}\sqrt{v^2_1 - 2r_{12}v_1v_2 + v^2_2} \qquad (10.32)$$

If we have four variables, the following formula for the correlation of indexes will yield a good approximation:

$$r_{(X_1/X_3)(X_2/X_4)} = \frac{r_{12}v_1v_2 - r_{14}v_1v_4 - r_{23}v_2v_3 + r_{34}v_3v_4}{\sqrt{v^2_1 + v^2_3 - 2r_{13}v_1v_3}\sqrt{v^2_2 + v^2_4 - 2r_{24}v_2v_4}} \qquad (10.33)$$

Although these formulas are very useful, their use is somewhat limited in that generally we cannot know whether the index distribution is normal, nor can we make a statement concerning linearity and homoscedasticity for the correlation between two indexes. Such information, if needed, must be obtained by first determining the numerical value of the indexes for each individual and then making distributions.

Several special cases can be deduced from formula (10.33). Thus the correlation between $X_1/X_3$ and $X_2$ is exactly equivalent to that between $X_1/X_3$ and $X_2/1$; i.e., $X_4$ is set equal to 1, which makes $v_4 = 0$, and therefore all terms involving the subscript 4 vanish. The correlation between $X_1/X_3$ and the reciprocal of a variable would be obtained by setting $X_2 = 1$, i.e., letting $1/X_4$ be the reciprocal; then $v_2 = 0$, whence the desired formula can be obtained by dropping all terms involving $v_2$. Likewise the correlation can be deduced for $1/X_3$ with $1/X_4$, for $1/X_3$ with $X_2$, and for $X_1/X_3$ with $X_2/X_3$. This last correlation is of particular interest because it is possible to find a relationship between these two indexes even though the three original variables are uncorrelated.

By substituting $X_3$ for $X_4$, i.e., replacing subscript 4 by 3, an expression for the correlation of indexes having a common variable denominator can readily be obtained. It will be

$$r_{(X_1/X_3)(X_2/X_3)} = \frac{r_{12}v_1v_2 - r_{13}v_1v_3 - r_{23}v_2v_3 + v_3^2}{\sqrt{v_1^2 + v_3^2 - 2r_{13}v_1v_3} \sqrt{v_2^2 + v_3^2 - 2r_{23}v_2v_3}} \quad (10.34)$$

If $r_{12} = r_{13} = r_{23} = 0$, this becomes

$$\frac{v_3^2}{\sqrt{v_1^2 + v_3^2} \sqrt{v_2^2 + v_3^2}}$$

and if the $v$s are equal, the value of the index correlation will be .50 even though there is no relationship between the original variables. This is known as *spurious correlation* due to indexes. There are instances, however, in which an analysis of the interrelations of ratios is of just as much import as the analysis of the variables from which the indexes are obtained, and therefore it does not follow that the correlation between ratios having a common denominator is necessarily misleading.

It has been asserted that the correlation between IQs derived from two tests or two forms of the same test will be spuriously high because of the common variable denominator, age. It can be shown, however, that such a correlation will not be spurious unless the two sets of IQs are correlated with age. If the IQ-vs.-age correlations are both positive or both negative, the index correlation will be spuriously high; if one is negative and the other positive, spuriously low. Thus, rather than make a blanket statement to the effect that the correlation between IQs is spuriously high, we should say that it can be spuriously high or low or not spurious at all, according to the IQ-vs.-age correlations. It should be remembered that, even though the IQs based on an ideal (properly constructed and standardized) test will be uncorrelated with age, a nonzero relationship might be produced for a single school-grade group by the selective factors that operate in age-grade location. Within a single grade group in a school system where acceleration is permitted, the younger children are likely to be the brighter, i.e., have the higher IQs, thus producing negative correlations for sets of IQs with age, and consequently a spuriously high correlation between IQs.

## PART-WHOLE CORRELATION

Another type of spurious correlation arises when a total score is correlated with a subscore which is a part of the total score. Suppose that a total score is made up of three parts, $X_T = X_1 + X_2 + X_3$ and that we correlate $X_1$ against $X_T$. Ordinarily in such situations the components will themselves be correlated positively. It should be obvious that the extent to which $X_1$ correlates with $X_T$ is more or less dependent on the fact that $X_T$ includes $X_1$. It does not follow, however, that a high value for $r_{1T}$ is not meaningful, even though spurious. For instance, a high value for $r_{1T}$ would, regardless of spuriousness, justify the use of $X_1$ in lieu of the battery of three subtests. There are times when we may wish to know how highly a subtest correlates with a total, based on any number of parts, *minus* the subtest. This correlation is given by

$$r_{1(T-1)} = \frac{r_{1T}S_T - S_1}{\sqrt{S^2_T + S^2_1 - 2r_{1T}S_1S_T}} \qquad (10.35)$$

## HETEROGENEITY WITH RESPECT TO A THIRD VARIABLE

We have already discussed the influence on $r$ of heterogeneity with regard to one or both the variables being correlated. Suppose variables $X_1$ and $X_2$ are two different traits, each of which is related to age as the third variable. Then an older individual will tend to be higher on both tests than a younger individual. In other words, heterogeneity with respect to age will tend to produce correlation between $X_1$ and $X_2$, and our present problem is to develop a method for correcting $r_{12}$ so that we can estimate what the correlation between $X_1$ and $X_2$ would be if age were constant. Suppose $r_{12}$, $r_{13}$, $r_{23}$, and the several means and standard deviations are known; then let us visualize the three scatter diagrams. The scatter for $r_{12}$ will be somewhat elongated as a result of the influence of age, since variation in both $X_1$ and $X_2$ are here supposed to be partly due to age variation. What is needed is the correlation, between measures of $X_1$ and $X_2$, which has been freed from the influence of age. If we were to express each $X_1$ in the first array of the scatter for $r_{13}$ as a deviation from the mean of this array and were to do the same for all other $X_1$s in the scatter—each as a deviation from the mean of the array in which it falls—we would have scores expressed as deviations from the means of the several ages. These deviations will be independent of age. As an example, suppose an 8-year-old individual scores 28 and the mean of 8-year-olds is 25, and a 14-year-old individual scores 54 and the mean of 14-year-olds is 51. The second individual scores higher than the first because he is older, but each would have a deviation (from his own age mean) of plus 3. Obviously, if we also expressed the $X_2$ scores as deviations from the averages for the several ages, they too would be independent of age influences. Now, if we correlated these deviations (from age means) we would be correlating sets of $X_1$ and $X_2$ scores which would be free from age, and hence we would arrive at a correlation, between variables $X_1$ and $X_2$, which would not be affected by age heterogeneity.

**Partial correlation.** The task of determining the correlation between two variables, with the influence of a third eliminated, can always be accomplished by actually computing all the deviations and then making a scatter diagram from which the $r$ can be determined. However, in those cases in which we can assume linearity of regression for $X_1$ on $X_3$ and $X_2$ on $X_3$, it is possible to set up a method for determining the desired correlation from the three correlation coefficients between the three variables. If linearity exists, we can correlate the deviations from the two regression lines instead of from the array means (or means for several ages if age is the third variable). Since

$$x'_1 = r_{13}\frac{S_1}{S_3}x_3 \quad \text{and} \quad x'_2 = r_{23}\frac{S_2}{S_3}x_3$$

the two sets of deviation-from-regression scores will be

$$x_1 - x'_1 = x_1 - r_{13}\frac{S_1}{S_3}x_3 \quad \text{and} \quad x_2 - x'_2 = x_2 - r_{23}\frac{S_2}{S_3}x_3$$

The correlation of these deviation scores, which is designated by the symbol $r_{12\cdot3}$ (read: the correlation between $X_1$ and $X_2$ with $X_3$ held constant) and known as the *partial correlation coefficient*, becomes

$$r_{12\cdot3} = \frac{\Sigma(x_1 - x'_1)(x_2 - x'_2)}{NS_{x_1-x'_1}S_{x_2-x'_2}}$$

$$= \frac{\Sigma\left(x_1 - r_{13}\frac{S_1}{S_3}x_3\right)\left(x_2 - r_{23}\frac{S_2}{S_3}x_3\right)}{NS_{x_1-x'_1}S_{x_2-x'_2}}$$

Multiplying and summing the numerator, and noting that the $S$s in the denominator are nothing more than the errors of estimate, $S_{1\cdot3}$ and $S_{2\cdot3}$, we have

$$r_{12\cdot3} = \frac{\Sigma x_1 x_2 - r_{23}\dfrac{S_2}{S_3}\Sigma x_1 x_3 - r_{13}\dfrac{S_1}{S_3}\Sigma x_2 x_3 + r_{13}r_{23}\dfrac{S_1 S_2}{S_3^2}\Sigma x_3^2}{N S_1\sqrt{1 - r_{13}^2}\, S_2\sqrt{1 - r_{23}^2}}$$

Dividing by $N$, cancelling $S$s, and collecting like terms, we get

$$r_{12\cdot3} = \frac{r_{12} - r_{13}r_{23}}{\sqrt{1 - r_{13}^2}\sqrt{1 - r_{23}^2}} \qquad (10.36)$$

This formula definitely assumes the linearity of the two regression lines for predicting $X_1$ and $X_2$ from $X_3$. Whether we correlate deviations from array means or use formula (10.36), we end with a correlation which has been freed of the influence of the third, or eliminated, variable. If, for example, age is the third variable, the partial correlation coefficient represents an estimate of what the correlation would be if we held age constant by the use of individuals of *any* one of the several age levels present in the original group.

The difference between $r_{12\cdot3}$ and $r_{12}$ indicates how much of the correlation between variables 1 and 2 is due to the influence of heterogeneity of a third variable. Obviously, if the third variable is unrelated to $X_1$ and $X_2$, the partial $r$ will equal $r_{12}$, and if either $r_{13}$ or $r_{23}$ is negative and $r_{12}$ positive, "partialing out" $X_3$ will raise the correlation. Is this reasonable?

The difficulties encountered in determining the direction of causation make it necessary to be careful in the use of the partial correlation technique. When it is said that heterogeneity with respect to a third variable ($X_3$) has in part (or entirely) produced correlation between $X_1$ and $X_2$, we must ask how the influence of $X_3$ comes about. Now if it can be argued that variation in $X_3$ is a cause of variation in $X_1$ and $X_2$, it is readily seen that $r_{12}$ is at least in part attributable to the fact that $X_1$ and $X_2$ have a common source of variation. The partial, $r_{12\cdot3}$, tells us the degree of correlation between $X_1$ and $X_2$ which would exist provided variation in $X_3$ were controlled. But if it cannot be claimed that $X_3$ produces variation in $X_1$ and $X_2$, the interpretation of the partial $r$ is far from clear. Suppose $X_1$ precedes $X_3$ in a temporal sense so that we know variation on $X_3$ couldn't possibly contribute to variation in $X_1$. Does it make sense to interpret $r_{12\cdot3}$ as the correlation between $X_1$ and $X_2$ with the influence of $X_3$ nullified when we know that $X_3$ could not influence $X_1$? Stated differently, the only way that $X_3$ can produce or contribute to the correlation between $X_1$ and $X_2$ is by way of $X_3$ producing variation in $X_1$ and $X_2$.

The technique can be extended for "partialing out" or eliminating more than one variable. Thus, to obtain an estimate of $r_{12}$ with $X_3$ and $X_4$ held constant, we can use

$$r_{12\cdot34} = \frac{r_{12\cdot3} - r_{14\cdot3}r_{24\cdot3}}{\sqrt{1 - r^2_{14\cdot3}}\sqrt{1 - r^2_{24\cdot3}}}$$

which is in terms of first-order partials calculable by formula (10.36).

The sampling error of the partial coefficient may be handled by the $z$ transformation. The standard error of the corresponding $z$ will be $1/\sqrt{N-4}$ when only one variable has been eliminated, and $1/\sqrt{N-5}$ when two variables have been eliminated.

The partial correlation coefficient based on a small sample can also be tested for significance by the $t$ technique. If one variable has been eliminated, we have

$$t = \frac{r_{12\cdot3}}{\sqrt{\dfrac{1 - r^2_{12\cdot3}}{N-3}}}$$

with $df = N - 3$. An additional degree of freedom is lost for each additional variable eliminated.

A perplexing and often-recurring question with regard to the inter-relations of three variables is this: Are the correlations consistent among themselves, or, if $r_{12}$ and $r_{13}$ are known, what are the possible limits for $r_{23}$? If $r_{12} = $ unity and $r_{13} = $ unity, $r_{23}$ must also equal unity, but, if $r_{12} = 0$ and $r_{13} = 0$, does it follow that $r_{23} = 0$? It can be shown that the limits for the correlation $r_{23}$ will always be $r_{12}r_{13} \pm \sqrt{1 - r^2_{12} - r^2_{13} + r^2_{12}r^2_{13}}$.

EXAMPLES:

When $r_{12}$ and $r_{13}$ each equal .90, the limits for $r_{23}$ are $+.62$ and $+1.00$;
When $r_{12}$ and $r_{13}$ each equal .50, the limits for $r_{23}$ are $-.50$ and $+1.00$;
When $r_{12}$ and $r_{13}$ each equal .25, the limits for $r_{23}$ are $-.875$ and $+1.00$.

**Part correlation.** There are times when we may wish to have the correlation between variables $X_1$ and $X_2$ with the influence of $X_3$ "removed" from $X_2$ only. For example, we may wish to calculate the correlation between intelligence and incidental memory with general memory partialled out of the incidental memory; or we may wonder what the correlation is between reading ability and academic achievement with the influence of college aptitude "taken out" of academic achievement; or we may wish to determine the correlation between intelligence and a set of final scores, obtained after extensive practice, with initial level partialled out of the final variance.

In symbols, if we seek the correlation between $X_1$ and $X_2$ with $X_3$ partialled out of $X_2$, we would in effect be correlating $X_1$ with the residual, $X_{2\cdot3}$. From the derivation of $r_{12\cdot3}$ it can be easily deduced that an appropriate formula is

$$r_{1(2\cdot3)} = \frac{r_{12} - r_{13}r_{23}}{\sqrt{1 - r^2_{23}}} \tag{10.37}$$

which is referred to as a *part* correlation coefficient. The part correlation will, of course, be most useful when it can be argued that $X_3$ causes variation in $X_2$ but not in $X_1$. Its significance is testable by way of formula (15.21).

## LACK OF NORMALITY

In Chapter 9 an attempt was made to specify the requisite assumptions for each possible interpretation of $r$. It will be recalled that it was for only the theoretical interpretation in terms of the normal bivariate scatter that normality for the distribution of $X$ and $Y$ was assumed. It was, however, mentioned that the lack of normality, particularly skewness, for the marginal distributions should alert one to look for curvilinearity, heterogeneity of array variances, and skewed array distributions. The presence of any of these may invalidate the chosen interpretation. In this chapter we have been considering the factors that might affect the magnitude of a correlation coefficient. Is it necessary to regard nonnormality for $X$ and/or $Y$ as such a factor?

Let's consider an extreme case in which $X$ shows marked negative skewness while $Y$ is markedly skewed in the positive direction. Now imagine the scattergram for this situation. With a piling up of cases at the bottom of the $Y$ distribution and at the top of the $X$ distribution it follows that a sizable number of tallies must fall in the lower-right quadrant of the scattergram. This means, of course, that it is impossible to have all the tallies in a lower-left to upper-right diagonal, a requisite for perfect positive correlation; hence a restriction on the upper positive limit for $r$. This simply means that the comparability of degrees of relationship is a function of skewness of the marginal distribution—an $r$ of $+.80$ could very well be the maximum possible for one scattergram but fairly far from the maximum possible either when both margins are normal or when the skewness is in the same direction for both $X$ and $Y$.

If the reader does not readily see that skewness in opposite directions imposes a limit on positive correlation, he should try to assign tallies so as to have all of them in a lower-left to upper-right diagonal, yet have the margins with opposite skewness. Then he should attempt to assign the tallies in an upper-left to lower-right diagonal, and thereby see that skewness in opposite directions does not preclude having all the tallies in such a diagonal, hence permitting perfect negative correlation. If you have now "got with" this type of thing, you should be able to demonstrate that marked skewness for one variable and normality for the other also imposes an upper limit on the numerical value of $r$.

At the intuitive level it seems reasonable to say that any effect of skewness will disturb the comparability of $r$s more when the relationships are relatively high than when either moderate or low. Fortunately, something can be done about the skewness: apply the normalizing $T$-score transformation to the variates. But whether you base your $r$ on the original skewed variates or on the transformed scores, you should be alert to the form of the regressions, which will likely be changed either from curvilinear to linear or vice versa by the transformation(s).

## SUMMARY

In this chapter, consideration has been given to factors which have a bearing on the magnitude of the correlation coefficient. If any of these is operative in the case of a particular coefficient, it is the responsibility of the investigator to qualify his conclusions accordingly. Published reports of correlational studies should include the following.

1. A definition of the population being sampled and a statement of the method used in drawing the sample.

2. The size of the sample and an adequate treatment of sampling by means of nonantiquated formulas.

3. The means and particularly the standard deviations of the variables being correlated, with some indication as to whether the sample is typical as regards heterogeneity with respect to the variables under consideration.

4. The reliability coefficients for the measures and the method of determining reliability.

5. A statement relative to the homogeneity of the sample with respect to possibly relevant variables such as age, sex, race.

6. A defense or precise interpretation of any reported correlations involving indexes or of any part-whole correlations.

7. At least a rough indication and direction of skewness, if present, for the marginal distributions.

The researcher who is cognizant of the assumptions requisite for a given interpretation of a correlation coefficient and who is also fully aware of the many factors which may affect its magnitude will not regard the correlational technique as an easy road to scientific discovery.

# Chapter 11

# MULTIPLE CORRELATION

So far our discussion of correlation has been concerned chiefly with the prediction of one variable from another or the attribution of a portion of the variance of one variable to the action of a second variable. We shall next consider the case where it is desired to predict one variable by using several other variables as a team of predictors, or where, if causation can be assumed, an attempt is made to analyze the variance for one variable into components or parts attributable to the action of two or more other variables. There is a close connection between the predicting and the analyzing problems; let us first consider the method of predicting one variable on the basis of other variables.

## THE THREE-VARIABLE PROBLEM

For simplicity, consider the problem of predicting $X_1$ from a knowledge of $X_2$ and $X_3$. The $X_1$ variable is frequently called the criterion, or dependent variable. If we had $X_1$ to be predicted from $X_2$ alone, we would have exactly the same situation as predicting $Y$ from $X$. That is, the linear prediction equation (in gross score form)

$$Y' = BX + A$$

becomes

$$X'_1 = BX_2 + A$$

and the deviation form

$$y' = bx + a$$

becomes

$$x'_1 = bx_2 + a$$

It will be recalled that the values of the constants, $B$ and $A$, or $b$ and $a$, were so determined as to give the maximum predictability, and that $B$ and

188

$A$ turned out to be functions of the correlation coefficient between the two variables and of the means and standard deviations for the variables. The equation which resulted from giving $A$ and $B$ specific values was said to be the equation of the best-fitting line—the error of prediction was minimized.

Now, if we wish to predict $X_1$ from $X_2$ and $X_3$, we start with an equation of the form

$$X'_1 = B_2 X_2 + B_3 X_3 + A \tag{11.1}$$

which can be written in deviation units as

$$x'_1 = b_2 x_2 + b_3 x_3 + a$$

Either of these forms represents the equation of a plane. It can be shown that $B_2 = b_2$ and $B_3 = b_3$. In fact, this is rather obvious when we consider the meaning of these $B$ or $b$ coefficients. They represent the slope of the plane; $B_2$ is the slope which the plane makes with the $x_2$ axis, and $B_3$ the slope with regard to the $x_3$ axis. When we shift from raw to deviation scores, we are merely shifting the origin, or point of reference, to the intersection of the means, and this point in terms of deviation scores becomes zero. This shift of the frame of reference does not change the position or angle of the plane; hence $B_2 = b_2$ and $B_3 = b_3$. (The student will recall that, for the ordinary two-variable problem, the slope of the line was equal to $B$ or $b$.)

It remains to attach meaning to $A$ and $a$. In the equation $Y' = BX + A$, it was noted that the constant $A$ was the $Y$ intercept, i.e., the value of $Y$ where the line cut the $y$ axis. It was also found that $a = 0$; i.e., that in the deviation form the line cut the $y$ axis at the origin. Perhaps the student has already anticipated, by analogy, that the $A$ in our three-variable equation is the value of $X_1$ where the plane cuts the $x_1$ axis, and that the value of $a$ will become zero.

Before going farther, it might be well to take a look at the problem geometrically. In the case of two variables, after plotting the $X$ and $Y$ values in a scattergram, we can readily picture the meaning of $B$ and $A$, and also obtain some notion of why certain values of $B$ and $A$ will lead to better predictions than those obtained by other values. In the case of three variables, $X_1$, $X_2$, and $X_3$, we have a trio instead of a pair of measurements. In order to draw up a plot of $N$ such sets of measurements, we will need to use a three-dimensional scheme. Instead of placing a tally mark in a cell defined by an interval along the $x$ axis and one along the $y$ axis, we now have to consider a cell as defined by intervals on the $x_1$, the $x_2$, and the $x_3$ axes. Instead of a square cell, we have a cubical cell.

Suppose an individual's three scores fall in intervals $i_1$, $i_2$, and $i_3$; then his "tally" will be placed in the cubicle formed at the intersection of these three intervals. The total number of cubicles will be the product of the number of intervals on each axis, and an individual's location in the "box" will depend on all three of his scores. The student may be at a loss to know just how he could make such a three-dimensional scattergram. Actually, this diagram is not necessary, but it is of interest to imagine what such a three-way distribution would look like. If the correlations, $r_{12}$, $r_{13}$, and $r_{23}$, are fairly high (and positive), and if we think of the frequencies in the several cubicles as being represented by dots (or different degrees of density), then the swarm of dots will extend from the lower-left front to the upper-right back of the box. The greatest density will be at the center of this swarm, and the density or frequency will fall off in all directions from the center. The swarm will have the general shape and appearance of a watermelon (ellipsoidal).

Imagine that a plane is to be cut through this swarm. Our job is to so locate the plane that, when we start upward vertically from any point on the bottom of the box, say the spot defined by any pair of values for $X_2$ and $X_3$, we will find that the altitude, i.e., the distance along the $x_1$ axis at which the plane is reached, will constitute the best estimate of $X_1$ for individuals having any given $X_2$ and $X_3$ scores. With a little reflection, the reader can see that, of many ways of placing the plane, some positions will obviously give very poor estimates, whereas others will lead to better estimates. What we need is that plane which for the given $N$ sets of $X_1$, $X_2$, and $X_3$ scores will yield the best possible estimates.

The criterion of "best" is a least square affair—the sum of the squares of the errors of estimate shall be a minimum. The task is really that of determining the values of $A$, $B_2$, and $B_3$ in formula (11.1) so that

$$\Sigma(X_1 - X'_1)^2$$

is a minimum. That is, we are to assign to $A$, $B_2$, and $B_3$ those values which will permit the best possible estimate of an unknown $X_1$ when we know the $X_2$ and $X_3$ values for the individual. The principle to be used is exactly the same as that employed to obtain the optimum value for $B$ and $A$ for the two-variable problem, but the present problem is more complicated because we have to determine the values for three constants.

**Derivation of regression equations.** Our task is simplified if deviation scores are used, and we assume $a = 0$ (if we carried $a$ along, it would prove to be zero). It is simplified somewhat more if we transform all three sets of scores into standard score form, i.e., if we set $z = (X - M)/S$. Then our equation becomes

$$z'_1 = \beta_2 z_2 + \beta_3 z_3 \tag{11.2}$$

It should be noted that, since we are changing the size of our unit of measure, it cannot be argued that $\beta_2$ will equal $B_2$ or $b_2$. The task now is to determine the value of the *beta coefficients*, $\beta_2$ and $\beta_3$, so as to have the best possible estimate of $z_1$, or so that the average of the squared errors, or

$$\frac{1}{N}\Sigma(z_1 - z'_1)^2$$

shall be a minimum. Since $z_1 - z'_1 = z_1 - \beta_2 z_2 - \beta_3 z_3$, the function, $f$, to be minimized is

$$f = \frac{1}{N}\Sigma(z_1 - \beta_2 z_2 - \beta_3 z_3)^2$$

The calculus is used to determine the values of $\beta_2$ and $\beta_3$ which will make this function a minimum. We take the partial derivative of the function first with respect to $\beta_2$, then with respect to $\beta_3$. Thus,

$$\frac{\partial f}{\partial \beta_2} = \frac{-2\Sigma z_2}{N}(z_1 - \beta_2 z_2 - \beta_3 z_3)$$

$$\frac{\partial f}{\partial \beta_3} = \frac{-2\Sigma z_3}{N}(z_1 - \beta_2 z_2 - \beta_3 z_3)$$

These two derivatives are to be set equal to zero and then solved simultaneously for the two unknowns, $\beta_2$ and $\beta_3$. Performing the indicated multiplications, summing, and dividing each equation by 2, we get

$$\frac{-\Sigma z_1 z_2}{N} + \beta_2 \frac{\Sigma z^2_2}{N} + \beta_3 \frac{\Sigma z_2 z_3}{N} = 0$$

$$\frac{-\Sigma z_1 z_3}{N} + \beta_2 \frac{\Sigma z_2 z_3}{N} + \beta_3 \frac{\Sigma z^2_3}{N} = 0$$

Since we are dealing with standard scores, we can now capitalize on certain properties thereof, namely, that the sum of their squares divided by $N$ is unity, whereas any sum of cross products divided by $N$ is the correlation between the two variables involved in the cross products. Thus, we have

$$-r_{12} + \beta_2 + \beta_3 r_{23} = 0$$

$$-r_{13} + \beta_2 r_{23} + \beta_3 = 0$$

or

$$\beta_2 + r_{23}\beta_3 - r_{12} = 0$$

$$r_{23}\beta_2 + \beta_3 - r_{13} = 0$$

Since the $r$s in the equations are determinable for any given sample of data, they are in effect knowns, whereas the $\beta$s are unknowns. We therefore have two simultaneous equations with two unknowns. These can readily be solved by a number of methods which the student will find in an algebra textbook. Straightforward solution gives

$$\beta_2 = \frac{r_{12} - r_{13}r_{23}}{1 - r^2_{23}}$$

$$\beta_3 = \frac{r_{13} - r_{12}r_{23}}{1 - r^2_{23}}$$

As soon as we have computed the $r$s, we can easily determine the $\beta$s. The obtained numerical values can then be substituted in the prediction equation

$$z'_1 = \beta_2 z_2 + \beta_3 z_3$$

so that for a given pair of $z_2$ and $z_3$ values we can predict the standard score on the criterion variable. However, in practice it is ordinarily more convenient to deal with raw scores; hence we need our prediction equation in raw score form. Obviously, if we replace the $z$s in the preceding equation by their values in terms of raw scores, means, and standard deviations, we will have

$$\frac{X'_1 - M_1}{S_1} = \beta_2 \frac{X_2 - M_2}{S_2} + \beta_3 \frac{X_3 - M_3}{S_3}$$

or

$$\frac{X'_1}{S_1} - \frac{M_1}{S_1} = \beta_2 \frac{X_2}{S_2} - \beta_2 \frac{M_2}{S_2} + \beta_3 \frac{X_3}{S_3} - \beta_3 \frac{M_3}{S_3}$$

Multiplying by $S_1$ and rearranging terms, we have

$$X'_1 = \beta_2 \frac{S_1}{S_2} X_2 + \beta_3 \frac{S_1}{S_3} X_3 + \left( M_1 - \beta_2 \frac{S_1}{S_2} M_2 - \beta_3 \frac{S_1}{S_3} M_3 \right) \quad (11.3)$$

from which we see that our original $B_2$ must equal $\beta_2(S_1/S_2)$, $B_3 = \beta_3(S_1/S_3)$, and $A =$ the parentheses term. Thus we can readily determine the numerical values of $B_2$, $B_3$, and $A$ and thereby have the constants for the prediction equation. Actually, the values of $B_2$ and $B_3$ are the optimum weights to be assigned to $X_2$ and $X_3$ in order to predict $X_1$.

**Error of estimate.** The accuracy of the prediction of $X_1$ by the best combination of $X_2$ and $X_3$ can be ascertained by examining the error term, i.e., $X_1 - X'_1$ or $S_1(z_1 - z'_1)$. The sum of the squares for the errors divided by $N$ will yield the variance of the errors. The square root would correspond to the standard error of estimate. Let $S_{z_{1\cdot23}}$ be this error (in $z$ units), then

$$S^2_{z1\cdot23} = \frac{\Sigma(z_1 - z'_1)^2}{N}$$

$$= \frac{\Sigma(z_1 - \beta_2 z_2 - \beta_3 z_3)^2}{N}$$

$$= \frac{\Sigma z^2_1}{N} + \beta^2_2 \frac{\Sigma z^2_2}{N} + \beta^2_3 \frac{\Sigma z^2_3}{N} - \frac{2\beta_2\Sigma z_1 z_2}{N} - \frac{2\beta_3\Sigma z_1 z_3}{N} + \frac{2\beta_2\beta_3\Sigma z_2 z_3}{N}$$

$$= 1 + \beta^2_2 + \beta^2_3 - 2\beta_2 r_{12} - 2\beta_3 r_{13} + 2\beta_2\beta_3 r_{23}$$

which by algebraic manipulation reduces to

$$S^2_{z1\cdot23} = 1 - (\beta_2 r_{12} + \beta_3 r_{13}) \tag{11.4}$$

in terms of standard scores. Then $S^2_1$ times this would give the error variance for raw scores.

**Multiple r.** We next define the *multiple correlation coefficient* as the correlation between $z_1$ and the best estimate of $z_1$ from a knowledge of $z_2$ and $z_3$. In symbols,

$$r_{1\cdot23} = r_{z_1 z'_1} = \frac{\Sigma z_1 z'_1}{N S_{z_1} S_{z'_1}}$$

$$= \frac{\Sigma z_1(\beta_2 z_2 + \beta_3 z_3)}{N S_{z'_1}} \tag{11.5}$$

Note that, although $S_{z_1} = 1$, it does not follow that $S_{z'_1} = 1$. In order to evaluate this last $S$, we write

$$z_1 = z'_1 + z_{1\cdot23}$$

That is, we think of $z_1$ as being made up of two parts—that which we can estimate plus a residual. It can easily be shown that these two parts are independent of each other; hence by the variance theorem we have

$$S^2_{z_1} = S^2_{z'_1} + S^2_{z1\cdot23}$$

or

$$1 = S^2_{z'_1} + S^2_{z1\cdot23}$$

then

$$S^2_{z'_1} = 1 - S^2_{z1\cdot23}$$

But $S^2_{z1\cdot23}$ is nothing more than the variance of the prediction errors as given by (11.4); therefore

$$S_{z'_1} = \sqrt{\beta_2 r_{12} + \beta_3 r_{13}}$$

Then, by substituting in formula (11.5), we have

$$r_{1\cdot23} = \frac{\Sigma z_1(\beta_2 z_2 + \beta_3 z_3)}{N\sqrt{\beta_2 r_{12} + \beta_3 r_{13}}}$$

$$= \frac{\beta_2 \Sigma z_1 z_2 + \beta_3 \Sigma z_1 z_3}{N\sqrt{\beta_2 r_{12} + \beta_3 r_{13}}} = \frac{\beta_2 r_{12} + \beta_3 r_{13}}{\sqrt{\beta_2 r_{12} + \beta_3 r_{13}}}$$

$$= \sqrt{\beta_2 r_{12} + \beta_3 r_{13}} \qquad (11.6)$$

It can also be shown that

$$r_{1\cdot23} = \sqrt{\frac{r^2_{12} + r^2_{13} - 2r_{12}r_{13}r_{23}}{1 - r^2_{23}}}$$

We thus see that, as soon as the $\beta$s are determined, we can write the regression equation for predicting $z_1$ from $z_2$ and $z_3$ and can also specify the degree of correlation and calculate the error of estimate. This error obviously can be written from formulas (11.4) and (11.6) as

$$S_{1\cdot23} = S_1\sqrt{1 - r^2_{1\cdot23}} \qquad (11.7)$$

which is in terms of raw scores.

Formula (11.7) has been used frequently to define the multiple correlation coefficient. Stated explicitly,

$$r^2_{1\cdot23} = 1 - \frac{S^2_{1\cdot23}}{S^2_1} = 1 - S^2_{z1\cdot23}$$

Then, by substituting from (11.4), we again arrive at (11.6).

The student will note the similarity of formula (11.7) to the ordinary error of estimate for the bivariate situation. Thus the multiple correlation coefficient can be interpreted, in terms of reduction in the error of estimate, in exactly the same manner as the ordinary bivariate correlation coefficient. The only difference is that we are now determining the regression coefficients, or weights for two variables as a team, so as to get the best possible prediction of a third variable, whereas in the bivariate situation only one regression coefficient is necessary. A multiple correlation coefficient of .60 has, aside from minor qualifications to be discussed later, the same meaning in a predictive sense as an ordinary correlation of .60. Furthermore, the interpretation in terms of contribution to variance also holds for the multiple correlation coefficient; i.e., if causation can be assumed, it may be said that a multiple $r$ of .60 indicates that 36 per cent of the variance in the criterion or dependent variable can be attributed to variation in the two independent variables.

**Relative weights.** The question arises as to the relative importance of the two variables as contributors to variation in the criterion variable. The $B$ coefficients in the regression equation have, at times, been misinterpreted as indicating the relative contribution of the two independent variables. The reader need only be reminded that the two $B$ coefficients usually involve different units of measurement (one may be in terms of feet and the other in pounds); hence they are not comparable at all. If $B_2$ is numerically twice $B_3$, it does not follow that $X_2$ is twice as important as $X_3$. In order to get around this difficulty, we must think in terms of standard scores; these will be comparable, and hence the $\beta$ coefficients in the standard score form of the regression equation will be comparable.

Since

$$S^2_{z_1} = S^2_{z'_1} + S^2_{z_{1 \cdot 23}}$$

or

$$1 = S^2_{z'_1} + S^2_{z_{1 \cdot 23}}$$

and

$$1 - S^2_{z_{1 \cdot 23}} = r^2_{1 \cdot 23}$$

it follows that

$$r^2_{1 \cdot 23} = S^2_{z'_1}$$

That is, $r^2_{1 \cdot 23}$, which corresponds to the percentage of variance explained, is equal to $S^2_{z'_1}$, or the variance of the predicted standard scores. This variance could be determined by actually making $N$ predictions of $z_1$ from the $N$ pairs of values of $z_2$ and $z_3$ and then computing the $S$ for the distribution of these predicted values. This is not done in practice, since the value of this $S$ squared is $r^2_{1 \cdot 23}$, which is easily calculated once the $\beta$s have been determined.

But note that, since

$$z'_1 = \beta_2 z_2 + \beta_3 z_3$$

we can indicate the value of $S^2_{z'_1}$ as

$$S^2_{z'_1} = \frac{\Sigma(z'_1)^2}{N} = \frac{\Sigma(\beta_2 z_2 + \beta_3 z_3)^2}{N}$$

$$= \frac{\beta^2_2 \Sigma z^2_2 + \beta^2_3 \Sigma z^2_3 + 2\beta_2 \beta_3 \Sigma z_2 z_3}{N}$$

which becomes

$$S^2_{z'_1} = \beta^2_2 + \beta^2_3 + 2\beta_2 \beta_3 r_{23} \qquad (11.8)$$

In other words, the predicted variance, which corresponds to the "explained" variance, can be broken down into three additive components. We thus see that the relative importance of the variables $X_2$ and $X_3$ in "explaining" or "causing" variation in $X_1$ can be judged by the magnitude of the squares of the $\beta$ coefficients. The third term in formula (11.8) represents a joint contribution which, it will be seen, is a function of the amount of correlation between the two predicting variables.

In summary, it can be said that the fundamental problem in multiple correlation is that of obtaining the optimum weighting to be assigned to independent variables ($X_2$ and $X_3$) in predicting or explaining variation in a dependent variable, $X_1$. That is, we determine the value of $B_2$, $B_3$, and $A$ in the equation

$$X'_1 = B_2X_2 + B_3X_3 + A$$

so as to get the best possible estimate of $X_1$. This is resolved by working with the prediction equation in standard score form with $\beta$ coefficients. The value of each $\beta$ is determinable from the intercorrelations among the three variables. Once the $\beta$s are calculated, we can: (1) readily compute the $B$ coefficients needed in the raw score form of the prediction equation; (2) determine the value of the multiple correlation coefficient and the error of estimate; (3) ascertain the relative importance of the independent variables as predictors or, if causation can be assumed, as contributors to the variance of the dependent or criterion variable. It is important to note that the multiple correlation coefficient represents the maximum correlation to be expected between the dependent variable and a linearly additive combination of $X_2$ and $X_3$.

## MORE THAN THREE VARIABLES

Suppose that we have a dependent variable and four independent variables which might be used as predictors or which might be thought of as causes of variation in the dependent variable. The cause and effect, as opposed to concomitant, relationship among variables is a logical problem which must be faced by the investigator as a logician rather than as a statistician. Whether we resort to the multiple correlation technique as an aid in predicting or as an aid in analysis will depend entirely on the problem being attacked; the mechanical solution is the same, but the investigator must choose the interpretation which best suits his purpose.

For a five-variable problem, we need the constants in the regression or prediction equation,

$$X'_1 = B_2X_2 + B_3X_3 + B_4X_4 + B_5X_5 + A$$

which can be written in standard score form as

$$z'_1 = \beta_2z_2 + \beta_3z_3 + \beta_4z_4 + \beta_5z_5$$

As in the three-variable situation, the problem is that of determining the optimum values of the $B$s or the $\beta$s so as to get the best possible prediction of $X_1$ or $z_1$, i.e., so that

$$\frac{\Sigma(X_1 - X'_1)^2}{N}$$

or

$$\frac{\Sigma(z_1 - z'_1)^2}{N}$$

shall be as small as possible. The mathematical solution is easier by way of the standard score form of the regression equation. We have the function

$$f = \frac{\Sigma(z_1 - z'_1)^2}{N} = \frac{\Sigma(z_1 - \beta_2 z_2 - \beta_3 z_3 - \beta_4 z_4 - \beta_5 z_5)^2}{N} \quad (11.9)$$

which is to be minimized by assigning proper values to the $\beta$s. These values are obtained by taking the derivative of the function with respect to, and in order for, each of the $\beta$s. This will yield four derivatives which when set equal to zero will give us four equations involving the four unknown $\beta$s. These equations can then be solved as simultaneous equations in order to determine the values of the $\beta$s. The obtained $\beta$s will be such that the sum of the squares of $z_1 - z'_1$ will be the least possible; i.e., we will have the best possible estimate of $z_1$ from an additive combination of the four independent variables.

The student of the calculus can readily verify that the four equations obtained by taking derivatives of formula (11.9) will take the following form (when set equal to zero):

$$\left. \begin{array}{l} \beta_2 \quad + \beta_3 r_{23} + \beta_4 r_{24} + \beta_5 r_{25} - r_{12} = 0 \\ \beta_2 r_{23} + \beta_3 \quad + \beta_4 r_{34} + \beta_5 r_{35} - r_{13} = 0 \\ \beta_2 r_{24} + \beta_3 r_{34} + \beta_4 \quad + \beta_5 r_{45} - r_{14} = 0 \\ \beta_2 r_{25} + \beta_3 r_{35} + \beta_4 r_{45} + \beta_5 \quad - r_{15} = 0 \end{array} \right\} \quad (11.10)$$

These equations result from steps exactly parallel to those used for the three-variable problem. The four $\beta$s are unknowns, whereas, for any given batch of data, the $r$s take on specific numerical values.

The extension of multiple correlation to include any number of variables involves the same principles as utilized here for the three- and the five-variable problem. For $n$ variables, formula (11.2) becomes

$$z'_1 = \beta_2 z_2 + \beta_3 z_3 + \cdots + \beta_n z_n \quad (11.11)$$

The extension of (11.3) as the gross score equation should be obvious. Formula (11.6) for the multiple correlation coefficient becomes

$$r_{1 \cdot 23 \cdots n} = \sqrt{\beta_2 r_{12} + \beta_3 r_{13} + \cdots + \beta_n r_{1n}} \quad (11.12)$$

To solve for the unknown $\beta$s, the student may resort to any of the schemes given in algebra textbooks for solving simultaneous equations. One method is by way of determinants and Cramer's rule. The coefficients of the unknowns are the intercorrelations among the four independent variables, whereas the constants in these equations are the respective correlations of the dependent with the independent variables. In the application of Cramer's rule, these constants are thought of as being on the right-hand side of the equation, i.e., shifted to the right of the equality mark, with the consequent change of sign. The student should keep in mind, however, the fact that the original sign of any of the computed correlation coefficients must be considered.

Solution by Cramer's rule becomes quite tedious and burdensome for a problem involving more than four or five variables. Indeed, this determinantal solution is practically impossible for problems involving a large number of variables. Fortunately, there is available a simplified solution, but before turning to it, we would like to indicate some algebraic manipulations in terms of determinants.

It will be noted from the foregoing simultaneous equations that all the intercorrelations among the five variables are involved. These correlations can be conveniently arranged in a table, or in determinantal form. Thus we can define a major determinant as

$$D = \begin{vmatrix} 1 & r_{12} & r_{13} & r_{14} & r_{15} \\ r_{12} & 1 & r_{23} & r_{24} & r_{25} \\ r_{13} & r_{23} & 1 & r_{34} & r_{35} \\ r_{14} & r_{24} & r_{34} & 1 & r_{45} \\ r_{15} & r_{25} & r_{35} & r_{45} & 1 \end{vmatrix}$$

If we were to delete the first row and first column, the minor which remains would involve the intercorrelations among the four independent variables. This minor might be conveniently symbolized as $D_{11}$; i.e., we have deleted the column and the row which involve the subscript 1. If we were to delete the row which involves the subscript 1 and the column involving the subscript 2 throughout, we would symbolize the resulting minor as $D_{12}$.

Now it can be shown that

$$\beta_2 = \frac{D_{12}}{D_{11}}$$

or any $\beta$, say $\beta_p$, will be

$$\beta_p = (-1)^p \frac{D_{1p}}{D_{11}}$$

where the quantity $(-1)^p$ is an indicator of either a positive or a negative sign, but the ultimate sign of $\beta_p$ is also dependent on whether the numerical values of the determinants are positive or negative. It can also be shown that the multiple correlation coefficient can be written as a function of determinants, thus

$$r^2_{1 \cdot 2345} = 1 - \frac{D}{D_{11}}$$

The student who is interested in following a treatment of multiple correlation in terms of determinants is referred to T. L. Kelley's *Statistical method*.*

## NUMERICAL SOLUTION

On a desk calculator, the solution of the simultaneous equations for the unknown $\beta$s can best be accomplished by resort to Doolittle's method.

**Table 11.1. Schema for arranging $r$s for Doolittle solution**

| $X_2$ | $X_3$ | $X_4$ | $X_5$ | $X_1$ |
|-------|-------|-------|-------|-------|
| 1 | $r_{23}$ | $r_{24}$ | $r_{25}$ | $-r_{12}$ |
|  | 1 | $r_{34}$ | $r_{35}$ | $-r_{13}$ |
|  |  | 1 | $r_{45}$ | $-r_{14}$ |
|  |  |  | 1 | $-r_{15}$ |

This method is applicable to the solution of any simultaneous equations involving a major determinant which, like $D$, is symmetrical about the diagonal. It is also applicable to problems involving less or more than five variables. The first step is to write down the intercorrelations (coefficients of the unknown $\beta$s) in the form indicated in Table 11.1, in which the right-hand column contains the correlation of each variable with the criterion or dependent variable. Negative signs are attached to these coefficients· because, in essence, we are dealing with equations (11.10). Obviously, if the original sign of an $r$ were negative, it would be preceded by a plus sign in an arrangement like that in Table 11.1.

As a numerical example, we shall use data from the Minnesota study of mechanical ability. † The sample size is 100.

* Kelley, T. L., *Statistical method*, New York: Macmillan, 1924.
† Paterson, D. G., *et al.*, *Minnesota mechanical ability tests*, Minneapolis: University of Minnesota Press, 1930.

Let

$X_1$ = Criterion (mechanical performance-quality).
$X_2$ = Minnesota assembling test.
$X_3$ = Minnesota spatial relations test.
$X_4$ = Paper form board.
$X_5$ = Interest analysis blank.

Since the several means and standard deviations will be needed, these are recorded in Table 11.2.

**Table 11.2. Means and $S$s (Minnesota data)**

|   | $X_1$ | $X_2$ | $X_3$ | $X_4$ | $X_5$ |
|---|-------|-------|-------|-------|-------|
| $M$ | 14.94 | 127.56 | 1422.90 | 46.60 | 107.00 |
| $S$ | 2.09 | 25.32 | 296.39 | 19.45 | 18.00 |

Table 11.3 gives the Doolittle solution for the $\beta$ coefficients. Once these are known, the regression equation in raw score form can be written, and the multiple $r$ and the error of estimate can be determined. The table includes an indication of the calculation of these values. The student will have to study the schema of the Doolittle solution carefully in order to grasp the necessary steps. We shall not attempt a complete exposition of the steps since the procedure of each step is indicated in the left-hand side of the table. A few remarks, however, will be of aid to the student.

As already specified, the correlations are written down in an order corresponding to equations (11.10) except that values to the left and below the diagonal are omitted. The first thing we do is to set up a check column. The first entry, 1.92, is obtained by summing, algebraically, the first row of correlations (including the diagonal 1.00); the second figure, 2.12, is the sum of the second row plus .56; the third entry, 1.99, is the sum of the third row plus .49 and .63; and the 1.63 is the sum of the fourth row plus .42, .46, and .39. The rule being followed should now be obvious: the $j$th entry in the check column is obtained by summing the 1.00 in the $j$th row with the values above it and to its right. The student should satisfy himself that this is equivalent to summing the correlations for the respective equations in (11.10). Since the check column will provide, at intervals, an automatic check on our computations, this summing should be done at least twice to insure accuracy.

Line 1 of the solution is obtained by copying down line $a$, the first row of $r$s; and line 2 consists of the line 1 values with the signs changed. The second part of the solution begins with line 3, which is obtained by

### Table 11.3. Computation of multiple $r$

|  | $X_2$ | $X_3$ | $X_4$ | $X_5$ | $X_1$ | ck |
|---|---|---|---|---|---|---|
| (a) | 1.00 | .56 | .49 | .42 | −.55 | 1.92 |
| (b) |  | 1.00 | .63 | .46 | −.53 | 2.12 |
| (c) |  |  | 1.00 | .39 | −.52 | 1.99 |
| (d) |  |  |  | 1.00 | −.64 | 1.63 |
| (1): line (a) | 1.00 | .56 | .49 | .42 | −.55 | 1.92 |
| (2) | −1.00 | −.56 | −.49 | −.42 | .55 | −1.92 |
| (3): line (b) |  | 1.000 | .63 | .46 | −.53 | 2.12 |
| (4): (1)(−.56) |  | −.314 | −.274 | −.235 | .308 | −1.075 |
| (5): (3) + (4) |  | .686 | .356 | .225 | −.222 | 1.045 ck |
| (6): (5)(−1/.686) |  | −1.000 | −.519 | −.328 | .324 | −1.524 ck |
| (7): line (c) |  |  | 1.000 | .39 | −.52 | 1.99 |
| (8): (1)(−.49) |  |  | −.240 | −.206 | .270 | −.941 |
| (9): (5)(−.519) |  |  | −.185 | −.117 | .115 | −.542 |
| (10): (7) + (8) + (9) |  |  | .575 | .067 | −.135 | .507 ck |
| (11): (10)(−1/.575) |  |  | −1.000 | −.116 | .235 | −.882 ck |
| (12): line (d) |  |  |  | 1.000 | −.64 | 1.63 |
| (13): (1)(−.42) |  |  |  | −.176 | .231 | −.806 |
| (14): (5)(−.328) |  |  |  | −.074 | .073 | −.343 |
| (15): (10)(−.116) |  |  |  | −.008 | .016 | −.059 |
| (16): (12) + (13) + (14) + (15) |  |  |  | .742 | −.320 | .422 ck |
| (17): (16)(−1/.742) |  |  |  | −1.000 | .431 | −.569 ck |

Back solution

| From (17) | $.431 = \beta_5$ |
|---|---|
| From (11) | $(.431)(−.116) + .235 = \beta_4 = .185$ |
| From (6) | $(.185)(−.519) + (.431)(−.328) + .324 = \beta_3 = .087$ |
| From (2) | $(.087)(−.56) + (.185)(−.49) + (.431)(−.42) + .55 = \beta_2 = .230$ |

Final checks

$$(.230)(1.00) + (.087)( .56) + (.185)( .49) + (.431)( .42) − .55 = .000$$
$$(.230)( .56) + (.087)(1.00) + (.185)( .63) + (.431)( .46) − .53 = .001$$
$$(.230)( .49) + (.087)( .63) + (.185)(1.00) + (.431)( .39) − .52 = .001$$
$$(.230)( .42) + (.087)( .46) + (.185)( .39) + (.431)(1.00) − .64 = .000$$

From formula (11.3)

$$B_2 = (.230)\frac{2.09}{25.32} = .0190, \qquad B_3 = (.087)\frac{2.09}{296.39} = .0006,$$

$$B_4 = (.185)\frac{2.09}{19.45} = .0199, \qquad B_5 = (.431)\frac{2.09}{18.00} = .0500, \qquad A = 5.40$$

Then

$$X'_1 = .0190X_2 + .0006X_3 + .0199X_4 + .0500X_5 + 5.40$$
$$r^2_{1\cdot2345} = (.230)(.55) + (.087)(.53) + (.185)(.52) + (.431)(.64) = .54465$$
$$r_{1\cdot2345} = .738, \qquad S_{1\cdot2345} = 2.09\sqrt{1 − (.738)^2} = 1.40$$

copying down the $b$ row of correlations. Line 4 is obtained by multi-plying entries in line 1 by −.56, which figure is found in line 2 directly above the 1.000 of line 3. As indicated at the left, line 5 results from summing lines 3 and 4, i.e., 1.000 + (−.314) equals .686, etc.

At this point we have our first automatic check: summing line 5 across should yield 1.045, already obtained by vertical summing of values in the check column. To be a satisfactory check, these two sums should agree within limits consistent with errors imposed by rounding off to three decimal places. Acceptable discrepancies will be of the order $\pm.001$, $\pm.002, \cdots \pm.005$, seldom larger.

Line 6 is obtained by multiplying line 5 by the negative reciprocal of its first entry. The correctness of the reciprocal used is evidenced by the fact that, when multiplied by .686, unity results. The ck attached to $-1.524$ indicates that summing the entries in line 6 yields the same value as 1.045 multiplied by the negative reciprocal of .686, thus providing a further check. This completes the second part of the solution.

The third part begins with a copying of row $c$ of the correlation table. The student should now be able to follow the steps; in particular, he should note that a multiplier is secured from the last line of each preceding part of the solution; that each multiplier is applied in turn to the values in the line just above it; that, when all such multipliers have been utilized, the lines are summed (summing across again provides a check), and the resulting line is, as before, multiplied by the negative reciprocal of its first entry, thus completing the third part of the solution.

The fourth part involves similar operations. If we had five independent variables, we would proceed in like fashion, with an additional or fifth part. The schema can be extended to any number of variables.‡. There will be as many parts to the solution as there are independent variables. The last part always consists of three columns of figures, and the bottom figure in the middle column *is* the value for $\beta_n$. In our example, $\beta_n = \beta_5 = .431$.

The other $\beta$s are determined by a "back"solution, which always involves a substitution of the value or values already found into the last line of the various parts (lines 11, 6, and 2 in our illustration). This back solution is given in Table 11.3. As a final check on all the computations, the four $\beta$s obtained must be substituted into the four simultaneous equations with which we began. This check appears next in Table 11.3.

In order to put our results into useful form, we ordinarily require the multiple regression equation in raw score form, and for this we need the $B$ coefficients and $A$ as called for in formula (11.3) extended for more variables. To get the multiple correlation coefficient, the $\beta$s and appropriate $r$s are substituted in formula (11.12), and from (11.7) we obtain the standard error appropriate for judging the accuracy of predictions made by the calculated regression equation. Table 11.3 includes these additional values.

If the problem involves analysis rather than prediction, there is no

‡ For more than five or six variables, the computations are more economically accomplished by electronic computers.

need to set up the regression equation or calculate the error of estimate. Appropriate interpretations will depend on the $\beta$s and $r^2_{1 \cdot 2345 \cdots n}$.

## SAMPLING ERRORS

The classical formula for the standard error of a multiple correlation involving $n$ variables is

$$S_{r_{1 \cdot 23 \cdots n}} = \frac{1 - r^2_{1 \cdot 23 \cdots n}}{\sqrt{N}} \tag{11.13}$$

This formula came into being before much was known about the random sampling distribution of multiple correlation coefficients, which we will now designate as $R$ without the necessary subscripts. It was falsely presumed that the distribution is normal, hence that the formula could be used to test not only whether an obtained $R$ deviates significantly from zero but also whether it deviates from some a priori specified nonzero value. Actually the sampling distribution of $R$ is a complicated affair, even under null conditions, so that it is not safe to use (11.13) unless $N$ is indeed very large. Even then it is questionable.

Consider the null condition: multiple $R$ is zero in a defined population. The only way that $R_{pop}$ can equal zero is for *all* $r_{1j}, j = 2 \cdots n$, to equal zero in the population. This proposition does not depend on the intercorrelations among the predictors. If, for example, $r_{12}$ is nonzero, $\beta_2$ can be zero only if $r_{12} - r_{13}r_{23} = 0$. If this were so, $r_{13}$ would have to be nonzero. If $r_{13}$ is nonzero, can $\beta_3$ be zero? Yes, if $r_{13} - r_{12}r_{23} = 0$. But aside from the trivial (nonexistent in practice) condition that $r_{23} = 1$ exactly, it is impossible to have $r_{12} - r_{13}r_{23}$ and $r_{13} - r_{12}r_{23}$ simultaneously equal to zero with either $r_{12}$ or $r_{13}$ or both being nonzero. Accordingly, at least one of the $\beta$s will be nonzero, and the value of $R$ will exceed zero. $R$ is always positive and higher numerically than the highest $r_{1j}$ value.

Now suppose the condition that all $r_{1j}$ are zero in the population, which means that the $R_{pop}$ must also be zero. When we draw a sample, it would indeed be a rarity to find all the observed $r_{1j}$ to be exactly zero. It follows that for a sample the computed $R$ will rarely if ever be zero, and since $R$ is always positive, the mean of sample values of $R$ will certainly not tend to equal the population value, $R_{pop} = 0$. It has been shown mathematically that for $n$ variables and the null condition the mean value of sample $R^2$s is $(n - 1)/(N - 1)$. For example, when $n = 11$ and $N = 101$, the sampling distribution of $R^2$ will center at .10, or $R$ on the average will be nearly $\sqrt{.10} = .316$ even though $R_{pop}$ is zero. A corollary is obvious: an $R^2$ must deviate upward significantly from $(n - 1)/(N - 1)$, rather than from zero, in order to possess statistical significance.

Any student who is puzzled as to how under null conditions (meaning that all $r_{1j} = 0$ in the population) a multiple based on 11 variables and $N = 101$ will tend to have a value of .316 (chance will produce still higher values—about 5 per cent will exceed .42) should ponder the fact that with $n - 1 = 10$ predictors and $N = 101$ any one or more of the $r_{1j}$ can easily exceed .15 or .20 by chance; several such chance departures from zero can build up to a "sizable" $R$.

An adequate method for testing the significance of a sample $R$, with due allowance for the foregoing expected mean value, is by the $F$ distribution used in the analysis of variance. The $F$ test can also be utilized to test the significance (from zero) of the beta coefficients. These tests are discussed near the end of Chapter 15.

The fact that the mean of sample $R^2$s under null conditions is $(n - 1)/(N - 1)$ instead of zero certainly implies that an $R$ computed by (11.12) is a biased estimator. Perhaps we can see intuitively what is going on here by considering geometrically an extreme case. For the ordinary bivariate correlation, it is evident on a moment's reflection that if $N = 2$ the correlation between the two variables must be perfect positive or perfect negative (it would be indeterminate if for either variable the two scores were the same); the regression line will pass through both plotted points on the scatter diagram, which means that there will be no error of estimate (for the two cases). Now for the three-variable situation and $N = 3$, it is always possible to pass a plane through all three plotted points in the three-dimensional space, hence no error of estimate and $R = 1$. In general, if $n = N$ we will always get a perfect multiple correlation. Obviously $N$ must be greater than $n$ before meaning can be attached to $R$. As $n$ approaches $N$, $R$ always approaches unity. The difference between $N$ and $n$ must have something to do with degrees of freedom for the sum of squares about the regression plane. This in turn suggests that perhaps in calculating $R$, as an estimator, we should somehow bring in the number of degrees of freedom.

When we examine formula (11.12) it is not readily seen that degrees of freedom can be utilized. Suppose that we think of $R$ as being calculated by formula (11.7) which leads to

$$r^2_{1 \cdot 23 \cdots n} = 1 - S^2_{1 \cdot 23 \cdots n}/S^2_1 \tag{11.7a}$$

with $S^2_{1 \cdot 23 \cdots n}$ calculated by (11.4) extended to the $n$ variable case and $S^2_1$ as a multiplier to get the error of estimate in terms of raw scores, i.e., on the $X_1$ scale rather than the standard score scale. In the foregoing formula for $R$ we have the ratio of two biased estimates of variances. We can readily change these to unbiased estimates by recalling the general proposition that $\Sigma x^2 = NS^2_x$ always, hence the unbiased estimate becomes

$$s^2 = \frac{\Sigma x^2}{N-1} = \frac{NS^2}{N-1}$$

What of the necessary degrees of freedom for converting our two $S^2$ values to unbiased estimators? That for the estimate of the variance for the criterion, or $X_1$, is old hat: $N-1$. That for the error of estimate variance will be $N-n$ because we are dealing with deviations from the regression plane the equation of which involves $n$ constants computed from the data—the intercept (or $A$) and $n-1$ $B$ coefficients.

Accordingly, $s^2_1 = NS^2_1/(N-1)$ and $s^2_{1\cdot23\cdots n} = NS^2_{1\cdot23\cdots n}/(N-n)$ $= NS^2_1(1 - r^2_{1\cdot23\cdots n})/(N-n)$. When we substitute these unbiased estimates into formula (11.7a) for the multiple correlation coefficient, we will have changed our estimate of the coefficient. Let's designate the new estimate as $\hat{R}$, with $R$ as the original estimate:

$$\hat{R}^2 = 1 - \frac{NS^2_1(1-R^2)/(N-n)}{NS^2_1/(N-1)} = 1 - (1-R^2)\frac{N-1}{N-n} \quad (11.14)$$

This estimate is usually referred to as the "shrunken" multiple $R$ because it is always less than the original $R$ based on biased variances or computed by (11.12), which give identical values. The procedure is sometimes referred to as correction for shrinkage. Obviously, if $N$ is very large relative to $n$, the correction will not shrink the estimate very much. But when $n$ approaches $N$ the story is quite different. As an example, consider the recently reported finding that $R = .947$ between IQs and a battery of motor skills tests. This .947 is startlingly high. Any explanation? Well, $N = 40$ and there were 36 motor skills tests, so we have an $n = 37$ variable situation, a situation in which the use of unbiased estimators, as in (11.14), may make a difference. By direct substitution we have

$$\hat{R}^2 = 1 - (1 - .947^2)\frac{39}{3} = 1 - 1.3415 = -.3415$$

which immediately seems to be as ridiculous as the starting .947. Of course, the square root of a negative number is said to be an "imaginary" number, so we could have $R$ as such an imaginary. Why such an absurdity? To answer this we ask what we would expect $R^2$ (not $\hat{R}^2$) to be, under null conditions, when we have $N = 40$ and $n = 37$. From the previously given expectation we see that under the null condition the mean of sample $R^2$s for the given $N$ and $n$ is $(n-1)/(N-1) = 36/39 = .9230$. But the original $R^2 = .947^2 = .8968$ is *below* the expected value. This in and of itself does not help us in an effort to interpret $\hat{R}^2 = -.3415$, but it does say that the estimate $\hat{R}^2$ can be (and will be) a negative number when the observed $R^2$ is less than the mean expected value of $(n - 1)/(N - 1)$. Obviously, if $R^2$ does not exceed the value expected under the null condition, it will not possess statistical significance, a fact that can be verified by applying the $F$ test. Thus we are spared the agony of trying to explain an estimate that, being negative for the square of a number, defies any statistical or practical explanation.

Even so, it may be worthwhile to re-examine $\hat{R}^2$ as an estimator. The factor in (11.14) that leads to the "shrunken" estimate is, of course, $(N - 1)/(N - n)$. Now

$$\frac{N - 1}{N - n} = \frac{1}{\dfrac{N - n}{N - 1}} = \frac{1}{\dfrac{N - 1 - n + 1}{N - 1}} = \frac{1}{1 - \dfrac{n - 1}{N - 1}} = \frac{1}{1 - \bar{R}^2}$$

in which $\bar{R}^2$ is used to designate the expected value of $R^2$ under the null condition; i.e., the mean of the random sampling distribution of $R^2$s. When we substitute into (11.14) we have

$$\hat{R}^2 = 1 - (1 - R^2)\frac{1}{1 - \bar{R}^2}$$

from which we see that when a sample $R^2 = \bar{R}^2$, the value of $\hat{R}^2$ is exactly zero. That is, the mean of the sample values of $R^2$ transforms into zero, which is $R^2{}_{pop}$ under the null condition. Stated differently, the random sampling distribution of $\hat{R}^2$ as an estimator centers at zero, the population value for the multiple correlation coefficient squared; hence it follows that under the null condition $\hat{R}^2$ is an unbiased estimator. With a sampling distribution centering at zero, it is not at all surprising to have estimates that are negative in sign.

It is not argued that under nonnull conditions formula (11.14) will yield an unbiased estimate. It won't, but it will yield a far less biased estimate than $R^2$, hence it should be used unless $n$ is very small relative to $N$. For example, when $N = 100$ and $n = 11$ formula (11.14) yields an estimate that is trivially biased—of the order .005 too low as an estimate of $R^2{}_{pop}$. It is of interest to note that the betas, as computed, are unbiased estimates of their respective population values.

## CAUTIONS AND REMARKS

As already indicated, there are two principal uses for the multiple correlation technique: (1) it yields the optimum weighting for combining a series of variables in predicting a criterion and provides an indication of the accuracy of subsequent predictions; (2) it permits the analyzing of variation into component parts. There are certain more or less obvious pits into which the unwary user of the multiple regression and correlation method may fall. For example, it is possible to write a multiple regression equation for predicting school achievement $(X_1)$ from a knowledge of age $(X_2)$ and mental age $(X_3)$. In standard score form it might be $z'_1 = .27z_2 + .67z_3$, from which it might be inferred that school achievement depends on age to a certain extent but on mental age to a greater extent. However, it is entirely possible to argue that mental age depends partly on school achievement. One could also use the same data to write the regression for age on mental age and school achievement; thus $z'_2 = .56z_1 + .06z_3$, from which the unwary might conclude that age depends on school achievement and mental age.

Multiple correlation may be particularly deceptive when we have available several variables, each of which yields a rather low correlation with the criterion and from which those yielding the higher correlations with the criterion are selected for the prediction equation. Such selecting tends to capitalize on correlations which might be high because of sampling fluctuations. For example, the author was once requested to compute the multiple $r$ for an 11-variable problem. None of the 10 variables showed a very high correlation with the criterion, the highest being .27. The resulting multiple was .44, which was statistically significant for the sample of 89 cases. When it was learned that 10 variables out of 40 had been selected as the most promising, i.e., because they showed the highest correlations with the criterion, the real significance of the multiple $r$ of .44 was questioned. That it really was misleading was clearly evidenced by the fact that for a second and similar sample the variable originally yielding the highest $r$ (.27) now produced an $r$ of $-.11$. That is, the supposedly best single predictor was actually of very doubtful value, and this, coupled with a tendency for the next highest $r$s to drop appreciably, meant that predictions by the regression equation could not be as good as was inferred from the multiple of .44.

The misleading value of a multiple $R$ based on predictor variables selected, from a pool of possible predictors, because of their relatively high $r$s with the criterion can be comprehended by a somewhat different approach. Recall that $R$ is the correlation that holds between the criterion and the predicted values for the individuals in the sample upon which the optimal weightings in the regression equation were obtained. The question can be raised as to whether or not predictions by this regression equation will show the same degree of correlation with the criterion for a second random sample from the population under consideration. This is, of course, a highly relevant question when, as is usual, the prediction equation is to be used with future cases.

Now consider the preceding example for which the "best" single predictor yielded a correlation of .27 with the criterion. From this we would expect said variable to have a positive regression weight, hence a high score would contribute to a high predicted value. Now the fact that this variable produced a negative $r$ of .11 on the second sample definitely leads to doubt that a high score thereon is either indicative or predictive of success on the criterion.

Another way of selecting variables from a pool would be to determine (easily done with a high-speed electronic computer) the regression equation, in standard score form, for the entire pool of potential predictors and then proceed to select as a subset, those predictors with the largest beta weights (or drop those with the smallest betas). A new regression equation based on the retained predictors will lead to an $R$ which, presumably, indicates the potential predictive value of the variables so selected and weighted. This procedure will tend to capitalize on chance high betas in the regression equation (for the pool) and yield a highly misleading $R$ for the subset.

When predictors have been chosen because they show promise for a sample at hand, it is imperative that we proceed to a second sample in order to secure a more dependable estimate of the predictive worth of the selected variables. The check may be performed by using the regression equation weights (or regression equation) based on the first sample as a basis for calculating predicted values for the individuals in the second sample. The correlation of these predicted values with the criterion measures of these second-sample individuals will tell the story as to the worth or worthlessness of the multiple regression equation. This process is referred to as *cross-validation*. The drop in validity from the first to the second (or cross-validation) sample should not be confused with the shrinkage in an $R$ which is calculable by formula (11.14). These are two entirely different issues: cross-validation allows for the capitalization on chance that tends to occur when there is *selection* of predictors based on statistics derived from the first, or original sample; the predictable shrinkage allows for the failure to have $N$ large relative to $n$, thus making it more important to have unbiased estimates of criterion and residual variances. If the selection of predictors does not depend on statistics based on the (first) sample at hand, cross-validation is not necessary.

The usually advocated procedure in cross-validating is to calculate a predicted score for each person in the second sample by substituting his scores into the regression equation derived from the first sample (a tedious job), then compute the correlation between these predicted values and the criterion scores. This will certainly provide an estimate as to the usefulness of the predictive index obtained by weighting the variables (predictors selected at first-sample stage) according to the regression coefficients based on the original sample. But are the weights in this regression the best possible to use when making predictions for new cases? Recall the situation: from among, say, $m_1$ possible predictors, $m_2$ have been selected either because they yielded relatively high $r$s with the criterion or because they had the higher beta weights in a regression equation involving all $m_1$ possible predictors. This selection depends on statistics obtained from a first sample; then weights for a second regression equation involving the $m_2$ selected variables are calculated from statistics (correlations) based on the selfsame sample that was used in the selection of predictors. It should be obvious that sample estimation of the optimal weights for some of the chosen predictors will be biased upward if our selection of predictors has in any way predetermined that they be ones with relatively high weights; and any selection that may capitalize on chance high $r$s (or $\beta$s) does just that.

We thus arrive at the conclusion that the usually advocated procedure should be replaced by a different strategy: use the second (cross-validation) sample as a basis for deriving a new set of regression weights and $R$ for the $m_2$ selected predictors. These weights derived from the second sample will not only be optimal for the second sample, they will also constitute far better (and unbiased) estimates of the unknown population regression weights in that at this second stage there is no selection that can capitalize on chance. Correction of the obtained $R$ for shrinkage is in order unless the second sample $N$ is very large relative to $m_2$. There is no need for further cross-validation simply because no selection of predictors is involved at the second-sample stage.

We make three further observations.

1. Cross-validation is not replication, and vice versa.

2. Although the need for cross-validation is obvious in connection with the predictive worth of a regression equation based on the type of selection being discussed, it should be noted that selection will also disrupt the outcome of a study designed to analyze the variance of a dependent variable into sources (independent variables). It cannot be safely said that the square of the $R$, based on the selected variables and calculated for the sample used in selecting the variables, represents the proportion of the variance that can be explained. The variables selected at the first-sample stage should be used with a new sample from which the correlations necessary for calculating a new set of beta coefficients and a new $R$ can be obtained. This $R$, corrected for shrinkage, will provide a far better basis for saying what proportion of the variance is attributable to the (selected) independent variables. Since in this case we are not concerned with predictive validity, it would be a misnomer to call the second stage

cross-validation, although the procedure is exactly the same as in cross-validating a prediction equation.

3. Cross-validation is needed in one situation that does not involve multiple regression. When, in test construction, from a pool of items we select those that either discriminate between two groups or show a relationship with a criterion, a capitalization on chance is very apt to occur so that the resulting test will yield a deceptively large differentiation between the two groups (used in selecting the items) or an illusory, high correlation with the criterion for the sample upon which item selection was based. Hence the need for cross-validating on additional groups or on a second sample when a criterion is involved.

Nothing has been said as yet concerning the principal assumption and consequent limitation in the use of multiple regression equations, namely, that regressions for the first-order correlations must be linear. There are methods for handling multiple correlation for curvilinear regressions. The reader is referred to M. Ezekiel's *Methods of correlation analysis*.§

It is not obvious from our discussion that, in general, the increase in the multiple correlation which results from adding variables beyond the first five or six is very small. This phenomenon of diminishing returns would not, of course, operate if we were to find an additional variable which correlated much more highly with the criterion than any of those already utilized.

Another fact which may not be apparent to the reader is that we can expect the multiple $r$ to be higher when the intercorrelations among the predictors are low instead of high. This point can be easily demonstrated by computing the multiples for, say, $r_{12} = .50$, $r_{13} = .50$, and varying values for $r_{23}$.

An interesting paradox of multiple correlation and an exception to the fact mentioned in the previous paragraph is that it is possible to increase prediction by utilizing a variable which shows no, or low, correlation with the criterion, *provided* it correlates well with a variable which does correlate with the criterion. Thus, if $r_{12} = .400$, $r_{13} = .000$, and $r_{23} = .707$, the regression equation will be $z'_1 = .800z_2 - .566z_3$, and $r_{1 \cdot 23}$ will equal .566. It is thus seen that, when $z_3$ is combined with $z_2$, an appreciable gain in prediction occurs even though when taken alone $z_3$ is worthless as a predictor of $z_1$.

Such a variable has been termed a "suppressant." We do not quickly see just how a suppressant variable, showing no correlation with the criterion, can increase the accuracy of prediction. Perhaps this point can be explained by reasoning by way of the notion that correlation can be

§ Ezekiel, M., *Methods of correlation analysis*, New York: John Wiley and Sons, 1959.

thought of in terms of common elements. Suppose that $X_1$ is composed of 10 elements, $X_2$ of 10, $X_3$ of 5, and suppose that $X_1$ and $X_2$ have 4 elements in common, $X_2$ and $X_3$ have 5 elements in common, and $X_1$ and $X_3$ have no overlapping elements. Diagrammatically, the variables and elements would be

$$
\begin{array}{cc}
\underline{\quad\quad x_1 \quad\quad} & \underline{\quad x_3 \quad} \\
\multicolumn{2}{c}{\underline{a\,a\,a\,a\,a\,a\,b\,b\,b\,b\,c\,d\,d\,d\,d}} \\
\multicolumn{2}{c}{x_2}
\end{array}
$$

By substituting in the common element formula (9.15) for correlation, we find $r_{12} = .400$, $r_{13} = .000$, $r_{23} = .707$. These lead to $z'_1 = .800z_2 - .566z_3$, and $r_{1\cdot23} = .566$. Variable $X_3$ has a negative regression weight, i.e., by use of $X_3$ something is being subtracted or suppressed. As set up here for illustrative purposes, all the elements of $X_3$ are contained in $X_2$; these elements are not related to $X_1$ and hence their presence in $X_2$ must tend to lower the correlation between $X_1$ and $X_2$; if these elements could be suppressed, the correlation between $X_1$ and $X_2$ minus the irrelevant (so far as $X_1$ is concerned) elements of $X_2$ should be higher than $r_{12}$. Actually, if we think of the "$d$" elements of the diagram as being nonexistent, we would have variation in $X_2$ dependent on only 5 elements, 4 of which overlap with $X_1$. The correlation between $X_1$ and the abridged $X_2$ would be $4/\sqrt{10(5)}$ or .566, which has exactly the same value as the multiple $r$ obtained previously. This exact correspondence to $r_{1\cdot23}$ will be obtained only when all the $X_3$ elements are contained in $X_2$. If $X_3$ contains other elements, its use as a suppressant will aid in predicting $X_1$, but the resulting $r_{1\cdot23}$ will not correspond to an $r$ deducible from the common element formula. The reason for this is left as an exercise.

The student, by resort to the notion of common elements, may secure a better understanding of the proposition that a higher multiple is obtainable when the correlations with the criterion are high and the correlations between the predictors low or zero. The reader should be warned, however, that such a condition is hard to realize in practice, as is also the finding of variables which will qualify as suppressants.

## NOTE ON NOTATION

The symbol $r_{1\cdot23}$ has been used to represent the correlation (multiple) between $X_1$ and the best combination of $X_2$ and $X_3$. This should not be confused with $r_{12\cdot3}$, which indicates the correlation (partial) between $X_1$ and $X_2$ with the effect of $X_3$ ruled out or held constant. The symbol $S_{y\cdot x}$, it will be recalled, stood for the standard error of estimate of $Y$ as estimated from $X$; $S_{1\cdot2}$ would be the error of $X_1$ when estimated from $X_2$; and $S_{1\cdot23}$ would be the standard error of estimate of $X_1$ when estimated from $X_2$ and $X_3$ by means of the multiple regression equation.

In the foregoing discussion, $\beta_2$ has been used as the symbol for the regression weight of $X_2$. A more formal, albeit cumbersome, notation would be $\beta_{12\cdot345}$, which would be read as the regression of $X_1$ on $X_2$, i.e., the coefficient for $X_2$, when used in combination with $X_3$, $X_4$, and $X_5$. It is not an accident that the subscript pattern resembles that for the partial correlation coefficient. If we were dealing with a three-variable problem, $\beta_2$ could be written as $\beta_{12\cdot3}$. This notation really means that we have the net regression of $X_1$ on $X_2$ when $X_3$ is held constant. Hence the coefficients are sometimes spoken of as *partial* regression coefficients. As a matter of fact, these partial or multiple regression coefficients can be computed by way of partial correlation coefficients, but the method is not nearly so straightforward and self-checking as the Doolittle procedure.

The meaning of the notation can be tied more specifically to the meaning of a beta value. Suppose we deal with standard scores and write the equation for predicting $z_1$ from $z_3$ as $z'_{13} = r_{13}z_3$. Then $z_{1\cdot3} = z_1 - z'_{13}$ $= z_1 - r_{13}z_3$ represents the error, or residual. Now none of these residual values can be predicted by further use of $z_3$ simply because, as we showed in deriving formula (9.12), $r_{z_3z_{1\cdot3}} = 0$. Suppose we find another variable, $z_2$, that might be useful in predicting the residuals represented by $z_{1\cdot3}$. First it will be noted that that part of $z_2$ which can be estimated from $z_3$ will be uncorrelated with $z_{1\cdot3}$. This correlation can be written as $r_{z_{1\cdot3}z'_{23}}$ $= r_{z_{1\cdot3}(r_{23}z_3)}$ since $z'_{23} = r_{23}z_3$. We have already shown that $z_{1\cdot3}$ and $z_3$ are uncorrelated; multiplying $z_3$ by the constant $r_{23}$ will not change the correlation of $z_3$ with any other variable, hence $r_{z_{1\cdot3}z'_{23}} = 0$. Since that part of $z_2$ which is predictable from $z_3$ is useless in predicting $z_{1\cdot3}$ we next raise the question as to whether that part of $z_2$ not predictable from $z_3$ will help. Thus we turn to $z_{2\cdot3}$ as a possible predictor of $z_{1\cdot3}$ and proceed to write the needed regression equation,

$$z'_{1\cdot3} = r_{z_{1\cdot3}z_{2\cdot3}} \frac{S_{z_{1\cdot3}}}{S_{z_{2\cdot3}}} z_{2\cdot3}$$

The $r$ in this is a partial correlation coefficient and the two $S$s are errors of estimate, hence we have

$$z'_{1\cdot3} = \frac{r_{12} - r_{13}r_{23}}{\sqrt{1 - r^2_{13}}\sqrt{1 - r^2_{23}}} \cdot \frac{\sqrt{1 - r^2_{13}}}{\sqrt{1 - r^2_{23}}} z_{2\cdot3}$$

$$= \frac{r_{12} - r_{13}r_{23}}{1 - r^2_{23}} z_{2\cdot3} = \beta_{12\cdot3}z_{2\cdot3}$$

In words, $\beta_{12\cdot3}$ as a regression coefficient is seen to involve the prediction of that part of $z_1$ which is independent of $z_3$ by that part of $z_2$ which is independent of $z_3$. The fact that the partial correlation coefficient comes into the picture ties in with the designation of the weights in a multiple regression equation as partial regression coefficients.

## FURTHER NOTES ON INTERPRETATIONS

Given that $n - 1$ predictors have been selected independently of the data on which an $R^2$ has been calculated, we may regard $\hat{R}^2$ obtained by (11.14) as a very nearly unbiased estimate of the proportion of the criterion variance predictable by the $n - 1$ variables. The error of estimate for future individual predictions can be taken as $s_1\sqrt{1 - \hat{R}^2}$. Now, groupwise, the set of regression weights derived from sample A will not be optimal for sample B; a new set based on sample B will not be optimal for sample C; and so on, ad infinitum. To avoid this impasse we simply regard the $\beta$s and $\hat{R}^2$, based on the largest available single sample, as providing satisfactory estimates of population values.

The *relative* importance of the $n - 1$ variables can be judged by the $\beta$s or by ascertaining the reduction in $R^2$ that results from dropping out, in turn, each predictor while retaining the remaining $n - 2$ predictors; for example, if $n = 5$, we would have the following differences between $R^2$s (note the subscripts):

$$r^2_{1\cdot2345} - r^2_{1\cdot345}, \; r^2_{1\cdot2345} - r^2_{1\cdot245}, \; r^2_{1\cdot2345} - r^2_{1\cdot235}, \; r^2_{1\cdot2345} - r^2_{1\cdot234}$$

as the net contributions of the predictors within the given $n$-variable system. Since the foregoing differences are needed for testing the significance of the $\beta$s, it is not surprising that these differences are nearly proportional to $\beta^2_2, \beta^2_3, \beta^2_4$, and $\beta^2_5$. Therefore it is safe to gauge the relative importance of the $n - 1$ predictors in terms of the $\beta^2$ values, irrespective of the type of "joint" contribution indicated in formula (11.8).

It must be clearly understood that in no sense does multiple regression analysis permit us to specify the *absolute* contribution of a source variable. Also, the relative effect of, say, $X_3$ and $X_5$ is a function of the other predictors that happen to have been included in the analysis. However, none of these interpretive considerations forces any qualifications on the use of a multiple regression equation as a basis for making practical predictions.

# Chapter 12

## OTHER CORRELATION METHODS

The product moment correlation measure is applicable only when the two variables are graduated, is restricted by the assumption of linearity of regression, and needs careful qualifying if either or both variables yield skewed distributions. There are, therefore, many problems for which it is inappropriate. In general, the majority of the situations which are met in practice can be handled by some type of correlational technique.

There are no general rules to follow in the case of variables yielding skewed distributions. Frequently, we can use a logarithmic transformation of such a variable and thereby secure scores which are at least approximately normal; or we may deliberately normalize the distribution by converting the raw scores into $T$ scores. When we consider the arbitrary units involved in most psychological measurement, such a procedure would seem not only permissible but also defensible in that the correlational description of the relationship need not be qualified because of skewness.

The situations arising most frequently in practice, for which measures of correlation are apt to be needed, can be subsumed under the following six headings: (1) graduated measures for one variable, dichotomized or two-category information for the second variable; (2) both variables dichotomized; (3) three or more categories for one variable and two or more for the second; (4) three or more categories for one variable and a graduated series of measures for the other; (5) both variables graduated, with curvilinear relationship; (6) when data are rank-orders.

An estimate of the degree of correlation for each of the foregoing situations can be obtained provided certain assumptions concerning the

214

variables can be regarded as tenable. Ordinarily the graduated variable can be thought of either as being continuous or as progressing in a sufficient number of discrete steps so as to give the appearance of continuity. The approach to normality for such series can, obviously, be specified. The nature of the categorized variable, whether discrete or continuous, can ordinarily be ascertained on logical grounds, but the question of whether a continuous variable for which we have only a distribution by categories would yield a normal distribution if we had some measuring stick for the trait is not easy to answer.

## BISERIAL CORRELATION

When one variable is measured in a graduated fashion and the other is in the form of a dichotomy, we have the so-called biserial situation, for which there are two measures of correlation: biserial $r$ and point biserial $r$. The difference between these two measures depends essentially on the type of assumption which is made concerning the nature of the dichotomized variable.

The most typical example of situations calling for one or the other of these measures is to be found in the test (mental and personality) field: the correlation between an item scored as pass or fail (yes or no, like or dislike, etc.) and a graduated criterion variable (or a total score on all of a set of items). We need to know each individual's score on the graduated variable and the dichotomy to which he belongs. Then we can make a distribution or scattergram with from 12 to 20 intervals for the graduated variable along the $y$ axis, and with two intervals for the two categories along the $x$ axis. Such a correlation scattergram is given in Table 12.1, which involves pass-fail on "abstract words" vs. composite IQ on Forms L and M of the 1937 Stanford-Binet. It is obvious that there is a tendency for those who fail the item to have lower IQs than those who pass—performance on the item is related to IQ.

**Biserial coefficient, $r_b$.** If it can be assumed that underlying the dichotomy there is a continuous variable, we can obtain a measure of correlation which is an estimate of what the product moment correlation would be in case the dichotomous variable were measured in such a way as to produce a normal distribution. This estimate is given by

$$r_b = \frac{(M_2 - M_1)(p_1 p_2)}{h S_y} \tag{12.1}$$

or by the exact equivalent

$$r_b = \frac{(M_2 - M_y)p_2}{h S_y} \tag{12.2}$$

**Table 12.1 Biserial table for "abstract words" as $X$ and Binet IQ as $Y$**

| IQ | Item Fail (1) | Item Pass (2) | Totals |
|---|---|---|---|
| 145–149 | | 1 | 1 |
| 140–144 | | | |
| 135–139 | | 1 | 1 |
| 130–134 | | 3 | 3 |
| 125–129 | | 4 | 4 |
| 120–124 | | 6 | 6 |
| 115–119 | | 10 | 10 |
| 110–114 | | 7 | 7 |
| 105–109 | 1 | 8 | 9 |
| 100–104 | 1 | 5 | 6 |
| 95–99 | 4 | 9 | 13 |
| 90–94 | 7 | 6 | 13 |
| 85–89 | 9 | 2 | 11 |
| 80–84 | 3 | 1 | 4 |
| 75–79 | 4 | | 4 |
| 70–74 | 5 | | 5 |
| 65–69 | | | |
| 60–64 | 3 | | 3 |
| Totals | 37 | 63 | 100 |
| Means | 84.43 | 109.86 | 100.45 |

$S_y = 17.69$

$p_1 = .37$

$p_2 = .63$

$h = .378$

$$r_b = \frac{(109.86 - 84.43)(.37)(.63)}{(.378)(17.69)}$$

$$= .89$$

Or by formula (12.2):

$$r_b = \frac{(109.86 - 100.45)(.63)}{(.378)(17.69)}$$

$$= .89$$

$$r_{pb} = \frac{(109.86 - 100.45)}{17.69}\sqrt{\frac{.63}{.37}}$$

$$= .69$$

in which

$p_1 =$ proportion of cases in the first category.

$p_2 =$ proportion of cases in the second category.

$M_1 =$ mean of $Y$s for cases in the first category.

$M_2 =$ mean of $Y$s for cases in the second category.

$M_y =$ mean of all the $Y$ scores.

$S_y = S$ of all the $Y$ scores.

$h =$ ordinate for the unit normal curve at the point where $p_1$ (or $p_2$) cases are cut off; it is determined by entering $p_1$ or $p_2$, whichever is smaller, as a $q$ value in Table A, then reading off the adjacent ordinate value in the fourth column of the table (interpolating if necessary).

Formula (12.2) is the more convenient when each of a series of items is to be correlated against the same graduated variable. The computations are illustrated in Table 12.1.

In the derivation of $r_b$ it is assumed not only that a normal distribution underlies the dichotomy but also that the regressions would be linear if the dichotomized variable were measured. The latter assumption cannot be checked; it is apt to hold for ability variables but may be violated for personality traits. The former assumption has troubled many. Actually, the main issue is the question of continuity. Consider the pass-fail dichotomy; it is obvious that failing a test item represents anything from a dismal failure up to a near pass, whereas passing the item involves barely passing up to passing with the greatest of ease. Such a line of reasoning is certainly presumptive evidence for continuity, and a similar argument can be advanced as regards yes-no, like-dislike, and similar categories. Given a continuous trait, it is usually (if not always) possible to construct a test thereof which yields a normal distribution, and consequently we need not worry about the mathematical assumption of normality when using $r_b$. We can justify the use of $r_b$ with obviously continuous variables by saying, as pointed out earlier, that the obtained coefficient represents what we would expect the product moment correlation to be if we had a measuring scale, for the dichotomized trait, which yielded a normal distribution.

The sampling error of biserial $r$ is given approximately by

$$S_{r_b} = \frac{\dfrac{\sqrt{p_1 p_2}}{h} - r_b^2}{\sqrt{N}} \tag{12.3}$$

As an exercise, the student should compare the magnitude of the sampling error of biserial $r$ for various cuts ($p$ values) with that of the product moment $r$ as given by the analogous classical form, $S_r = (1 - r^2)/\sqrt{N}$. It might be anticipated that the sampling error will be large when dichotomies are extreme, i.e., involve cuts yielding very high (and low) $p$s. Thus, if $N = 100$, and we have a .95–.05 cut, it follows that one of the means used in computing $r_b$ by formula (12.1) will be based on only five cases and therefore will be subject to rather large sampling fluctuation, which incidentally will not be counterbalanced entirely by the relatively greater stability of the other mean. It may occur to the reader that the use of formula (12.2) would overcome this difficulty, since it is always possible to arrange to use the mean for the category having the larger number of cases, thereby avoiding the unstable mean. This appears plausible enough; its refutation is left to the student.

The fact that the sampling error for biserial $r$ is large when extreme dichotomies are involved should serve as a warning. Unless $N$ is fairly large, we should not place much confidence in a biserial $r$ based on cuts more extreme than .10 (or .90).

Since no $r$ to $z$ transformation is available for use with biserial $r$, the difficulty of skewed sampling distributions for high $r_b$s cannot be overcome. In testing the null hypothesis (that no correlation exists), the $r$ term in formula (12.3) may be dropped. For $N$ small, a more adequate test of the significance of $r_b$ is possible by way of the $t$ test for the difference between $M_2$ and $M_1$.

Although $r_b$ is an estimate of a product moment $r$, there are limitations as to its interpretation. It is, of course, a measure of the degree of relationship between the two variables. It does not, however, enter into prediction formulas, nor does it lead to an error of estimate. If we know to which $X$ category an individual belongs, the predicted $Y$ is simply the mean of the $Y$ scores for that category, and the error of such an estimate is the standard deviation of the $Y$ scores in the given category. This error of estimate would not equal $S_y\sqrt{1 - r_b^2}$.

If we have a $Y$ score to use in predicting an individual's $X$ category, we estimate on the basis of the tendency for those possessing $Y$ scores in a given interval to fall predominantly into the first or second category on $X$. The error for such a prediction must depend on the relative frequencies in these two categories for individuals possessing a given $Y$ score. Thus, if the frequencies in the first and second categories were 18 and 6 (for a given $Y$ interval), the error might be stated something like this: the odds are 3 to 1 that the given individual's $X$ position is in the first category; i.e., 75 per cent of the time the prediction would be correct. But such a percentage statement might itself be subject to grave sampling error since it is based on a small $N$; and such a statement of error might need to be qualified according to the $p$s. Why?

**Point biserial, $r_{pb}$.** If the dichotomous trait is truly discrete, an appropriate measure of correlation is given by

$$r_{pb} = \frac{(M_2 - M_1)\sqrt{p_1 p_2}}{S_y} \tag{12.4}$$

or its equivalent

$$r_{pb} = \frac{M_2 - M_y}{S_y}\sqrt{\frac{p_2}{p_1}} \tag{12.5}$$

Actually, $r_{pb}$ is the product moment correlation between $Y$ and the $X$ categories scored as either 0 or 1 (scoring as 1 and 2, or as 4 and 10, or any other two values will yield the same correlation). The value of $r_{pb}$ for the data of Table 12.1 is .69, compared to an $r_b$ of .89. The magnitude of $r_{pb}$ tends to be less than that of $r_b$ for the same set of data, as can be seen by examining the following connection between the two coefficients:

$$r_{pb} = \frac{h}{\sqrt{p_1 p_2}} (r_b)$$

For a 50–50 dichotomy, $h = .3989$ and $r_{pb} = .798r_b$, and as the dichotomy departs farther and farther from 50–50 the discrepancy between $r_{pb}$ and $r_b$ increases. For a 10–90 cut we have $r_{pb} = .585r_b$. The maximum degree of correlation between a dichotomous variable and a normally distributed variable will occur when there is no overlap between the $Y$ distributions for the two categories. For such a situation $r_b$ will be either $+1.00$ or $-1.00$ regardless of the cut, whereas $r_{pb}$ will be $\pm.798$ for a 50–50 cut and only $\pm.585$ for a 10–90 cut. These two coefficients are not on the same scale; they will agree only when there is exactly no relationship between the two variables. Even if the dichotomous variable were a genuine point variable, $r_{pb}$ as an expression of the degree of relationship would not be comparable either to $r_b$ or to the product moment $r$ between two variables measured in a graduated fashion.

Despite the fact that true point variables are practically nonexistent in psychology and despite the difficulties of interpreting $r_{pb}$ as a terminal descriptive statistic, $r_{pb}$ has a rightful place in certain analytical and practical work where the two categories are arbitrarily, for convenience, assigned point scoring values of, say, 0 and 1. For example, if a dichotomized variable with point scoring were included in an $n$ variable multiple regression equation, point biserial $r$s would be the correct values for the correlation of the dichotomized variable with the remaining $n - 1$ variables.

For the large sample situation the significance of $r_{pb}$ (as a deviation from zero) may be tested by using $\sigma_{r_{pb}} = 1/\sqrt{N}$ as its standard error. For small samples, the $t$ test for the difference, $M_2 - M_1$, is appropriate.

A troublesome difficulty with the biserial coefficient, $r_b$, is that it occasionally exceeds unity. The usually given explanation for this is that the assumption of normality for the dichotomous variable is not tenable, but it seems more likely that when such $r$s occur it is because the graduated variable, for the combined categories, is either platykurtic or bimodal in distribution.

**Biserial illustrations.** In order for the reader to have some impression as to the distributions for varying degrees of biserial relationships for differing dichotomous cuts, the normal bivariate frequencies in Figs. 9.2, 9.3, and 9.4 have been used to generate Table 12.2, with frequencies rounded to yield $N = 100$. The effects of the rounding and grouping errors are such that actually computed values for $r_b$ will not correspond exactly with the given values, which are theoretical $r$s for the given cuts. Ordinarily, $r_{pb}$ values would not be calculated when it is known that a

# Table 12.2. Biserial distributions

| | | | | | | | | |
|---|---|---|---|---|---|---|---|---|
| 12 | | 1 | | 1 | | 1 | | 1 |
| 11 | | 2 | | 2 | | 2 | 1 | 1 |
| 10 | | 4 | | 4 | 1 | 3 | 2 | 2 |
| 9 | 1 | 8 | 2 | 7 | 5 | 4 | 7 | 2 |
| 8 | 2 | 13 | 7 | 8 | 11 | 4 | 14 | 1 |
| 7 | 7 | 12 | 13 | 6 | 17 | 2 | 19 | |
| 6 | 12 | 7 | 17 | 2 | 18 | 1 | 19 | |
| 5 | 13 | 2 | 14 | 1 | 15 | | 15 | |
| 4 | 8 | 1 | 9 | | 9 | | 9 | |
| 3 | 4 | | 4 | | 4 | | 4 | |
| 2 | 2 | | 2 | | 2 | | 2 | |
| 1 | 1 | | 1 | | 1 | | 1 | |
| | 50 | 50 | 69 | 31 | 83 | 17 | 93 | 7 |
| $r_b$ | .80 | | .80 | | .80 | | .80 | |
| $r_{pb}$ | .64 | | .61 | | .54 | | .42 | |
| 12 | | 1 | | 1 | | 1 | | 1 |
| 11 | | 2 | | 2 | | 2 | 1 | 1 |
| 10 | 1 | 3 | 1 | 3 | 2 | 2 | 3 | 1 |
| 9 | 2 | 7 | 4 | 5 | 6 | 3 | 7 | 2 |
| 8 | 5 | 10 | 8 | 7 | 12 | 3 | 14 | 1 |
| 7 | 8 | 11 | 13 | 6 | 16 | 3 | 18 | 1 |
| 6 | 11 | 8 | 15 | 4 | 17 | 2 | 19 | |
| 5 | 10 | 5 | 13 | 2 | 14 | 1 | 15 | |
| 4 | 7 | 2 | 8 | 1 | 9 | | 9 | |
| 3 | 3 | 1 | 4 | | 4 | | 4 | |
| 2 | 2 | | 2 | | 2 | | 2 | |
| 1 | 1 | | 1 | | 1 | | 1 | |
| | 50 | 50 | 69 | 31 | 83 | 17 | 93 | 7 |
| $r_b$ | .50 | | .50 | | .50 | | .50 | |
| $r_{pb}$ | .40 | | .38 | | .34 | | .27 | |
| 12 | | 1 | | 1 | 1 | | 1 | |
| 11 | 1 | 1 | 1 | 1 | 2 | | 2 | |
| 10 | 2 | 2 | 2 | 2 | 3 | 1 | 4 | |
| 9 | 4 | 5 | 5 | 4 | 7 | 2 | 8 | 1 |
| 8 | 7 | 8 | 10 | 5 | 12 | 3 | 13 | 2 |
| 7 | 9 | 10 | 13 | 6 | 16 | 3 | 17 | 2 |
| 6 | 10 | 9 | 14 | 5 | 16 | 3 | 18 | 1 |
| 5 | 8 | 7 | 11 | 4 | 13 | 2 | 14 | 1 |
| 4 | 5 | 4 | 7 | 2 | 8 | 1 | 9 | |
| 3 | 2 | 2 | 3 | 1 | 4 | | 4 | |
| 2 | 1 | 1 | 2 | | 2 | | 2 | |
| 1 | 1 | | 1 | | 1 | | 1 | |
| | 50 | 50 | 69 | 31 | 85 | 15 | 93 | 7 |
| $r_b$ | .20 | | .20 | | .20 | | .20 | |
| $r_{pb}$ | .16 | | .15 | | .13 | | .11 | |

continuous variable has been dichotomized, and if we were to make believe
that these distributions hold for truly point dichotomous variables, we
would not calculate $r_b$. But these considerations do not lessen the worth of
Table 12.2 as illustrations of the general appearance of biserial distributions that underlie differing degrees of relationship, whether computed
as $r_b$ or as $r_{pb}$. Notice how the distributions depend on the cuts, and note
the dependence of $r_{pb}$ on the cuts.

## TETRACHORIC CORRELATION

When both variables yield only dichotomized information, as, for
example, two items scored as passed or failed, it is possible to secure an
estimate of what the correlation would be if the underlying traits were
continuous and normally distributed or if they were so measured as to give
normal distributions. The measure of relationship for such a situation
is known as the *tetrachoric correlation coefficient*, usually designated as $r_t$.
It is not feasible to derive here the formula for tetrachoric correlation, but
perhaps a few words will help us understand the reasoning back of the
formula.

Let us suppose that we have before us a scattergram for the correlation
between height and weight; let us further assume that this scatter exhibits
all the characteristics of a normal correlational surface as defined by
equation (9.16). That is, the two marginal distributions and all the vertical
and horizontal array distributions are normal; the regressions are linear;
and the arrays homoscedastic. For such a normal plot, it is possible,
knowing the degree of correlation and the $M$s and $S$s of the two variables,
to specify how many or what proportion of the cases will fall in any given
segment of the scatter plot. This can be done by mathematical manipulation of formula (9.16) or by the aid of Table VIII of Pearson's *Tables for
statisticians and biometricians, part II.**

Now, of course, if the student had placed before him a scatter for height
vs. weight and were asked how many cases fell in that portion of the table
below 120 pounds *and* also below 68 inches, he would simply count them.
But suppose he were told that, when the two axes were cut at 120 pounds
and 68 inches, the frequencies in each of the four quadrants so formed were
as shown in Table 12.3. The purpose of tetrachoric correlation is to
ascertain the degree of correlation which would permit the observed
frequencies in such a fourfold table. A more rigorous statement would be:
Given the four frequencies, what should be the true correlation—for the

* Pearson, Karl, *Tables for statisticians and biometricians, part II*, Cambridge:
Cambridge University Press, 1931.

**Table 12.3. Correlation for height and weight dichotomized**

|  | Below 120 lb. | Above 120 lb. |  |
|---|---|---|---|
| Above 68 in. | 10 | 80 | 90 |
| Below 68 in. | 60 | 50 | 110 |
|  | 70 | 130 | 200 |

scatter underlying the fourfold table—in order to make the obtained four frequencies most likely?

In order to secure this estimate it is necessary to convert into a proportion each of the four frequencies and each of the marginal totals by dividing by $N$. For the fourfold table we may symbolize the frequencies as in Table 12.4, the proportions as in Table 12.5. Then, the tetrachoric coefficient can be obtained from the following rather forbidding equation:

$$\frac{c - qq'}{h_x h_y} = r + xy\frac{r^2}{2} + (x^2 - 1)(y^2 - 1)\frac{r^3}{6}$$
$$+ (x^3 - 3x)(y^3 - 3y)\frac{r^4}{24} + \cdots \quad (12.6)$$

in which it is assumed that both $q$ and $q'$ are less than .50. The general rule is to choose whichever is smaller, $p$ or $q$, to pair with whichever is smaller, $p'$ or $q'$. This determines, logically, whether $a$ or $b$ or $c$ or $d$ becomes a part of the formula. Thus one can have $c - qq'$ (as given) or $b - pp'$, each of which will yield a positive $r$ for positive correlation or a negative $r$ for negative correlation, or one can have $a - q'p$ or $d - qp'$, each of which will yield an $r$ with sign opposite to its true sign. (It is, of course, here assumed that reading to the right on the $x$ axis and up on the $y$ axis means *more* of the traits.)

We must next specify the meaning of the $x$, $y$, and $h$ in formula (12.6). As for biserial $r$, $h_y$ is the ordinate of the unit normal curve where $q$

**Table 12.4. Frequencies**

|  | − | + |  |
|---|---|---|---|
| + | A | B | A + B |
| − | C | D | C + D |
|  | A + C | B + D | N |

**Table 12.5. Proportions**

|  | − | + |  |
|---|---|---|---|
| + | a | b | p |
| − | c | d | q |
|  | q' | p' | 1.0 |

proportion of the cases are cut off; $h_x$ has a similar meaning for $q'$. The $y$ represents the value on the base line of the unit normal curve where $q$ cases are cut off, i.e., the $x/\sigma$ in Table A of the Appendix, and $x$ is similarly determined from a knowledge of $q'$.

Additional terms may be added to equation (12.6), which will result in a closer approximation at the expense of a greater, if not an impossible, amount of computation. For the given formula, the solution for $r$ involves determining the roots of a fourth-degree or quartic equation. Either Horner's or Newton's methods, as described in college algebra texts, will do the trick. The fourth-degree equation will yield satisfactory approximations except when $r$ is high.

The solution of a quartic equation is not difficult, nor is it so easy as to lead to mass production of tetrachoric $r$s. Fortunately, it is no longer necessary to go through this tedious method for getting an approximation to the value of $r_t$. Diagrams† are available which enable us to determine quickly the value of $r_t$ for any given table of proportionate frequencies. Anyone having as many as a half-dozen tetrachorics to compute will find it economical to possess a copy of these diagrams.

The tetrachoric $r$ is particularly useful in estimating the degree of correlation between variables for which we have only dichotomized information, but it can also be used instead of biserial $r$ or the product moment $r$, since situations for which these two methods apply can readily be converted into fourfold tables by simply dichotomizing the graduated variables. The advantage of so estimating correlation is that tetrachoric $r$ is much easier to determine (by using the computing diagrams) than is calculating either biserial $r$ or the product moment $r$. Indeed, this fact of computational economy has led a number of investigators to use $r_t$ when product moment $r$s could be determined. That such a practice may be short-sighted economy becomes quite evident when we turn to the sampling fluctuation of $r_t$.

The standard error of $r_t$ is closely approximated by

$$S_{r_t} = \frac{\sqrt{pqp'q'}}{h_x h_y \sqrt{N}} \sqrt{(1 - r^2)\left[1 - \left(\frac{\sin^{-1} r}{90°}\right)^2\right]}  \tag{12.7}$$

When this is compared to the classical formula for the standard error of a product moment, $r$, i.e., to $S_r = (1 - r^2)/\sqrt{N}$, it will be seen that the tetrachoric $r$ has a much larger sampling error. To illustrate the difference, the errors for four $r$s for two different dichotomies are presented in Table 12.6 along with the errors (by the classical formula) of the corresponding product moment $r$s for $N = 100$.

† Chesire, L., Saffir, M., and Thurstone, L. L., *Computing diagrams for the tetrachoric correlation coefficient*, Chicago: University of Chicago Bookstore, 1933.

Table 12.6. Sampling errors of $r_t$ and $r$ compared

| $r$ or $r_t$ | $p$ | $p'$ | $S_{r_t}$ | $S_r$ |
|---|---|---|---|---|
| .00 | .50 | .50 | .157 | .100 |
| .00 | .80 | .80 | .204 | .100 |
| | | | | |
| .40 | .50 | .50 | .130 | .084 |
| .40 | .80 | .80 | .182 | .084 |
| | | | | |
| .60 | .50 | .50 | .115 | .064 |
| .60 | .80 | .80 | .150 | .064 |
| | | | | |
| .80 | .50 | .50 | .073 | .036 |
| .80 | .80 | .80 | .095 | .036 |

It can readily be seen from this table that $r_t$ is much less stable than $r$; in fact, even for the most favorable comparison (.50–.50 cuts, low $r$s), the standard error of the tetrachoric coefficient is more than 50 per cent greater than that for the product moment coefficient. This means that we must have more than twice as many cases to attain the same degree of sampling stability for a tetrachoric as for a product moment correlation coefficient. For .80–.20 cuts and low correlations, four times as many cases are needed to have comparable sampling errors. For high correlations and also for more extreme cuts, $r_t$ compares still less favorably with $r$.

The foregoing discussion and further study of formula (12.7) lead to two obvious conclusions.

First, the increasing sampling instability of $r_t$ as the dichotomies become more extreme warns us that, unless $N$ is large, we cannot place much reliance on $r_t$ for cuts more extreme than .10–.90; seldom will $N$ be large enough to warrant confidence in a tetrachoric based on cuts more extreme than .05–.95.

Second, in using $r_t$ instead of the product moment $r$ when the latter is calculable, we are always throwing away the equivalent of more than half the available data. Thus the computational economy may be an expensive luxury—it is very doubtful whether the calculation of a product moment $r$ for $N$ cases will ever require anything but a fraction of the expense of securing data on the additional $N$ cases needed to counterbalance the greater sampling error incurred in using the tetrachoric coefficient.

As in the case of $r_b$, no $r$ to $z$ transformation exists for handling the sampling errors of high tetrachorics. For testing the null hypothesis, that $r_t$ for the universe is zero, we may use a simpler expression for its standard error, namely, $S_{r_t} = \sqrt{pqp'q'}/h_x h_y \sqrt{N}$. Another method for judging the significance of the correlation computed from a fourfold table will be presented in the next chapter.

The use of tetrachoric $r$ is circumscribed by an assumption that the underlying correlational surface is of the normal type. Among other things this implies (1) that the dichotomized traits are continuous and normally distributed, and (2) that the regressions are linear. Although, as discussed in connection with biserial $r$, we are usually ignorant of the tenability of 1, this ignorance can be partially overcome if the correlation is regarded as that which would obtain if the traits were normalized; i.e., it can be argued that the use of tetrachoric $r$ automatically normalizes the distributions. It is not so easy to dispose of assumption 2, since the normalizing of variables will not necessarily lead to linearity of regression. The only consolation here is that *measured* psychological traits are usually linearly related, if related at all.

## FOURFOLD POINT CORRELATION

If we can safely assume point distributions for both dichotomous variables, a descriptive measure of correlation can be obtained from a fourfold table (Table 12.4) by

$$r_p = \frac{BC - AD}{\sqrt{(A + B)(C + D)(A + C)(B + D)}} \tag{12.8}$$

or from the table of proportionate frequencies (Table 12.5) by the exact equivalent

$$r_p = \frac{c - qq'}{\sqrt{pqp'q'}} \tag{12.9}$$

The fourfold point correlation coefficient is frequently referred to as the *phi* coefficient and designated by $\phi$. Actually, it is the product moment correlation between the two variables each scored in a point fashion (say, 0 and 1). Unlike the point biserial, $r_p$ can be unity but only when $p = p'$. Otherwise (i.e., in nearly all situations) $r_p$ and $r_t$ from the same table will differ in value, with $r_p$ being lower, and the difference between the two becomes greater as the dichotomy for either variable, or both, varies farther and farther from 50–50.

An examination of the fourfold tables in Table 12.7 may be instructive. All but one of the 12 tables illustrate the fact that when $r_t$ and $r_p$ are each computed for the same data, $r_p$ tends to be less than $r_t$, and that the discrepancy tends to increase as we move away from situations with 50–50 cuts (on the margins).

Aside from the contrasts of $r_p$ and $r_t$, it is of interest to note how, in the

### Table 12.7. Fourfold tables: $r_t$ vs. $r_p$

| | | | | | | | | | | | | | | | |
|---|---|---|---|---|---|---|---|---|---|---|---|---|---|---|---|
| 10 | 40 | 50 | | 3 | 47 | 50 | | 9 | 60 | 69 | | 1 | 30 | 31 | |
| 40 | 10 | 50 | | 28 | 22 | 50 | | 22 | 9 | 31 | | 30 | 39 | 69 | |
| 50 | 50 | | | 31 | 69 | | | 31 | 69 | | | 31 | 69 | | |
| $r_t$ | .81 | | | | .81 | | | | .80 | | | | .77 | | |
| $r_p$ | .60 | | | | .54 | | | | .58 | | | | .40 | | |

| | | | | | | | | | | | | | | | |
|---|---|---|---|---|---|---|---|---|---|---|---|---|---|---|---|
| 17 | 33 | 50 | | 3 | 47 | 50 | | 10 | 74 | 84 | | 0 | 16 | 16 | |
| 33 | 17 | 50 | | 13 | 37 | 50 | | 6 | 10 | 16 | | 16 | 69 | 84 | |
| 50 | 50 | | | 16 | 84 | | | 16 | 84 | | | 16 | 84 | | |
| $r_t$ | .49 | | | | .52 | | | | .46 | | | | ?? | | |
| $r_p$ | .32 | | | | .27 | | | | .26 | | | | .19 | | |

| | | | | | | | | | | | | | | | |
|---|---|---|---|---|---|---|---|---|---|---|---|---|---|---|---|
| 22 | 28 | 50 | | 6 | 44 | 50 | | 2 | 14 | 16 | | 1 | 15 | 16 | |
| 28 | 22 | 50 | | 10 | 40 | 50 | | 14 | 70 | 84 | | 15 | 69 | 84 | |
| 50 | 50 | | | 16 | 84 | | | 16 | 84 | | | 16 | 84 | | |
| $r_t$ | .19 | | | | .20 | | | | .10 | | | | .31 | | |
| $r_p$ | .12 | | | | .11 | | | | .04 | | | | .12 | | |

top tables, differing combinations of cell frequencies lead to practically the same tetrachoric $r$s. This also is seen in the first three of the middle tables and the first two of the bottom tables. Varying combinations along with differing cuts do not preclude the obtaining of similar tetrachorics, but it will be noted that such is not the case for fourfold point $r$s. Furthermore, 31–69 and 69–31 cuts reduce $r_p$ more than 31–69 and 31–69 cuts (see top right-hand tables). The middle right-hand tables show a similar effect, but for more extreme cuts.

Since the value of a calculated $r_p$ seems to be in part a function of the cuts, one might ask about the maximum possible $r_p$ for certain cuts. Consider the table that led to an $r_p$ of .58; for the given cuts the maximum (positive) correlation would be obtained if all the frequencies were in the southwest and northeast cells, i.e., 31 and 69, respectively. The computed $r_p$ for such a fourfold table is +1.00. Next, consider the table that yielded $r_p = .40$; for the given cuts the maximum positive correlation will be obtained when the $(BC - AD)$ part of formula (12.8) is as large as possible. Moving the $A = 1$ to the $B$ cell and moving one of the 39 in the $D$ cell to the $C$ cell will maintain the cuts on the margin. Thus $A$, $B$, $C$, and $D$ become 0, 31, 31, and 38, which lead to $r_p = .449$ as the highest possible positive fourfold point correlation for the given cuts.

The fact that $r_t$ is not influenced by the cuts is a strong argument for its use as a descriptive statistic when (as is usual) the variables underlying the dichotomies are continuous. Even so, there is one situation, most apt to occur with extreme cuts and when the relationship is relatively high, which proves to be an insurmountable difficulty in computing $r_t$. That cell frequency of zero in one of the middle tables is a real stumbling block: none of the four roots in equation (12.6) will be less than 1.00, hence $r_t$ is indeterminate. Obviously, the correlation is not perfect in the given example. Actually, the "observed" frequencies resulted from reducing a bivariate scattergram (Fig. 9.3) to a fourfold table and rounding the frequencies to $N = 100$. The underlying normal bivariate $r$ was .50.

As a matter of fact the "observed" frequencies for all the fourfold tables in Table 12.7, except for the one yielding $r_t = .10$, were produced from normal bivariate scattergrams—the underlying $r$s were .80 for the top four, .50 for the middle four, and .20 for all except the third one of the bottom four. The rounding to $N = 100$ accounts for frequencies which do not yield $r_t$ values in exact agreement with the underlying normal bivariate coefficients. The greatest disagreement occurs for the table that gives $r_t = .31$. The frequencies of 1 and 69 involved downward roundings and the 15s upward roundings, hence an increase in $r_t$. The adjacent table differs in that just two fewer cases fall in the southwest-northeast diagonal cells (the two cases were "moved" sidewise to the other cells). The resulting $r_t$ is .10: that small difference in the frequencies in the two fourfold tables produces $r_t$s of .10 and .31, a happening that ties in with the sampling instability of $r_t$ when extreme cuts are present.

Despite the fact that $r_p$ is not interpretable on the same scale as $r_t$ (or $r$ or $r_b$) as a measure (terminal descriptive statistic) of the degree of relationship, it is useful and necessary in certain analytic work. If variable $U$ and variable $V$ were dichotomous and each scored as 0 and 1, then $r_p$ would be the appropriate value to use in formula (9.9) to obtain the variance of $W$, defined as $U + V$. If formula (5.5) for the standard error of the difference between correlated proportions were written analogously to formula (6.8), $r_p$ would be used. It is also used in the statistical theory of mental tests.

For testing whether $r_p$ deviates significantly from zero we may safely use $1/\sqrt{N}$ as its standard error when $N$ is not small.

## CONTINGENCY COEFFICIENT

The *contingency coefficient* is a measure of the degree of association or correlation which exists between variables for which we have only categorical information. The number of categories can be such as to provide a

2 by 2 table (as for tetrachoric correlation) or a 2 by 3, or a 3 by 3, or a 3 by 4, or a 4 by 4, or a $k$ by $l$ table. This coefficient is stated in terms of a quantity known as $\chi^2$ (*chi square*) thus

$$C = \sqrt{\frac{\chi^2}{N + \chi^2}} \tag{12.10}$$

where

$$\chi^2 = \Sigma \frac{(O - E)^2}{E} \tag{12.11}$$

in which $O$ is the observed frequency (not percentage) and $E$ is the expected frequency for a given cell. In a 2 by 3 table there would be six cells, hence six values summed to get $\chi^2$. The expected cell frequencies for the contingency situation are those frequencies which would exist if there were no association or relationship between the given variables. It can thus be anticipated that, the larger the discrepancy between expected and observed frequencies *relative* to the expected, the larger the value of $\chi^2$ and consequently the higher the value of $C$.

An example will help to clarify the preceding. Suppose that we have two variables, each of which yields three categories or classifications, and that the observed frequencies are as given in Table 12.8, which also contains the expected frequencies in parentheses. (Fictitious data; marginal frequencies arranged so as to simplify exposition.) In order to ascertain the expected frequencies needed in the computation of $\chi^2$, we ask what cell frequencies would be expected if there were no relationship, or zero association, between the two variables. Consider the 100 classified as college; if no association existed, we would expect that these 100 would be distributed according to a 1, 3, 1 ratio, i.e., in the same ratio as the

#### Table 12.8. Contingency table

|              | Low       | Medium      | High      |     |
|--------------|-----------|-------------|-----------|-----|
| College      | 5 (20)    | 45 (60)     | 50 (20)   | 100 |
| High school  | 50 (40)   | 110 (120)   | 40 (40)   | 200 |
| Grade school | 45 (40)   | 145 (120)   | 10 (40)   | 200 |
|              | 100       | 300         | 100       | 500 |

marginal frequencies at the bottom. Thus the expected cell frequencies for the top row of cells would be 20, 60, 20. The expected frequencies for the middle and bottom rows of cells should also be in a 1, 3, 1 ratio. Both these rows would have expected frequencies of 40, 120, 40.

It will be noted that (1) the expected frequencies for the *columns* follow, as they should, the ratio of 1, 2, 2, i.e., the ratio of 100, 200, 200 for the marginal frequencies on the right; (2) the expected frequencies sum to the same marginal totals as the observed frequencies; and (3) the expected frequencies actually exhibit a zero relationship between the two characteristics.

In practice, the computation of the expected frequencies can readily be accomplished by either of two schemes: (1) express the marginal totals along the bottom as proportions of the total $N$, then multiply each of the frequencies on the right margin by each proportion in turn, entering the resulting product in the cell common to the two marginal figures involved in the multiplication; or (2) multiply any frequency on the bottom margin by any frequency on the right margin, and then divide this product by $N$; the result is the expected frequency for the cell common to the two marginals involved in the products.

The computation of $\chi^2$ is now a routine matter. We simply take each cell in turn, square the difference between the observed and expected value, and divide by the expected frequency. Thus we have

$$(5 - 20)^2/20 = 11.25$$
$$(45 - 60)^2/60 = 3.75$$
$$(50 - 20)^2/20 = 45.00$$
$$(50 - 40)^2/40 = 2.50$$
$$(110 - 120)^2/120 = .83$$
$$(40 - 40)^2/40 = .00$$
$$(45 - 40)^2/40 = .62$$
$$(145 - 120)^2/120 = 5.21$$
$$(10 - 40)^2/40 = 22.50$$

The sum of these quantities, 91.66, is $\chi^2$. To get $C$, the coefficient of contingency, the value of $\chi^2$ is substituted in formula (12.10), thus

$$C = \sqrt{\frac{91.66}{500 + 91.66}} = .39$$

This strength of association is not to be interpreted as indicating the same degree of relationship as an ordinary (or biserial or tetrachoric) coefficient of the same magnitude. One reason for this is that the upper limit for the contingency coefficient is a function of the number of categories. The upper limit for a 2 by 2 table is $\sqrt{\frac{1}{2}}$; for a 3 by 3 table, $\sqrt{\frac{2}{3}}$;

for a 4 by 4 table, $\sqrt{\frac{3}{4}}$; for a 5 by 5 table, $\sqrt{\frac{4}{5}}$; for a $k$ by $k$ table, $\sqrt{(k-1)/k}$. The exact upper limits for rectangular tables, such as 2 by 3, 2 by 4, 3 by 4, are unknown. (As an exercise, the student might demonstrate to his own satisfaction the upper limit for 2 by 2 and 3 by 3 tables.) The reader will also note that $C$ can never be negative.

Despite having varying maximal values, contingency coefficients have a decided advantage over other measures of relationship; no assumptions involving the nature of the variables need be met—continuous or discrete variables, normal or skewed or any shaped distributions for underlying traits, ordered or unordered series, and combinations thereof are permissible.

Disadvantages are that any two contingency coefficients are not comparable unless derived from tables of the same size, that they are noncomparable to product moment $r$s (and estimates thereof) unless certain corrections are applied, and that the formula for sampling error is unwieldy. The necessary corrections and the sampling error formula may be found in Kelley,‡ but before consulting Kelley, the reader might bear in mind the following comments.

In regard to the corrections, the first is for number of categories. The additional correction to make $C$ an estimate of $r$ involves the assumption that the underlying traits are continuous and normal in distribution. Furthermore, this correction is very tedious to make. It is suggested that, if the assumption of normally distributed continuous variables is tenable, we are justified in reducing a contingency table of more than four cells to a 2 by 2 table and then determining the value of tetrachoric $r$. When reducing to a fourfold table, we should combine adjacent categories so as to have dichotomies as near to .50–.50 proportions as possible. The combination should *not* be made on the basis of the pattern of cell frequencies, since this is likely to involve a capitalization or decapitalization on chance. We might take several or all possible fourfold combinations, thus securing several tetrachoric $r$s which may then be averaged.

As to the unwieldy sampling error formula for $C$, it is suggested that insofar as we wish simply to test the null hypothesis, i.e., that there is no relationship between the two given variables, we need only enter the value of $\chi^2$ into an appropriate probability table to test its significance. If $\chi^2$ is significant, the relationship is significantly greater than zero. This use of $\chi^2$ will be discussed in Chapter 13.

Chi square for a fourfold table can be readily obtained by formula without first computing expected frequencies. Thus for a set of frequencies

‡ Kelley, T. L., *Statistical method*, pp. 266–271, New York: Macmillan, 1924.

like that of Table 12.4 we have

$$\chi^2 = \frac{N(BC - AD)^2}{(A + B)(C + D)(A + C)(B + D)}$$

This resembles formula (12.8). In fact, there is a relationship between the fourfold point coefficient $(r_p)$, $\chi^2$, and $C$:

$$r^2{}_p = \frac{\chi^2}{N} \quad \text{and} \quad C = \sqrt{\frac{r^2{}_p}{1 + r^2{}_p}}$$

Other measures of association or of correlation between attributes have been advocated. This is not the place to argue the pros and cons of these other measures. It seems to the author that the measures we have discussed are the more defensible.

## THE CORRELATION RATIO OR $\eta$ (ETA)

It will be recalled that one way of understanding the product moment correlation coefficient is to note from the relationship $r^2 = 1 - S^2{}_{y \cdot x}/S^2{}_y$ (or $r^2 = 1 - S^2{}_{x \cdot y}/S^2{}_x$) that the degree of correlation is a function of the error of estimate variance relative to the total variance of the variable being predicted by a linear regression line. If the array means fail to fall on a straight line, it can rightly be argued that better prediction can be made by using a curve which really "fits" the means or by using the means themselves. The latter procedure would entail an error of estimate which would be a function of the variance within the arrays about the array means. An over-all variance about the means of the vertical arrays can be calculated by squaring the deviations about the mean of each array, summing these for all arrays, and then dividing by $N$. The resulting variance for the vertical arrays may be labeled $S^2{}_{ay}$, for the horizontal arrays, $S^2{}_{ax}$.

The *correlation ratio*, $\eta$, in terms of the accuracy with which $Y$s can be predicted from $X$s is defined as

$$\eta^2{}_{yx} = 1 - \frac{S^2{}_{ay}}{S^2{}_y} \tag{12.12}$$

and for $X$s predicted from $Y$s, we have

$$\eta^2{}_{xy} = 1 - \frac{S^2{}_{ax}}{S^2{}_x} \tag{12.13}$$

Are two $\eta$s necessary? We have elsewhere proved that the variance about the mean is smaller than about any other point. It follows that $S_{ay}$ will be less than $S_{y \cdot x}$ and that $S_{ax}$ will be less than $S_{x \cdot y}$; hence both $\eta$s will exceed $r$, but to varying degrees, depending on the extent to which the array means fail to fall on a straight line. Since it is possible, and likely, that the means for the vertical arrays will not exhibit the same departure from linearity as those for the horizontal arrays, it is not reasonable to expect the two $\eta$s to agree.

The $\eta$s indicate the relative accuracy with which we can predict on the basis of array means, and accordingly they are useful measures of the extent of correlation when the regressions are curvilinear. The correlation ratio can also be utilized when the regression is linear; hence it is more generally applicable than the product moment coefficient which is useful only in the special case where the assumption of linearity is tenable. The correlation ratio, however, does not enter into the regression equation constants.

Even if the regressions were exactly linear for some defined population, a given sample would show deviations from linearity, and therefore $\eta$s for successive samples would show chance sampling deviations from $r$. By how much must $\eta$ exceed $r$ before we suspect curvilinearity? The only adequate statistical test for answering this question involves the analysis of variance technique and hence is postponed to Chapter 15.

Another definition of $\eta$ can be had by starting with the proposition that the variance $S^2_y$ can be broken down into components, a predictable and an unpredictable part, or $S^2_y = S^2_{m_y} + S^2_{a_y}$, in which $S^2_{m_y}$ is the variance of the array means weighted for the number of cases in the several arrays. Then we have $\eta$ defined as $\eta^2_{yx} = S^2_{m_y}/S^2_y$ and also as $\eta^2_{xy} = S^2_{m_x}/S^2_x$. These are analogous to $r^2 = S^2_{y'}/S^2_y$ and $r^2 = S^2_{x'}/S^2_x$, and accordingly we may interpret $\eta^2_{yx}$ as the proportion of $Y$ variance explained by or associated with variation in $X$.

Since the $\eta$s are most readily computed by methods to be developed later (pp. 315–16), no illustration will be given here.


## RANK CORRELATION

Rank-ordering by judges is frequently resorted to when no measuring instrument is available for a trait. One measure of relationship between variables for which we have individuals ranked is given by $\rho$ (rho), the Spearman rank-difference correlation coefficient:

$$\rho = 1 - \frac{6\Sigma D^2}{N(N^2 - 1)} \qquad (12.14)$$

in which $D$ is the difference between an individual's two ranks (for the two traits). When we have ranks for one variable and scores for the other we can use the scores as a basis for setting up ranks for the latter, and then compute rho.

Whenever rankings on a given variable involve ties (the judges fail to distinguish between two or more individuals or the scores used for ranking are such that two or more persons have the same score), the ranks are split between individuals who are in tie positions. Suppose three ranks have been assigned and that two individuals are tied for the fourth position. If they were distinguishable, they would use up ranks 4 and 5, so we assign

**Table 12.9. Computation of rank-difference correlation coefficient**

| Persons | Ranks 1st | 2nd | Differences D | $D^2$ |
|---|---|---|---|---|
| A | 3 | 1 | 2 | 4 |
| B | 4 | 2 | 2 | 4 |
| C | 10 | 10 | 0 | 0 |
| D | 8 | 4.5 | 3.5 | 12.25 |
| E | 5 | 6 | −1 | 1 |
| F | 9 | 11 | −2 | 4 |
| G | 1 | 3 | −2 | 4 |
| H | 2 | 7 | −5 | 25 |
| I | 13 | 13 | 0 | 0 |
| J | 11 | 4.5 | 6.5 | 42.25 |
| K | 7 | 8.5 | −1.5 | 2.25 |
| L | 6 | 8.5 | −2.5 | 6.25 |
| M | 12 | 12 | 0 | 0 |
| | | | 0 | $105.00 = \Sigma D^2$ |

$$\rho = 1 - \frac{6(105)}{13(169 - 1)} = .71$$

each a value of 4.5. Had three persons tied for this position, we would split ranks 4, 5, and 6, giving each a rank of 5. Then when we proceed to the remaining individuals we must remember that rank position 6 has been used.

The computation of rho is illustrated in Table 12.9. The fact that the algebraic sum of the $D$s must be zero can be utilized as a means of checking the $D$ column values.

Rho for ranks based on scores for two normally distributed variables tends to be slightly (less than .02) lower than the product moment $r$ computed from the scores; hence rho is comparable with $r$ as a measure of the strength of relationship.

To test the significance of rho, for $N$ of 10 or more, we may safely use

$$t = \rho\sqrt{\frac{N-2}{1-\rho^2}}$$

which approximates the $t$ distribution with $N-2$ degrees of freedom.

Rho does not possess the mathematical advantages inherent in $r$, and therefore has merit only when the observations on one or both variables are ranks instead of measures. Because of judgmental difficulties in assigning ranks for $N$ large, rank-order data are apt to be confined to small samples, but for $N$ less than 10 the $t$ test of the significance of rho is not satisfactory. Kendall§ has proposed another measure, designated $\tau$ (tau), for use with ranks; it is superior to rho insofar as testing significance is concerned when $N$ is very small. As a measure of the degree of relationship, tau, like rho, has the property of being unity for a perfect relationship; for zero and near zero correlation these two measures tend to be alike numerically, but for other degrees of association tau tends to be lower than rho—at times only two-thirds the magnitude of rho. Thus tau is not comparable with rho (and $r$), and furthermore there seems to be no specifiable way of estimating one from the other. For a much more adequate discussion of both tau and rho, the reader is referred to Kendall.

## THE DISCRIMINANT FUNCTION

Suppose we have two or more variables (measured in a graduated fashion) which we wish to combine into a total score for the purpose of discriminating between two groups. The question arises as to how best weight the variables so as to obtain maximum difference between the total score means for the two groups. This difference must be considered *relative* to the within-groups variability; otherwise we could easily produce a large numerical difference by the simple operation of summing the scores and multiplying by a large constant, whereas the real purpose is to have score distributions with the least amount of overlap for the two groups. We want the difference to be maximal relative to the spread of scores within the groups.

The simplest way to determine the weights for the several variables is to compute the $\beta$s, thence the $B$s, as in the multiple regression problem. For this purpose, the product moment correlations among the two or more independent variables are calculated, and the *point* biserial $r$ is calculated between each independent variable and $X_1$, the dependent variable (membership in one or the other of the two groups, with one of the groups consistently designated as corresponding to the first category for the biserial setup).

§ Kendall, M. G., *Rank correlation methods*, London: Griffin, 1948.

Actually, since the problem here is that of ascertaining optimum relative weights rather than fitting a regression plane, we need not calculate the $A$ of the regression equation nor worry about $S_1$ ($= \sqrt{p_1 p_2}$ of the biserial setup). The weights may be taken simply as $\beta_2/S_2$, $\beta_3/S_3$, etc., all multiplied by a constant so chosen as to have weights which exceed, say, 10—thereby avoiding decimals. Some of the weights may be negative, according to the sign of the corresponding $\beta$. If all or a majority of the weights are negative, the signs of all may be reversed. The relationship of the total of the optimally weighted scores to group membership is describable by the multiple $r$ computed by equation (11.12). Such a multiple $r$ is the point biserial between the total weighted scores and belonging to one or the other of the two groups. Or we may compute the weighted scores for all $N$ cases and then make distributions for the two groups separately in order to scrutinize the amount of differentiation (or overlap) present.

## CORRELATION OF SUMS (OR AVERAGES)

There are times when it is useful to have a formula for the correlation between two variables, each of which is made up as the sum (or average) of two or more variables. As an introduction to the problem, consider the situation in which one variable is obtained by summing three parts, the other by summing two. In deviation units, let

$$x = x_a + x_b + x_c \quad \text{and} \quad y = y_A + y_B$$

Then

$$r_{xy} = \frac{\Sigma xy}{NS_x S_y} = \frac{\Sigma(x_a + x_b + x_c)(y_A + y_B)}{NS_x S_y}$$

$$= \frac{\Sigma x_a y_A + \Sigma x_a y_B + \Sigma x_b y_A + \Sigma x_b y_B + \Sigma x_c y_A + \Sigma x_c y_B}{NS_x S_y}$$

Each term in the numerator when divided by $N$ will yield an $r$ times the product of two $S$s, and the value of $S_x$ and $S_y$ will each be given by the square root of the variance of a sum of correlated scores, hence we have

$$r_{xy} = \frac{r_{aA}S_a S_A + r_{aB}S_a S_B + r_{bA}S_b S_A + r_{bB}S_b S_B + r_{cA}S_c S_A + r_{cB}S_c S_B}{\sqrt{S^2_a + S^2_b + S^2_c + 2r_{ab}S_a S_b + 2r_{ac}S_a S_c + 2r_{bc}S_b S_c}}$$

$$\times \sqrt{S^2_A + S^2_B + 2r_{AB}S_A S_B}$$

as a formula in terms of the $S$s for the component parts, the correlations among the parts entering into $x$ and ditto $y$, and the cross-correlations of the $x$ and $y$ components (in the six numerator terms). This means that if the given $r$s and $S$s are available it is possible to obtain the correlation between the specified sum scores without ever computing a sum score for any of the $N$ individuals.

Suppose $x$ is made up of $m$ variables and $y$ of $M$ variables ($m$ not necessarily equal to $M$):

$$x = x_a + x_b + \cdots + x_i + \cdots + x_m$$

$$y = y_A + y_B + \cdots + y_I + \cdots + y_M$$

A little thought will indicate that the expression for the $r$ between the two sums will involve a numerator containing $mM$ terms of the type $r_{iI}S_iS_I$ with $i = a, \cdots, m$ and $I = A, \cdots, M$. These are the cross-correlations. Under one radical sign we will have $m$ variances of the type $S^2_i$ plus either $m(m-1)/2$ terms of the type $2r_{ij}S_iS_j$ with $i \neq j$ or $m(m-1)$ terms like $r_{ij}S_iS_j$, and under the other radical we will have $M$ variances of the type $S^2_I$ and $M(M-1)$ terms of the form $r_{IJ}S_IS_J$ with $I \neq J$. Instead of using $+$ signs to indicate the addition of the various terms, we can use $\Sigma$ to indicate the adding process. Accordingly, a general formula for the correlation of two sums (or averages) can be written as

$$r_{xy} = r_{\Sigma x \Sigma y} = \frac{\Sigma r_{iI}S_iS_I}{\sqrt{\Sigma S^2_i + \Sigma r_{ij}S_iS_j}\sqrt{\Sigma S^2_I + \Sigma r_{IJ}S_IS_J}} \begin{matrix} (i \neq j) \\ (I \neq J) \end{matrix} \quad (12.15)$$

in which each $r_{ij}$ term and each $r_{IJ}$ term is added twice in the summing.

The $r$ between $\Sigma x/m$ and $\Sigma y/M$ will also be given by (12.15), since dividing by a constant does not change the degree of correlation. That is, the correlation between averages is the same as the correlation between the sums entering into the averages.

Formula (12.15) can be written in a different way if it is noted that any sum can be replaced by the number of items being summed, multiplied by their mean; e.g., $\Sigma X = N$ times the mean of the $X$s. Thus each of the sums in (12.15) can be replaced by the appropriate mean times the number of terms being summed. Using an overhead bar to indicate a mean, we have

$$r_{\Sigma x \Sigma y} = \frac{mM\overline{r_{iI}S_iS_I}}{\sqrt{m\overline{S^2_i} + m(m-1)\overline{r_{ij}S_iS_j}}\sqrt{M\overline{S^2_I} + M(M-1)\overline{r_{IJ}S_IS_J}}} \quad (12.16)$$

Consider now a first special case of formula (12.16). Suppose that all $m$ of the $X_i$ are parallel measures of trait $X$ and that all $M$ of the $Y_I$ are parallel measures of trait $Y$. By definition of parallel measures, the following hold:

(1) $S_a = S_b = \cdots = S_i = \cdots = S_m$

(2) $S_A = S_B = \cdots = S_I = \cdots = S_M$

(3) all $r_{ij}$ are equal to one another

(4) all $r_{IJ}$ are equal to one another

and we would expect

(5) all $r_{iI}$ to have the same value

Under these conditions, the $S_i$ will cancel, the $S_I$ will cancel, $\bar{r}_{ij}$ will be the same as any one $r_{ij}$, $\bar{r}_{IJ}$ will equal any $r_{IJ}$, and $\bar{r}_{iI}$ will equal any $r_{iI}$. This leads to

$$r_{\Sigma x \Sigma y} = \frac{mM r_{iI}}{\sqrt{m + m(m-1)r_{ij}}\sqrt{M + M(M-1)r_{IJ}}} \qquad (12.17)$$

Dividing both numerator and denominator by $mM$, we find

$$r_{\Sigma x \Sigma y} = \frac{r_{iI}}{\sqrt{\frac{1}{m} + \left(1 - \frac{1}{m}\right)r_{ij}}\sqrt{\frac{1}{M} + \left(1 - \frac{1}{M}\right)r_{IJ}}}$$

Next consider what happens as we allow both $m$ and $M$ to approach infinity. Both $1/m$ and $1/M$ become zero, leaving 1 times $r_{ij}$ and 1 times $r_{IJ}$ under the radicals. Thus the correlation between two sums (or averages), each based on an infinite number of parallel measures, becomes simply

$$r_{\Sigma x \Sigma y} = r_{(\Sigma x/m)(\Sigma y/M)} = \frac{r_{iI}}{\sqrt{r_{ij}}\sqrt{r_{IJ}}} \qquad (12.18)$$

But since both $r_{ij}$ and $r_{IJ}$ are correlations between parallel measures, each is a reliability coefficient; and $r_{iI}$ is simply $r_{xy}$, the correlation between any $X_i$ and $Y_I$. Therefore, for $m$ and $M$ infinitely large,

$$r_{(\Sigma x/m)(\Sigma y/M)} = \frac{r_{xy}}{\sqrt{r_{xx}}\sqrt{r_{yy}}} \qquad (12.19)$$

which is, as might have been anticipated, the correction for attenuation, or formula (10.20).

As a second special case of (12.16), suppose $M = m$ and the $X_i$ and $Y_I$ are all parallel measures of just one trait, say $X$. Again the $S$s will cancel, but now $r_{ij} = r_{IJ} = r_{iI} = r_{xx}$, and with $M = m$ formula (12.17) becomes

$$r_{\Sigma x \Sigma y} = \frac{m^2 r_{xx}}{m + m(m-1)r_{xx}} = \frac{m r_{xx}}{1 + (m-1)r_{xx}} \qquad (12.20)$$

If $m = 2$, this becomes $2r_{xx}/(1 + r_{xx})$, or the previously derived Brown-Spearman formula (10.17) for the reliability of a test doubled in length. In other words, formula (12.20) is the generalized Brown-Spearman formula for a test increased $m$-fold in length. If a test of given length has reliability $r_{xx}$, and we wish to know how much we would need to lengthen the test to achieve a reliability of, say, .90 we simply solve (12.20) for $m$, with $r_{\Sigma x \Sigma y}$ set to .90:

$$m = \frac{.90(1 - r_{xx})}{r_{xx}(1 - .90)}$$

## QUASI INDICATORS OF CORRELATION

Here we shall discuss (a) a fallacious measure of correlation derived from a fourfold table, (b) a frequently used method for inferring corre-

**Table 12.10. Fourfold tables with differing marginal cuts**

| A | B | 50 |   | A | B | 90 |
|---|---|----|---|---|---|----|
| C | D | 50 |   | C | D | 10 |

|   | 50 | 50 |   |   | 10 | 90 |

lation, a sort of quasi-correlational technique, (c) a situation that is quasi-correlational for which it would be desirable to have a standard measure of association, and (d) a hit-miss interpretation of predictions.

**Percentage agreement.** Suppose we have two fourfold tables as shown in Table 12.10. The frequencies in these tables might have to do with whether two judges have agreed in their classification of 100 patients into one of two diagnostic categories. It is easy to say that agreement should be expressed in percentage form. Thus $100(B + C)/100$, or more generally, $100(B + C)/N$, is taken as a percentage measure of agreement. The difficulty with such a measure is that it depends upon the marginal splits in that for the 50–50 cuts (both margins) chance would be expected to produce 50 per cent agreement whereas for the 10–90 cuts chance would tend to produce 82 per cent agreement. Note that if both judges happen to assign 90 to the same group, there automatically ensues a forced agreement of not less than 80 per cent. Note also that the percentage agreement tends to increase with the departure (in the same direction) of the two marginal cuts from 50–50. Note further that if the margins were 10–90 and 90–10 we would have a situation in which agreement could not possibly exceed 20 per cent, with 18 per cent as the chance expected value.

Thus it would seem the percentage agreement is far from satisfactory as a measure. If used, it has little meaning unless accompanied with the agreement expected by chance for the given marginal frequencies.

**Use of extremes.** When measures are available or easily obtainable for a pool of $N_p$ subjects on variable $X$ and we wish to test the hypothesis that variable $Y$ is related to $X$ but because of time or economic or facility factors we can measure only $N_s$ of the potential $N_p$ subjects on $Y$, we might proceed by either of two commonly used routes: (*a*) draw a random sample from the pool, secure $Y$ measures for the $N_s$ subjects so drawn, compute a product moment $r$, and test it for significance or (*b*) take the $N_s/2$ subjects at the top and the $N_s/2$ subjects at the bottom of the $X$ distribution of the pool, measure these $N_s$ cases on $Y$, then run a $t$ test for the difference between the $Y$ means for these two extreme (as regards $X$) groups. The chief advantage of this usage of extreme groups is that for fixed $N_s$ the $t$ test by way of the difference between extremes is more powerful than the $t$ test for $r$ computed on $N_s$ cases. For example, a $t$ test based on 17 cases in each of the extreme groups (top and bottom 10% of $N_p = 170$) is as apt to yield significance as an $r$ based on near 100 cases. The more extreme the groups on $X$, the greater will be the group difference on $Y$ if there is nonzero correlation in the population, but the attainment of more extreme groups for fixed $N_s$ cannot be achieved unless $N_p$ has somehow been made to increase (not likely). One could for fixed $N_p$ pick greater extremes by taking $N_s$ smaller and smaller, but as $N_s$ becomes smaller the value of $t$ will tend to decrease because the increase in the difference between means does not offset the increase in the estimated standard error of the difference. The extremeness of the selected groups will, of course, depend upon $N_s$ relative to $N_p$.

There are two, usually ignored, difficulties in inferring correlation by use of extremes. First, no information emerges concerning whether the form of the relationship is linear or curvilinear. Now the researcher who says that he is not interested in the form of the relationship cannot escape the fact that if there is a genuine curvilinear relationship of, say, the U or inverted U type, the use of extremes may all too frequently lead to accepting falsely the null hypothesis when indeed correlation exists. To guard against overlooking a possible nonlinear relationship, it would be wise to take $N_s/3$ cases as extremes with $N_s/3$ chosen from the middle of the $X$ distribution.

The second difficulty with deductions from extreme groups is that a significance test between means tells us nothing about the strength, or degree, of relationship between $Y$ and $X$. For example, it can be shown that using the top and bottom 10 per cent from a pool of $N_p = 170$ subjects can easily lead to a difference between means that is significant at the .05 level, two-tailed test, when the underlying $r$ is only .20. The exuberance over finding a "significant" difference in this situation is somewhat dampened when one examines Fig. 9.2 long enough to have some appreciation of the trivial amount of relationship represented by $r = .20$. Take a second example: the use of the highest 44 and lowest 44 from a pool of 440 possible subjects will tend to produce a one-tailed, .05 level of significance when the underlying $r$ is a mere .10, which is not something that calls for shouting about "confirmation" of a theory that led to the hypothesis that $Y$ is related to $X$. Related, yes, but to a piddling degree. If we are to know what the shouting is about when extremes yield a significant difference, we should have an estimate of the degree of relationship.

**Estimation, based on extremes, of $r$.** When highs and lows on $X$ are measured for $Y$, the available or computable statistics will be as follows: $M_x$ and $S_x$ for the pool of $N_p$ cases; differences, $D_{M_x}$ and $D_{M_y}$, in means for the high and low groups; $s_{w_y}$ for within groups (needed in calculating $t$ for testing $D_{M_y}$). We will not know either $M_y$ or $S_y$ for the $N_p$ cases. We can compute a quantity here designated $k$ and defined as

$$k = \frac{D_{M_y}}{D_{M_x}} \cdot \frac{S_x}{s_{w_y}}$$

Then $k^2/(1 + k^2)$ will provide an estimate of $r$, the sign of which will depend on the sign of $k$ (in getting the difference between means the direction of subtraction should be the same for both variables). The number of subjects in the high and low groups need not be exactly the same. The above will yield a reasonably good approximation when each extreme group is less than 15 per cent of $N_p$, and for low relationships, say $r$ less than .50. It is for low degrees of correlation that an estimated value is most needed if we are to avoid making too much fuss over a claimed relationship that is, unbeknownst to us, of paltry magnitude.

**Quasi-correlational situations.** Given the distribution for the scores on, say, $Y$ for the two sex groups, ordinarily the question of sex difference is examined by way of the significance of the difference between the two means. The resulting $t$ (or $z$) is in part a function of sample sizes and therefore is not a measure of the magnitude of the difference. The actual numerical difference between the means does not readily provide an answer to the query as to whether the sex difference on variable $Y$ is larger or smaller than that for any other variable simply because the two variables typically do not involve the same metric, or measure unit. What is needed is a "standard" index for gauging differences.

One solution, advocated by some, is in terms of the percentage overlap of the two distributions. Another solution is the difference between the two means divided by the average of the two standard deviations. Both of these solutions have some merit. The percentage overlap could be taken as the percentage in one group that exceeds the mean or the median of the other group or it could be the area under the two distribution curves which is under both curves, expressed relative to the total area for either group as 100 per cent. What meaning this overlap has in case of unequal variances and/or skewed distributions is somewhat ambiguous. For normal distributions, having equal or nearly equal standard deviations, it is possible to set up tables for specifying the percentage of overlap (common area) as a function of $(M_1 - M_2)/S_y$, where $S_y$ is based on averaging the two variances. If this ratio is zero, the overlap is 100 per cent; if the ratio is 5.15, the overlap is 1 per cent; if 8 or 10 or 11 or higher, no overlap. Now the failure of this percentage overlap to distinguish between differing large differences in means should not be too bothersome simply because such large differences do not arise in practice.

A third solution can be stated in quasi-correlational terms: Is scoring performance associated with sex? Or does performance depend on group membership? Thus we could regard sex as a point variable, $X$, with scoring as 0 or 1 for males vs. females (or vice versa), then proceed to calculate the point biserial correlation as a measure of the degree of correlation between $Y$ and $X$. Such an $r_{pb}$ need not have an upper limit of .798, which holds when the $Y$s for the two groups combined form a normal distribution. With variation on $Y$ within each group, the upper limit for $r_{pb}$ will not be 1.00, although as the difference between the means becomes larger and larger the calculated value of $r_{pb}$ will approach 1.00 if the variation within the groups has not also increased. Thus $r_{pb}$ would seem to provide a reasonably satisfactory "standard" index for gauging group differences on a scale that would have similar descriptive meaning for group comparisons on any and all measurable variables.

Need we place any qualifications on the use of $r_{pb}$ as a descriptive measure of the relationship between a measured variable and group membership? Let's consider the difference between the sexes in fear of snakes, for which we presume there is a reasonably good test available. Two M.A. theses get under way, one at Stanford and one at Indiana University. In the tradition of campus-bound research psychologists, each M.A. candidate draws a random sample from the general psychology courses in his institution. The Indiana sample yields 64 males and 64 females whereas the Stanford sample consists of 96 males and 32 females (there is a difference in the sex ratio at these universities). Now to illustrate a point we shall say that all four distributions are normal with equal variances and that the difference between the sex means is the same at both universities. If the difference between the sex means is .89 times the common within-groups standard deviation, the $r_{pb}$ will be .41 in the Indiana study and .36 for that at Stanford. This is nothing more or less than an earlier mentioned undesirable characteristic of $r_{pb}$ as a descriptive measure: it is a function of the split on the dichotomous variable, hence does not necessarily lead to comparable indexes. Aside from lack of comparability, there is another issue that is illustrated by the foregoing sex difference example. Shall we regard .41 or .36 as the better measure of the association between fear of snakes and sex? The answer is the $r$ that is based on the split that approximates most closely the sex ratio in the population. Take another example: in order to learn whether the scores on a newly constructed personality test are related to the presence or absence of schizophrenia, a researcher might test 200 schizophrenics and 200 normals, run a significance check on the difference between the two means, and compute $r_{pb}$ as an index of association. This index would be quite different if the researcher had cuts as between normals and schizo-phrenics nearer the extreme cuts, of the order of .9998 to .0002, in the population. A feasible and justifiable method for estimating $r_{pb}$ for the population would be a simple blowup of the frequencies in the distribution for normals so that the split is near the ratio of the proportions in the population. This, in our example, could be achieved by multiplying the frequencies for the normals by 4999, thus inflating the $N$ for normals to 999,800. This will not change the difference between the means, but $S_y$ will need to be recalculated in order to substitute in (12.4) to get the adjusted $r_{pb}$. The significance test must, of course, be based on the original $N$s, not the inflated one for the normal group.

A fourth solution to the problem of a standard expression has been proposed: biserial eta. This is simply the limiting situation for eta—two arrays, or columns, of scores for the two groups. Apparently the advocates of biserial eta are not aware of the fact that, algebraically, biserial eta is precisely the same as the point biserial $r$. It is therefore no better or worse than $r_{pb}$ as a "standard" index.

**Relationship in terms of "hits."** Given the following steps in "validating" a new test designed to differentiate between delinquent and nondelinquent boys. The test is a short one, the items having been selected because each yielded a highly significant response difference between groups of delin-quents and nondelinquents. The three authors of the test, who had collaborated in constructing the test while employed at the University of Chicago, realized that its worth must be judged on new sample groups (the cross-validation problem, p. 210). At this juncture, one author accepted a position at UCLA and another author a position at Columbia, and this happening made it easy to collect cross-validation data in three cities. It was decided that the budget would allow the testing of 256 in each city.

Whether or not the 256 were to be split evenly as regards being delinquent or nondelinquent was to depend on testing feasibility. In due course the three investigators met to discuss their separate results. It had been agreed that each would examine his delinquent-nondelinquent score distribution for a cutting score which would produce the maximum number of hits in predicting to which group a boy belongs. The results were as follows (percentages based on $N = 256$):

New York, 78 per cent hits, cutting score of 8.5
Chicago, 70 per cent hits, cutting score of either 7.5 or 8.5
Los Angeles, 78 per cent hits, cutting score of 7.5
New York, 42 of 256 predicted as delinquent
Chicago, either 84 or 114 of 256 predicted as delinquent
Los Angeles, 171 of 256 predicted as delinquent

It was quickly inferred from the differing optimal cutting scores that the over-all means likely differed from city to city, and it was suspected that validity might depend on locale. The puzzling finding was the big difference in the numbers predicted as delinquent.

That such diverse outcomes could hold when the means do not differ from city to city and when the difference between the means for delinquents and nondelinquents is precisely the same for the three cities is illustrated by the contrived artificial distributions in Table 12.11, from which the "results" given were derived. What we see here is: (a) that validity in the sense of hits is, like $r_{pb}$, a function of the split on the dichotomous variable; (b) that the optimal cutting score on the test is likewise a

Table 12.11. Artificial data to illustrate "hit" percentages

| Group* | New York | | Chicago | | Los Angeles | | N.Y. adj. | |
|---|---|---|---|---|---|---|---|---|
| | I | II | I | II | I | II | I | II |
| 12 | | 1 | | 2 | | 3 | | 1 |
| 11 | | 6 | | 12 | | 18 | | 6 |
| 10 | 3 | 15 | 2 | 30 | 1 | 45 | 6 | 15 |
| 9 | 18 | 20 | 12 | 40 | 6 | 60 | 36 | 20 |
| 8 | 45 | 15 | 30 | 30 | 15 | 45 | 90 | 15 |
| 7 | 60 | 6 | 40 | 12 | 20 | 18 | 120 | 6 |
| 6 | 45 | 1 | 30 | 2 | 15 | 3 | 90 | 1 |
| 5 | 18 | | 12 | | 6 | | 36 | |
| 4 | 3 | | 2 | | 1 | | 6 | |
| $N$s | 192 | 64 | 128 | 128 | 64 | 192 | 384 | 64 |
| $M$s | 7 | 9 | 7 | 9 | 7 | 9 | 7 | 9 |

* Referred to in text as nondelinquent and delinquent, respectively.

function of the split on the dichotomous variable and also a function of the direction of the split from 50–50; and (c) that the variation in the numbers predicted as delinquent is very much dependent on the splits.

How does one avoid such discrepant results? Obviously, if the splits as regards group sizes had been the same for each city, the discrepancies would not have arisen. But is that the whole answer? One statistical consideration would lead us to say that splitting the 256 cases 50–50 would be optimum for getting a significant difference between the group means (see p. 121) or for getting a significant $r_{pb}$. Is such a split best in the sense of providing the best cutting score and also yielding a good estimate of the percentage of hits to be expected in further use of the test? The answer to both these questions is no if there is not an equal (or nearly equal) split in the population of boys as to delinquency or nondelinquency. The split shown for New York is likely more realistic than a 50–50 split, whereas the split in the Los Angeles "situation" is certainly highly unrealistic. It is not here argued that the establishment of validity in terms of hits requires the use of samples proportionate in size to the split in the population; rather, it is argued that what should be done is to adjust the frequencies so as to have splits corresponding to the *base rate* (the population proportions, known or estimated).

To pursue the matter a bit further, let us assume the delinquency rate is .141, or 1 to 6 (perhaps unrealistic). Then adjustments to the distributions in Table 12.11 would require multiplying the nondelinquent frequencies in the distribution for New York by a factor of 2; those for Chicago by a factor of 6; and those for Los Angeles by a factor of 18. These adjustments will lead to distributions for nondelinquents and delinquents that will be exactly the same for the three cities when further adjusted to equalize the total $N$s (no longer 256 for each city). We may therefore take any one of the possible adjusted tables as the basis for further discussion; that for New York, the right-hand part of Table 12.11, will be used. From these adjusted values, it is easily deduced that the cutting score that maximizes the number of hits is 9.5, and the percentage of hits becomes 94 as compared to 70 and 78 per cent deduced from the original unadjusted distributions.

That boost to 94 per cent hits may hit the reader as a bit of statistical chicanery, but if the split in the population is 1 to 6, then our hit rate should be based on frequencies that hold for the population. Otherwise, we have hit percentages that fortuitously depend on the numbers of cases in two groups that an investigator chose because of feasibility. Some of that magical increase to 94 per cent hits seems less magical if it is noted that when we have a 6 to 1 split we can achieve hits 86 per cent of the time by predicting nondelinquency for all regardless of test scores!

# Chapter 13

# FREQUENCY COMPARISON: CHI SQUARE

The quantity chi square ($\chi^2$), defined in the last chapter as

$$\chi^2 = \Sigma \frac{(O - E)^2}{E} \qquad (12.11)$$

or as the sum of the squared discrepancies, between observed and expected frequencies, each divided by the expected frequency, is a statistic very useful in a variety of problems involving frequencies. Let us begin by an examination of what might be expected to happen if a penny were tossed 100 times. The expected frequency for heads is 50, and for tails is also 50. If for a particular series of tosses we secured 55 heads and 45 tails, the discrepancies would be $+5$ and $-5$. When these discrepancies are squared, each becomes $+25$, and dividing each squared discrepancy by the expected value we would have $.5 + .5 = 1.0$ as the value for $\chi^2$. Had we obtained 40 heads and 60 tails, the discrepancies of $-10$ and $+10$, when squared and divided by $E$, would give $2 + 2 = 4$ as $\chi^2$.

Three things are readily apparent from the aforementioned: first, the greater the discrepancy relative to $E$, the greater the contribution to $\chi^2$; second, the two parts being summed to obtain $\chi^2$ are *not independent*—when the absolute discrepancy for heads is known, that for tails can be inferred to be the same; and third, the squaring process means that $\chi^2$ is always a positive quantity regardless of the direction of the discrepancies. A fourth fact becomes apparent if we recall what happens when a series of tosses is repeated. The number of heads (or tails) secured will vary from one series of 100 tosses to the next; hence the amount of discrepancy will vary, and therefore the magnitude of $\chi^2$ will vary from series to series. In other words, successive sampling will yield varying values for $\chi^2$. If we knew the sampling distribution for $\chi^2$, we could specify the probability of securing by chance as large a value as any obtained $\chi^2$, and thereby we could judge whether a given amount of discrepancy is significantly large enough to warrant the conclusion that the coin is biased.

Situations similar to this arise in research work. We may, on the basis of a hypothesis that a certain proportion of individuals possess a given characteristic, state how many of a sample of $N$ cases would be expected to show the characteristic. Observations on $N$ cases will provide an observed number. If the hypothesis is tenable, the discrepancy between observed and expected should be no larger than might arise on the basis of chance. If the obtained discrepancy is too large, i.e., not apt to arise by chance, the hypothesis becomes suspect. The student who recalls that the standard error of a proportion can be used in comparing observed with expected proportions may wonder whether another technique is necessary. The answer will be forthcoming.

## CHI SQUARE AND THE BINOMIAL DISTRIBUTION

Perhaps some insight regarding the sampling distribution of $\chi^2$ can be obtained by a re-examination of the binomial distribution, which was discussed in Chapter 5. Suppose we consider the binomial distribution, $(p + q)^{10}$ with $p = q = \frac{1}{2}$ as yielding the chance distribution of number of heads when 10 unbiased coins are tossed (see Table 13.1). When 10 coins are tossed we expect to get 5 heads and 5 tails, that is, the $E$s are 5 and 5, but for a particular toss we will have an observed number of heads (and

Table 13.1. The binomial and $\chi^2$ when 10 coins are tossed

| Number of Heads | $f_b$ | $\chi^2$ | $f$ for $\chi^2$ |
|---|---|---|---|
| 10 | 1 | 10.0 | 2 |
| 9 | 10 | 6.4 | 20 |
| 8 | 45 | 3.6 | 90 |
| 7 | 120 | 1.6 | 240 |
| 6 | 210 | 0.4 | 420 |
| 5 | 252 | 0.0 | 252 |
| 4 | 210 | 0.4 | |
| 3 | 120 | 1.6 | |
| 2 | 45 | 3.6 | |
| 1 | 10 | 6.4 | |
| 0 | 1 | 10.0 | |
| | 1024 | | 1024 |

tails) which may differ from 5 and 5. The observed values, or $O$s, could be 10 heads and zero tails; 9 heads, 1 tail; and so on to zero heads, 10 tails. If we obtained 9 heads and 1 tail, we could write $\chi^2 = (9 - 5)^2/5 + (1 - 5)^2/5 = 6.4$. Similarly, if we compute $\chi^2$ for 10 heads and no tails we get a value of 10.0; for 8 heads and 2 tails we get 3.6; etc. Note that for each $\chi^2$, $\Sigma E = \Sigma O = 10$.

The third column of Table 13.1 gives the values of $\chi^2$ for various possible sets of observed frequencies for number of heads and tails. All the given numerical values of $\chi^2$, except 0, appear twice: 9 tails and 1 head will obviously lead to the same $\chi^2$ as 9 heads and 1 tail. Now the probability of obtaining 9 heads and 1 tail is 10/1024 and the $P$ for 1 head and 9 tails is also 10/1024; hence the $P$ for obtaining a $\chi^2$ of 6.4 is 20/1024. Likewise, we may combine the appropriate binomially derived chance frequencies ($f_b$) so as to write the chance frequencies for the several $\chi^2$ values. These appear as the fourth column of the table. We have thus established the chance or probability distribution of $\chi^2$ for a specified coin-tossing situation. A plot of these frequencies against the $\chi^2$ values will reveal a highly skewed distribution.

The probability of a $\chi^2$ as large as 6.4 will be 20/1024 + 2/1024, or 22/1024, a value which obviously represents the probability of a discrepancy, between $O$ and $E$, as great as 4 in either direction (at least 9 heads or at least 9 tails). The $P$ of 22/1024 involves 1 tail of the distribution of $\chi^2$ values, but both tails of the binomial contribute thereto. This fact will need to be recalled below when we discuss one- vs. two-tailed tests of hypotheses.

Before we leave Table 13.1, it might be well to point out a connection between $\chi^2$ and $x/\sigma$. Consider again an obtained frequency of 9 heads. If we express 9 as a deviation from the mean of the binomial, $np = 5$, relative to the $\sigma$ of the binomial, $\sqrt{npq} = 1.581$, we have 4/1.581, which when squared gives 6.401 or the corresponding value of $\chi^2$ (within limits of rounding error). This agreement is not accidental; as will be seen shortly, under specifiable conditions $\chi^2 = (x/\sigma)^2$. Another characteristic of $\chi^2$ is obvious from Table 13.1: for the 10-coin situation no values of $\chi^2$ other than those given can be obtained because the possible number of heads (and tails) is a discrete series. This lack of continuity imposes a restriction on the use of $\chi^2$ which will receive more attention as we proceed.

The $\chi^2$ values in Table 13.1 are for possible discrepancies of observed frequencies from an expected frequency of 5 for a *single* toss of 10 coins. Suppose that we have, as shown in Table 13.2, an observed distribution of frequencies obtained by tossing 7 coins 1000 times, and that we wish to compare these observed frequencies with those expected on the basis of the binomial expansion. We are not concerned this time with a single toss

**Table 13.2.** $\chi^2$ **for discrepancies of expected and observed frequencies when 7 coins were tossed 1000 times**

| Number of Heads | $E$ | $O$ | $O - E$ | $\dfrac{(O - E)^2}{E}$ |
|---|---|---|---|---|
| 7 | 8 | 4 | −4 | 2.00 |
| 6 | 55 | 55 | 0 | .00 |
| 5 | 164 | 157 | −7 | .30 |
| 4 | 273 | 283 | 10 | .37 |
| 3 | 273 | 267 | −6 | .13 |
| 2 | 164 | 177 | 13 | 1.03 |
| 1 | 55 | 45 | −10 | 1.82 |
| 0 | 8 | 12 | 4 | 2.00 |
| Sums | 1000 | 1000 | 0 | 7.65 |
|  | ($N$) | ($N$) |  | ($\chi^2$) |

for which the expectation would be 3.5, but rather with the results expected when a large number of tosses are made. Note that both the $E$ column and the $O$ column sum to 1000 (or $N$) and that the $(O - E)$s sum to zero. The several contributions to $\chi^2$ are given in the last column, which sums to 7.65, or the $\chi^2$ for the entire table. Two other series of 1000 tosses made by students in the author's classes yielded $\chi^2$ values of 12.52 and 15.02. Two of these values for $\chi^2$ are larger than any of the values in Table 13.1, and one reason for this is the fact that more $(O - E)^2/E$ terms are being summed—8 such values instead of 2. Thus, the possible magnitude of a $\chi^2$ would seem to be a function of two things: the size of the squared discrepancies (relative to their respective $E$s) and the number of categories or possibilities for discrepancy. Actually, the chance or sampling distribution of $\chi^2$ is only indirectly a function of the number of discrepancies; it is a direct function of the number of *independent* discrepancies or the *degrees of freedom*, which we shall next discuss.

## DEGREES OF FREEDOM

We have seen that the $\chi^2$ of 6.4 in Table 13.1 involves two $(O - E)^2/E$ values: $(9 - 5)^2/5$ and $(1 - 5)^2/5$, or two discrepancies of exactly the same absolute magnitude. This means that the two discrepancies are not independent—as soon as one is calculated, the other can be written down at once without any further calculation; hence 1 degree of freedom exists. If we study the data of Table 13.2, we see that, since the discrepancies must sum to zero, all eight cannot be independent or vary freely. As soon as seven are known, the eighth is determined. This means that there are 7 degrees of freedom for this situation. If we were to roll a die 600 times and then compare the observed frequency for 6 spots, 5 spots, etc., with the number expected on the basis of a perfectly homogeneous (unloaded) cube, we would have five possible independent discrepancies, or 5 degrees of freedom. In each of these situations the expected frequencies are determinable on the basis of some a priori principle, and the only restriction is that the total expected frequency must be the same as the total observed frequency, i.e., $N_E$ must equal $N_O$. In all such cases the number of degrees of freedom $(df)$ is 1 less than the number of categories.

The $df$ for other situations in which the $\chi^2$ technique is applicable will follow the same principles as to the number of independent discrepancies,

### Table 13.3. $\chi^2$ and fourfold table

(Expected frequencies in parentheses)

|          | No        | Yes         | Totals        |
|----------|-----------|-------------|---------------|
| Group 1  | 50  (40)  | 50   (60)   | $100 = N_1$   |
| Group 2  | 70  (80)  | 130  (120)  | $200 = N_2$   |
| Totals   | 120       | 180         | $300 = N$     |
|          | $N_n$     | $N_y$       |               |

but not the rule just laid down. Suppose we consider a 2 by 2 or fourfold table such as that given in Table 13.3 (which contains fictitious data for purpose of ease in exposition). The expected frequencies are set up on the assumption that there is no difference between the two groups (the null hypothesis). If this were the case, we would expect that the 180 yeses would be distributed in the 1 to 2 ratio of the right-hand totals; likewise the 120 noes. Note that the expected frequencies reading across, i.e., 40 and 60, and 80 and 120, are proportional to the marginal totals at the bottom. In determining the $df$, we can observe either of two things: first that all four discrepancies have the same absolute value, so that when one is known the other three can be written down at once; or second, that in setting up the expected frequencies, we are restricted by the requirement that the two top-row values must sum to $N_1$, the next two must sum across to $N_2$, the left-hand column must sum to $N_n$, and the next column to $N_y$; as soon as the value 40 has been ascertained, the remaining three expected values become fixed. Either way we look at the situation, we see that there is but 1 degree of freedom even though there are four cells or four discrepancies.

The fundamental question is: How many of the discrepancies are independent? In practice this can be answered by determining how many categories or cells can be filled in at will before the others become fixed because of the restrictions imposed. If we turn back to Table 12.8, we see that the restrictions for a 3 by 3 table are similar to those for a 2 by 2 table: the expected frequencies must add across and down to the observed marginal totals. The student should ponder Table 12.8 long enough to see that the proper $df$ is 4. The general rule-of-thumb for ascertaining the degrees of freedom for all contingency-type tables of $k$ rows and $l$ columns, where the marginal totals are utilized in setting up the expected frequencies, is to take $df = (k - 1)(l - 1)$. Thus for the fourfold table we have $(2 - 1)(2 - 1) = 1$, and for the 3 by 3 table, $(3 - 1)(3 - 1) = 4$, etc. Such tables need not be square; in fact, very often the psychologist wishes to compare two groups on the basis of $k$ possible responses to a question. For this $k$ by 2 table, the $df$ becomes $(k - 1)(2 - 1)$, or simply $k - 1$.

## SAMPLING DISTRIBUTION OF $\chi^2$

Before discussing further the applications of $\chi^2$, we turn again to the sampling distribution of this statistic. It is easy enough to see from the coin-tossing situations which we have considered previously that chance leads to discrepancies between observed and expected frequencies. In those situations wherein we wish to compare groups, we know from the discussion of sampling in Chapter 5 that differences in responses or characteristics can and will arise as a result of chance sampling even though the two universes do not differ. Likewise, contingency tables involving the possible relationship between two categorized variables will yield varying chance values of $\chi^2$ even though no real association exists. Knowing the chance sampling distribution of $\chi^2$ for various degrees of freedom, we can specify the probability of obtaining a $\chi^2$ as large as any value and conclude therefrom, according to the situation, that observations do not agree with hypothesized frequencies or that two or more groups differ significantly or that a real association exists.

We have already suggested that, for 1 degree of freedom, the distribution of $\chi^2$ is the same as for $(x/\sigma)^2$. The general equation for the $\chi^2$ distribution* involves an $n$ or the $df$, and therefore there is no one $\chi^2$ distribution but a very large number of distributions, one for each value of $n$. It happens that practical work seldom involves more than 30 degrees of freedom, so that we need not concern ourselves with all possible distributions. Curves for the distribution of $\chi^2$ can be drawn for various $n$s with $\chi^2$ along the abscissa and the ordinates as the $h$ values obtained by the equation in the

$$* \qquad h = \frac{1}{2^{n/2} \cdot \Gamma(n/2)} (\chi^2)^{(n-2)/2} e^{-\chi^2/2}$$

in which $\Gamma$ indicates the gamma function as defined in texts in advanced calculus.

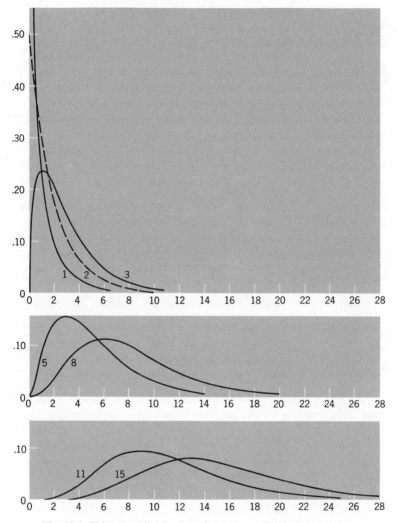

**Fig. 13.1** Chi square distributions for various *df*s. $\chi^2$ along abscissa.

footnote. The area under each curve will be 1 unit, as in the unit normal curve. Figure 13.1 contains curves for seven different values of *n* or *df*, so drawn as to be comparable. Note that the shapes of these curves and their general locations along the abscissa vary with *n*.

For *n* = 1, or for 1 degree of freedom, the curve starts very high (strictly speaking, it is asymptotic to the ordinate and hence starts at infinity) and drops quite rapidly. For this curve the height or *h* value at $\chi^2 = .16$ is .92 (not shown). At $\chi^2 = .01$, the height is more than four times greater than .92. By the time we reach a $\chi^2$ of 1.00, the height is .242 (what $x/\sigma$ value does this height correspond to when the unit normal curve is considered?). Then the curve trails off until, at $\chi^2 = 6.25$, the height is about .007. Regardless of $n$, the right-hand parts of the curves never reach the base line; i.e., they are asymptotic. If we think of the total area under any curve as unity, the area between ordinates erected at any two base-line points, or the area beyond any point, can be expressed as a proportion of the total. Thus, for $n = 1$, .99 of the area is beyond (to the right of) a $\chi^2$ value of .000157, and only .05 is beyond 3.841. Stated differently, the probability of obtaining a $\chi^2$ value as large as 3.841 is .05; for $\chi^2$ as large as 6.635, $P = .01$; and the $P = .001$ point is at a $\chi^2$ of 10.827. These hold only for $df = 1$.

The curve for $n = 2$ starts at a height of .50 and then descends, but less rapidly than that for $n = 1$. It is readily seen that large values for $\chi^2$ occur more frequently when $n = 2$ than when $n = 1$. The $P = .05$ point is at 5.991; i.e., the probability of obtaining by chance a $\chi^2$ value as great as 5.991 is .05. The .01 point is at 9.210, and the .001 point is at 13.815.

For $n = 3$, the distribution curve begins at zero height, rises sharply to a maximum (modal value) at $\chi^2 = 1$, and then falls off so that the $P = .01$ point is at $\chi^2 = 11.341$. As $n$ is taken larger and larger, the distributions become less and less skewed and move farther and farther to the right. The mean of a given distribution always corresponds to a $\chi^2$ equal to $n$, and except for $n = 1$ the modal value is at a $\chi^2$ of $n - 2$. The variance is $2n$.

The distributions of $\chi^2$ for varying $n$s are theoretical probability distributions. They may be interpreted as random sampling distributions, and by them we can judge the statistical significance of discrepancies. Their use is exactly analogous to testing the significance of the difference between means, which it will be recalled involves setting up the null hypothesis: if there is no real difference between two universe means, the $D/S_D$ values for successive samples will form a normal curve with center at zero and with unit variance. If a found difference is 1.96 times its standard error, the null hypothesis becomes suspect; if 2.58 times its standard error, the hypothesis of no difference can fairly safely be rejected; if $D/S_D = 3.00$, rejection is more definitely indicated. These three $z$s, it will be recalled, correspond to the .05, the .01, and the .003 levels of significance, for two-tailed tests.

Now $\chi^2$ can likewise be used to test the null hypothesis. The essential difference between the $D/S_D$ and the $\chi^2$ techniques is that the latter involves skewed probability distributions; but, knowing the distribution for a given $n$, we can ascertain the necessary value of $\chi^2$ for the .05, the .01, the .001, or other levels of significance. The statement of the null hypothesis in connection with $\chi^2$ may vary slightly according to the given situation. If the frequencies in the universe agree with the a priori expected frequencies, if the frequencies in two or more universes are the same, if there is zero association in the universe between two classifications or variables—if any such conditions hold for the universe or universes, then successive samplings will yield $\chi^2$ values which will distribute themselves in a determinable manner, thus permitting us to specify the probability of obtaining by chance a $\chi^2$ value as large as any given or obtained value. When this probability is small, say .01 or less, the null hypothesis is rejected, and its rejection implies that there are real discrepancies or real differences exist or there is a real association.

Since the random sampling distribution of $\chi^2$ depends on the $df$, which varies from situation to situation, it is not feasible to give a rule-of-thumb criterion in terms of the magnitude of $\chi^2$ which would be deemed significant. If we adopt $P = .01$ as the level of significance we wish to attain, we need to refer to available tables of $\chi^2$ in order to find how large $\chi^2$ must be to correspond to this level; likewise for any other chosen level of significance. Probability tables for $\chi^2$ are available in two forms. One form, Fisher's (see Table D of the Appendix), gives the values of $\chi^2$ which will be exceeded by chance a specified number of times, such as .10, .05, .01, and .001. Elderton's table† gives the probabilities for obtaining chi squares as large as specified values expressed as integers, such as 1, 2, 3, $\cdots$, 21, 22. Both tables include varying degrees of freedom. Because of an early erroneous notion as to the meaning of degrees of freedom, Elderton's table must be entered with $df$ equal to 1 less than his $n'$ values, e.g., use $n' = 4$ when $n$ or $df = 3$. Elderton's table has one advantage over that given in our Appendix: $P$ values as small as .000001 can be ascertained.

For $n$s larger than 30, the expression $\sqrt{2\chi^2} - \sqrt{2n - 1}$ will have a sampling distribution which will follow very closely the unit normal curve. The probability is accordingly .05 that this expression will exceed $+1.64$, and .01 that it will exceed $+2.33$, by chance.

Before the applications of $\chi^2$ are summarized, a word should be said about the underlying assumptions which restrict its usage. In the derivation of the equation for the $\chi^2$ distribution(s) it is assumed that the sample discrepancies, or distribution of observed from expected, follow a normal distribution. In practical applications this assumption can easily be violated in two senses: skewed distribution for $O - E$ values and lack of continuity. If $E$ is small, say, equal to 2, the $O$s are restricted on one side of $E$ to zero and 1, whereas on the other side the possible values may

---

† Table XII in Pearson, Karl, *Tables for statisticians and biometricians, part I*, Cambridge: Cambridge University Press, 1931.

be 3, 4, 5, and upward. Such a curtailment ordinarily leads to a skewed distribution of the observed frequencies (if the other side were restricted to just 3 and 4, symmetry could exist). Now it is obvious that when $E$ is small we have a greater degree of discontinuity, hence the sampling distribution of the observed frequencies (and therefore, of $O - E$ values) will be discrete instead of continuous as required for the normal curve. Even for the situation involving a symmetrical distribution of $O$s about $E$, such as that in Table 13.1, the discontinuities in the possible observed $\chi^2$ values are marked. But that situation was for $df = 1$, and, as in the approximation of binomial probabilities by use of the normal curve, a correction for continuity will better the approximation when the $\chi^2$ curve is used to obtain the probability for as great a discrepancy as the observed one. Even so, with $df = 1$ and with the correction for continuity (correction formulas yet to be given herein), it is not safe to use $\chi^2$ if an $E$ is less than 5.

However, the effect of small $E$s in producing discontinuities is not as marked when $df$ is 2 or more. Consider the dice-rolling situation. If we rolled 12 dice once we would have 2 as the expected frequency for each of the possible outcomes: 1, 2, 3, 4, 5, 6 spots. For successive rolls, or trials, we would have $O$s for each outcome that would vary from zero up to a possible (though unlikely) 12. In calculating $\chi^2$ for each trial we would be summing six $(O - E)^2/E$, or $(O - 2)^2/2$, ratios with each of the six involving marked discontinuity over successive trials. The $df$ would, of course, be 5. But over a large number of successive trials, the possible combinations of $O$ values per trial are so numerous that the calculated $\chi^2$s, hence the sample $\chi^2$s, will show relatively little discontinuity. In Table 13.1 we saw, for a situation involving $E = 5$ and $df = 1$, that $\chi^2$ can have just six values, whereas for the foregoing dice situation the summing of six $(O - E)^2/E$ ratios permits a large number of possible values for calculated $\chi^2$s, or a greater approach to continuity. This is somewhat analogous to the approximation of the discontinuous binomial by the continuous normal distribution, which approximation becomes increasingly better as $n$ increases (or as the $n + 1$ possible outcomes increase).

Although discontinuity as an aspect of the violation of the assumption of normality for $O$s about small $E$ is not serious when the $df$ is more than 5 or 6, there remains the question of the effect of possible skewness when $E$s are small. There is evidence that, when $df$ is not small, $E$s as low as 2 will not produce misleading $\chi^2$ values.

A second assumption is that the observations be independent of one another. This assumption is violated when the total of the observed frequencies exceeds the total number of persons in the sample(s). Such an inflation of $N$ occurs when multiple observations are made on each person and each person is counted more than once.

## APPLICATIONS

The chief situations for which it is permissible to use $\chi^2$ may be classified into three types.

1. The discrepancy of observed frequencies from frequencies expected on the basis of some a priori principle. Such situations are most frequently found in genetics, wherein it is hypothesized that certain crossings should lead to the presence, in a certain proportion of offspring, of some defined characteristic or variation thereof. The frequency table for such situations is 1 by $k$, with $k - 1$ degrees of freedom, since the only restriction is that the expected frequencies must sum to $N$. This type of situation does not arise often in research in the social sciences.

2. Contingency tables. Here we have two types of situations which differ only in the methods of classifying.

*a.* We may have a contingency table which is analogous to a correlation table in that both classifications are based on continuous or ordered discrete variables for which we have only categorized information for $N$ individuals. The two variables might be in dichotomy (fourfold table), or one might be a dichotomy and the other manifold, or both might involve multiple categories. For these contingency tables it is meaningful to speak of the correlation between the two variables, and the degree of correlation might be appropriately specified by the tetrachoric $r$ or the fourfold point $r$ or the contingency coefficient (corrected or uncorrected); which measure is used depends upon meeting the requisite assumptions. Insofar as we are concerned only with $\chi^2$, we have the means for testing the significance of the correlation or association as a chance departure from zero or no relationship, and the significance test can be used without knowledge of the degree of correlation. Such a test of significance is sometimes spoken of as a test of independence—are the two classifications independent? If so, $\chi^2$ should be no larger than would arise by chance. If we have evidence for correlation or a lack of independence from the $\chi^2$ technique, we can proceed to calculate an appropriate coefficient for measuring the degree of correlation or the strength of association. The student should, as an exercise, convince himself that $\chi^2$ *per se* is not a measure of association.

*b.* The other contingency-type situation involves classification into categories for one variable vs. classification into unordered groups for the other, or one unordered grouping vs. another. The fundamental problem is apt to be that of comparing two or more groups with regard to multiple responses; i.e., we want a test of the difference between groups rather than a measure of correlation, which would not be entirely meaningful except in the loose sense that a particular response is associated more often with a particular group. As previously stated, the $df$ for a $k$ by $l$ contingency table is $(k - 1)(l - 1)$.

3. Goodness of fit. If we wish to check on whether it is reasonable to believe that a given frequency distribution is, within the limits of chance sampling, of the normal or some other specified type, a frequency curve having the same basic constants (e.g., $N$, $M$, and $S$ for the normal curve) as those computed from the observed frequency distribution can be fitted to the data. If a normal curve is being fitted, the table of normal curve functions is used to set up the theoretical or expected frequencies for the

**Table 13.4. Setup for computing $\chi^2$ from a fourfold table by means of a formula**

|   |   |   |
|---|---|---|
| $A$ | $B$ | $A + B$ |
| $C$ | $D$ | $C + D$ |
| $A + C$ | $B + D$ | $N$ |

several grouping intervals. Then $\chi^2$ can be computed in the usual manner. The $df$ will correspond to the number $(k)$ of grouping intervals less the number of constants derived from the data and used in the fitting process. For the normal curve the observed and theoretical distributions are made to agree as to $N$, $M$, and $S$; hence $df = k - 3$. An attempt will be made later to explain the reasoning back of the determination of $df$ when checking the goodness of fit of frequency curves.

**Fourfold contingency tables.** For illustrative purposes, let us first apply $\chi^2$ to a couple of 2 by 2 contingency tables for which the tetrachoric $r$, as well as the contingency coefficient, is an appropriate measure of the degree of correlation. Before we do this, it might be well to recall that $\chi^2$ for a fourfold table can be computed by a simple formula which does not require calculation of the four expected frequencies. Let the fourfold frequencies and marginal totals be set up as in Table 13.4. Chi square can be computed from

$$\chi^2 = \frac{N(BC - AD)^2}{(A + B)(C + D)(A + C)(B + D)} \tag{13.1}$$

This is simpler than calculation from the discrepancies between observed and expected frequencies. The requisite that no *expected* frequency shall

Table 13.5. $\chi^2$ applied to contingency (fourfold) tables

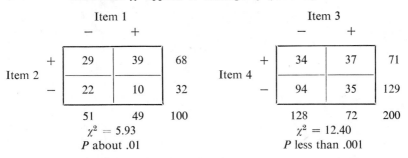

| | Item 1 − | Item 1 + | | | | Item 3 − | Item 3 + | |
|---|---|---|---|---|---|---|---|---|
| Item 2 + | 29 | 39 | 68 | Item 4 + | | 34 | 37 | 71 |
| Item 2 − | 22 | 10 | 32 | Item 4 − | | 94 | 35 | 129 |
| | 51 | 49 | 100 | | | 128 | 72 | 200 |

$$\chi^2 = 5.93$$
$$P \text{ about } .01$$

$$\chi^2 = 12.40$$
$$P \text{ less than } .001$$

be less than 5 still holds. A quick check on this can be obtained by multiplying the smaller right-hand marginal frequency by the smaller frequency on the bottom margin and dividing the product by $N$. This will yield the smallest expected frequency. In Table 13.5 will be found two fourfold tables for Stanford-Binet items. Direct substitution into formula (13.1) yields the two chi squares at the bottom of the table. The $P$ values are approximately .01 and less than .001, respectively. We can be reasonably sure that there is some correlation between the first two items, and fairly certain that items 3 and 4 are correlated. The value of the tetrachoric $r$ is .40 for each table, and the contingency coefficient (with no corrections) is .24 for each table. Thus we see that the $\chi^2$ $Ps$ associated with the same degree of correlation can be different. Why? Would it be possible for two fourfold tables to yield the same $\chi^2$ $P$, yet differ in the degree of relationship?

Another application of $\chi^2$ to fourfold tables is given in Table 13.6, in which the sexes at four age levels are compared in performance on a Stanford-Binet item. None of the $\chi^2$ values reaches 6.635, the value corresponding to the .01 level of significance, but three of them are large

Table 13.6. $\chi^2$ used to test sex differences in passing ( + ) or failing ( − )
a Binet item

| Age | 6 − | 6 + | | 7 − | 7 + | | 8 − | 8 + | | 9 − | 9 + | |
|---|---|---|---|---|---|---|---|---|---|---|---|---|
| B | 84 | 18 | 102 | 66 | 36 | 102 | 58 | 44 | 102 | 37 | 66 | 103 |
| G | 93 | 8 | 101 | 80 | 20 | 100 | 62 | 39 | 101 | 52 | 49 | 101 |
| | 177 | 26 | 203 | 146 | 56 | 202 | 120 | 83 | 203 | 89 | 115 | 204 |
| $\chi^2$ | | 4.30 | | | 5.89 | | | .43 | | | 5.02 | |
| $P$ | | <.05 | | | <.02 | | | <.50 | | | <.05 | |

enough to suggest a real sex difference. That a real difference may exist is also suggested by the fact that the boys are consistently superior at all four age levels. This brings us to an important property of $\chi^2$. The several chi squares for independent (i.e., based on different samples) tables may be summed to a total $\chi^2$, with $df$ equal to the sum of the $df$s for the chi squares being summed. Thus for Table 13.6 we have $4.30 + 5.89 + .43 + 5.02 = 15.64$ as a $\chi^2$ based on 4 degrees of freedom, by which we can judge the significance of the over-all sex differences shown in the four tables. With $\chi^2 = 15.64$ and $n = 4$, we find (from Table D) that $P$ is less than .01 (for $n = 4$, a $\chi^2$ of 13.28 corresponds to the .01 level). From Elderton's tables it can be ascertained that $P$ is about .004. In other words, as great a sex difference, considering all four age groups, would arise 4 times in 1000 by chance; hence it would be concluded that a real difference does exist for this item.

This combinatorial property of $\chi^2$ is important for all situations where frequency data from different groups cannot first be legitimately combined because of age or other differences. It is most useful when consistency is present among several comparisons, none of which taken singly possesses statistical significance. However, neither consistency nor insignificance for single comparisons constitutes a requisite for using the sum of chi squares as an over-all test of significance or as a means of arriving at one summary probability figure.

A rigorous proof of the proposition that a sum of independent $\chi^2$s will yield a new $\chi^2$ that follows the $\chi^2$ distribution with $df$ equal to the sum of the $df$s for the $\chi^2$s being summed is beyond the level of this book. Perhaps the reasonableness of the proposition can be seen from the following argument. Let

$$\chi^2_a = \Sigma \frac{(O_a - E_a)^2}{E_a} \quad \text{and} \quad \chi^2_b = \Sigma \frac{(O_b - E_b)^2}{E_b}$$

If we added the two $\chi^2$ values to get $\chi^2_t = \chi^2_a + \chi^2_b$, we would have exactly the same total $\chi^2$ as we would obtain by summing all the possible ratios of the form $(O_a - E_a)^2/E_a$ along with those of the form $(O_b - E_b)^2/E_b$ without regard to their source. That is, if there were six ratios involving the subscript $a$ and four ratios involving the subscript $b$, and we simply added these ten ratios without first computing $\chi^2_a$ and $\chi^2_b$, we would get precisely the same value as $\chi^2_t = \chi^2_a + \chi^2_b$. Now in determining the degrees of freedom for the $\chi^2_t$ obtained by summing the ten ratios, we need to look at the restrictions within the separate sets in order to specify how many of the $(O - E)$ deviations are independent. The number of independent deviations in set $a$ plus the number in set $b$ would, of course, be the total number of independent deviations in the two sets combined.

But this total number is simply the sum of the $df$s for the separate sets. Stated differently, $\chi^2$ is not conscious of how it was computed, whether by first getting partial sums which are then added or by proceeding directly to a total sum.

The single age comparisons in the foregoing example could, of course, be made by means of proportions. This could be done by formula (5.6), the discussion of which should be reviewed at this time. Let us examine the connection between the $\chi^2$ technique and the $D/S_D$ for proportions

**Table 13.7. Schema for comparing groups via $\chi^2$ and via difference between proportions (or percentages)**

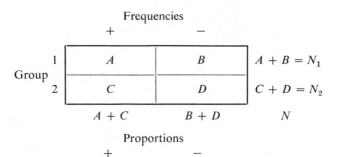

method of testing the significance of the difference between two groups, the individuals of which have been classified as either passing or failing, saying either yes or no, possessing or not possessing a characteristic, etc. All such comparisons begin with a fourfold frequency table of the type symbolized in Table 13.4, or an equivalent (the frequencies may have been recorded for only one category of the dichotomy, say the yeses, from which the frequencies for the other category may be readily inferred by subtraction). Table 13.7 contains the basic table of frequencies for the presence ($+$) or absence ($-$) of a characteristic for groups 1 and 2, and the basic table of proportions obtained by dividing the frequencies by the proper $N$s is indicated. Note that the $p$ and $q$ values on the bottom margin are the proportions to use in formula (5.6) for the standard error of the difference between $p_1$ and $p_2$. Note also that $p_1 = A/N_1 = A/(A + B)$ and that $p_2 = C/N_2 = C/(C + D)$.

In order to avoid carrying along a square root sign or radical, and for another reason which if not now obvious will soon become so, let us write the square of the expression for the critical ratio of the difference between the two proportions, $p_1$ and $p_2$, thus,

$$\frac{D^2}{S^2_D} = \frac{(p_1 - p_2)^2}{\dfrac{pq}{N_1} + \dfrac{pq}{N_2}}$$

When we replace all the proportions by their equivalents involving frequencies and the proper $N$s and also substitute frequencies for $N_1$ and $N_2$, we have

$$\frac{D^2}{S^2_D} = \frac{[A/(A + B) - C/(C + D)]^2}{\dfrac{[(A + C)/N] \cdot [(B + D)/N]}{A + B} + \dfrac{[(A + C)/N] \cdot [(B + D)/N]}{C + D}}$$

$$= \frac{\dfrac{(AC + AD - AC - BC)^2}{[(A + B)(C + D)]^2}}{\dfrac{(A + C)(B + D)(C + D) + (A + C)(B + D)(A + B)}{N^2(A + B)(C + D)}}$$

$$= \frac{(AD - BC)^2 N^2}{\left\{ \begin{array}{l} (A + B)(C + D)[(A + C)(B + D)(C + D)] \\ \quad + (A + C)(B + D)(A + B)] \end{array} \right\}}$$

$$= \frac{(AD - BC)^2 N^2}{(A + B)(C + D)(A + C)(B + D)(A + B + C + D)}$$

$$\frac{D^2}{S^2_D} = \frac{(AD - BC)^2 N}{(A + B)(C + D)(A + C)(B + D)}$$

which equals $\chi^2$ as given by formula (13.1) for the fourfold table. This confirms a fact already mentioned, that for 1 degree of freedom $\chi^2$ is the same as the square of the critical ratio. Since formula (5.6) is applicable only for comparing proportions based on independent samples, it follows that $\chi^2$ is similarly restricted. That is, $\chi^2$ as computed from a fourfold table by (13.1) does not allow for any correlational factor which might be introduced because the two groups consist of paired or matched individuals or for the correlational factor which would be present if $p_1$ and $p_2$ (or the corresponding frequencies) were based on the *same* individuals as in a pretest, intervening experience, posttest situation.

**Significance of changes.** The student should carefully note that although the application of $\chi^2$ to fourfold tables of frequencies like that of Table 5.2, which is here reproduced with minor changes as Table 13.8, provides

**Table 13.8. Fourfold table of frequencies and proportions for a first set vs. a second set of responses from the *same* individuals**

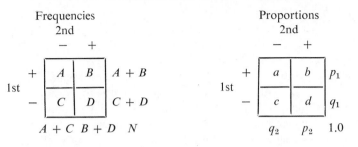

a means of testing the significance of the association or correlation between two sets of responses, such an application does not test the significance of change from the first to the second set of responses. This latter test can be made by means of formula (5.5). It is also possible to test the significance of any found change by the use of $\chi^2$. To do this, we first note that a net change for the group must necessarily involve the difference between the frequencies, $A$ and $D$, since the $B$ and $C$ cases represent those who showed no change. The null hypothesis would be that the universe frequencies are not different; i.e., for a given sample, $A$ and $D$ would differ only as a result of chance sampling. Since $A + D$ represents the total number of individuals who changed (the $A$s from $+$ to $-$, and the $D$s from $-$ to $+$), in setting up the null hypothesis concerning the net change it would seem appropriate to say that, if $A + D$ individuals changed, $(A + D)/2$ would change in one direction and $(A + D)/2$ in the other direction. Thus $(A + D)/2$ would become the expected frequency; then $A - (A + D)/2$ and $D - (A + D)/2$ would become the discrepancies between observed and expected (on the basis of the null hypothesis) frequencies. If $A = D$, both discrepancies would become zero. Squaring each discrepancy and dividing by $E$ and then summing the two quotients or doubling either one will give a $\chi^2$ which is based on 1 degree of freedom (why 1 degree of freedom?). A little algebraic manipulation shows that

$$\chi^2 = \frac{(A - D)^2}{A + D} \tag{13.2}$$

for the particular situation in which we wish to test the significance of over-all changes.

Comparison of formula (13.2) with formula (5.3) shows that we again have a $\chi^2$, with 1 degree of freedom, which equals the square of an $x/\sigma$. The reasoning back of the statement given on p. 58 that formulas (5.3), (5.4), and (5.5) are inapplicable unless $A + D$ equals 10 or more should now be clearer to the reader. If $A + D$ were less than 10, the two $E$s would be less than 5, an acceptable though none too conservative lower limit for $E$. A correction (for continuity) needed when the $E$s are smaller than 10 will be given shortly. One thing which may puzzle the reader at this time is the fact that formula (13.2) does not contain a total $N$. Its algebraic equivalent, $(D/\sigma_D)^2$, with $\sigma_D$ calculated by formula (5.5), does contain $N$, so the absence of $N$ from (13.2) is more apparent than real.

The advantage of the $\chi^2$ over the $D/\sigma_D$ technique for testing the significance of net changes in responses lies in the fact that $\chi^2$ values for two or more groups which have been used in an experiment can be summed to a new $\chi^2$ with $n$ equal to the sum of the separate $df$s; in this case $n$ equals the number of chi squares being summed.

Formula (13.2) is, of course, not restricted to situations involving changes in responses. If we have the same individuals giving, say, yes or no responses to two different questions and we desire to test the significance of the difference between the frequencies (or proportions) of yeses or noes, formula (13.2) is applicable. Or suppose we wish to know whether there is a significant difference in the difficulty of two test items which have been administered to the same group. For example, in Table 13.5 we have 49 and 68 individuals passing items 1 and 2 respectively. Since $N = 100$, the proportions are .49 and .68 (or 49 and 68 per cent). By formula (13.2) we have $\chi^2 = (29 - 10)^2/(29 + 10) = 9.26$, which for 1 degree of freedom falls between the .01 and .001 levels of significance; hence, it would be concluded that the two items are different in difficulty. If we use formula (5.5), we get $z = (p_2 - p_1)/\sigma_D = (.68 - .49)/\sqrt{(.10 + .29)/100} = .19/.0624 = 3.04$, which leads to the same probability figure as that for a $\chi^2$ of 9.26. Either method may be used. Both make due allowance for the correlation which is present because the frequencies or proportions being compared are based on the *same* individuals.

**Correction for continuity.** As pointed out earlier, when $df = 1$ the use of $\chi^2$ is suspect if any one $E$ is less than 5. For fourfold contingency tables, an allowance for discontinuity can be made by applying Yates' correction, which should always be used when any one $E$ is 5 to 10. A small $E$ is most likely to occur either when the total $N$ is small or when one or both of the marginal totals involve extreme dichotomies. It is easy to determine the smallest $E$ by dividing the product of the two smaller marginal frequencies by the total $N$. Yates' correction can be incorporated in formula (13.1), which becomes

$$\chi^2 = \frac{N(|BC - AD| - N/2)^2}{(A + B)(C + D)(A + C)(B + D)} \tag{13.3}$$

and indicates that the absolute difference between $AD$ and $BC$ is to be reduced by $N/2$. Formula (13.2) can also be written to include a correction for continuity. The corrected form

$$\chi^2 = \frac{(|A - D| - 1)^2}{A + D} \qquad (13.4)$$

involves decreasing the absolute value of the difference between $A$ and $D$ by 1. Formula (13.4) is to be preferred to (13.2) when $A + D$ is less than 20. The reasoning back of Yates' correction is precisely the same as that given on p. 47 of Chapter 5.

**One-tailed vs. two-tailed test.** It will be recalled from our discussion of the sampling distribution of $\chi^2$ that the $P$s obtainable from Table D are the probabilities of the chance occurrence of as large a $\chi^2$ as that observed; that is, levels of significance such as $P = .05$ or .01 or .001 are based on *one* (the right-hand) tail of the sampling distribution of $\chi^2$. Does this mean that it is a one-tailed test in the hypothesis testing sense discussed earlier (pp. 64–65)? Let us recall a couple of facts. First, when using the binomial to indicate something of the nature of the $\chi^2$ distribution we saw that both tails of the binomial were combined as one tail of the $\chi^2$ distribution. Second, for 1 degree of freedom $\chi^2 = (x/\sigma)^2$. Now an $x/\sigma$ of 1.96 corresponds to the $P = .05$ level as a two-tailed test. The square of 1.96 gives a $\chi^2$ of 3.84, which we can see from Table D also corresponds to the .05 level. Hence the $P$s, for 1 degree of freedom, read from Table D are equivalent to those based on the two-tailed test despite the fact that only one tail of the $\chi^2$ distribution is involved.

If the decision to be made or the hypothesis to be tested calls for a one-tailed test, the $P$s from Table D need to be halved: a $\chi^2$ of 5.41 (instead of 6.64) is required for the .01 level, and a $\chi^2$ of 2.71 (instead of 3.84) gives the .05 level. Incidentally, for 1 degree of freedom, a $\chi^2$ $P$ obviously can be obtained by entering its square root into the normal curve table—whether such a $P$ from $x/\sigma$ is based on one or both tails of the normal distribution depends on the hypothesis being tested. As we proceed, the student should convince himself that the notion of direction of differences, hence the idea of a one-tailed test, does not make sense in other applications of $\chi^2$.

**Comparison of two or more correlated proportions.** Formula (13.2) has been extended to provide a method for testing whether three or more nonindependent proportions (or sets of frequencies) differ significantly among themselves. For example, we may have pass-fail (or yes-no, or some other dichotomous) information on $C$ items (or questions) for $N$ individuals; or we may have only one item with responses from $N$ persons under $C$ different conditions; or one item with responses from $N$ sets of $C$ matched persons each, that is, $C$ matched groups.

Data from such situations can be arranged in a table consisting of $N$ rows and $C$ columns. The total number of passes (yeses) in a given column divided by $N$ will, of course, be the proportion of passes (or yeses) in that column. Do these $C$ proportions (or the totals) differ significantly in an over-all sense? The null hypothesis is that all the proportions are the same except for chance. To test the null hypothesis we will need to obtain not only the column totals (number of passes) but also a similar total for each of the $N$ rows. Let $T$ stand for the total in any column and $X$ stand for the total in any row. This $X$ is a sort of "score" for the person—his number of passes (or yeses) on the $C$ items. The sampling distribution of the quantity

$$Q = \frac{(C - 1)[C\Sigma T^2 - (\Sigma T)^2]}{C\Sigma X - \Sigma X^2} \tag{13.5}$$

follows the $\chi^2$ distribution with $C - 1$ degrees of freedom for $N$ large ($N > 30$, presumably).

The computation of $Q$ is so easy that it need not be illustrated. If an obtained $Q$ exceeds the $\chi^2$ required for a chosen level of significance, we conclude that the (correlated) proportions do differ in an over-all sense, that is, they are not homogeneous.

**Chi square for 2 by $k$ tables.** The calculation of $\chi^2$ from a table with two rows and $k$ columns (or two columns and $k$ rows) can be accomplished by way of expected cell frequencies calculated as previously suggested from the marginal totals or by means of

$$\chi^2 = \frac{N^2}{A_t B_t}\left[\Sigma \frac{B^2_i}{A_i + B_i} - \frac{B^2_t}{A_t + B_t}\right] \tag{13.6}$$

in which the $A$s and $B$s have the meanings indicated in Table 13.9, wherein will be found the frequencies for two groups classified according to five response categories. The necessary computations required by formula (13.6) are also included in the table. Note that, as usual, the marginal totals are first found by summing across and down. Column D is obtained by dividing the entries in column B by the adjacent values in column C, and column E results from multiplying the D column values by the B column figures. These same operations, when applied to the last (or totals) line, lead to the column E entry of 49.44, which is the value of the $B^2_t/(A_t + B_t)$ term in formula (13.6). Summing the first five figures in column E yields 50.83, or the $\Sigma$ term of (13.6), and the difference between 50.83 and 49.44 is 1.39, the value of the bracketed part of the formula. When this is multiplied by $N^2/A_t B_t$, we have $\chi^2$, which for a $df$ of 4 yields a $P$ of about .16. In other words, once in six trials differences as large as those in Table 13.9 would occur by chance; hence we have insufficient

**Table 13.9. The calculation of $\chi^2$ from a 2 by $k$ table: two groups and $k$ ($= 5$) responses**

| | Col. A Group I | Col. B II | Col. C $A_i + B_i$ | Col. D $\dfrac{B_i}{A_i + B_i}$ | Col. E $\dfrac{B^2_i}{A_i + B_i}$ |
|---|---|---|---|---|---|
| 1 | $27(= A_1)$ | $15(= B_1)$ | 42 | .3571 | 5.36 |
| 2 | $26(= A_2)$ | $16(= B_2)$ | 42 | .3810 | 6.10 |
| 3 | $247(= A_3)$ | $110(= B_3)$ | 357 | .3081 | 33.89 |
| 4 | $41(= A_4)$ | $8(= B_4)$ | 49 | .1633 | 1.31 |
| 5 | $39(= A_5)$ | $15(= B_5)$ | 54 | .2778 | 4.17 |
| | | | | | 50.83 |
| Totals | $380(= A_t)$ + | $164(= B_t)$ = | $544(= N)$ | .3015 | 49.44 |
| | | | | | 1.39 |

$$\frac{544^2}{(380)(164)} = 4.75; \qquad \chi^2 = (4.75)(1.39) = 6.60$$
$$n = 4, \qquad P = .16$$

evidence for concluding that the universes from which these two samples were drawn differ in regard to their responses to the asked question.

If we had to depend on the $D/S_D$ technique for testing the significance of the group differences in Table 13.9, five $z$ ratios would result—for each category there is a possible difference in proportions or percentages with a standard error for each difference. The five $z$s might, and usually would, lead to five different $P$ values with a consequent predicament as to interpretation. Offhand, it might be argued that, if any $z$ so determined reached an acceptable level of significance, we would be justified in concluding that the difference between the groups was real rather than chance. That such an argument may be fallacious is well illustrated by the data of Table 13.9, which are actual data. When these data first came to the author's attention, the table was in percentage form with a $z$ worked out only for the category showing the largest difference. This $z$, based on formula (5.6), was 2.54, which is near the $P = .01$ level of significance, and it had accordingly been concluded that a real difference had been found. Now, when we consider the $\chi^2$ $P$ of .16 for the over-all comparison, we are not justified in placing much confidence in such a conclusion.

Why the apparent inconsistency between two tests of significance? Since most investigators are looking for group differences rather than group similarities, there is the tendency to single out a category for comparison, not because of intrinsic a priori interest in that category but because it happens to yield the largest difference. By this a posteriori selection, there is a tendency to capitalize on differences which may be

266          PSYCHOLOGICAL STATISTICS

large mainly as a result of chance. A similar situation occurs when we have
the means for several groups—the largest of the possible differences may
be the largest partly or entirely as a result of chance. Thus the use of $\chi^2$
for such situations as are exemplified in Table 13.9 not only provides an
over-all single index of significance, but also helps us avoid false conclusions.

**Application to $k$ by $l$ tables.** Consider the data of Table 13.10, which
contains a contingency-type table involving three groups and three possible
opinion responses. To test the significance of the differences between the
groups by use of the $D/S_D$ technique would involve comparing the percentages for group I vs. II, I vs. III, and II vs. III, for each of the three

**Table 13.10. Table of frequency of three possible responses for three groups of
individuals—percentages in the parentheses add downward to 100\***

| Motivation of Conscientious Objectors | Group | | | Total |
|---|---|---|---|---|
| | I | II | III | |
| Not cowards | 24(27.0) | 56(53.8) | 71(69.6) | 151 |
| Partly cowards | 30(33.7) | 23(22.1) | 19(18.6) | 72 |
| Cowards | 35(39.3) | 25(24.0) | 12(11.8) | 72 |
| $N$s | 89(100.0) | 104(99.9) | 102(100.0) | 295 |

\* Data from Leo Crespi, *J. Psychol.*, 1945, **19**, p. 285.

responses—a total of nine $z$s. Straightforward computation gives $\chi^2$
$= 36.58$, which for $df = 4$ is double the value of the $\chi^2$ needed for the
$P = .001$ point. From Elderton's table we find that $P$ is about .000001;
hence Table 13.10 as a whole exhibits highly significant differences between
the groups.

Perhaps a better understanding of the extent of the differences can be had
by considering the percentages given in parentheses in the table. Membership in group III means a greater tendency to the "not cowards" response.
Group I tends more to give the "cowards" response. Now it happens that
the three groups, I, II, and III, can be (and are) placed in an ordered series
for amount of education: grammar school, high school, and college,
respectively. Thus the association shown in the table is in the direction of
less disparagement of conscientious objectors by those in the higher
educational level. The strength of association or degree of correlation is
represented by a contingency coefficient of .33, which may seem rather
low in light of the highly significant $\chi^2$ $P$. This illustrates a point which
most readers will already have grasped: high statistical significance and a

high degree of association are far from synonymous. Consideration of the data of Table 13.10 readily indicates the difficulty of predicting responses when the extent of association is represented by a $C$ of .33.

As in the 2 by $k$ table, so here it is better to calculate an over-all $\chi^2$ before examining by the $z$ technique any of the possible separate comparisons. Unless the $\chi^2$ $P$ is significant, it is unwise to proceed with such comparisons.

The calculation of $\chi^2$ for a $k$ by $l$ table is greatly facilitated by the

**Table 13.11. Schema for calculating $\chi^2$ from a $k$ by $l$ table**

|   | 1 | 2 | 3 | Total |
|---|---|---|---|-------|
| 1 | $f_{11}$ $f^2_{11}/N_1$ | $f_{12}$ $f^2_{12}/N_2$ | $f_{13}$ $f^2_{13}/N_3$ | $n_1$ $\sum_c f^2_{1c}/N_c$ |
| 2 | $f_{21}$ $f^2_{21}/N_1$ | $f_{22}$ $f^2_{22}/N_2$ | $f_{23}$ $f^2_{23}/N_3$ | $n_2$ $\sum_c f^2_{2c}/N_c$ |
| 3 | $f_{31}$ $f^2_{31}/N_1$ | $f_{32}$ $f^2_{32}/N_2$ | $f_{33}$ $f^2_{33}/N_3$ | $n_3$ $\sum_c f^2_{3c}/N_c$ |
| Total | $N_1$ | $N_2$ | $N_3$ | $N$ |

$$\chi^2 = N \left[ \frac{\sum_c f^2_{1c}/N_c}{n_1} + \frac{\sum_c f^2_{2c}/N_c}{n_2} + \frac{\sum_c f^2_{3c}/N_c}{n_3} - 1 \right]$$

following procedure. Let the observed frequencies be represented by $f$s as in Table 13.11. Divide the square of each cell frequency by $N_c$, the total $N$ for its column; sum these quotients across, one sum for each row; divide each of these sums by the respective total row frequency, $n_r$; add *these* quotients, deduct 1, and multiply by the grand total $N$. The result is $\chi^2$. The first set of quotients ($kl$ in number) should be carried to two decimals, and the second set of quotients should be carried to three decimals. The computational process is given in symbols in Table 13.11. The first subscript to $f$ indicates the row and the second the column. $\sum_c$ means sum with $c$ taking on in turn the column designations, 1, 2, 3.

**Goodness of fit.** The use of $\chi^2$ in testing the goodness of fit of a theoretical curve to an observed frequency distribution is illustrated in Table 13.12. We start with an actual distribution, usually with more grouping intervals than in our example, and the descriptive statistical measures therefor. In fitting the normal curve to the distribution of Table 13.12, we need $N$, $M$, and $S$. To set up for each interval the frequency which would hold for the best-fitting normal curve, we go

**Table 13.12. Goodness of fit of normal curve to Stanford-Binet IQs, form M**

| IQ | $O$ | $x/S$ | Proportionate Area | $E$ | $O - E$ | $(O - E)^2/E$ |
|----|-----|-------|--------------------|-----|---------|----------------|
| 160 | 3⎫ | | | | | |
| 150 | 13⎬16 | | .0041 | 12 | 4 | 1.33 |
| | | 2.645 | | | | |
| 140 | 55 | | .0158 | 47 | 8 | 1.36 |
| | | 2.057 | | | | |
| 130 | 120 | | .0512 | 152 | −32 | 6.74 |
| | | 1.468 | | | | |
| 120 | 330 | | .1186 | 352 | −22 | 1.38 |
| | | .879 | | | | |
| 110 | 610 | | .1958 | 582 | 28 | 1.35 |
| | | .291 | | | | |
| 100 | 719 | | .2316 | 688 | 31 | 1.40 |
| | | −.298 | | | | |
| 90 | 592 | | .1950 | 579 | 13 | .29 |
| | | −.886 | | | | |
| 80 | 338 | | .1177 | 350 | −12 | .41 |
| | | −1.475 | | | | |
| 70 | 130 | | .0506 | 150 | −20 | 2.67 |
| | | −2.064 | | | | |
| 60 | 48 | | .0155 | 46 | 2 | .09 |
| | | −2.652 | | | | |
| 50 | 7⎫12 | | .0040 | 12 | 0 | .00 |
| 40 | 4⎬ | | | | | |
| 30 | 1⎭ | | | | | |
| | 2970 = N | | .9999 | 2970 | 0 | 17.02 = $\chi^2$ |

$$M = 104.56 \qquad df = 11 - 3 = 8; \qquad P = .03$$
$$S = 16.99$$

through the tedious process of determining the proportionate area under the theoretical curve for each interval. Once the proportions are known, each is multiplied by $N$ to secure the expected frequencies. The proportions are ascertained by calculating the $x/S$ value of the boundary limits of the intervals. For example, the 110–119 interval may be thought of as running from 109.5 to 119.5, since IQs are rounded to the nearest integer. Then $(109.5 - 104.56)/16.99 = .2907$ as the $x/S$ for the lower limit, and $(119.5 - 104.56)/16.99 = .8793$ as the $x/S$ for the upper limit of the 110–119 interval. Of course, .8793 is also the lower limit of the 120–129 interval. Now the difference, $.8793 - .2907 = .5886$, is the same as $10/16.99$ or $i/S$, which is the interval width expressed in $x/S$ units. Adding

.5886 once to .2907 gives .879 (it is sufficient to retain three decimals); adding it twice gives 1.468; and so on. Then subtracting .5886 once from .2907 gives $-.298$; subtracting twice gives $-.886$; etc.

When the boundary limits in terms of $x/S$ have been set up, the proportionate area for a given interval is found by using the table of normal curve areas. The two top intervals have been combined, and likewise the three bottom intervals, so as to have no expected frequencies less than 10. The proportionate areas, .0041 and .0040, represent the areas beyond given points, and the $E$s at top and bottom are the number of cases expected beyond these same points. Note that the sum of the proportions should be unity within limits of rounding errors, and that the sum of the expected frequencies should be the same as the sum of the observed frequencies. Perhaps it is unnecessary to point out that the expected frequencies form an exactly (within limits of rounding errors and for the given intervals) normal distribution which will yield the same $M$ and $S$ as the observed distribution with which we started.

Straightforward calculation gives a $\chi^2$ of 17.02. With $df = 11 - 3$ (number of intervals minus the number of constants used in the fitting), $P = .03$; i.e., only 3 times in 100 would as large a $\chi^2$ arise by chance, or only 3 times in 100 would we get a worse fit if the universe of IQs were distributed as a normal curve. This would lead us to question whether IQs, as measured by Form M of the 1937 Revision of the Stanford-Binet, are distributed in the normal curve fashion. The same data with intervals of size 5 give a $\chi^2$ $P$ of .003, and the degree of kurtosis (by moments) is thrice its standard error; therefore it can be concluded that the observed distribution is not a chance departure from a normal distribution.

Thus the $\chi^2$ technique provides us with a test by means of which we can judge that the frequencies of a given distribution do not follow the frequencies of a theoretical curve closely enough to be regarded as chance departures therefrom. Note that a smaller value for $\chi^2$ for the example of Table 13.12 would not prove that the universe is normal even though the $P$ were as large as .90 or .95. This would merely indicate that the given data were consistent with the normal distribution. As a matter of fact, so-called excellent fits leading to $P$s of .99 or more are suspect. When $P = .01$, it is said that chance sampling would lead to a worse fit only once in 100 times; when $P = .99$, it is said that chance sampling would lead to a better fit only once in 100 times. In other words, if $P$ is between .05 and .01, the hypothesis that the universe distribution is of the normal type (or whatever type was fitted) is questionable; if $P$ is .01 or less, this hypothesis is rejected; if $P$ is between .95 and .99, we may suspect the fit as being too good; if $P$ is .99 or more, we should definitely look for an error in calculation or for some type of restraint on the operation of chance. Too good a fit is as open to question as too poor a fit. If $P$ is between .05 and .95, the fit is said to be satisfactory.

When the goodness of fit of frequency curves is being tested, the $df$ depends on the number of grouping intervals and on the number of restrictions imposed or the ways in which the expected distribution is made to agree with the observed distribution. The general principle behind the determination of $df$ for $\chi^2$ as a test of fit may be illustrated for the case of testing the goodness of fit of the normal curve. The expected and observed distributions are made to agree with respect to $N$, $M$, and $S$. Suppose that we have $k$ grouping intervals and that we let $f_i$ stand for the frequency in the $i$th interval and $X_i$ for its score value (midpoint), and that $x_i$ represents the corresponding deviation score value for this midpoint. Then the following equations will hold:

$$f_1 + f_2 + f_3 + \cdots + f_i + \cdots + f_k = N$$
$$f_1 X_1 + f_2 X_2 + f_3 X_3 + \cdots + f_i X_i + \cdots + f_k X_k = NM$$
$$f_1 x^2_1 + f_2 x^2_2 + f_3 x^2_3 + \cdots + f_i x^2_i + \cdots + f_k x^2_k = NS^2$$

Now, if all the $f$ values were known except $f_1$, $f_2$, and $f_3$, those parts to the right of the $f_3$ term in the first of these equations could be added numerically. The resulting sum could be shifted to the right of the equality sign and then combined numerically with $N$, giving an equation of the type $f_1 + f_2 + f_3 = A$, where $A$ equals $N$ minus the sum of all the frequencies save the first three. Likewise, the parts beyond the $f_3$ term in each of the other two equations could be summed numerically, shifted to the right, and combined numerically with the constant, $NM$ for the second and $NS^2$ for the third equation.

This procedure will lead to three simultaneous equations with $f_1$, $f_2$, and $f_3$ as the unknowns.

$$f_1 + f_2 + f_3 = A$$
$$f_1 X_1 + f_2 X_2 + f_3 X_3 = B \text{ (say)}$$
$$f_1 x^2_1 + f_2 x^2_2 + f_3 x^2_3 = C \text{ (say)}$$

It is a well-known principle of algebra that three equations in three unknowns will be satisfied (if solvable) by just one set of values for the unknowns. For our particular problem, this means that, as soon as the frequencies for all but three (any three) intervals are known, these three remaining frequencies are not "free to vary"; they are fixed because of the requirements that the frequencies or functions thereof must add to $N$, $NM$, and $NS^2$. We accordingly lose 3 degrees of freedom, and therefore when we are testing the fit of a normal curve to a distribution with $k$ intervals, the $df$ is $k - 3$.

Although the Kolmogorov-Smirnov (K-S) test does not involve $\chi^2$, we include it here since it also provides a test of goodness of fit. The procedure is relatively simple. The $k$ observed frequencies are converted to cumulative frequencies, which are divided by $N$ to secure cumulative proportions. For the given $M$ and $S$, the proportions per interval expected on the basis of the normal curve are calculated (e.g., the proportionate area column of Table 13.12), then cumulated. We thus have two sets of cumulative proportions. The $k$ pairs of values are examined to find the largest pair difference, $D$; that is, the largest discrepancy between observed and expected. (This $D$ is, of course, in proportion, not percentage, units.)

The sampling distribution of $D$ is such that for $N$ greater than 35, $D$ must reach:

$$1.14/\sqrt{N} \text{ for significance at the } P = .10 \text{ level}$$

$$1.36/\sqrt{N} \text{ for significance at the } P = .05 \text{ level}$$

$$1.63/\sqrt{N} \text{ for significance at the } P = .01 \text{ level}$$

The advantages of the K-S test over the $\chi^2$ test for goodness of fit are twofold: the K-S test is applicable for $N$ smaller, and it is a more powerful test than the $\chi^2$ test. The latter advantage means that departure from normal form is more apt to be detected by the K-S test. Stated differently, compared to the $\chi^2$ test the K-S test is less apt to mislead us into accepting the hypothesis of normality of distribution.

Although the method of fitting set forth in Table 13.12 should aid in comprehending the meaning of goodness of fit—an observed frequency contrasted with an expected frequency obtained via the concept of area under a normal curve for the interval—there is a computationally shorter method for calculating the $E$ values. The $x/S$ value of the *midpoint* of each interval is determined, and then the ordinate (or $h$ value) for each $x/S$ value is written down from Table $A$. The products, each $h$ times $iN/S$, provide the series of $E$s for the intervals. For the K-S test, $h$ times $i/S$ will yield the expected proportions, which are then cumulated.

The $\chi^2$ test can be used to test the difference between two observed distributions (this becomes a 2 by $k$ situation, with $k - 1$ as the $df$), and the K-S test has been extended for the same purpose. But for comparing two observed distributions, both the $\chi^2$ test and the K-S test have serious drawbacks: significance may reflect difference in location parameters, or in variances, or in distribution shape, or in any combination of these.

In this chapter we have discussed the essential nature of $\chi^2$ and have pointed out typical applications. By now the student should appreciate the advantages of $\chi^2$ over percentage comparisons and have some insight into the use of $\chi^2$ as a means of testing hypotheses.

## EXACT OR DIRECT PROBABILITIES

The $\chi^2$ $P$s obtainable from Table D are approximations in that areas under a continuous curve are taken as estimates of values which form a point distribution. Even with Yates' correction for continuity, the approximation is none too good when $E$ values are less than 5. This raises the question as to the criterion for judging the closeness of such approximations, and the answer is that for situations involving 1 degree of freedom it is possible to specify exact probabilities. How?

First, consider the problem of deciding on the basis of a specified number of successes whether a chap can distinguish between two cigarette brands. We learned in Chapter 5 that the exact $P$ for the probability of as many correct identifications can be obtained by the binomial distribution; hence we need not use the normal curve or the $\chi^2$ approximation. But such approximations not only are very convenient computationally for $N$ (or $n$) large, but also are accurate enough. In checking a $\chi^2$ $P$ against an exact $P$ derived from the binomial, we must bear in mind the possibility of confusing one- and two-tailed tests; both methods should be alike in this regard.

Second, consider the $\chi^2$ test of the significance of change (or difference between two correlated frequencies or proportions) given by formula (13.2). We can obtain an exact $P$ for this situation by resorting to the binomial (see p. 56). Again, in calculating the binomial $P$, we must give consideration to whether we had intended a one-tailed or a two-tailed test.

Third, consider the fourfold table for which formula (13.1) is appropriate in testing either the significance of association or the significance of the difference between two groups. For this situation the binomial is not easily applicable (except when the frequencies are equal on one, or both, of the margins). Exact $P$s can be obtained for such tables by a rather tedious procedure which we shall now describe. It can be shown that the probability for a particular observed set of frequencies, $A$, $B$, $C$, and $D$, for fixed margins is

$$P = \frac{(A + B)!\,(C + D)!\,(A + C)!\,(B + D)!}{N!\,A!\,B!\,C!\,D!}$$

To have a test comparable to the usual significance test, we would also need the $P$s for all sets of frequencies deviating farther than the observed set from the null values of no association. This can be made clearer by an example. Table 13.13 shows an observed set (part I) and sets showing higher association (parts II and III). Note that each part is derived from the preceding part by subtracting 1 from both $A$ and $D$ and adding 1 to

**Table 13.13. Series of fourfold table frequencies required for calculating $P$ directly and exactly**

| I | − | + | |
|---|---|---|---|
| + | 3 | 9 | 12 |
| − | 6 | 2 | 8 |
| | 9 | 11 | 20 |

| II | − | + | |
|---|---|---|---|
| + | 2 | 10 | 12 |
| − | 7 | 1 | 8 |
| | 9 | 11 | 20 |

| III | − | + | |
|---|---|---|---|
| + | 1 | 11 | 12 |
| − | 8 | 0 | 8 |
| | 9 | 11 | 20 |

both $B$ and $C$. This process is continued until $A$ or $D$ or both become zero. Note that the marginal frequencies remain the same.

Application of the foregoing formula to each table in turn will yield the probability for each set of frequencies, and the sum of these $P$s will be the probability of as great association (in the given direction) as that indicated by the starting (observed) set of frequencies. We have

$$P_{\rm I} = \frac{(12!)(8!)(11!)(9!)}{(20!)(3!)(9!)(6!)(2!)} = .0367$$

$$P_{\rm II} = \frac{(12!)(8!)(11!)(9!)}{(20!)(2!)(10!)(7!)(1!)} = .0031$$

$$P_{\rm III} = \frac{(12!)(8!)(11!)(9!)}{(20!)(1!)(11!)(8!)(0!)} = .0001$$

The sum of these separate probabilities gives $P = .0399$, or .04 (to two decimals) as the probability of obtaining sets as extreme (in one direction) as the set observed in part I of Table 13.13. If the situation calls for a two-tail test, the mere doubling of the calculated one-tailed $P$ will give the exact probability of as large a difference (or as great an association) irrespective of direction *only* when the marginal frequencies are identical; that is, in a setup such as Table 13.4, only when $A + B = B + D$. Otherwise, exactly the same degree of association in the opposite direction cannot occur, and the doubling of the one-tailed $P$ will only approximate the required two-tailed $P$. Consider again the left-hand fourfold table in Table 13.13. Association in the opposite direction would call for a majority of the cases in the upper-left and lower-right cells, but how many cases for as great a negative association as the observed positive association? One criterion would be to say that for the negative association the frequencies $A$, $B$, $C$, $D$ should be such that, with the margins unchanged, the value of $(BC - AD)$ must equal $(9 \times 6 - 3 \times 2)$ or 48 except for having the opposite sign. [Note: the value of $(BC - AD)$ enters into both the fourfold point $r$ and the contingency coefficient for the $2 \times 2$ table.] If we try

$$
\begin{array}{ccc}
8 & 4 & 12 \\
1 & 7 & 8 \\
9 & 11 &
\end{array}
$$

we get $(BC - AD) = -52$, which would indicate a greater degree of association than the 48. If we try 7 for $A$, we would have $(BC - AD) = (5 \times 2 - 7 \times 6) = -32$, which leads to a lesser degree of association than the 48. It is simply impossible with nonsymmetrical marginal frequencies to have the same degree of negative as positive association.

**Table 13.14. Sets of frequencies for negative association as strong as the positive association in I of Table 13.13**

|  | I | | | | II | |
|---|---|---|---|---|---|---|
| + | 8 | 4 | 12 | + | 9 | 3 | 12 |
| − | 1 | 7 | 8 | − | 0 | 8 | 8 |
|  | 9 | 11 | 20 | | 9 | 11 | 20 |

The best that we can do to obtain a two-tailed $P$ is to consider all possible sets of frequencies that give rise to negative association as great as the observed positive association. For this, only the two tables given in Table 13.14 qualify. For I of Table 13.14 we have

$$
P_{\mathrm{I}} = \frac{12!\,8!\,11!\,9!}{20!\,8!\,4!\,1!\,7!} = .0224
$$

and for II we have

$$
P_{\mathrm{II}} = \frac{12!\,8!\,11!\,9!}{20!\,9!\,3!\,0!\,8!} = .0026
$$

which when summed give a $P$ of .0250. When this is added to the $P$ of .0399 for the positive association we get a $P$ of .0649 as the $P$ for the two-tailed test, a value which does not correspond to *either* of the one-tailed $P$s doubled.

The argument regarding the effect of asymmetry on the reciprocity of one- and two-tailed $P$s also holds for those situations involving fairly sizable $N$s when either or both margins of the fourfold table depart markedly from a 50–50 split. If the smallest $E$ is less than, say, 10, the $\chi^2$ $P$ (even with Yates' correction) when halved does not lead to an entirely valid one-tailed $P$ value.

The computation of the separate $P$s, laborious even with an ordinary table of logarithms, is greatly facilitated by a table of the logarithms of factorials, such as Table XLIX of Part I of Pearson's *Tables for statisticians and biometricians*. For $N$s up to 28, special tables are available (see Table I of S. Siegel's *Nonparametric statistics*, New York: McGraw-Hill, 1956) for judging whether the exact probabilities reach certain commonly used levels of significance.

# *Chapter 14*

# INFERENCES ABOUT
# VARIABILITIES

We now return to the problem of statistical inference based on measures for continuous variables. This chapter is concerned with inferences regarding variances and differences between variances, and presents a basic theoretical distribution which will serve in later chapters when we again discuss tests based on means.

First, a digression from our main interest in this chapter to present a simple algebraic development of the standard error of the mean formula, which will make it unnecessary to "take on faith" a part of the next derivation concerning variance estimation. Let $\bar{X}$ (the symbol which we will use henceforth) represent a sample mean. Suppose a random sample of size $N$, and that we pull $R$ successive samples (with replacements if $R$ is large and the population size is finite). We can arrange each set of $N$ scores to form a row, with random assignment to the 1st, 2nd, $\cdots$, $N$th position. We will have $R$ rows and we can regard the positions as forming columns, $N$ in number. Our table will consist of $R$ rows and $N$ columns of scores with no particular meaning attached to the columns because assignment thereto is a chance matter. With $r$ standing for the $r$th row and $c$ for the $c$th column, we will have a table like Table 14.1.

Summing across columns leads to a $\Sigma X$ $(= N\bar{X})$ for each row. These row sums will, of course, vary from row to row (we long ago learned that means vary from sample to sample). Each of the $R$ sums is the result of summing $N$ scores and thus may be regarded as a sort of "total score" for the row obtained by adding $N$ parts, each column contributing a part, or a component. With variation within each column, this summing is as though $N$ variables were being summed to get a total score. Since the

**Table 14.1. Score layout for $R$ replications (successive samples) of size $N$**

|   | 1 | 2 | $\cdots c \cdots$ | | $N$ | |
|---|---|---|---|---|---|---|
| 1 | $X$ | $X$ | $X$ | $X$ | $_1\Sigma X = N\bar{X}_1$ |
| 2 | $X$ | $X$ | $X$ | $X$ | $_2\Sigma X = N\bar{X}_2$ |
| . | . | . | . | . | . |
| $r$ | $X$ | $X$ | $X$ | $X$ | $_r\Sigma X = N\bar{X}_r$ |
| . | . | . | . | . | . |
| $R$ | $X$ | $X$ | $X$ | $X$ | $_R\Sigma X = N\bar{X}_R$ |

assignment to columns depends on chance, there will be no correlation between columns when $R$ is infinitely large; hence the sums for rows will have a variance given by the well-known variance theorem for a sum:

$$\sigma^2_{r\Sigma X} = \sigma^2_{N\bar{X}_r} = \sigma^2_1 + \sigma^2_2 + \cdots + \sigma^2_c + \cdots + \sigma^2_N$$

We have used the symbol $\sigma$ instead of $S$ because, when we regard $R$ as infinitely large, we are dealing with theoretical rather than observed variances. Under the condition of infinite $R$, all columns will have the same variance, $\sigma^2_x$, and their sum will simply be $N\sigma^2_x$. That is, $\sigma^2_{N\bar{X}} = N\sigma^2_x$. But this is the variance of $N$ times the row means; if we divide this variance by $N^2$ we have the variance of the means themselves. (If this last step is not immediately clear, recall that for any variable, $Y$, we have $\sigma_{ay} = a\sigma_y$, where $a$ is a positive constant.) Thus

$$\frac{\sigma^2_{N\bar{X}_r}}{N^2} = \frac{N\sigma^2_x}{N^2} = \frac{\sigma^2_x}{N}$$

the square root of which is $\sigma_x/\sqrt{N}$, the familiar formula for the standard error of the mean. In practice, we need to estimate $\sigma_x$; by $S_x$ if the sample is large and by $s_x$ if small. No claim is made that this rather simple derivation would satisfy the step-by-step rigor required by contemporary mathematical statisticians.

**Estimation of variance.** To show that $s^2 = \Sigma x^2/(N-1)$ is an unbiased estimate of $\sigma^2$, we need to show that the average of a very large (infinite) number of $s^2$ values is $\sigma^2$. Such a mean value is called an expected value. If the expected value of a measure corresponds to the population value, the measure is said to be an unbiased estimate.

Suppose $s^2_r$ is the estimate based on the $r$th sample and that we let $x_r = X_r - \bar{X}_r$ and $x'_r = X_r - \mu$ be deviation scores from the sample mean, $\bar{X}_r$, and the population mean, $\mu$, respectively. If we subtract $x_r$

from $x'_r$ we have

$$x'_r - x_r = (X_r - \mu) - (X_r - \bar{X}_r) = \bar{X}_r - \mu$$

hence

$$x'_r = x_r + (\bar{X}_r - \mu) \quad \text{and} \quad x'^2_r = [x_r + (\bar{X}_r - \mu)]^2$$

Then

$$\Sigma x'^2_r = \Sigma x^2_r + \Sigma(\bar{X}_r - \mu)^2 + 2(\bar{X}_r - \mu)\Sigma x_r$$

Now $\Sigma x_r = 0$ and $\Sigma(\bar{X}_r - \mu)^2 = N(\bar{X}_r - \mu)^2$, thus leading to

$$\Sigma x'^2_r = \Sigma x^2_r + N(\bar{X}_r - \mu)^2$$

or

$$\Sigma x^2_r = \Sigma x'^2_r - N(\bar{X}_r - \mu)^2$$

This permits us to write

$$s^2_r = \frac{\Sigma x^2_r}{N - 1} = \frac{\Sigma x'^2_r}{N - 1} - \frac{N(\bar{X}_r - \mu)^2}{N - 1}$$

Suppose we have $R$ replications (samples) and that we average the $R$ estimates:

$$\frac{\Sigma s^2_r}{R} = \frac{\Sigma \dfrac{\Sigma x'^2_r}{N - 1}}{R} - \frac{\Sigma \dfrac{N(\bar{X}_r - \mu)^2}{N - 1}}{R}$$

$$= \frac{\Sigma \dfrac{\Sigma x'^2_r}{N - 1}}{R} - \frac{N}{N - 1} \frac{\Sigma(\bar{X}_r - \mu)^2}{R}$$

$$= \frac{N}{N - 1} \frac{\Sigma\Sigma x'^2_r}{NR} - \frac{N}{N - 1} \frac{\Sigma(\bar{X}_r - \mu)^2}{R}$$

Now as $R$ becomes infinitely large (or large enough to exhaust the population, if finite in size), the first term will involve the sum of the squares of all the scores in the population about the population mean, and hence will be the population variance, $\sigma^2$, multiplied by the $N/(N - 1)$ factor. With $R$ infinitely large, the second term yields $N/(N - 1)$ times the variance of an infinite number of sample means about the population mean, or the sampling variance of means, which is $\sigma^2/N$. Thus, this term becomes

$$\frac{N}{N - 1} \sigma^2_{\bar{x}} = \frac{N}{N - 1} \frac{\sigma^2}{N} = \frac{1}{N - 1} \sigma^2$$

Hence, as $R \rightarrow \infty$,

$$\frac{\Sigma s^2_r}{R} = \frac{N}{N - 1} \sigma^2 - \frac{1}{N - 1} \sigma^2$$

Factoring out $\sigma^2/(N - 1)$, the mean or expected value of $s^2$, becomes

$$\frac{\sigma^2}{N - 1}(N - 1) = \sigma^2$$

and therefore $s^2$ is an unbiased estimate of the population variance.

If the student follows through the foregoing development with $N$ instead of $N - 1$ as the divisor, he will discover that $S^2 = \Sigma x^2/N$ is a biased estimator.

**Variance and $\chi^2$.** The student who peruses the statistical literature may encounter the relationship, $\chi^2 = NS^2/\sigma^2 = (N - 1)s^2/\sigma^2 = \Sigma x^2/\sigma^2$. Since $\chi^2$ is a random value varying from sample to sample, this implies that the random sampling distribution of $S^2$ and of $s^2$ is related to the random sampling distribution of $\chi^2$. We will now attempt to build up the connection between the two.

Consider the binomial situation with $n$ elements ($n$ coins, $n$ dice, etc.) with $p$ the probability of success on a single element. The mean number of successes, $np$, may be regarded as the expected frequency of success and the mean number of failures, $nq$, may be regarded as the expected frequency of failure. A trial toss (or roll) will lead to an observed number (frequency) of successes, $O_s$, and an observed frequency of failures, $O_f$. (If, e.g., 8 of 10 coins show heads, $O_s = 8$ and $O_f = 2$; $O_s + O_f = n$.) We have

$$\chi^2 = \Sigma \frac{(O - E)^2}{E} = \frac{(O_s - np)^2}{np} + \frac{(O_f - nq)^2}{nq}$$

which may be rewritten as

$$\chi^2 = \frac{q(O_s - np)^2}{npq} + \frac{p(O_f - nq)^2}{npq}$$

But for this $\chi^2$ with 1 degree of freedom the numerical (absolute) value of $(O_f - nq) = (O_s - np)$; that is, the discrepancy is the same except for sign, which sign disappears in the squaring. We may, therefore, replace the $(O_f - nq)^2$ of the second term by its exact equivalent, $(O_s - np)^2$, thus giving

$$\chi^2 = \frac{q(O_s - np)^2}{npq} + \frac{p(O_s - np)^2}{npq} = \frac{(O_s - np)^2(q + p)}{npq}$$

Since $q + p = 1$, we get

$$\chi^2 = \frac{(O_s - np)^2}{npq}$$

The $(O_s - np)$ is the deviation of a random value from a (theoretical) mean value, hence may be regarded as an $x$; and $npq = \sigma^2$ is nothing more than the theoretical (population) variance of these $x$s. Accordingly, $\chi^2 = x^2/\sigma^2$, a relationship which we previously had in another context for $\chi^2$ when the $df$ is 1.

Next, suppose for the foregoing general binomial situation we have individuals $A, B, C, \cdots , N$, each making a single trial. For each we would have two $O$ values and the corresponding $E$ values. The $\chi^2$ for $A$'s outcome could, according to the preceding argument, be expressed (letting $O_A$ represent his frequency of success) as

$$\chi^2_A = \frac{(O_A - np)^2}{npq} = \frac{x^2_A}{\sigma^2_x}$$

and

$$\chi^2_B = \frac{(O_B - np)^2}{npq} = \frac{x^2_B}{\sigma^2_x}$$

$$\cdots \cdots \cdots \cdots \cdots \cdots$$

$$\chi^2_N = \frac{(O_N - np)^2}{npq} = \frac{x^2_N}{\sigma^2_x}$$

A summing of these separate $\chi^2$ values leads to a total chi square, $\chi^2_t$, with $df = N$, the sum of the $df$s for the $\chi^2$s being summed. Thus,

$$\chi^2_t = \sum_{i=A}^{N} \chi^2_i = \Sigma \frac{x^2_i}{\sigma^2_x} = \frac{\Sigma x^2}{\sigma^2_x}$$

Since the sampling values of $\chi^2_t$ follow the chi square distribution with $N$ degrees of freedom, the values of $\Sigma x^2/\sigma^2_x$ also follow the chi square distribution with $N$ $df$. But this $\Sigma x^2$ differs from the usual $\Sigma x^2$ in that the $x$s here used are deviations from a theoretical mean instead of (as usual) from an observed mean. Since there are no restrictions on these $x$s, the $df$ for this $\Sigma x^2$ is $N$. For the usual $\Sigma x^2$ we have $x = X - \bar{X}$, and the $df$ is $N - 1$. With $df = N - 1$ for the usual $\Sigma x^2$, it follows that for the usual situation the $df$ associated with $\Sigma x^2/\sigma^2_x$ is also $N - 1$; therefore we can say that $\Sigma x^2/\sigma^2_x$ is a $\chi^2$ variable with $df = N - 1$. With $x$ having its usual meaning (i.e., $x = X - \bar{X}$), we have the two variance estimates, $S^2 = \Sigma x^2/N$ and $s^2 = \Sigma x^2/(N - 1)$, from which we see that $\Sigma x^2 = NS^2 = (N - 1)s^2$, and therefore

$$\frac{\Sigma x^2}{\sigma^2_x} = \frac{NS^2}{\sigma^2_x} = \frac{(N - 1)s^2}{\sigma^2_x} = \chi^2 \tag{14.1}$$

with $df = N - 1$. That is, both $NS^2/\sigma^2$ and $(N - 1)s^2/\sigma^2$ have sampling distributions that follow the chi square distribution with $N - 1$ degrees of freedom. (Note that since $\sigma$, $S$, and $s$ pertain to the same variable, $X$, we drop the subscript to $\sigma$.)

**Inference based on a single variance.** The relationship set forth in (14.1) permits us to use either $S^2$ or $s^2$ as the basis for testing a hypothesis about a population variance and also for establishing confidence limits for $\sigma^2$. Although situations rarely arise in psychological research where logic leads to a hypothesis regarding $\sigma^2$, we will illustrate the procedure, with a deliberately cooked up example that will help us understand the more important task of setting a confidence interval for $\sigma^2$.

Suppose an $S^2$ of 100 or an $s^2$ of 105 based on $N = 21$ and the hypothesis that $\sigma = 16$, or $\sigma^2 = 256$. If the hypothesis is true, we have

$$\chi^2 = \frac{\Sigma x^2}{\sigma^2_h} = \frac{(21)(100)}{256} = \frac{(21-1)(105)}{256} = 8.21$$

with $N - 1 = 20$ degrees of freedom. When we turn to Table D we find that the $P$ associated with a $\chi^2$ of 8.21, $df = 20$, is .99, which is to be interpreted by saying that a $\chi^2$ greater than 8.21 will occur .99 of the time by chance; hence, a value as small as 8.21 has a $P$ of .01. That is, if $\sigma^2_h = 256$ is true, then only .01 of the time would we get a sample $S^2$ as low as 100 or a sample $s^2$ as low as 105. Accordingly, we would reject the hypothesis at the .01 level of significance.

Next, let us suppose the sample $S^2$ is 457.96 or $s^2$ is 480.86. This gives $\chi^2 = (21)(457.96)/256 = (21-1)(480.86)/256 = 37.57$, which falls at the $P = .01$ point. The probability of an $S^2$ as great as 457.96, or an $s^2$ as great as 480.86, as a deviation from $\sigma^2_h = 256$ is .01; again, we would reject the hypothesis.

Turning now to the setting of confidence limits, the student will recall that when ascertaining the .95 confidence interval for a population mean, the procedure was to find $t$ for the .05 significance level for the given $df$. Call this $t_{.05}$, then the limits are given by $\overline{X} \pm t_{.05}s_{\bar{x}}$. The $t_{.05}$ is of course the $t$ that cuts off .025 at each end of the $t$ distribution. This suggests that when setting the .95 confidence interval for $\sigma^2$ we would need to find the $\chi^2$ value that cuts off the top .025 area and the $\chi^2$ that cuts off the bottom .025, for the given $df$. Or if we had, for the $df$, the values for $\chi^2$ that cut off the top and the bottom .01 area, we would have values to use in setting the .98 confidence limits. For the latter limits, our rule-of-thumb procedure will be to find, for the given $df$, $\chi^2_{.99}$ as the $\chi^2$ that cuts off the lower .01 area and $\chi^2_{.01}$ as the $\chi^2$ that cuts off the upper .01 area.

Returning now to the example with $S^2 = 100$ (or $s^2 = 105$) and $df = 20$, we find from Table D that $\chi^2_{.99}$ is 8.26 and $\chi^2_{.01}$ is 37.57. Next we ask what values of $\sigma^2$ will yield these two $\chi^2$s. Using 8.26, we have

$$\chi^2 = 8.26 = \frac{(21)(100)}{\sigma^2} = \frac{(21 - 1)(105)}{\sigma^2} = \frac{2100}{\sigma^2}$$

as an equation to solve for $\sigma^2$. Thus, $\sigma^2 = 2100/8.26 = 254.24$. Using $\chi^2_{.01}$ we have $\chi^2 = 37.57 = 2100/\sigma^2$ from which we get $\sigma^2 = 55.89$. Note that the lower $\chi^2_{.99}$ point leads to the upper limit, whereas the higher $\chi^2_{.01}$ leads to the lower limit, for $\sigma^2$. If we take the square roots of the two $\sigma^2$s, we get 7.47 and 15.94 as the limits for the .98 confidence interval for the population standard deviation.

## DIFFERENCES BETWEEN VARIATIONS

Formulas (6.10) and (6.11) for the standard error of the difference between standard deviations, for $N$ not small, have been given earlier in this book. When testing the difference between two standard deviations or two variances we must, as always, distinguish between situations involving correlated values and situations in which the measures are independent (or based on independent samples). The methods about to be presented are applicable for both small and large samples and are based on differences between variances rather than differences between standard deviations.

**Differences between correlated variances.** Correlated variabilities arise when we have two forms of a psychological test administered to the same group with an $S$ or $s$ for each form, or when we have the $S$ for a first trial vs. the $S$ for a later trial for the same sample, or $S$s for the performance of one group under different experimental conditions, or $S$s based on two groups ($N$ pairs of individuals) related by blood or related by matching. For such situations the difference between variations can be tested by

$$t = \frac{(s^2_1 - s^2_2)\sqrt{N - 2}}{\sqrt{4s^2_1 s^2_2(1 - r^2_{12})}} \qquad (14.2)$$

or its exact equivalent with $s^2_1$ and $s^2_2$ replaced by $S^2_1$ and $S^2_2$. This $t$ follows the $t$ distribution with $N - 2$ degrees of freedom.

**Differences between independent variances.** For the purpose of testing the difference between uncorrelated $S$s or $s$s, Professor R. A. Fisher developed the mathematics of the sampling distribution of a function designated by $z$ and defined as

$$z = \log_e s_1 - \log_e s_2 \qquad (14.3)$$

If successive samples are drawn from a single universe or from two universes having the same variance, the sampling variation of $z$ will center at zero and depend on $n_1$ and $n_2$, the two $dfs$. Note that the sampling distribution is independent of the universe value of the variance or standard deviation. In other words, we do not require an estimate of a standard error which uses information from the samples, as required for the standard error of the difference between $S$s. Probability tables for the $z$ function are available by which we can, for given $dfs$, i.e., $n_1$ and $n_2$, find how large $z$ must be for the .05, the .01, and the .001 levels of significance.

The $z$, defined by formula (14.3), has one disadvantage: logarithms must be used. Since (14.3) can be written in the equivalent form

$$z = \frac{1}{2} \log_e \frac{s^2_1}{s^2_2} \qquad (14.4)$$

it is seen that, instead of the difference between two logarithms, we have $z$ as a function of the ratio of the two estimated variances. From the sampling distribution of one-half the log of a ratio, the sampling distribution of the ratio itself can be inferred. For $n_1 = 5$ and $n_2 = 16$, the value of $z$, which will be exceeded 1 per cent of the time by chance (the .01 probability level), is .7450. This is one-half the log of the ratio of the two variances, and hence the log of the ratio would be 1.4900; by reference to a table of natural logarithms the antilog of 1.4900 is found to be 4.44. That is, as large a ratio as 4.44 would occur .01 time by chance. In order to avoid the necessity of using logs, Professor George W. Snedecor has developed tables for the *variance ratio*, which is defined as

$$F = \frac{s^2_1}{s^2_2} \qquad (14.5)$$

The equation* of the sampling distribution of $F$ contains two $n$s: $n_1$ for the $df$ upon which $s_1$ is based, and $n_2$ as the $df$ for $s_2$. This means that there is a sampling distribution curve of $F$ for each possible combination of $n_1$ and $n_2$. The probability table for $F$ must accordingly be entered with $n_1$ and $n_2$ in order to learn what level of significance a given $F$ reaches. To use Table F of the Appendix, we take the larger of the two variance estimates as the numerator in computing $F$, and the $df$ for this larger estimate is symbolized as $n_1$ regardless of any system of subscripts that may have been used to designate the two groups. Thus the $F$ that is used with the table is always unity or greater, even though the sampling distribution of $F$

*
$$h = \frac{\Gamma\left(\dfrac{n_1 + n_2}{2}\right) n_1^{n_1/2} n_2^{n_2/2}}{\Gamma\left(\dfrac{n_1}{2}\right) \Gamma\left(\dfrac{n_2}{2}\right)} \cdot \frac{F^{(n_1-2)/2}}{(n_1 F + n_2)^{(n_1+n_2)/2}}$$

involves values less than unity. That is, if we were drawing successive samples from groups $A$ and $B$ and each time took $F$ as $s^2_a/s^2_b$, regardless of which was the larger estimate, the sampling distribution of $F$ would obviously involve values below unity as well as above unity. The table, however, is set up in terms of the greater-than-unity side of the sampling distribution.

If we wish to judge whether two samples, either large or small, yield a difference in variability which is large enough to warrant concluding that the two population variabilities differ, we set up the null hypothesis that no difference exists in the two population variances. Then, instead of dealing as usual with the difference between the two estimates, we take their ratio. Obviously, the departure of this ratio, $F$, from unity reflects or depends on the difference between the two variance estimates. If the value of $F$, computed with the larger estimate in the numerator, is so large that it is not reasonable to believe it a chance deviation from a true value of unity, the null hypothesis is rejected, and it is concluded that the two populations do not have the same variance. If $F$ is small, i.e., near unity, the null hypothesis is accepted.

Now it happens that, although the $F$ values given in Table F for the .05, the .01, and the .001 levels of significance hold for the major and very extensive uses of the $F$ table to be discussed in Chapters 15–18, these values are *not* applicable to the simple case where we wish to ascertain the probability of as great a difference (irrespective of direction, i.e., a hypothesis or decision requiring a two-tailed test) between the variances for two groups. For this particular case, an $F$ which falls at, say, the .01 level signifies that as large a difference in *one* direction would occur 1 per cent of the time by chance. This is so because in placing the larger estimate in the numerator we are considering only one tail of the $F$ distribution. In asking whether two variance estimates of, say, 10 and 25 based on two groups differ, i.e., lead to an $F$ which departs significantly from unity (no difference), we should consider not only the probability of securing an $F$ as large as 25/10 but also the probability of obtaining one as small as 10/25. This, it will be observed, is exactly analogous to considering both positive and negative values for the $z$ of formula (14.3) and then raising the question as to the probability of obtaining on a chance basis as large a difference, irrespective of direction. If we had this last probability, we would halve it to obtain the $P$ for one direction only; conversely, if we had an $F$ which fell at the $P = .01$ level in the table, we would need to double .01 to secure the probability for as large a difference irrespective of direction. In other words, for this particular case, that of testing the significance between the variability for two groups, an $F$ at the .01 point of the table means significance at the .02 level; an $F$ at the .05 level means significance at the .10 level; and an $F$ at the .001 level indicates significance at the .002 level. We will *not* have to make this type of adjustment when we come to the principal uses of $F$ in connection with the analysis of variance.

For example, suppose that 50.21 and 147.62 are variance estimates available for two samples of eight and nine cases respectively. The respective $df$s would be 7 and 8. In computing $F$ we have 147.62/50.21 = 2.94, and $n_1$ becomes 8, with $n_2 = 7$. Turning to Table F, we see that $F$ would need to be 3.73 for the .05 level, which for this type of problem is the .10 level. Therefore the null hypothesis is not rejected. If we take the square roots of the two variance estimates, we get $s$s of 7.09 and 12.15. By the $F$ test, we are in effect saying that the difference between these two $s$s is not significant. As usual, this does not prove the null hypothesis—it becomes acceptable because we cannot with sufficient certainty reject it.

If the research hypothesis being tested or the decision to be made calls for a one-tailed test, the $F$ values in Table F are applicable without further ado. As a matter of fact, if the null hypothesis is to be accepted unless $s^2_a$ is significantly larger than $s^2_b$, we would not bother to compute $F$ if $s^2_a$ turned out to be smaller than $s^2_b$.

**Differences between several independent variances.** We have seen in Chapter 13 that $\chi^2$ can be used to provide an over-all test of the difference between several independent proportions (p. 266) for $C$ groups and also between $C$ correlated proportions (p. 263). In the next chapter we shall see how an over-all test can be made for the differences between several means, either correlated or independent. We shall consider now an over-all test of the difference between three or more variance estimates. This test is not applicable when the variances are correlated (based on the same group or matched groups).

Suppose we have $k$ variance estimates, $s^2_1, s^2_2, \cdots, s^2_i, \cdots, s^2_k,$ based on $m_1 - 1, m_2 - 1, \cdots, m_i - 1, \cdots, m_k - 1$ degrees of freedom respectively. Let $N$ be the sum of the $m$s. Compute the products: each $s^2$ times its $df$. Sum these $k$ products (the equivalent of summing the $k$ sums of squares of deviations). Let $s^2_w$ stand for this sum divided by $N - k$. Determine the log of each of the $k$ $s^2$ values, then calculate the products: each $\log s^2$ times the $df$ for the given $s^2$. Sum these products, that is, $\sum_i (m_i - 1) \log s^2_i$ in which $i$ takes on values from 1 to $k$. Determine the log of $s^2_w$, and compute

$$C = 1 + \frac{1}{3(k-1)} \left( \sum_i \frac{1}{m_i - 1} - \frac{1}{N - k} \right)$$

Finally, calculate the quantity

$$V = \frac{2.3026}{C} [(N - k) \log s^2_w - \sum_i (m_i - 1) \log s^2_i] \qquad (14.6)$$

The sampling distribution of $V$ follows the $\chi^2$ distribution with $k - 1$ degrees of freedom. If $V$ reaches the $P = .05$ or $P = .01$ or any a priori chosen level of significance, the differences between the $k$ variances may be regarded as nonchance, hence the conclusion that the $k$ groups have not been drawn from populations having equal variances. If $V$ is not significant, we accept the hypothesis that the groups have been drawn from populations having equal variances. The variances are said to be homogeneous. The procedure just described is known as Bartlett's test for the homogeneity of variances. It is appropriate for testing the assumption of homoscedasticity in bivariate correlation scattergrams.

## $F$, $\chi^2$, $t$, AND $z$ (NORMAL DEVIATE)

Since $F$ involves the ratio of two variance estimates and since there is a connection between a variance and $\chi^2$, it is possible to write $F$ as a function of two $\chi^2$s. Recall that $(N - 1)s^2/\sigma^2 = \chi^2$ with $df$ of $N - 1$. Solving for $s^2$ we have $s^2 = \sigma^2\chi^2/(N - 1)$. Thus $s^2_1 = \sigma^2_1\chi^2_1/n_1$ and $s^2_2 = \sigma^2_2\chi^2_2/n_2$ in which $n_1$ and $n_2$ are the $df$s. Then

$$F = \frac{s^2_1}{s^2_2} = \frac{\sigma^2_1\chi^2_1/n_1}{\sigma^2_2\chi^2_2/n_2} \qquad (14.7)$$

Under the null hypothesis condition that $\sigma^2_1 = \sigma^2_2$ we see that

$$F = \frac{\chi^2_1/n_1}{\chi^2_2/n_2} \qquad (14.8)$$

or that $F$ is the ratio of two $\chi^2$ variables, each divided by its $df$. This provides a more fundamental definition of the $F$ ratio and also serves as the basis for the derivation of the sampling distribution of $F$ under the null condition, which in turn leads to the values of $F$ for levels of significance (Table F). If $F$ is significantly large, we simply say that the sample value of $F$ is not consistent with the null hypothesis, or that $\sigma^2_1$ exceeds $\sigma^2_2$.

If the estimate $s^2_2$ is based on an infinitely large $df$ (i.e., $n_2 =$ infinity) it will equal $\sigma^2_2$ or if $\sigma^2_2$ happens to be a known theoretical variance (e.g., the variance of a binomial), we can write $F$ as

$$F = \frac{s^2_1}{\sigma^2_2} = \frac{\sigma^2_1\chi^2_1/n_1}{\sigma^2_2} \qquad (14.9)$$

in which we did not replace the denominator by a $\chi^2$ equivalent because it is a constant, hence can not possibly vary as a $\chi^2$ variable. Again, under the null condition that $\sigma^2_1 = \sigma^2_2$, the two variances cancel, leaving $F = \chi^2_1/n_1$ when $n_2 = \infty$. This means that in the $\infty$ lines of Table F, each

$F$ is a $\chi^2$ divided by $n_1$. If you had a $\chi^2$ and no $\chi^2$ table available, you could divide it by its $df$ and enter the quotient in the $\infty$ lines of Table F to learn whether it reached one of three levels of significance.

If $n_2 = \infty$ and $n_1 = 1$, we see that $F = \chi^2$. But a $\chi^2$ with 1 $df$ equals an $x^2/\sigma^2$, or a $z^2$ where $z$ is a unit normal deviate; hence $F = z^2$ when $n_2 = \infty$ and $n_1 = 1$. The first column entries of the $\infty$ lines of Table F are the squares of $x/\sigma$ values.

If $n_1 = 1$ and $n_2$ varies, we have

$$F = \frac{\sigma^2_1 \chi^2_1 / 1}{s^2_2}$$

in which we deliberately did not replace $s^2_2$ by its $\chi^2$ equivalent. The numerator involves a $\chi^2$ with 1 $df$, hence $\chi^2_1 = x^2_1/\sigma^2_1$ in which $x_1$ is a normal deviate with variance $\sigma^2_1$. Substituting for $\chi^2_1$, we have

$$F = \frac{\sigma^2_1 x^2_1 / \sigma^2_1}{s^2_2} = \frac{x^2_1}{s^2_2} \tag{14.10}$$

If $\sigma^2_1 = \sigma^2_2 = \sigma^2$, we may regard $s^2_2$ as an estimate of the common variance, $\sigma^2$. Now since the $x_1$ values are random normal deviates from the assumed common population, we may drop the subscripts and have $F = x^2/s^2$ with $n_1 = 1$ and $n_2 = n$ where $n$ is the $df$ for $s^2$. The square root of $F$ becomes $x/s$, or the ratio obtained by dividing a normally distributed variate by an unbiased estimate of its standard deviation. Since this corresponds to one definition of $t$, we have $F = t^2$ when $n_1 = 1$ and $n_2 = n$. All the entries in the first column of Table F, with the exception of $\infty$ lines, are $t^2$ values. The exceptions are $z^2$, or $(x/\sigma)^2$, values of the unit normal distribution.

# Chapter 15
## ANALYSIS OF VARIANCE: SIMPLE

The $F$ or variance ratio defined in Chapter 14 is applicable in a wide variety of situations. The general requirement is that we have two independent estimates of variance, which estimates are, on the basis of the null hypothesis, regarded as estimates of the same population value. If $F$ is sufficiently large, the null hypothesis becomes suspect, and we draw a positive conclusion, the nature of which depends on the given situation. Each application in this and the following chapter requires an *assumption of normality* and an *assumption of homogeneity of certain variances;* normality of what, and homogeneity of which variances, will need to be specified for each type of situation.

Although these assumptions are incorporated in the mathematical derivation of the $F$ distribution, there is ample evidence that marked skewness, departures from normal kurtosis, and extreme differences in variance (of the order 1 to 4 to 9—it is not the numerical differences but the relative sizes of the variances that are pertinent) do not greatly disrupt the $F$ test as a basis for judging significance in the analysis of variance. The error introduced by violations of assumptions is such that $F$s may reach the .05 level about .06 or .07 of the time, and the .01 level about .02 of the time.* If the investigator wishes to have some assurance that he is not risking the making of the type I error more often than his chosen level for judging significance, he may wish to adopt a somewhat more rigorous level: requiring a computed $F$ to reach the .01 level provides a very safe base for claiming significance at the .02 level.

* See the study by Norton, reported in Lindquist, E. F., *Design and analysis of experiments*, New York: Houghton Mifflin, 1953.

It will be recalled that under certain circumstances the squared correlation coefficient is interpretable in terms of the proportion of variance "explained." The idea is that variation can be broken down into component parts in such a way as to permit specification of the relative importance of the component sources. Back of this is the fact that variances are additive to a total variance, as shown when we derived formula (9.10), which is basic to the so-called variance theorem. Although this theorem is fundamental to the analysis of variance technique, it is not our aim to consider methods of estimating the proportion or percentage of variance due to a given source but rather to discuss ways of testing whether a possible source is contributing to the total variance to a statistically significant degree.

## BREAKDOWN OF SUM OF SQUARES

Let us begin with the simple situation in which the total variation for a set of scores based on $N$ individuals is possibly due in part to the fact that the total group is heterogeneous with respect to some factor, such as socioeconomic level or age or racial origin or type of treatment or method used in memorizing or varying level of illumination—any factor which permits breaking down the total group into subgroups. In other words, the individuals or their scores can be classified into subgroups, or the total group can be regarded as made up of specified subgroups. For simplicity, let us assume that the subgroups are of the same size, say $m$ cases per group, and that we have $G$ groups. Let $g$ stand for any subgroup; i.e., $g$ takes on values of $1, 2, 3, \cdots, G$, and let the mean score for the groups be specified as $\bar{X}_1, \bar{X}_2, \cdots, \bar{X}_g, \cdots, \bar{X}_G$, with $\bar{X}$ as the mean for all groups combined (total mean). Although it is possible to use a precise notation, such as $X_{ig}$, to denote the score of any, the $i$th, person in group $g$, we shall in this chapter simply use $X$ as the score for any individual.

We are now in a position to write an individual's score as a deviation from the total mean in terms of the deviation of his score from his group mean and the deviation of the group mean from the total mean. Thus, for a score in group $g$,

$$(X - \bar{X}) = (X - \bar{X}_g) + (\bar{X}_g - \bar{X}) \qquad (15.1)$$

which indicates two sources of variation: the variation of a group mean from the total mean and the variation of an individual's score from his group mean.

If we rewrite formula (15.1) specifically for group 1, we have

$$(X - \bar{X}) = (X - \bar{X}_1) + (\bar{X}_1 - \bar{X})$$

Squaring both sides gives

$$(X - \bar{X})^2 = (X - \bar{X}_1)^2 + (\bar{X}_1 - \bar{X})^2 + 2(\bar{X}_1 - \bar{X})(X - \bar{X}_1)$$

as the squared deviation, from the total mean, of any score in group 1. Each of the $m$ persons in the group will have such a squared deviation score. We may indicate the sum of the squares for the $m$ cases as

$$\Sigma(X - \bar{X})^2 = \Sigma(X - \bar{X}_1)^2 + \Sigma(\bar{X}_1 - \bar{X})^2 + 2(\bar{X}_1 - \bar{X})\Sigma(X - \bar{X}_1)$$

Note that in the last term the constants 2 and $(\bar{X}_1 - \bar{X})$ have been taken from under the summation sign, and that $\Sigma(X - \bar{X}_1)$, being the sum of deviations of a set of scores about their own mean, will be exactly zero. Therefore, the last term vanishes. Note also that the second right-hand term involves summing a constant, which is the same as multiplying it by the number of cases involved in the summation, i.e.,

$$\Sigma(\bar{X}_1 - \bar{X})^2 = m(\bar{X}_1 - \bar{X})^2.$$

Thus we see that we may write the sum of squares (of deviations) for the first group and by analogy for the other groups as follows:

1st group:  $\Sigma(X - \bar{X})^2 = \Sigma(X - \bar{X}_1)^2 + m(\bar{X}_1 - \bar{X})^2$

2nd group:  $\Sigma(X - \bar{X})^2 = \Sigma(X - \bar{X}_2)^2 + m(\bar{X}_2 - \bar{X})^2$

gth group:  $\Sigma(X - \bar{X})^2 = \Sigma(X - \bar{X}_g)^2 + m(\bar{X}_g - \bar{X})^2$

Gth group:  $\Sigma(X - \bar{X})^2 = \Sigma(X - \bar{X}_G)^2 + m(\bar{X}_G - \bar{X})^2$

If we summed the left-hand parts of the foregoing, we would obviously have the sum of squares of deviations for the entire set of $N = mG$ cases. This summing of sums, or double summation, can be conveniently indicated by using two summation signs, or $\Sigma\Sigma(X - \bar{X})^2$. We may sum the right-hand terms separately. The first term on the right involves summing sums, and the result can be indicated symbolically by $\Sigma\Sigma(X - \bar{X}_g)^2$, which implies that we first sum for each group, then sum over all groups. The first summation sign indicates that the subscript $g$ takes in turn values running from 1 to $G$. The sum of the other right-hand terms can be written as $m \underset{g}{\Sigma}(\bar{X}_g - \bar{X})^2$.

Since adding of equations leads to an equation, we have

$$\Sigma\Sigma(X - \bar{X})^2 = \underset{g}{\Sigma\Sigma}(X - \bar{X}_g)^2 + m\underset{g}{\Sigma}(\bar{X}_g - \bar{X})^2 \qquad (15.2)$$

as a means of expressing the fact that the total sum of squares (of deviations) can be broken down into two components, the first of which has to do with variation about group means, i.e., *within* groups, and the second of which involves variation of group means about the total mean, i.e., *between* groups. In other words, the total sum of squares is made up of two additive parts. If we divide both sides by $N$ or $mG$, we have the total variance broken into additive components, but for our present purposes we shall need unbiased estimates of variance, and hence it becomes necessary to divide through by degrees of freedom.

The correct *df* can be ascertained by examining the three sums of squares. For the total sum of squares we have one restriction, the total mean, and as seen in Chapter 7 the *df* will be $N - 1$ or $mG - 1$. The within-groups sum is based on $N$ or $mG$ squares, but since these are about $G$ different means there are $G$ restrictions, or $mG - G$ ($= N - G$) degrees of freedom. The last or between-groups sum involves $G$ means, varying more or less about the total mean; thus, aside from the $m$ factor, it contains $G$ squares with one restriction, and the *df* becomes $G - 1$. In other words, the $G$ means are analogous to varying scores, and obviously the mean of these means will equal the total mean.

We may indicate the division of the three sums of squares by the proper *df*s as follows:

$$\frac{\Sigma\Sigma(X - \bar{X})^2}{mG - 1}, \quad \frac{\Sigma\Sigma_g(X - \bar{X}_g)^2}{mG - G}, \quad \frac{m\Sigma_g(\bar{X}_g - \bar{X})^2}{G - 1}$$

Notice that we are no longer dealing with an equation. Why? Each division will result in a variance estimate, but these are not directly additive, which means that we cannot specify what proportion of the estimated total variance is due to the between-groups variation. The reader should note, however, that the *df*s *are* additive:

$$(mG - 1) = (mG - G) + (G - 1)$$

Before examining the meaning of these three variance estimates, let us label them: $s^2$ for the estimate of total variance, $s^2_w$ for that based on the *within*-groups sum of squares, $s^2_b$ for that based upon *between* groups. Variance estimates are sometimes referred to as "mean squares."

It is of interest to note that $s^2_w = \dfrac{\Sigma\Sigma_g(X - \bar{X}_g)^2}{mG - G}$ can be written as

$$\Sigma_g \frac{\dfrac{\Sigma(X - \bar{X}_g)^2}{m - 1}}{G}$$

which indicates explicitly that $s^2_w$ may be regarded as the average of $G$ estimates of the within-groups variance.

## MEANING OF VARIANCE ESTIMATES

Insofar as we think of the total $mG$ cases as a sample drawn from one population, $s^2$ will be the best unbiased estimate of the variance of the population, $\sigma^2$. If we think of the $m$ cases for each of our $G$ groups as samples from $G$ possibly different populations, then $s^2_w$ will be a composite estimate of the several population variances, a sort of average which makes sense if the population variances are equal; if the $G$ groups have been drawn from just one population, this within-groups variance estimate or $s^2_w$ will differ little from, but be somewhat smaller than, $s^2$. Note that $s^2$ and $s^2_w$ *cannot* be regarded as *independent* estimates because the two estimates are based on practically the same deviations: extreme scores, in either direction, will tend to make both $s^2$ and $s^2_w$ large. If $m$, or the number of cases per group, is taken larger and larger and if the groups are regarded as belonging to the same population or populations differing in some respects but having the same mean and variance for the given trait or variate, $s^2$ and $s^2_w$ will tend to the same value, $\sigma^2$.

Let us next look at $s^2_b$. The division of $m \sum_g (\bar{X}_g - \bar{X})^2$ by its $df$ may be accomplished by dividing the sum factor by $G - 1$. In making this division we are dividing a sum of squares by degrees of freedom; hence the result will be a variance estimate. Let us use $s^2_{\bar{x}_b}$ as a symbol for this estimate. Then

$$s^2_b = \frac{m\sum_g(\bar{X}_g - \bar{X})^2}{G - 1} = ms^2_{\bar{x}_b}$$

In order to understand the meaning of $s^2_{\bar{x}_b}$, we may regard our $G$ means as a sample of sample means from an indefinitely large supply of possible sample means for groups drawn from the same population. The variance for this universe of sample means is given by the standard error of mean formula, i.e., $\sigma^2_{\bar{x}_b} = \sigma^2/m$. If we were given the value of $\sigma^2_{\bar{x}_b}$ and told to determine the universe trait variance or $\sigma^2$, we would simply solve $\sigma^2_{\bar{x}_b} = \sigma^2/m$ for $\sigma^2$. Thus, $\sigma^2 = m\sigma^2_{\bar{x}_b}$. If we had only an estimate of $\sigma^2_{\bar{x}_b}$, such as $s^2_{\bar{x}_b}$, we could use this estimate as a basis for estimating the trait variance; i.e., $ms^2_{\bar{x}_b}$ can be taken as an estimate of $\sigma^2$. Since $ms^2_{\bar{x}_b} = s^2_b$, we have $s^2_b$ and $s^2_w$ (see previous paragraph) as estimates of the same population variance.

These estimates should agree within the limits of chance, and being independent estimates of the same variance the sampling distribution of their ratio is that of the $F$ distribution. When an obtained $F$, or $s^2_b/s^2_w$, is larger than expected on the basis of chance sampling, the implication is that $s^2_b$ is greater than expected on the basis of chance sampling, hence that there are real differences among the $G$ means. If the null hypothesis of no difference is true, we expect that in the long run $s^2_b$ will tend to equal $\sigma^2$; that is, the average or expected value of $s^2_b$ is $\sigma^2$. Suppose that the null hypothesis is *not* true and we ask about the mean or expected value of $s^2_b$. An expected value is defined as the mean of an infinite number of obtained values.

Now

$$s^2_b = \frac{m\Sigma(\overline{X}_g - \overline{X})^2}{G - 1}$$

in which $\overline{X}_g$ is a sample mean from the gth group. It seems reasonable to say that the variation among $\overline{X}_1, \overline{X}_2, \cdots, \overline{X}_g, \cdots, \overline{X}_G$ will have two sources under nonnull conditions: the extent of the variation among the corresponding $G$ population means, $\mu_1, \mu_2, \cdots, \mu_g, \cdots, \mu_G$, plus a random sampling component because each $\overline{X}_g$ is based on a sample of $m$ cases. If we symbolized the sampling error by $E_g$, we would have $E_g = \overline{X}_g - \mu_g$, which leads to $\overline{X}_g = \mu_g + E_g$. (This is similar to the conception of a score $X$ as being composed of a true part $X_t$ and an error $E$ so that $X = X_t + E$.)

We seek an expression for the deviation $(\overline{X}_g - \overline{X})$ which will incorporate the two sources of variation. The random error part can be expressed as $(\overline{X}_g - \mu_g)$ and the possible differences among the population means as $(\mu_g - \mu)$ in which $\mu$ is the mean of the population means. Thus we could write

$$(\overline{X}_g - \overline{X}) = (\mu_g - \mu) + (\overline{X}_g - \mu_g) - (\overline{X} - \mu) \qquad (15.3)$$

in which the third term on the right is a (nuisance) variable which must be incorporated to make the equation balance, i.e., to provide an identity. To obtain an expression for $s^2_b$ (aside from the $m$ factor) we could square and sum either side of the foregoing identity. If we did this we would discover that the $(\overline{X} - \mu)$ term is really a nuisance. There is a trick that will help, but before considering the squaring and summing further, let us set up a scheme which will eventually aid in ascertaining the mean (expected) value of $s^2_b$. Let us imagine that we have $R$ replications or sets of data, each set leading to $G$ means based on $m$ cases. We might arrange the obtained means in a table (Table 15.1) consisting of $R$ rows and $G$ columns, each mean having two subscripts, the first of which designates the row, the second the column. Thus $\overline{X}_{36}$ would be the mean in the third row (set or replication) and the sixth column (group). Note the notation for the mean of means ($M$ of $M$s) along the bottom and right margin. The $M$ of $M$s on the right are obtained by summing across columns (groups)—the dot replaces the second subscript. For the $M$ of $M$s at the

**Table 15.1. Means for $G$ groups, with $R$ replications**

|         | 1 | 2 | $g$ | $G$ | $M$ of $Ms$ |
|---------|-----------|-----------|------------|------------|------------|
| 1       | $\bar{X}_{11}$ | $\bar{X}_{12}$ | $\bar{X}_{1g}$ | $\bar{X}_{1G}$ | $\bar{X}_{1.}$ |
| 2       | $\bar{X}_{21}$ | $\bar{X}_{22}$ | $\bar{X}_{2g}$ | $\bar{X}_{2G}$ | $\bar{X}_{2.}$ |
| $r$     | $\bar{X}_{r1}$ | $\bar{X}_{r2}$ | $\bar{X}_{rg}$ | $\bar{X}_{rG}$ | $\bar{X}_{r.}$ |
| $R$     | $\bar{X}_{R1}$ | $\bar{X}_{R2}$ | $\bar{X}_{Rg}$ | $\bar{X}_{RG}$ | $\bar{X}_{R.}$ |
| $M$ of $Ms$ | $\bar{X}_{.1}$ | $\bar{X}_{.2}$ | $\bar{X}_{.g}$ | $\bar{X}_{.G}$ | $\bar{X}_{..}$ |
| Pop. $Ms$ | $\mu_{.1}$ | $\mu_{.2}$ | $\mu_{.g}$ | $\mu_{.G}$ | $\mu_{..}$ |

bottom the summing is over rows (replications), hence the dot replaces the first subscript. Averaging the $M$ of $Ms$ along the right or those at the bottom leads to the $\bar{X}..$ at the lower-right corner. The $\mu$s at the bottom hold for the $G$ populations.

To rewrite (15.3) for our first set (or row) we need only insert the subscript 1 to designate the first row, along with appropriate dots as part of the subscript notation. Thus,

$$(\bar{X}_{1g} - \bar{X}_{1.}) = (\mu_{.g} - \mu_{..}) + (\bar{X}_{1g} - \mu_{.g}) - (\bar{X}_{1.} - \mu_{..}) \quad (15.4)$$

or more generally, we have

$$(\bar{X}_{rg} - \bar{X}_{r.}) = (\mu_{.g} - \mu_{..}) + (\bar{X}_{rg} - \mu_{.g}) - (\bar{X}_{r.} - \mu_{..}) \quad (15.5)$$

as a sort of model for expressing the deviation of the $g$th mean in the $r$th replication $\bar{X}_{rg}$ from the over-all mean of the $r$th replication $\bar{X}_{r.}$, given on the left of the equality sign, in terms of the three components on the right side. (Note that writing the mean of the $g$th population as $\mu_{.g}$ indicates that its value is independent of the first subscript.)

When we square both sides of (15.4) and sum over groups we will obtain for the right side three terms involving squares and three involving cross-products. One of these cross-product terms, $-2\sum_{g}(\bar{X}_{1g} - \mu_{.g})(\bar{X}_{1.} - \mu_{..})$, turns out to be unmanageable. We can avoid this difficult term by shifting the nuisance component, $(\bar{X}_{1.} - \mu_{..})$, to the left, rewriting (15.4) as

$$(\bar{X}_{1g} - \bar{X}_{1.}) + (\bar{X}_{1.} - \mu_{..}) = (\mu_{.g} - \mu_{..}) + (\bar{X}_{1g} - \mu_{.g})$$

Squaring both sides and summing over $g$, we have

$$\sum_{g}(\bar{X}_{1g} - \bar{X}_{1.})^2 + G(\bar{X}_{1.} - \mu_{..})^2 + 2(\bar{X}_{1.} - \mu_{..})\sum_{g}(\bar{X}_{1g} - \bar{X}_{1.})$$

$$= \sum_{g}(\mu_{.g} - \mu_{..})^2 + \sum_{g}(\bar{X}_{1g} - \mu_{.g})^2 + 2\sum_{g}(\mu_{.g} - \mu_{..})(\bar{X}_{1g} - \mu_{.g})$$

Note that since the second term on the left side involves summing a constant $G$ times, the summation sign was replaced by $G$; note also that in the next term, a constant is taken from under the summation sign. Note further that in this same third term, the expression $\sum_g(\bar{X}_{1g} - \bar{X}_{1.})$ is zero, hence this cross-product term vanishes.

Omitting the vanishing term and shifting the second term on the left to the right-hand side, we may write the equation for the first set, and by analogy for any set (the $r$th) and then the $R$th (for a total of $R$ replications):

$$\sum_g(\bar{X}_{1g} - \bar{X}_{1.})^2 = \sum_g(\mu_{.g} - \mu_{..})^2 + \sum_g(\bar{X}_{1g} - \mu_{.g})^2$$

$$- G(\bar{X}_{1.} - \mu_{..})^2 + 2\sum_g(\mu_{.g} - \mu_{..})(\bar{X}_{1g} - \mu_{.g})$$

$$\sum_g(\bar{X}_{rg} - \bar{X}_{r.})^2 = \sum_g(\mu_{.g} - \mu_{..})^2 + \sum_g(\bar{X}_{rg} - \mu_{.g})^2$$

$$- G(\bar{X}_{r.} - \mu_{..})^2 + 2\sum_g(\mu_{.g} - \mu_{..})(\bar{X}_{rg} - \mu_{.g})$$

$$\sum_g(\bar{X}_{Rg} - \bar{X}_{R.})^2 = \sum_g(\mu_{.g} - \mu_{..})^2 + \sum_g(\bar{X}_{Rg} - \mu_{.g})^2$$

$$- G(\bar{X}_{R.} - \mu_{..})^2 + 2\sum_g(\mu_{.g} - \mu_{..})(\bar{X}_{Rg} - \mu_{.g})$$

The addition of these equations will lead to a new equation for which we will need double summation signs, summing over $g$ from 1 to $G$ and over $r$ from 1 to $R$. Thus,

$$\underset{r\ g}{\sum\sum}(\bar{X}_{rg} - \bar{X}_{r.})^2 = \underset{g}{R\sum}(\mu_{.g} - \mu_{..})^2 + \underset{r\ g}{\sum\sum}(\bar{X}_{rg} - \mu_{.g})^2$$
$$\quad\text{A}\qquad\qquad\text{B}\qquad\qquad\text{C}$$

$$- \underset{r}{G\sum}(\bar{X}_{r.} - \mu_{..})^2 + 2\underset{r\ g}{\sum\sum}(\mu_{.g} - \mu_{..})(\bar{X}_{rg} - \mu_{.g})$$
$$\qquad\text{D}\qquad\qquad\qquad\text{E}$$

To facilitate further consideration, the five terms have been designated by letters. It will be recalled that when we divide $m \sum_g(\bar{X}_g - \bar{X})^2$ by its $df$, $G - 1$, we get $s^2_b$ which may be thought of as $ms^2_{\bar{x}_b}$. When we divide the exact equivalent, $m \sum_g(\bar{X}_{rg} - \bar{X}_{r.})^2$, by $G - 1$ we get an $s^2_b$ for the $r$th set, and this $s^2_b$ might be written as $ms^2_{\bar{x}_{b(r)}}$. But note that in the foregoing $R$ equations we did not (and need not) have $m$, hence the division of $\sum_g (\bar{X}_{rg} - \bar{X}_{r.})^2$ by $G - 1$ gives $s^2_{\bar{x}_{b(r)}}$.

Our plan of attack is first to note that when each of the $R\sum_g$ parts of term **A** is divided by $G - 1$, we have a variance estimate, $s^2_{\bar{x}_{b(r)}}$, for each of the $R$ replications. If we next sum these (varying) estimates and divide by $R$ we will have the mean of the $R$ estimates. Then if we think of $R$ as approaching infinity or as infinitely large, we will have the mean (or expected) value of the estimate. This process of first dividing $\sum_g (\bar{X}_{rg} - \bar{X}_{r.})^2$ by $G - 1$, then summing over $r$, followed by a division by $R$ can be stated as follows:

$$\frac{\Sigma\Sigma(\bar{X}_{rg} - \bar{X}_{r.})^2}{R(G-1)} = \Sigma_r \frac{\dfrac{\Sigma_g(\bar{X}_{rg} - \bar{X}_{r.})^2}{G-1}}{R} = \frac{\Sigma_r s^2_{\bar{x}_{b(r)}}}{R}$$

By definition, as $R$ becomes infinitely large we have the expected value of the variance estimate, $s^2_{\bar{x}_{b(r)}}$. But we wish to express this in terms of two parts—one which reflects a random error in the observed means whereas the other reflects possible real differences among the $G$ means, $\mu_{.g}$. To do this we will need to work with the right-hand side of the above five-term equation in order to see what happens when the four right-hand terms are also divided by $R(G-1)$ and $R$ is allowed to approach infinity. Obviously, such a division will maintain the equation. When dividing, it is necessary that we break up $R(G-1)$ so as to evaluate a given term as $R$ becomes infinitely large.

For the **B** term, we have

$$\frac{R\,\Sigma_g(\mu_{.g} - \mu_{..})^2}{R(G-1)} = \frac{\Sigma_g(\mu_{.g} - \mu_{..})^2}{G-1}$$

which is a fixed quantity regardless of $R$.

For the **C** term, we may write

$$\frac{\Sigma\Sigma_{rg}(\bar{X}_{rg} - \mu_{.g})^2}{R(G-1)} = \Sigma_g \frac{\left[\Sigma_r(\bar{X}_{rg} - \mu_{.g})^2\right]\Big/R}{G-1}$$

This involves for any one of the $G$, say the $g$th group, the squared deviations of a series of sample means, each based on $m$ cases, about the population mean for the group. For $R$ infinitely large, the sum of these squared deviations divided by $R$ is the true (or theoretical population) variance of sample means, hence **C** becomes

$$\frac{\Sigma_g \sigma^2_{\bar{x}(g)}}{G-1} = \frac{\Sigma_g \sigma^2_g/m}{G-1} = \frac{G}{G-1} \cdot \frac{\sigma^2}{m}$$

in which the sample variance of the means from the $g$th group, $\sigma^2_{\bar{x}(g)}$, has been replaced by the familiar formula for the sampling variance of means in terms of population score variance, $\sigma^2_g$, and sample size, $m$. It may help the student to note that we are dealing with the distribution of sample means in any column, the $g$th, in Table 15.1. The last step in the foregoing procedure involves the assumption of homogeneity of variances. When $\sigma^2_1 = \sigma^2_2 = \sigma^2_g = \sigma^2_G$, the summing over $g$ of the $G$

variances, all equal, is nothing more than $G$ times $\sigma^2$, the common population variance.

For the **D** term, we have (ignoring its negative sign which is to be picked up later)

$$\frac{G\Sigma_r(\bar{X}_{r\cdot} - \mu_{\cdot\cdot})^2}{R(G-1)} = \frac{G}{G-1}\frac{\Sigma_r(\bar{X}_{r\cdot} - \mu_{\cdot\cdot})^2}{R}$$

which, as $R$ becomes infinitely large, involves the sampling variance of the means of means along the right-hand margin of Table 15.1, which variance we will symbolize as $\sigma^2_{\bar{x}(r)}$. Since each of these $R$ means is based on $mG$ cases, we could easily jump to the mistaken conclusion that this variance of sample means is a function of the variance of the $mG$ scores about $\mu_{\cdot\cdot}$ and would therefore depend on the variance within the $G$ groups plus possible variation among the $G$ groups. Note that although it is presumed that the $m$ cases of group $g$ have been randomly drawn from population $g$, it does not follow that the $mG$ cases have been drawn randomly from a grand total population made up by combining the $G$ subpopulations. Instead, the sampling process ensures that each group is equally represented by $m$ cases, which would not necessarily be true if $mG$ cases were randomly drawn from the grand total population without the provision of equal representation. (This involves the concept of stratified sampling, to be discussed briefly in Chapter 20.)

To evaluate the variance of the $\bar{X}_{r\cdot}$ about $\mu_{\cdot\cdot}$, let us note that

$$\bar{X}_{r\cdot} = (\bar{X}_{r1} + \bar{X}_{r2} + \cdots + \bar{X}_{rg} + \cdots + \bar{X}_{rG})/G$$

which indicates that $\bar{X}_{r\cdot}$ as a random variable is made up by adding $G$ variables of the form $\bar{X}_{rg}$, the means in the $r$th row of Table 15.1. With $R$ infinitely large the $\bar{X}_{r1}$, $\bar{X}_{r2}$, $\bar{X}_{rg}$, $\bar{X}_{rG}$ will be independent random variables (zero correlation between all possible pairs of columns in Table 15.1), hence the variance of the sum in the preceding parentheses will be given by the sum of the variances for the random variables being summed, and to get the variance of the averages, $\bar{X}_{r\cdot}$ we need only divide by $G^2$. Thus

$$\sigma^2_{\bar{x}(r)} = (\sigma^2_{\bar{x}_{r1}} + \sigma^2_{\bar{x}_{r2}} + \cdots + \sigma^2_{\bar{x}_{rg}} + \cdots + \sigma^2_{\bar{x}_{rG}})/G^2$$

but each of the $G$ variances is the sampling variance for the means for a particular column in Table 15.1, which for infinite $R$ will be nothing more than $\sigma^2/m$, with $\sigma^2$ being the (assumed) common variance for the $G$ populations. When we sum $G$ of these, we will have $G\sigma^2/m$, so that

$$\sigma^2_{\bar{x}(r)} = \frac{G\sigma^2/m}{G^2} = \frac{\sigma^2}{mG}$$

Hence the value for **D** term becomes

$$\frac{G}{G-1}\sigma^2_{\bar{x}(r)} = \frac{G}{G-1}\frac{\sigma^2}{mG} = \frac{\sigma^2}{m(G-1)}$$

As for the last, or **E**, term of our five terms, i.e.,

$$2\underset{r\,g}{\Sigma\Sigma}(\mu_{\cdot g} - \mu_{\cdot\cdot})(\bar{X}_{rg} - \mu_{\cdot g})$$

let us note that the contribution of group 1 to this double sum would be

$$2\underset{r}{\Sigma}(\mu_{\cdot 1} - \mu_{\cdot\cdot})(\bar{X}_{r1} - \mu_{\cdot 1})$$

which, since $(\mu_{\cdot 1} - \mu_{\cdot\cdot})$ is a constant, may be written as

$$2(\mu_{\cdot 1} - \mu_{\cdot\cdot})\underset{r}{\Sigma}(\bar{X}_{r1} - \mu_{\cdot 1})$$

Now with $R$ infinitely large, the mean of the $\bar{X}_{r1}$ will equal $\mu_{\cdot 1}$, hence we have the sum of a set of deviations about their own mean. It will be recalled that such a sum is always zero, a fact which will likewise hold for any and all values of $g$; therefore, the **E** term, when divided by infinite $R$ vanishes—we need not divide by $G - 1$ of the $R(G - 1)$ divisor.

We now bring together the results of dividing the five terms by $R(G - 1)$, with $R$ becoming infinitely large.

$$\frac{\underset{r}{\Sigma}s^2_{\bar{x}_{b(r)}}}{R} = \frac{\underset{g}{\Sigma}(\mu_{\cdot g} - \mu_{\cdot\cdot})^2}{G-1} + \frac{G}{G-1}\frac{\sigma^2}{m} - \frac{\sigma^2}{m(G-1)} + 0$$

Multiplying both sides by $m$ gives

$$\frac{m\underset{r}{\Sigma}s^2_{\bar{x}_{b(r)}}}{R} = \frac{m\underset{g}{\Sigma}(\mu_{\cdot g} - \mu_{\cdot\cdot})^2}{G-1} + \frac{G\sigma^2}{G-1} - \frac{\sigma^2}{G-1}$$

We place the $m$ of the left-hand side under the summation sign, we note that the last two terms combined become $\sigma^2$, and we rearrange, giving

$$\frac{\underset{r}{\Sigma}ms^2_{\bar{x}_{b(r)}}}{R} = \sigma^2 + \frac{m\underset{g}{\Sigma}(\mu_{\cdot g} - \mu_{\cdot\cdot})^2}{G-1}$$

Thus the mean of an infinite number of $ms^2_{\bar{x}_{b(r)}}$ (or of the exact equivalent, $s^2_b$) values is given on the right-hand side of the last equation. That is, the right side gives the mean or expected value of $s^2_b$. Stated differently, we may say that $s^2_b$ as an estimate includes the two components on the right. If we let $\rightarrow$ stand for "is an estimate of," then

$$s^2_b \rightarrow \sigma^2 + \frac{m\Sigma(\mu_{\cdot g} - \mu_{\cdot\cdot})^2}{G - 1}$$

and also

$$s^2_w \rightarrow \sigma^2$$

which helps us see that $F = s^2_b/s^2_w$ is a test for the presence of real differences among the $\mu_{\cdot g}$, or the $G$ population means.

Next, we distinguish between two differing situations as regards the term involving the $\mu_{\cdot g}$ values. If the $G$ groups represent, say, $G$ schools drawn at random from a possible population of schools, we have the so-called *random* model, whereas if the $G$ groups are the two sex groups or five defined socioeconomic groups or groups working under two or more experimental conditions, etc., we have the *fixed effects* model. Note that for the latter model, the number of groups is typically small and "fixed." Even though we were defining $G$ experimental conditions as, for example, $G$ differing degrees of illumination, we would not draw the $G$ levels at random from the theoretically possible large number of levels. Instead, we would deliberately *select* $G$ levels so they would be spaced along the illumination continuum. If we were interested in the effect of sense modality on reaction time, the number of possible sense modalities which we could use is fixed. The fixed effects model is sometimes called the fixed constants model because the $G$ values of $\mu_{\cdot g}$ are constants—with no sampling of groups, exactly the same population means are involved for each replication. This would not be the case when replication of the experiment involved, for example, drawing another sample of schools.

For this chapter we need not worry further about the models except to note that for the random model it makes sense to replace the term containing the random $\mu_{\cdot g}$ by $m\sigma^2_{\mu_{\cdot g}}$ because the sum of squares is being divided by $(G - 1)$ degrees of freedom and hence is an unbiased estimate of the variance of the population means based on a sample of size $G$. The use of such a symbol when only a few (2 or 3, or more) definitely fixed $\mu_{\cdot g}$ are involved does not provide a very meaningful description of the variation among them.

The student who got lost in the foregoing rather tedious, though rigorous, method for determining the expected value of $s^2_b$ might like a more intuitive and more easily understandable approach, which is restricted to the random model but has similar implication for the fixed effects model. Recall that when considering the reliability of measurement problem we regarded the variable score, $X$, as made up of two variable parts, $X_t$ and $E$

so that $X = X_t + E$, whence we had $S^2_x = S^2_t + S^2_e$. By analogy we may say that $\bar{X}_g$ as a random variable is made up of two parts, $\mu_g$ and sampling error, $E_g \, (= \bar{X}_g - \mu_g)$. Thus $\bar{X}_g = \mu_g + E_g$, and if we had a sample mean $\bar{X}_g$ from each of the possible populations, the variance of the distribution of these means would be

$$\sigma^2_{\bar{x}_g} = \sigma^2_{\mu_g} + \frac{\sigma^2}{m}$$

in which the last term is the sampling error component (each mean is based on a sample of $m$ cases). Multiplying both sides by $m$ gives

$$m\sigma^2_{\bar{x}_g} = \sigma^2 + m\sigma^2_{\mu_g}$$

Therefore, since $m$ times the variance of the means, when all populations are represented, can be broken into two components, it follows that the estimate, $ms^2_{\bar{x}_g} \, (= s^2_b)$, may also be subject to the same two sources of variation.

In practice we do not have a priori knowledge as to whether the second component, expressed either as $m \sum_g (\mu._g - \mu..)^2/(G - 1)$ or as $m\sigma^2_{\mu._g}$, is or is not zero. What we have are two estimates of variance, $s^2_w$ and $s^2_b$ $(= ms^2_{\bar{x}_b})$. If $s^2_b$ is significantly larger than $s^2_w$, i.e., if $F = s^2_b/s^2_w$ is beyond the chosen level of significance, it can be argued that $s^2_b$ does involve a source of variation over and above that of random sampling errors in the observed means; hence the second component is real.

Although the table of $F$ requires that the larger of the two estimates be used as the numerator in computing the variance ratio, it should be noted that $s^2_w$ cannot be significantly larger than $s^2_b$ unless the operation of chance sampling has been restricted in some manner. In practical applications we are primarily and nearly always interested in the case in which $s^2_b$ is the larger of the two estimates. If it is smaller than $s^2_w$, it is ordinarily not necessary to compute $F$.

We may now summarize the foregoing. When we have scores on $G$ groups of $m$ cases each, the total sum of squares can be broken down into two additive parts, that for between and that for within groups. Dividing by the appropriate degrees of freedom, the within sum of squares gives $s^2_w$ as an estimate of the trait variance for the population, and $s^2_b \, (= ms^2_{\bar{x}_b})$ yields a second and independent estimate of the same population variance. The sampling variation of the ratio of these two estimates is that of the variance ratio, $F$, if the $G$ groups belong to the same population. If $s^2_b$ is significantly larger than $s^2_w$, which is an estimate of the population variance, $s^2_b$ must be regarded as an estimate of the same variance *plus* variation due to real, nonchance, differences between the $G$ groups. Again, letting $\rightarrow$

stand for "is estimate of" or "has expected value of," we have

$$s^2_w \rightarrow \sigma^2$$

$$s^2_b \rightarrow \sigma^2 + \frac{m\Sigma(\mu_{.g} - \mu_{..})^2}{G - 1} \qquad \text{(fixed model)}$$

$$s^2_b \rightarrow \sigma^2 + m\sigma^2_{\mu_{.g}} \qquad \text{(random model)}$$

The null hypothesis is that the second component in the expected value of $s^2_b$ is zero, and rejection of the null hypothesis because $s^2_b/s^2_w$, as an $F$ with $df$ or $n_1$ of $G - 1$ and $df$ or $n_2$ of $mG - G$ (or $N - G$), is significantly large implies that the second component is not zero. In other words, we have a technique that provides an over-all test of the significance of the differences between several means considered simultaneously.

For all the applications discussed in this chapter, it is *assumed* (1) that the $m$ cases constituting each group have been drawn from a normally distributed population of scores for the trait or variable as measured and (2) that the $G$ populations have the same variance. For large samples the first assumption can be checked by way of measures of skewness and kurtosis relative to their standard errors or by the chi square test of goodness of fit. Unfortunately neither of these checks is very sensitive for small samples. The second assumption may be evaluated, regardless of sample sizes, by Bartlett's test for the homogeneity of variances. The reader will have noted that these two assumptions have to do with the distribution of scores within groups, which lead to the denominator $s^2_w$ of $F$.

**Computational formulas.** The required arithmetical labor can be shortened by resort to the general principle for computing the sum of squares of deviations inherent in formula (3.6):

$$\Sigma(X - \bar{X})^2 = \Sigma X^2 - \frac{(\Sigma X)^2}{N} = \frac{1}{N}[N\Sigma X^2 - (\Sigma X)^2]$$

Thus we would have

$$\Sigma\Sigma(X - \bar{X})^2 = \frac{1}{N}[N\Sigma\Sigma X^2 - (\Sigma\Sigma X)^2] \qquad (15.6)$$

for total sum of squares, in which the double summation indicates that the summing is over all groups. It can be shown by easy algebra that

$$\Sigma\Sigma(X - \bar{X}_g)^2 = \frac{1}{m}[m\Sigma\Sigma X^2 - \Sigma(\Sigma X)^2] \qquad (15.7)$$

for within sum of squares and that

$$m\Sigma(\bar{X}_g - \bar{X})^2 = \frac{1}{mG}[G\Sigma(\Sigma X)^2 - (\Sigma\Sigma X)^2] \qquad (15.8)$$

for between sum of squares.

Accordingly, to compute the three sums of squares of deviations, we need to sum all the raw scores, $\Sigma\Sigma X$; sum the squares of all the raw scores, $\Sigma\Sigma X^2$; and sum the squares of the separate group sums, $\Sigma(\Sigma X)^2$. These sums can readily be obtained on a calculating machine by computing $\Sigma X$ and $\Sigma X^2$ for each group separately, squaring each $\Sigma X$, and then summing the several $\Sigma X$ values for $\Sigma\Sigma X$, the $\Sigma X^2$ values for $\Sigma\Sigma X^2$, and the $(\Sigma X)^2$ values for $\Sigma(\Sigma X)^2$.

### EXAMPLE: TESTING THE SIGNIFICANCE OF DIFFERENCES BETWEEN SEVERAL MEANS

To illustrate the application of the technique outlined previously we shall use unpublished data of Wright[†] on massed vs. distributed practice in the learning of nonsense syllables by the anticipation method. The essential comparison is based on the amount of learning shown in 34 minutes by five $(= G)$ groups of sixteen $(= m)$ cases each. The groups differed in length of rest intervals between trials and/or in the total number of trials, as indicated at the top of Table 15.2. The scores of all 80 subjects are included in this table, and the necessary sums are given at the bottom of the table, separately for each group. Summing across yields the required double sums. The group means are also given, although not actually needed in determining $F$.

The sums of squares (of deviations) are obtained by substituting in formulas (15.6), (15.7), and (15.8):

$$\Sigma\Sigma(X - \overline{X})^2 = \tfrac{1}{80}[80(7638) - (692)^2] = 1652.20$$

$$\Sigma\Sigma_g(X - \overline{X}_g)^2 = \tfrac{1}{16}[16(7638) - 110{,}778] = 714.38$$

$$m\Sigma_g(\overline{X}_g - \overline{X})^2 = \tfrac{1}{80}[5(110{,}778) - (692)^2] = 937.82$$

These sums of squares, along with the respective degrees of freedom and the resulting variance estimates are conveniently arranged in Table 15.3, usually referred to as a variance table. Note that the sums of squares for between and within groups add to the sum for the total, which provides a check on the arithmetic involved in substituting in the formulas. This does not check on the accuracy of the sums given in Table 15.2. Note also that the degrees of freedom add to the total $df$.

The variance ratio, or $F$, becomes $234.46/9.53$ or $24.60$. With $df$s of $n_1 = 4$ and $n_2 = 75$, we refer to the table of $F$ to learn whether $24.60$ is larger than expected on the basis of chance. That this $F$ is highly

[†]Wright, Suzanne T., *Spacing of practice in verbal learning and the maturation hypothesis*, Unpublished Master's Thesis, Stanford University, California, 1946.

**Table 15.2. Number of syllables correctly anticipated at the 34th minute of practice**

| Group | 1 | 2 | 3 | 4 | 5 | | |
|---|---|---|---|---|---|---|---|
| Rest interval | | | | | | | |
| (minutes) | 8 | 3.5 | 2 | 1.25 | 0 | | |
| Number of | | | | | | | |
| trials | 5 | 8 | 11 | 14 | 29 | | |
| | 5 | 8 | 9 | 11 | 17 | | |
| | 5 | 7 | 3 | 12 | 16 | | |
| | 1 | 4 | 9 | 15 | 18 | | |
| | 5 | 4 | 10 | 11 | 11 | | |
| | 8 | 7 | 5 | 10 | 15 | | |
| | 1 | 7 | 11 | 8 | 9 | | |
| | 2 | 5 | 9 | 13 | 18 | | |
| | 2 | 6 | 6 | 13 | 13 | | |
| | 2 | 8 | 7 | 5 | 12 | | |
| | 8 | 14 | 6 | 7 | 15 | | |
| | 4 | 8 | 16 | 11 | 8 | | |
| | 1 | 5 | 12 | 12 | 13 | | |
| | 3 | 1 | 11 | 12 | 7 | | |
| | 4 | 5 | 15 | 9 | 15 | | |
| | 4 | 8 | 13 | 16 | 15 | | |
| | 2 | 5 | 4 | 7 | 13 | | |
| $m$ | 16 | 16 | 16 | 16 | 16 | | |
| $\Sigma X$ | 57 + | 102 + | 146 + | 172 + | 215 = $\Sigma\Sigma X$ | = | 692 |
| $\Sigma X^2$ | 279 + | 768 + | 1,550 + | 1,982 + | 3,059 = $\Sigma\Sigma X^2$ | = | 7,638 |
| $(\Sigma X)^2$ | 3,249 + | 10,404 + | 21,316 + | 29,584 + | 46,225 = $\Sigma(\Sigma X)^2$ | = | 110,778 |
| Means | 3.56 | 6.38 | 9.12 | 10.75 | 13.44 | $\bar{X}$ = | 8.65 |

significant is immediately apparent when we note that for the given $df$s an $F$ of about 5.2 is significant at the .001 level. With the between-groups variance estimate significantly larger than that for within groups, we can conclude with high confidence that the five sets of scores have not been drawn from the same population of scores, or that amount of time spent in practice is a real source of variation. This is, of course, equivalent to

**Table 15.3. Variance table for data of Wright**

| Source | Sum of Squares | $df$ | Variance Estimate |
|---|---|---|---|
| Between | 937.82 | 4 | 234.46 = $s^2_b$ |
| Within | 714.38 | 75 | 9.53 = $s^2_w$ |
| Total | 1652.20 | 79 | |

saying that the several group means considered simultaneously differ significantly among themselves.

In the illustration just given the groups can be arranged in order before any of the data are seen, and additional credence can be placed in the results because the means follow this ordering. It should be understood, however, that the variance technique does not presuppose an a priori ordering of the several groups—it is generally applicable for testing the significance of the differences between group means regardless of prior considerations.

If only the $z$ or $t$ technique were available and we wished to compare the means for five groups, it would ordinarily be necessary to compute $t$ or $z$ for each possible difference, and five means would lead to $5 \times 4/2$ or 10 differences. Obviously, the variance method requires less computation, and furthermore it provides an over-all test of significance which is not subject to the fallacy inherent in singling out the comparison involving the largest obtained $t$ or $z$, a practice which is likely to capitalize on chance differences. This problem is discussed at the end of this chapter.

## SPECIAL CASE OF $F$ TEST WHEN $n_1 = 1$

If we had $G = 2$ groups, the testing of the between-groups variance would appear to be much like testing the difference between two means. Let us examine this case by starting with the expressions for the sum of squares for two groups:

$$\text{1st group:} \quad \Sigma(X - \bar{X})^2 = \Sigma(X - \bar{X}_1)^2 + m(\bar{X}_1 - \bar{X})^2$$
$$\text{2nd group:} \quad \Sigma(X - \bar{X})^2 = \Sigma(X - \bar{X}_2)^2 + m(\bar{X}_2 - \bar{X})^2$$

Instead of using double summation signs, we may indicate the within-groups sum of squares as $\Sigma(X - \bar{X}_1)^2 + \Sigma(X - \bar{X}_2)^2$, and the between-groups sum of squares as $m(\bar{X}_1 - \bar{X})^2 + m(\bar{X}_2 - \bar{X})^2$. The respective $df$s will be $2m - 2$ and 1. Indicating the division of the sums of squares by their $df$s, we can write the variance ratio as

$$F = \frac{\dfrac{m(\bar{X}_1 - \bar{X})^2 + m(\bar{X}_2 - \bar{X})^2}{1}}{\dfrac{\Sigma(X - \bar{X}_1)^2 + \Sigma(X - \bar{X}_2)^2}{2m - 2}}$$

Since the number of cases for the two groups is the same, it is readily seen that the mean for one group will be exactly as far above the general mean ($\bar{X}$) as the other group mean is below $\bar{X}$, or that $\bar{X}$ will bisect the distance between $\bar{X}_1$ and $\bar{X}_2$; therefore $(\bar{X}_1 - \bar{X})^2 = (\bar{X}_2 - \bar{X})^2 = \frac{1}{4}(\bar{X}_1 - \bar{X}_2)^2$. The numerator for $F$ becomes $(m/2)(\bar{X}_1 - \bar{X}_2)^2$. It will be noted that the denominator term, which defines $s^2_w$, is identical to the $s^2$ defined by (7.2) in connection with the $t$ test. Accordingly, we may write

$$F = \frac{\dfrac{m}{2}(\bar{X}_1 - \bar{X}_2)^2}{s^2}$$

Dividing both numerator and denominator by $m/2$, we have

$$F = \frac{(\bar{X}_1 - \bar{X}_2)^2}{s^2 \dfrac{2}{m}}$$

the square root of which is

$$\sqrt{F} = \frac{\bar{X}_1 - \bar{X}_2}{s\sqrt{\dfrac{2}{m}}} = \frac{\bar{X}_1 - \bar{X}_2}{s\sqrt{\dfrac{1}{m} + \dfrac{1}{m}}}$$

which is identical with a formula for $t$, p. 115. When $G = 2$ or two groups are being compared, then $F = t^2$. It can be shown that this is also true when the $N$s or $m$s for the two groups are unequal. In fact, it can be shown that, when $n_1 = 1$, the sampling distribution of $F$ becomes the same as that for $t^2$ provided the estimate based on between groups, i.e., that based on 1 degree of freedom, is used as the numerator regardless of which of the two estimates is the larger. It is thus seen that the $t$ test is a special case of the $F$ test. Note that $F$ involves the square of the difference between means; hence it provides a basis for judging whether a difference between means, irrespective of direction, is significant (cf. pp. 284–85). The $z$ technique for comparing the means of two large samples is also a special case of the more general $F$ test. That is, when $n_1 = 1$ and $n_2$ is not small, the square root of $F$ is $z$, interpretable via the normal curve table (Table A of the Appendix).

## GROUPS OF UNEQUAL SIZE

When the number of cases varies from group to group, we may let $m_1, m_2, \cdots, m_g, \cdots, m_G$ stand for the several $N$s. The sum of squares for the $g$th group would be written as

$$\Sigma(X - \bar{X})^2 = \Sigma(X - \bar{X}_g)^2 + m_g(\bar{X}_g - \bar{X})^2$$

and the double summation over all groups would be

$$\Sigma\Sigma(X - \bar{X})^2 = \underset{g}{\Sigma}\Sigma(X - \bar{X}_g)^2 + \underset{g}{\Sigma}m_g(\bar{X}_g - \bar{X})^2$$

which differs from formula (15.2) in that the varying $m$s must be left under the summation sign in the last term. In specifying the degrees of freedom, we must replace $mG$ by $N$, where $N$ is the total cases for all groups. The respective $df$s become $N - 1$, $N - G$, and $G - 1$. The computational formulas are changed to

$$\Sigma\Sigma(X - \bar{X})^2 = \Sigma\Sigma X^2 - \frac{(\Sigma\Sigma X)^2}{N} \quad \text{for total sum} \quad (15.9)$$

$$\Sigma\Sigma_g(X - \bar{X}_g)^2 = \Sigma\Sigma X^2 - \Sigma_g \frac{(\Sigma X)^2}{m_g} \quad \text{for within sum} \quad (15.10)$$

$$\Sigma_g m_g(\bar{X}_g - \bar{X})^2 = \Sigma_g \frac{(\Sigma X)^2}{m_g} - \frac{(\Sigma\Sigma X)^2}{N} \quad \text{for between sum} \quad (15.11)$$

Note that the second term for the within sum (and the first for the between) requires that for each group the square of the sum of its scores be first divided by its $m$; then the several quotients are summed. An additional row would be needed along the bottom of Table 15.2 for these quotients if the $m$s differed, or the $(\Sigma X)^2$ row might be replaced by $(\Sigma X)^2/m_g$ values.

A variance table (like Table 15.3) may be formed, and $F$ taken to equal $s^2_b/s^2_w$ as before. The same interpretation holds: if $F$ is significantly large, i.e., if $s^2_b$ is significantly larger than $s^2_w$, the variation of the several group means among themselves is larger than expected on the basis of sampling; hence nonchance differences exist between the groups. Although for the unequal $m$ situation the sampling values of $s^2_b/s^2_w$ follow the $F$ distribution, the use of unequal $m$s does not provide as sensitive (as powerful) a test as a test based on equal $m$s. If $N$ cases are to be divided into $G$ experimental groups, it is preferable to assign $m = N/G$ cases to each group, unless there is a cost factor that is differential from experimental to experimental condition. See the argument on p. 121.

Thus the $F$ technique may be applied as a test of the significance of the difference between two or more means based on large or small samples of equal or unequal size (per group) regardless of whether there is an a priori basis for arranging the groups in order. It might be said parenthetically that the scientific hypothesis being tested will specify the direction of differences if such are expected.

## TESTING THE SIGNIFICANCE OF THE CORRELATION RATIO

If the definitions of the correlation ratio, $\eta$ (pp. 231–32), are reexamined, it is readily seen that for one variable the within-arrays variance is the same as the within-groups variance, the grouping being made on the basis of intervals on another variable. Also the variance of array means is the same as between-groups variance. We recall, however, that the correlation ratio, as defined, does not involve the idea of variance estimates. It should be rather obvious that, unless the between-arrays (groups) variance is significantly larger than expected on the basis of sampling errors in the array means, a correlation ratio cannot be deemed significant.

For purposes of exposition we shall outline the procedure for testing the significance of $\eta_{yx}$, for which we shall use the simpler symbol $\eta$. The grouping will be on the basis of the intervals on the $X$ variable, and the required sums of squares will be in terms of $Y$. The sums of squares and their respective degrees of freedom will be

$$\Sigma\Sigma(Y - \overline{Y})^2 = \Sigma\Sigma(Y - \overline{Y}_g)^2 + \Sigma m_g(\overline{Y}_g - \overline{Y})^2$$
$$\underset{(N-1)}{} \qquad \underset{g \ \ (N-G)}{} \qquad \underset{g \ \ (G-1)}{}$$

for $G$ arrays with varying number, $m_g$, of cases per array. From the definition formula of the correlation ratio, we have

$$\eta^2 = 1 - \frac{S^2_{ay}}{S^2_y}$$

which becomes, in the notation of this chapter,

$$\eta^2 = 1 - \frac{\Sigma\Sigma(Y - \overline{Y}_g)^2/N}{\Sigma\Sigma(Y - \overline{Y})^2/N}$$

Since $N$ cancels, we see that the following holds:

$$\Sigma\Sigma(Y - \overline{Y}_g)^2 = (1 - \eta^2)\Sigma\Sigma(Y - \overline{Y})^2 = \text{within sum of squares} \quad (15.12)$$

From the alternate expression for $\eta$ we have

$$\eta^2 = \frac{S^2_{my}}{S^2_y}$$

which becomes

$$\eta^2 = \frac{\Sigma m_g(\overline{Y}_g - \overline{Y})^2/N}{\Sigma\Sigma(Y - \overline{Y})^2/N}$$

which leads to

$$\Sigma m_g(\overline{Y}_g - \overline{Y})^2 = \eta^2\Sigma\Sigma(Y - \overline{Y})^2 = \text{between sum of squares} \quad (15.13)$$

When we wish to divide the sum of squares of formula (15.12) or (15.13) by the proper $df$, we may choose either the left- or right-hand part as representing the sum of squares. Thus the between-arrays estimate may be written as

$$s^2_b = \frac{\eta^2\Sigma\Sigma(Y - \overline{Y})^2}{G - 1}$$

and that for within arrays as

$$s^2_w = \frac{(1 - \eta^2)\Sigma\Sigma(Y - \bar{Y})^2}{N - G}$$

The ratio, $F = s^2_b/s^2_w$, may be written as

$$F = \frac{\eta^2\Sigma\Sigma(Y - \bar{Y})^2/(G - 1)}{(1 - \eta^2)\Sigma\Sigma(Y - \bar{Y})^2/(N - G)} = \frac{\eta^2/(G - 1)}{(1 - \eta^2)/(N - G)} \quad (15.14)$$

It is accordingly seen that for fixed $df$s the value of $F$, even though computed from the sums rather than from their equivalents in terms of $\eta^2$, can be thought of as depending on the size of $\eta^2$; therefore a significant $F$ indicates a significant correlation ratio.

With the three sums of squares computed, we can readily determine whether any correlation in the sense of the correlation ratio exists, and we also have the necessary sums for calculating $\eta$ if it is desired to have this measure of the degree of correlation. A significant $F$ does not, however, mean a high correlation ratio; with $N$ large, a low $\eta$ can possess statistical significance.

The computation of the sums of squares is accomplished by means of formulas (15.9–15.11) with the $X$s replaced by $Y$s.

## SIGNIFICANCE OF LINEAR CORRELATION

An appreciable correlation between two variables which are linearly related implies that the slopes of the regression lines are not zero, which in turn implies that the variance of predicted values is large enough to have some kind of statistical significance. The variance technique may be used as a test of the significance of linear regression.

Suppose that we develop the argument in terms of the regression of $Y$ on $X$. We may write the linear equation for predicting $Y$ from $X$ as $Y' = BX + A$. If we think of this regression line as having been drawn on the scatter diagram, we can readily see the deviation of any person's $Y$ value from the mean of the $Y$s can be expressed in terms of its deviation from the regression line (or predicted value) plus the deviation of the predicted value from the mean of the $Y$s:

$$(Y - \bar{Y}) = (Y - Y') + (Y' - \bar{Y})$$

in which $Y'$ will vary from person to person in accordance with his $X$ score. If we square all such $(Y - \bar{Y})$ deviations and sum over all cases, we get

$$\Sigma\Sigma(Y - \bar{Y})^2$$
$$= \Sigma[(Y - Y') + (Y' - \bar{Y})]^2$$
$$= \Sigma(Y - Y')^2 + \Sigma(Y' - \bar{Y})^2 + 2\Sigma(Y - Y')(Y' - \bar{Y})$$

for which double summation signs are not needed for clarity even though the summing is over all cases. The last or cross-product term has to do with a possible relationship between predicted values and residuals, but, as was shown in Chapter 9, this correlation is always zero, and hence this last term vanishes.

Therefore the sum of squares can be broken down into two components: residuals or within arrays about the regression line and a part depending on the variation of the predicted values about the mean. If the correlation between $X$ and $Y$ were zero, this latter component would be zero because $\overline{Y}$ would be predicted for all cases. The departure of this sum of squares or of a variance estimate based thereon from zero might lead us to conclude that real correlation exists in the population being sampled if it were not for the fact that sampling errors ordinarily operate so as to prevent the obtaining of zero correlation.

Before attempting to understand the operation of chance sampling, we should consider the degrees of freedom associated with the sums of squares. As usual, the total sum of squares is based on $N - 1$ degrees of freedom. The $df$ for $\Sigma(Y - Y')^2$ may not be immediately obvious, but note that, if $N = 2$ and variation exists for both $X$ and $Y$, the regression line would necessarily pass through the two points defined by the pair of scores, $r$ would be unity, and $\Sigma(Y - Y')^2$ would be zero. In other words, with $N = 2$, there is no freedom for deviation from the regression line. From this it would be inferred that $N$ needs to be reduced by 2, or that $df = N - 2$, a deduction which is consistent with the fact that, in fitting a straight line, two constants are determined from the data, and hence two restrictions are imposed on the $N$ deviations of the type $(Y - Y')$.

Since the $df$s for the component sums of squares are additive to that for the total, we can determine the $df$ for the regression or $\Sigma(Y' - \overline{Y})^2$ term by subtracting the $df$ for residuals from that for the total: $(N - 1) - (N - 2) = 1$ as the $df$ for the regression term. But determination of a $df$ by subtraction does not permit the additive check on the correctness of the $df$s which is possible in case each $df$ is ascertained separately on the basis of some principle. By what principle could we determine that for the regression sum of squares the proper $df$ is 1? The value of $\Sigma(Y' - \overline{Y})^2$ will not be changed by shifting from gross scores to deviation scores, i.e., by moving the origin to the intersection of $\overline{X}$ and $\overline{Y}$. It will be recalled that the regression equation in deviation units is $y' = bx$ (where $b = B$ of the gross score form), and accordingly we may write

$$\Sigma(Y' - \overline{Y})^2 = \Sigma(y' - \bar{y})^2 = \Sigma(y' - 0)^2 = \Sigma(bx)^2 = b^2\Sigma x^2$$

which permits us to examine the source or sources of variation in the regression sum of squares. Its value depends on $b^2$ and $\Sigma x^2$, but the value of $\Sigma x^2$ does not depend on the degree of correlation. For a fixed set of $X$s, the freedom of $\Sigma(Y' - \overline{Y})^2$ to vary springs from $b$, i.e., from *one* value.

Now let us return to a brief consideration of sampling or the meaning of the variance estimates which result from dividing the sums of squares by their *df*s. On the basis of the null hypothesis, that the degree of linear correlation is zero for the population being sampled, the regression line for the population would pass through $\mu_y$, with zero slope or parallel to the $x$ axis. Hence $(Y - Y')$ will equal $(Y - \mu_y)$ and the variance of the residuals will equal the total variance of the $Y$s. A sample from the population will seldom yield zero correlation (line with zero slope), and therefore the residuals will tend to be somewhat reduced, or $\Sigma(Y - Y')^2$ will tend to be less than $\Sigma(Y - \overline{Y})^2$. Will $\Sigma(Y - Y')^2/(N - 2)$ give an unbiased estimate of the population variance, $\sigma^2_y$, when no correlation exists in the population? Now $s^2_{y \cdot x}$ is, of course, an unbiased estimate of $\sigma^2_{y \cdot x}$, the residual variance for the population regardless of the value of $r_{pop}$. But when $r_{pop} = 0$, $\sigma^2_{y \cdot x} = \sigma^2_y(1 - 0^2) = \sigma^2_y$, hence under the null condition of no correlation we see that the unbiased estimate, $s^2_{y \cdot x}$, of $\sigma^2_{y \cdot x}$, may be regarded as an unbiased estimate of $\sigma^2_y$ simply because $\sigma^2_{y \cdot x} = \sigma^2_y$ when $r_{pop} = 0$.

What does $\Sigma(Y' - \overline{Y})^2/1$ estimate under the null condition? Since

$$\Sigma(Y - \overline{Y})^2 = \Sigma(Y - Y')^2 + \Sigma(Y' - \overline{Y})^2$$

we have

$$\Sigma(Y' - \overline{Y})^2 = \Sigma(Y - \overline{Y})^2 - \Sigma(Y - Y')^2 \qquad (15.15)$$

Now on the average

$$\frac{\Sigma(Y - \overline{Y})^2}{N - 1} = \sigma^2_y$$

hence

$$\Sigma(Y - \overline{Y})^2 = (N - 1)\sigma^2_y$$

Also on the average, under the null condition,

$$\frac{\Sigma(Y - Y')^2}{N - 2} = \sigma^2_y$$

hence

$$\Sigma(Y - Y')^2 = (N - 2)\sigma^2_y$$

Substituting into (15.15) gives

$$\Sigma(Y' - \overline{Y})^2 = (N - 1)\sigma^2_y - (N - 2)\sigma^2_y = \sigma^2_y$$

Thus we have that on the average a sum of squares is equal to a population variance, but unbiased variance estimation always involves dividing a sum of squares by its *df*. We see at once that $df = 1$ is the only possible value that will preserve the equation, hence we may regard $\Sigma(Y' - \overline{Y})^2/1$ as an unbiased estimate of $\sigma^2_y$ under the null condition of no correlation.

Let $s^2_r$ stand for the estimate based on the residual sum of squares and $s^2_p$ stand for the estimate based on the predictions by a linear regression function. If $F = s^2_p/s^2_r$ is significant, the null hypothesis becomes suspect. Thus the $s^2_p$ estimate is larger than expected on the basis of chance sampling, from which it may be inferred that regression is a real source of variation in $\Sigma(Y' - \overline{Y})^2$, i.e., that the slope of the regression for the population is not zero, or that some correlation exists.

We have already noted that

$$\Sigma(Y' - \overline{Y})^2 = Nr^2S^2_y$$

Since $\Sigma(Y - Y')^2$ divided by $N$ equals the error of estimate variance, previously proved to equal $S^2_y(1 - r^2)$, it follows readily that

$$\Sigma(Y - Y')^2 = N(1 - r^2)S^2_y$$

Accordingly

$$s^2_p = \frac{\Sigma(Y' - \overline{Y})^2}{1} = \frac{Nr^2S^2_y}{1}$$

and

$$s^2_r = \frac{\Sigma(Y - Y')^2}{N - 2} = \frac{N(1 - r^2)S^2_y}{N - 2}$$

Therefore

$$F = \frac{Nr^2S^2_y/1}{N(1 - r^2)S^2_y/(N - 2)} = \frac{r^2}{(1 - r^2)/(N - 2)} \quad (15.16)$$

which is the square of the $t$, formula (10.2), for testing the significance of $r$. Thus, again we have $F = t^2$, when $n_1 = 1$.

The reader will have noted that, since the required sums of squares and the resulting $F$ can readily be expressed in terms of $r$, there is no need to worry further about a computational scheme for securing the sums of squares. The easier thing to do is simply to compute $r$. After that is done, either the $F$ or the $t$ test may be used for judging whether the correlation is significant. This discussion of the linear correlation problem here should help the student appreciate the generality of the analysis of variance technique and should also provide him with relevant concepts for understanding the test for curvilinearity of regression, to which we now turn.

## TESTING LINEARITY OF REGRESSION

We have seen that the correlation ratio is a general measure of the degree of correlation and that $r$ measures the degree of linear relationship.

Even though the regression of $Y$ on $X$ for a population be exactly linear, it will be found for a sample that the means of the arrays will show some deviation from a straight line; hence, as previously pointed out, the correlation ratio will tend to be larger than $r$. How large should the difference between $\eta$ and $r$ be before we suspect nonlinearity, or how much can the array means deviate from a straight line by chance? Before the development of the analysis of variance technique, the inadequate Blakeman criterion was used to answer the foregoing. In presenting the currently accepted method, we shall carry the argument through on the basis of the regression of $Y$ on $X$.

Imagine a scatter diagram with regression line drawn and the array mean located in each vertical array. For a score in the $g$th array, the deviation of $Y$ from $\overline{Y}$ can be thought of in terms of its deviation from the array mean, $\overline{Y}_g$, plus the deviation of the array mean from the predicted value, $Y'_g$, plus the deviation of the predicted value from the total mean. In symbols,

$$(Y - \overline{Y}) = (Y - \overline{Y}_g) + (\overline{Y}_g - Y'_g) + (Y'_g - \overline{Y})$$

Squaring and summing for the $m_g$ cases in each array and then summing over all $G$ arrays (equivalent to summing over all groups), we have

$$\Sigma\Sigma(Y - \overline{Y})^2 = \Sigma\Sigma_g(Y - \overline{Y}_g)^2 + \Sigma_g m_g(\overline{Y}_g - Y'_g)^2 + \Sigma_g m_g(Y'_g - \overline{Y})^2$$

the cross-product terms having vanished because the component parts are uncorrelated.

The first component is a sum of squares based on within-array variation with $N - G$ degrees of freedom. We encountered this in checking the significance of the correlation ratio, and we then labeled as $s^2_w$ the variance estimate based thereon.

The second sum involves deviations of array means from linear regression. Its $df$ will be $G - 2$ since there are $G$ means and two restrictive constants in $Y'_g$. If $G = 2$, the two means cannot vary from the fitted line. Let us use $s^2_d$ as a symbol for the variance estimate based on this sum of squares.

The third sum, which has to do with the part of the total variance predictable by means of linear regression, is very similar to that occurring a few pages earlier in connection with the $F$ test of the correlation coefficient. It differs only in that the same value is predicted for all cases within an array regardless of their location in the $X$ interval defining the array. This is equivalent to a linear prediction of the mean of the array. Actually, the numerical value of $\Sigma(Y' - \overline{Y})^2$ as calculated by $Nr^2S^2_y$, which equals $r^2\Sigma\Sigma(Y - \overline{Y})^2$, will be the same as $\Sigma_g m_g(Y'_g - \overline{Y})^2$ computed directly, provided $r$ was originally determined from a scatter diagram with the same intervals now being used to define the arrays. We have already seen that the $df$ for this sum is 1, and we have used $s^2{}_p$ as a symbol for the estimate based thereon.

It will be recalled that, in the scheme for testing the significance of the correlation ratio, the total sum of squares was broken down into a within-array and a between-array part. We now have a breakdown into within

**Table 15.4. Analysis of variance functions for bivariate correlation**

| Source of Variation | Sum of Squares | Equivalent | $df$ | Esti-mate |
|---|---|---|---|---|
| (a) Linear regression | $\sum_g m_g(Y'{}_g - \bar{Y})^2 = r^2\Sigma\Sigma(Y - \bar{Y})^2$ | | 1 | $s^2{}_p$ |
| (b) Deviation of means from line | $\sum_g m_g(\bar{Y}_g - Y'{}_g)^2 = (\eta^2 - r^2)\Sigma\Sigma(Y - \bar{Y})^2$ | | $G - 2$ | $s^2{}_d$ |
| (c) Between-array means | $\sum_g m_g(\bar{Y}_g - \bar{Y})^2 = \eta^2\Sigma\Sigma(Y - \bar{Y})^2$ | | $G - 1$ | $s^2{}_b$ |
| (d) Within arrays | $\sum_g \Sigma\Sigma(Y - \bar{Y}_g)^2 = (1 - \eta^2)\Sigma\Sigma(Y - \bar{Y})^2$ | | $N - G$ | $s^2{}_w$ |
| (e) Residual from line | $\sum_g \Sigma\Sigma(Y - Y'{}_g)^2 = (1 - r^2)\Sigma\Sigma(Y - \bar{Y})^2$ | | $N - 2$ | $s^2{}_r$ |
| (f) Total | $\Sigma\Sigma(Y - \bar{Y})^2$ | | $N - 1$ | |

array (as before) plus two additional parts—the sum $\sum_g m_g(\bar{Y}_g - \bar{Y})^2$ is broken into

$$\sum_g m_g(\bar{Y}_g - Y'{}_g)^2 + \sum_g m_g(Y'{}_g - \bar{Y})^2$$

It will also be recalled that

$$\sum_g m_g(\bar{Y}_g - \bar{Y})^2 = \eta^2\Sigma\Sigma(Y - \bar{Y})^2$$

and that

$$\sum_g m_g(Y'{}_g - \bar{Y})^2 = r^2\Sigma\Sigma(Y - \bar{Y})^2$$

By subtraction, we see that the new sum, $\sum_g m_g(\bar{Y}_g - Y'{}_g)^2$, is equivalent to $(\eta^2 - r^2)\Sigma\Sigma(Y - \bar{Y})^2$.

For convenience, we shall now assemble in an analysis of variance table the several symbolic expressions having to do with testing the significance of (1) the correlation ratio, (2) the linear regression coefficient, and (3) nonlinearity of regression. Table 15.4 gives the sources of variation, the sums of squares and their equivalents in terms of $r$ or $\eta$, the degrees of freedom, and a symbol for each of the variance estimates. Note, in review, that for the sums of squares, their equivalents, and the $dfs$, the following additions hold true:

$$(a) + (b) = (c)$$
$$(a) + (e) = (f)$$
$$(c) + (d) = (f)$$
$$(a) + (b) + (d) = (f)$$

The several useful and permissible $F$s, or ratios of independent and unbiased variance estimates, along with the proper $dfs$ ($n_1$ and $n_2$ values) for entering the table of $F$, may be stated in summary form:

$F_1 = s^2_b/s^2_w;$   $n_1 = G - 1,$ $n_2 = N - G$:   significance of correlation ratio

$F_2 = s^2_p/s^2_r;$   $n_1 = 1,$      $n_2 = N - 2$:   significance of linear correlation

$F_3 = s^2_d/s^2_w;$   $n_1 = G - 2,$ $n_2 = N - G$:   significance of curvilinearity

We have already discussed the first two of these $F$s. If we write the third in terms of sums and $dfs$, we have

$$F_3 = \frac{s^2_d}{s^2_w} = \frac{\sum_g m_g(\overline{Y}_g - Y'_g)^2/(G - 2)}{\sum\sum_g(Y - \overline{Y}_g)^2/(N - G)}$$

$$= \frac{(\eta^2 - r^2)\sum\sum(Y - \overline{Y})^2/(G - 2)}{(1 - \eta^2)\sum\sum(Y - \overline{Y})^2/(N - G)}$$

$$= \frac{(\eta^2 - r^2)/(G - 2)}{(1 - \eta^2)/(N - G)} \tag{15.17}$$

which indicates definitely that its value, for given $dfs$, is a reflection of the difference between the correlation ratio and the correlation coefficient. Therefore, in testing the significance of the variation of array means from linear regression, we are testing the significance of the difference between $\eta$ and $r$. If $F_3$ falls beyond the .01 probability level, the hypothesis of linear regression for the population being sampled is rejected. When this happens, it follows that the correlation coefficient and a linear regression function for $Y$ on $X$ are not appropriate measures to use in describing the relationship.

If we are also interested in testing the significance of the correlation ratio for $X$ on $Y$ and the linearity of the horizontal array means, the analysis is carried through with $X$s substituted for $Y$s. Since the number of grouping intervals on the two axes need not be the same, the value of $G$ may differ for the two analyses.

## ILLUSTRATIVE PROBLEM: $r$, $\eta$, AND CURVILINEARITY

The foregoing three tests of significance and the computations necessary thereto may be illustrated by the data of Table 15.5, which gives the bi-

Table 15.5. Bivariate scatter for initial and final scores of 92 boys on Koerth pursuit rotor

| $Y =$ Final Score Code | | $X =$ Initial Score | | | | | | | | $f_y$ |
|---|---|---|---|---|---|---|---|---|---|---|
| | | 0 | 30 | 60 | 90 | 120 | 150 | 180 | 210 | |
| 740 | 11 | | | | 1 | | | | | 1 |
| 700 | 10 | | 1 | 2 | 1 | 1 | | 2 | 2 | 9 |
| 660 | 9 | 1 | 1 | 1 | 4 | 3 | | 1 | 2 | 13 |
| 620 | 8 | 2 | 8 | 2 | 2 | 2 | | 1 | | 17 |
| 580 | 7 | 3 | 3 | 7 | 1 | 1 | | | 1 | 16 |
| 540 | 6 | 2 | 8 | 5 | | | | | | 15 |
| 500 | 5 | 2 | 5 | 3 | 1 | | | | | 11 |
| 460 | 4 | 3 | 1 | | | | | | | 4 |
| 420 | 3 | 2 | | | | | | | | 2 |
| 380 | 2 | | | | | | | | | |
| 340 | 1 | 3 | | | | | | | | 3 |
| 300 | 0 | 1 | | | | | | | | 1 |
| $f_x = m_g$ | | 19 | 27 | 20 | 10 | 7 | 0 | 4 | 5 | 92 $= N$ |
| $\Sigma Y$ | | 89 | 181 | 139 | 85 | 60 | 0 | 37 | 45 | 636 |
| $\Sigma Y^2$ | | 547 | 1269 | 1007 | 747 | 520 | 0 | 345 | 411 | 4846 |
| $(\Sigma Y)^2/m_g$ | | 416.89 | 1213.37 | 966.05 | 722.50 | 514.29 | 0 | 342.25 | 405.00 | 4580.35 |

variate distribution for the relationship between initial (sum of scores on trials 1–4) and final (trials 67–70) performance on the Koerth pursuit rotor. Since it is logical to be concerned with the prediction of final from initial score, or the regression of $Y$ on $X$, we shall be dealing with variations on the $Y$ variable.

In the first place, the correlation coefficient is computed from the scatter diagram by the method given in Chapter 8. Its value of .5687 is about .01 lower than the coefficient computed from a scatter with twice as many intervals. The use of so few intervals for the $X$ variable would obviously not be recommended for the computation of $r$, but in this illustration it is convenient because of page-space limitations. There is the additional consideration that for computing the correlation ratio we should avoid having too few cases per array, which if the sample is small may mean only a few intervals on the independent variable. At least twelve intervals should be used for the dependent variable. In checking on linearity, it is necessary that we calculate $r$ from a scatter with the same grouping intervals used in computing $\eta$, and no corrections for grouping error are needed.

For the computation of the correlation ratio and for the testing of its significance, we need the within-arrays, the between-arrays, and the total sum of squares. These may be computed from coded scores (deviations from an arbitrary origin in terms of step intervals), and the entire analysis may be carried through on the basis of coded scores, so that cumbersomely large figures are avoided. The reader who wishes to follow the computational procedure will need to note the following features of Table 15.5. The marginal frequencies on the right are for all the $Y$ scores, and the $f_x$s along the bottom margin are the $m_g$s, or cases per array. For each vertical array and for the right-hand margin, $\Sigma Y$ and $\Sigma Y^2$ are computed in terms of coded values (these correspond to $\Sigma d$ and $\Sigma d^2$ of Chapter 3). Summing across the $\Sigma Y$ and $\Sigma Y^2$ rows should yield the $\Sigma Y$ and $\Sigma Y^2$ obtained from the marginal distribution. For this problem, $\Sigma\Sigma Y = 636$ and $\Sigma\Sigma Y^2 = 4846$. The last row, containing the several values of $(\Sigma Y)^2/m_g$, is summed across for the needed $\displaystyle\sum_g \frac{(\Sigma Y)^2}{m_g}$, which is 4580.35 in this example. There is no check on this figure by calculations based on the margin.

In order to get the sums of squares of deviations, the values 636, 4846, and 4580.35 are substituted in formulas (15.9–15.11) with $X$ replaced by $Y$.

$$\Sigma\Sigma(Y - \overline{Y})^2 = 4846 - \frac{636^2}{92} = 449.30$$

$$\Sigma\Sigma(Y - \overline{Y}_g)^2 = 4846 - 4580.35 = 265.65$$

$$\Sigma m_g(\overline{Y}_g - \overline{Y})^2 = 4580.35 - \frac{636^2}{92} = 183.65$$

By formula (15.13) we now obtain

$$\eta^2 = \frac{183.65}{449.30} = .40874; \qquad \eta = .639$$

which is the correlation ratio for $Y$ on $X$.

The other sums of squares called for in schematic Table 15.4 may be calculated from their equivalents in terms of $r^2$ and/or $\eta^2$. Note that $r^2 = .5687^2 = .32342$.

$$\Sigma m_g(Y'_g - \overline{Y})^2 = (.32342)(449.30) = 145.31$$

$$\Sigma\Sigma(Y - Y'_g)^2 = (1 - .32342)(449.30) = 303.99$$

$$\Sigma m_g(\overline{Y}_g - Y'_g)^2 = (.40874 - .32342)(449.30) = 38.34$$

The several sums of squares and their respective degrees of freedom are set forth in Table 15.6, which contains also the variance estimates obtained by dividing the sums of squares by their $dfs$. From these variance estimates we have the following.

For testing the significance of the correlation ratio we have $F_1$ = 30.61/3.13 = 9.8, which for $n_1 = 6$ and $n_2 = 85$ is highly significant. The .001 level of significance requires an $F$ of about 4.0.

For testing the significance of linear correlation, i.e., $r$, we have $F_2$ = 145.31/3.38 = 43.0, which for $n_1 = 1$ and $n_2 = 90$ is likewise highly significant, the .001 level being at an $F$ of about 11.6.

**Table 15.6. Analysis of variance table for regression of final ($Y$) on initial score for data of Table 15.5**

| Source | Sum of Squares | $df$ | Variance Estimate |
|---|---|---|---|
| Linear regression | 145.31 | 1 | $145.31 = s^2_p$ |
| Deviation of means from line | 38.34 | 5 | $7.67 = s^2_d$ |
| Between-array means | 183.65 | 6 | $30.61 = s^2_b$ |
| Within arrays | 265.65 | 85 | $3.13 = s^2_w$ |
| Residual from line | 303.99 | 90 | $3.38 = s^2_r$ |
| Total | 449.30 | 91 | 4.94 |

For testing linearity of regression, i.e., the departure of the array means from a straight line, we have $F_3 = 7.67/3.13 = 2.5$, which for $n_1 = 5$ and $n_2 = 85$ is near the .05 level of significance. Thus the apparent departure from linearity in Table 15.5 is not sufficiently great to lead to rejection of the hypothesis of linearity; we would, however, question the hypothesis. This is an example of borderline significance which calls for drawing another sample or adding more cases before we set forth a conclusion. For the problem at hand, a second sample of 90 boys yields a scatter diagram much like that of Table 15.5, so we would reject the hypothesis of linearity of regression.

The student should keep in mind that the test for linearity can lead to the definite conclusion that the regression is curvilinear (if $F$ is large enough), whereas a low $F$ does not prove linearity. Why?

If the hypothesis of linearity is disproved, it follows that the correlation coefficient is not a suitable figure for describing the relationship. The correlation ratio can be used to describe the *degree* of association, but the *form* of the relationship should be described by a fitted curve or by a verbal description of the general curve tendency of the array means. Some readers will have noted that the correlation ratio cannot be considered very descriptive of the data of Table 15.5 because of heteroscedasticity.

## APPLICATION TO MULTIPLE CORRELATION

The reader may recall that the methods given in Chapter 11 for judging the significance of the multiple correlation coefficient involved unsatisfactory approximations. Insofar as we are interested in testing the deviation of a multiple $r$ from zero, the analysis of variance technique provides an exact test which is applicable when the sample is either small or large.

Let us suppose that $Y$ is a dependent variable which is to be predicted by a multiple regression equation containing $m$ independent variables designated by $X$s. The prediction equation may be written as

$$Y' = A + B_1 X_1 + B_2 X_2 + \cdots + B_m X_m$$

in which the $B$s are the regression coefficients. The deviation of any individual's $Y$ score from the mean $Y$ can be expressed as the sum of two parts: the deviation of his $Y$ from his predicted value plus the deviation of the predicted value from the mean of the $Y$s, thus,

$$(Y - \bar{Y}) = (Y - Y') + (Y' - \bar{Y})$$

If we square both sides and sum over all cases, we have

$$\Sigma\Sigma(Y - \bar{Y})^2 = \Sigma(Y - Y')^2 + \Sigma(Y' - \bar{Y})^2$$

which is exactly analogous to the breakdown used in connection with the test of the linear correlation coefficient. One part has to do with residuals about the regression *plane*, the other with variations in the predicted values. The cross-product term again vanishes—it can be shown that there is no correlation between residuals and predicted values.

As previously, we label the $\Sigma(Y - Y')^2$ as the residual sum of squares and $\Sigma(Y' - \bar{Y})^2$ as the regression sum of squares. The total sum of squares will, of course, have $N - 1$ degrees of freedom. The residual sum of squares will lose $df$s according to the number of constants in the regression equation. We have the constant $A$, and the number of $B$ constants is $m$; hence $df = N - (m + 1) = N - m - 1$ for the residual term. The reader who does not immediately see the reasonableness of this should consider the case of one dependent and two independent variables with varying scores on $N = 3$ cases. Imagine that the three scores for each case can be used to locate a point for each in three-dimensional space, and then think of fitting an ordinary plane to these three points. Obviously, the plane can be made to pass through all three; hence the prediction would be perfect, and there would be no freedom for any of the three points to vary from the plane. That is, with $N = 3$ (and with variation on all three variables), the multiple derived therefrom must be unity.

Now, as to the $df$ for the regression or prediction sum of squares, we note that for a fixed set of values for the $X$s the variation of this term must depend on the slopes of the regression plane or on the $B$s. There being $m$ $B$s, there are $m$ ways in which this sum can vary; therefore $df = m$. This is, it will be noted, an extension of the argument used to explain why $df = 1$ for testing the linear correlation coefficient. If our $df$ determinations are correct, we should have $(N - m - 1) + m$ adding to $N - 1$, which is seen to be the case.

In Chapter 11 it was pointed out that the multiple correlation coefficient can be defined as

$$r^2_{1 \cdot 23 \cdots} = 1 - \frac{S^2_{1 \cdot 23 \cdots}}{S^2_1}$$

in which $S^2_{1 \cdot 23 \cdots}$ represents the residual variance and $S^2_1$ is the variance for the dependent variable. Since the residual variance plus the predicted variance adds to the total, the multiple $r$ can also be expressed as the ratio of the predicted to the total variance. (Note that we are here speaking of variances, not estimates.) By definition, the residual variance is $\Sigma(Y - Y')^2/N$, the predicted variance is $\Sigma(Y' - \overline{Y})^2/N$, and the total variance is $\Sigma\Sigma(Y - \overline{Y})^2/N$. We may therefore write the multiple correlation coefficient, using $R$ in order to avoid subscripts, as

$$R^2 = 1 - \frac{\Sigma(Y - Y')^2/N}{\Sigma\Sigma(Y - \overline{Y})^2/N}$$

from which it is readily seen that

$$\Sigma(Y - Y')^2 = (1 - R^2)\Sigma\Sigma(Y - \overline{Y})^2$$

From the alternative way of regarding multiple correlation, we have

$$R^2 = \frac{\Sigma(Y' - \overline{Y})^2/N}{\Sigma\Sigma(Y - \overline{Y})^2/N}$$

which leads to $\Sigma(Y' - \overline{Y})^2 = R^2\Sigma\Sigma(Y - \overline{Y})^2$.

Thus the sums of squares have their equivalents in terms of $R$, and consequently they may be computed by way of $R$. The computation of these sums directly would be a hammer-and-tongs approach which would involve the laborious task of predicting by means of the regression equation the $Y$ for each individual.

The foregoing may be assembled in a schematic variance table, like Table 15.7. As in testing the significance of the ordinary correlation coefficient, we set the null hypothesis to the effect that the estimate based on the regression sum of squares will differ from that based on the residual sum only because of chance sampling errors. The null hypothesis implies

**Table 15.7.  Variance setup for testing significance of multiple correlation coefficient**

| Source | Sum of Squares | Equivalent | $df$ | Estimate |
|--------|----------------|------------|------|----------|
| Regression | $\Sigma(Y' - \bar{Y})^2 = R^2\Sigma\Sigma(Y - \bar{Y})^2$ | | $m$ | $s^2_p$ |
| Residual | $\Sigma(Y - Y')^2 = (1 - R^2)\Sigma\Sigma(Y - \bar{Y})^2$ | | $N - m - 1$ | $s^2_r$ |
| Total | $\Sigma\Sigma(Y - \bar{Y})^2$ | | $N - 1$ | |

that, if the entire population were measured, the correlation of the dependent variable with each independent variable would be zero.  Now, when a sample is drawn from such a population, the $r$s will vary more or less from zero with the result that the multiple $R$ will likewise differ from zero.  If the conditions of the null hypothesis hold true, the sampling distribution of $s^2_p/s^2_r$ follows that of the $F$ distribution with appropriate degrees of freedom.  Note that

$$F = \frac{s^2_p}{s^2_r} = \frac{\Sigma(Y' - \bar{Y})^2/m}{\Sigma(Y - Y')^2/(N - m - 1)}$$

$$= \frac{R^2\Sigma\Sigma(Y - \bar{Y})^2/m}{(1 - R^2)\Sigma\Sigma(Y - \bar{Y})^2/(N - m - 1)}$$

$$= \frac{R^2/m}{(1 - R^2)/(N - m - 1)} \tag{15.18}$$

hence $F$ is a ratio which depends on $R$ and the $df$s.  If the numerator is less than the denominator, we may conclude without reference to the table of $F$ that $R$ is insignificant.  When the numerator is the larger, we judge the significance of $F$ by entering the table of $F$ with $n_1 = m$ and $n_2 = N - m - 1$.  Once $R$ has been computed, the calculations involved in checking its significance are so simple that an example would be humdrum.

In the chapter on multiple correlation (Chapter 11), it was pointed out that $R$ as computed tends to have a positive bias, the extent of which could be judged by formula (11.14), which provides a very nearly unbiased estimate.  It should be stressed that neither the analysis of variance check on the significance of $R$ nor the improved estimate of $R$ allows for the fallacy involved in multiple correlation work when from among a large number of variables a few are chosen for inclusion in the analysis because they show correlation with the criterion.  Such selection tends to capitalize on $r$s which are among the highest partly because of chance errors.

A practical question of considerable importance arises when we wonder whether the inclusion of additional variables in the multiple regression equation leads to a significant increase in the accuracy of prediction or when we wish to know whether the dropping of certain variables results in a significant decrease in the amount of variance predicted. The inclusion of additional variables in the equation always tends to reduce the error of estimate somewhat and leads to an increase in $R$. Can it be said that the increase in $R$ possesses statistical significance?

Let $R_1$ be the multiple based on $m_1$ independent variables and $R_2$ be the value based on $m_2$ variables *selected from among* the $m_1$ variables. To test the significance of the difference between $R_1$ and $R_2$, we take

$$F = \frac{(R^2_1 - R^2_2)/(m_1 - m_2)}{(1 - R^2_1)/(N - m_1 - 1)} \qquad (15.19)$$

with $n_1 = m_1 - m_2$ and $n_2 = N - m_1 - 1$. If $F$ is sufficiently large, we can safely assume that the apparent gain in using the additional variable or variables possesses statistical significance.

An $F$ test by (15.19) has a special meaning when all but one of the $m_1$ variables are selected. Thus $m_2$ will equal $m_1 - 1$, and $m_1 - m_2 = 1$ will be the $df$ for the numerator. This is, of course, equivalent to omitting one variable. Suppose that variable is $X_4$; then $R^2_1 = r^2_{1\cdot2345\ldots n}$ and $R^2_2 = r^2_{1\cdot235\ldots n}$. Thus the difference $R^2_1 - R^2_2$ represents the decrease resulting from dropping $X_4$ (or the gain from its inclusion). If $F$ is significant it follows that the loss (or gain) is statistically significant. Actually this $F$ is a test of the significance of $\beta_4$. To test the significance of all the $m_1$ beta values, one would need to compute not only $R^2_1$ but also $m_1$ values of $R$ each based on $m_1 - 1$ predictors, or independent variables. This formidable task is easily accomplished by the electronic computer.

Consider the three-variable situation and that we wish to test the significance of $\beta_2$. We have $R^2_1 = r^2_{1\cdot23}$ and $R^2_2 = r^2_{13}$. That is, the dropping out of variable 2 simply reduces $R^2_2$ to a bivariate correlation. Then we have

$$F_{\beta_2} = \frac{r^2_{1\cdot23} - r^2_{13}}{(1 - r^2_{1\cdot23})/(N - 2 - 1)} \qquad (15.20)$$

Utilizing one of the formulas for $r_{1\cdot23}$ (see p. 194), we can write the numerator of this $F$ as

$$r^2_{1\cdot23} - r^2_{13} = \frac{r^2_{12} + r^2_{13} - 2r_{12}r_{13}r_{23}}{1 - r^2_{23}} - r^2_{13}$$

$$= \frac{r^2_{12} + r^2_{13} - 2r_{12}r_{13}r_{23} - r^2_{13} + r^2_{13}r^2_{23}}{1 - r^2_{23}}$$

$$= \frac{r^2_{12} - 2r_{12}r_{13}r_{23} + r^2_{13}r^2_{23}}{1 - r^2_{23}} = \frac{(r_{12} - r_{13}r_{23})^2}{1 - r^2_{23}}$$

which is the square of a part correlation, given by (10.37). Thus (15.20) becomes

$$F = \frac{r^2_{1(2\cdot3)}}{(1 - r^2_{1\cdot23})/(N - 3)} \qquad (15.21)$$

which provides a test of the significance of a part correlation coefficient. Note that this differs markedly from the $t^2$ $(= F)$, given on p. 185 for testing the significance of a partial correlation coefficient.

## INTRACLASS CORRELATION

Suppose we wish to specify the degree of resemblance of twins in terms of a correlation coefficient. We have measurements on just one variable, and if we attempt to make a scatter diagram we are faced with the problem of deciding which member of a pair, $A$ or $A'$, to assign to one axis and which to the other. This can be resolved by a double entry scheme: each pair is entered twice, $A$ as $X$ and $A'$ as $Y$, and then $A'$ as $X$ and $A$ as $Y$. An $r$ calculated from the double entry (symmetrical) table suffers from a slight bias, which may be avoided by using the formula given below.

In general, if we have $G$ families (or groups or classes) with $m$ cases per family, the degree of resemblance can be specified by the intraclass correlation coefficient, computable by

$$r' = \frac{s^2_b - s^2_w}{s^2_b + (m - 1)s^2_w} \qquad (15.22)$$

in which we have variance estimates for between families (groups or classes) and for within families. If $F = s^2_b/s^2_w$ is significant, we have evidence for a significant positive $r'$. Note that if there is no within-family variation, $r'$ becomes unity. Note also that $r'$ may be negative, but since in practice $s^2_w$ will rarely be significantly larger than $s^2_b$, one is seldom confronted with the necessity for trying to interpret a negative intraclass correlation.

When the number of cases per family varies, the average of the $m_g$ values is used in place of $m$ in the foregoing formula for $r'$. This does not affect the $F$ test as a way of judging the significance of the correlation.

The distinguishing characteristic of an intraclass correlation situation is that we have $G$ sets of scores on just one variable with no way of ordering the scores within a set (a sort of interchangeability). It is obvious that $r'$ can be used to describe group resemblance, regardless of how the groups have been defined.

## SELECTED CONTRASTS

When and only when $F$, as an over-all test, indicates significant differences among the $G$ groups may we safely make further tests to see whether two selected means differ significantly (here designated as a type $D$ contrast) or whether one mean (or the average of two or more group means) differs more than chance from the average of other group means (characterized here as a type $D'$ contrast or comparison). We need to distinguish between two motivations for making such additional significance tests: we may wish to do so because an a priori hypothesis calls for examining a given contrast, or, as "data snoopers," we may wish to make certain comparisons suggested by the data. For the former, a $t$ test is appropriate whereas for the latter the $t$ test may be misleading in that such snooping is apt to lead to the selection of those differences that are the largest, a process which tends to capitalize on chance differences with a resultant vitiating of the level of significance.

Regardless of motivation, we have to calculate either a $D$ or a $D'$ or both and an appropriate standard error. Suppose $G = 5$ groups, with means $\bar{X}_1$, $\bar{X}_2$, $\bar{X}_3$, $\bar{X}_4$, and $\bar{X}_5$. For equal $m_g$ we might have $D = \bar{X}_5 - \bar{X}_1$ or $D' = (\bar{X}_1 + \bar{X}_3 + \bar{X}_4)/3 - (\bar{X}_2 + \bar{X}_5)/2$, but with unequal $m_g$ the value of $D'$ should be based on weighted averages, thus

$$D' = \frac{m_1\bar{X}_1 + m_3\bar{X}_3 + m_4\bar{X}_4}{m_1 + m_3 + m_4} - \frac{m_2\bar{X}_2 + m_5\bar{X}_5}{m_2 + m_5}$$

For the sampling error variances we have

$$s^2_D = s^2_w\left(\frac{1}{m_1} + \frac{1}{m_5}\right)$$

and

$$s^2_{D'} = s^2_w\left(\frac{1}{m_1 + m_3 + m_4} + \frac{1}{m_2 + m_5}\right)$$

If all $m_g = m$, the latter simplifies to

$$s^2_{D'} = \frac{s^2_w}{m}\left(\frac{1}{3} + \frac{1}{2}\right) \quad \text{or as} \quad \frac{s^2_w}{m}\left(\frac{1}{a} + \frac{1}{b}\right)$$

in which $a$ and $b$ are the number of group means being averaged. Note that when $a = b = 1$, the latter yields $s^2_D$ as a special case. The required standard errors, or $s_D$ and $s_{D'}$ values, are obtained by taking the square roots of the foregoing variances.

For the a priori hypothesis situation, we have $D/s_D$ and $D'/s_{D'}$ as $t$ ratios with $\Sigma m_g - G$ degrees of freedom. The chosen level for judging significance carries the same connotation as that associated with any other ordinary $t$ test, provided the comparison was decided on before any of the data were scrutinized.

For the data-snooping situation, an allowance for the capitalization on chance-large differences can be made by any of several more or less satisfactory methods. We present here the Scheffé method because it does not require equal $m_g$, because it can be used for both $D$ and $D'$ types of contrasts, because it is robust under nonnormality and heterogeneous variance conditions, and because it is closely linked with the $F$ test and requires only the $F$ table.‡ The Scheffé method involves the computation of a quantity, which we designate as $K$, defined as the square root of the product of $(G - 1)$ times the $F$ required for the $\alpha$ level of significance for $n_1 = G - 1$ and $n_2 = \Sigma m_g - G$ degrees of freedom. For example, if we have adopted .01 as $\alpha$ and have $G = 5$ and $\Sigma m_g = 33$, we find $K$ as $\sqrt{4(4.07)}$, or 4.01. For any contrast to be regarded as significant at the $\alpha$ level, we must have $D/s_D$ (or $D'/s_{D'}$) equal to or greater than $K$. Stated differently, $D$ (or $D'$) must reach $Ks_D$ (or $Ks_{D'}$). $D \pm Ks_D$ (or $D' \pm Ks_{D'}$) will provide the $1 - \alpha$ confidence interval for the population values of the specified contrast.

Apparently this method is the best yet devised for contrasts or comparisons of the $D'$ type but for those of the $D$ type it is lacking somewhat in sensitivity. However, along with this lack of power we have the advantages listed earlier plus the satisfaction of knowing that its usage guards against the making of the type I error too frequently when testing differences suggested by the data. The error rate, using the ordinary $t$ test for such comparisons, increases astonishingly as $G$ increases.

‡ The frequently advocated Duncan *new multiple range test* is currently under suspicion by mathematical statisticians, and the unpublished Tukey test is valid only for equal $m_g$ and when equality of variance holds. The reader is referred to the not-easy-to-read discussion by Scheffé of his $S$-method and the Tukey $T$-method in H. Scheffé, *The analysis of variance*, New York: John Wiley and Sons, 1959.

# Chapter 16

# ANALYSIS OF VARIANCE:
# COMPLEX

In Chapter 15 an explanation of the fundamental idea of the analysis of variance technique was attempted, and applications to relatively simple situations were given. In general, these situations involved the testing of the significance of the over-all variation of the means for several groups, the groups differing on the basis of a single classificatory principle. Such setups are sometimes referred to as *single* variable experiments, by which is meant that groups differing in *one* known respect are compared on a dependent variable. For example, income might be considered a variable which is dependent in part on amount of education, which accordingly becomes the independent, single variable for classifying individuals into groups. Or it might be that the classificatory variable is subject to experimental manipulation, and we wish to determine whether variations thereof will lead to performance or response differences. The Wright experiment cited in Chapter 15 is an example of this.

There are times when it is not only feasible but advisable to design the experimental setup so as to make one set of data serve for the testing of hypotheses regarding the separate influence of two or more independent variables. This type of thing has been done for a long time in psychological research wherein it has been possible to classify a total group first one way, then another, and perhaps a third way. For example, in order to determine some of the possible correlates of measured intelligence, we may classify a group of children into urban, suburban, and rural groups; then, ignoring this basis for grouping, we may classify them as to sex or age. Such a procedure in which one variable is considered at a time is tantamount to the single variable setup, even though the same batch of data is made to answer questions about the "effects" of different independent variables.

Now it is obvious that, in studying factors associated with intelligence, we could make a double classification by classifying our cases simultaneously on two of the variables, or a triple classification by using three variables, etc. Consider for the moment a double classification based on the three rural-urban categories and on sex. This would lead to the assigning of the cases to six groups, each of which would have a mean IQ. Instead of having three means for groupings on the basis of the rural-urban characteristic, we would now have two sets of such means, one set for each sex. Instead of two means for the total group classified by sex, we would have three sets of sex means, a set for each of the three residence categories.

This type of breakdown and similar ones where percentages instead of means are involved were utilized in psychological research long before the advent of the analysis of variance technique. The further breakdown of each sex group for residence status (or of residence groups for sex) is made in order to see whether rural-urban differences hold for the sexes separately (or whether the sex differences are similar for each of the separate residence groups). Although researchers were not confined to the single variable approach before the invention of the variance technique, they were definitely limited in the possible statistical treatment of their data. Now that we have the analysis of variance method, we have an adequate statistical technique for checking such hypotheses as can be formulated concerning the "influence" of not only one but two or more variables. The advantages of using analysis of variance for such situations may be briefly mentioned.

First, as we have already seen, it provides an over-all test of the significance of the difference between two or more means when either large or small samples are involved.

Second, we shall soon see that it leads to a definitely improved estimate of sampling error when double or triple or higher-order classification is involved. For instance, when the older method is used to check the significance of the difference between the two sex means for the total group, the determination of the sampling error makes no allowance for likely heterogeneity in intelligence associated with residence status. The variance method permits a refined estimate of error by allowing for variation due to one or more variables when the differences between groups classified on the basis of some other variable are being tested.

Third, the variance technique provides a means of testing whether the influence of one independent variable on the dependent variable is similar for subgroups formed on the basis of a second independent variable. In a sex-by-residence analysis of IQs, the breakdown of each residence group by sex will likely show that the sex differences are not exactly the same for the three groups and that rural-surburban-urban differences are not exactly alike for the separate sex groups. Such inconsistencies as seem apparent from examination of the six cell means may not be real for the simple reason that random sampling errors are present. Before the development of the variance technique there was no way of testing such apparent inconsistencies, except when each classificatory characteristic led to just two categories.

This last point has to do with what has been termed *interaction*, a concept which is not easily understood. Rather than provide a detailed discussion now of what is meant by interaction, we will give a simple illustration. Suppose it has been found that one learning method has a distinct advantage over a second method, but that, when the data are broken down for two recall intervals, the superiority of the first method seems to hold only for those with the shorter recall interval. This failure of the first method to be consistently better becomes an example of interaction. Before concluding that there is evidence for real interaction, we need to apply a statistical test. For such a simple breakdown, we could compute the difference between the first and second method means, and the standard error of the difference, for those with the short recall interval; likewise, for those with the long interval; then we could determine the difference between the differences and its standard error and therefrom obtain either a $z$ or a $t$ as a test of inconsistency. But, when we think of a situation with three methods and three or four recall intervals, it is immediately obvious that such a simple test cannot be applied.

It is the purpose of this chapter to present the methods of analysis to be used when classification into groups is made on the basis of two or more variables. These extensions, which are somewhat restricted by the underlying assumptions of normality and homogeneity of certain variances, are applicable for either large or small samples and are particularly helpful with small samples when it seems imperative that we "get the most out of the available data."

## TWO-WAY CLASSIFICATION

Suppose that the individuals (or their scores) are classifiable into $C$ groups on the basis of one characteristic or variable and into $R$ groups on the basis of a second variable. This would lead to a table with $RC$ cells. Let us presume that we have $m$ scores per cell, with independence from cell to cell. It is convenient to let $X_{rc}$ stand for a score in the $r$th row and $c$th column of such a table. A score in the first row (from the top) and third column would be symbolized as $X_{13}$, and $\overline{X}_{13}$ would be the mean for the scores in this cell; $\overline{X}_{rc}$ would stand for the mean of any cell. The general patterning of labeling the scores is set forth in Table 16.1, which also includes symbols for the row and column means. Note that the first subscript denotes the row, and the second the column, to which a score belongs. We again use the "dot" notation to denote the means on the margins—the "dot" replaces the subscript over which we sum to get a mean. Thus $\overline{X}_{r.}$ is obtained by summing across the columns and dividing by $mC$, and $\overline{X}_{.c}$ is obtained by summing over rows and dividing by $mR$. We can have a second table in which we enter the cell means, or $\overline{X}_{rc}$ values.

**Table 16.1 Schema for labeling scores and means for groups, double classification**

|   | 1 | 2 | 3 | $c$ | $C$ |   |
|---|---|---|---|---|---|---|
| 1 | $X_{11}$ | $X_{12}$ | $X_{13}$ | $X_{1c}$ | $X_{1C}$ | $\overline{X}_{1.}$ |
| 2 | $X_{21}$ | $X_{22}$ | $X_{23}$ | $X_{2c}$ | $X_{2C}$ | $\overline{X}_{2.}$ |
| 3 | $X_{31}$ | $X_{32}$ | $X_{33}$ | $X_{3c}$ | $X_{3C}$ | $\overline{X}_{3.}$ |
| $r$ | $X_{r1}$ | $X_{r2}$ | $X_{r3}$ | $X_{rc}$ | $X_{rC}$ | $\overline{X}_{r.}$ |
| $R$ | $X_{R1}$ | $X_{R2}$ | $X_{R3}$ | $X_{Rc}$ | $X_{RC}$ | $\overline{X}_{R.}$ |
|   | $\overline{X}_{.1}$ | $\overline{X}_{.2}$ | $\overline{X}_{.3}$ | $\overline{X}_{.c}$ | $\overline{X}_{.C}$ | $\overline{X}$ |

Summing these across columns and dividing by $C$ will, of course, lead to the $\overline{X}_{r.}$ on the right-hand margin; summing over rows and dividing by $R$ will yield the $\overline{X}_{.c}$ values at the bottom. The average of the marginal means, either direction, will be the total mean, which consistent with the "dot" notation would be $\overline{X}_{..}$, but we will simply use $\overline{X}$ as a symbol for the total mean.

The deviation of any score, $X_{rc}$, from the total mean can be expressed in terms of the deviation of its row mean from the total mean, $(\overline{X}_{r.} - \overline{X})$, plus the deviation of its column mean from the total mean, $(\overline{X}_{.c} - \overline{X})$, plus its deviation from its cell mean, $(X_{rc} - \overline{X}_{rc})$. Thus we could write

$$(X_{rc} - \overline{X}) = (\overline{X}_{r.} - \overline{X}) + (\overline{X}_{.c} - \overline{X}) + (X_{rc} - \overline{X}_{rc})$$

which it will be seen is really not an equation. Something is missing. We simply ask, what can be added to make an equation or an identity? If we were to subtract the right-hand parts from the left-hand side, we would secure a sort of remainder which when included on the right-hand side would provide an equation. Simple algebraic manipulation leads to $(\overline{X}_{rc} - \overline{X}_{r.} - \overline{X}_{.c} + \overline{X})$ as the needed component. Thus we have

$$(X_{rc} - \overline{X}) = (\overline{X}_{r.} - \overline{X}) + (\overline{X}_{.c} - \overline{X})$$
$$+ (\overline{X}_{rc} - \overline{X}_{r.} - \overline{X}_{.c} + \overline{X}) + (X_{rc} - \overline{X}_{rc})$$

as an identity which specifies four possible sources of variation for the $X_{rc}$ scores. Later we will examine the meaning of the component obtained as a remainder.

With $r$ running from 1 to $R$, and $c$ taking on values from 1 to $C$, and $m$ scores per cell there will be $mRC$ ($=$ total $N$) individual deviations. We need the sum of their squares, which sum will be broken into four parts plus six cross-product terms when we square and sum the right-hand side of the identity. Since the limits of the printed page make it rather difficult to write out and manipulate a breakdown into $4 + 6$ terms, we will ask the reader to "take on faith" the fact that the six cross-product terms can

**Table 16.2. Variance schema for double classification with $m$ scores per cell**

| Source | Sum of Squares | $df$ | Variance Estimate |
|---|---|---|---|
| Rows | $mC\sum_{r}(\bar{X}_{r\cdot} - \bar{X})^2$ | $R - 1$ | $s^2_r$ |
| Columns | $mR\sum_{c}(\bar{X}_{\cdot c} - \bar{X})^2$ | $C - 1$ | $s^2_c$ |
| Interaction | $m\sum_{r}\sum_{c}(\bar{X}_{rc} - \bar{X}_{r\cdot} - \bar{X}_{\cdot c} + \bar{X})^2$ | $(R - 1)(C - 1)$ | $s^2_{rc}$ |
| Within cells | $\sum_{r}\sum_{c}(X_{rc} - \bar{X}_{rc})^2$ | $mRC - RC$ | $s^2_w$ |
| Total | $\sum_{r}\sum_{c}(X_{rc} - \bar{X})^2$ | $mRC - 1$ | |

be shown to vanish. Perhaps the reader has anticipated that any cross-product terms involving $(X_{rc} - \bar{X}_{rc})$ will vanish because this last term is a within-groups deviation and in one-way analysis of variance the cross-product term involving within-groups and between-groups deviations was proven to be exactly zero.

Before proceeding further, we admit that a more precise notation for the $i$th individual in the $r$th row and $c$th column would be $X_{irc}$ and hence $\bar{X}_{\cdot rc}$ would be the cell mean. An expression for the total sum of squares would be $\sum_{i}\sum_{r}\sum_{c}(X_{irc} - \bar{X})^2$. We will herein avoid the use of a subscript for an individual unless really needed for clarity. Accordingly, we will use the expression $\sum_{r}\sum_{c}(X_{rc} - \bar{X})^2$ as indicative of the total sum of squares for all $mRC$ individuals; it is to be understood that $mRC$ scores, not $RC$ scores, are involved in the summation.

The breakdown of the total sum of squares into four additive components is depicted in Table 16.2, along with the corresponding partitioning of the degrees of freedom. The $df$s for rows and for columns are precisely as for the $G$ group situation of the previous chapter. The $df$ for the within-cells term is analogous to that for the within-groups setup, i.e., we have a total of $mRC$ scores varying about $RC$ means, hence $RC$ restrictions. The new term, $\underset{r\,c}{\Sigma\Sigma}(\overline{X}_{rc} - \overline{X}_{r.} - \overline{X}_{.c} + \overline{X})^2$, which has to do with *interaction*, has $(R - 1)(C - 1)$ degrees of freedom. How does one arrive at this for the $df$? The situation here is analogous to the $\chi^2$ $df$ for a $k$ by $l$ contingency table. Imagine a table like Table 16.1, but with $\overline{X}_{rc}$ values in the cells. The marginal means constitute restrictions on the "deviations" which are involved in the interaction sum of squares—when $(R - 1)(C - 1)$ cell means are known the rest become determinable from the margins. Note that if you sum the degrees of freedom for the four parts you will get the total $df$.

Table 16.2 also contains symbols for the four variance estimates. We will now have three null hypotheses: (1) the row means are chance variations from one population mean, (2) the column means are also chance variations from the same population mean; and (3) there is no interaction. As will be seen later, the proper denominator for all three possible and legitimate variance ratios, $F_r$, $F_c$, and $F_{rc}$, is $s^2_w$. It will be noticed that the denominator, or "error" term, for these $F$s is based on a variance estimate that is independent of each of the three possible effects. Stated differently, the error term is freed of variation produced or associated with the variables or characteristics that permit us to classify the scores.

There are no restrictions on the nature of the bases for classification (however, see next paragraph). Either the rows or the columns or both may stand for groupings that are qualitatively different or that are based on chosen levels on a quantified variable. Categories, $R$ and/or $C$ in number, might be based on "natural" classification, such as sex, race, eye color, age; or they might be based on sociocultural status, such as residence locale (urban, suburban, rural), occupation, religion, political affiliation, educational level, socioeconomic level; or they might be based on qualitative groupings produced in the laboratory, such as dosages, type of therapy, teaching method, kinds of rewards; or the classification may involve a quantitative variable that is manipulable so as to permit the measuring of subjects on a dependent variable, $X$, under differing levels on such variables as size, distance, illumination, amount of reward, etc. In laboratory studies involving independent variables for which the subject does not bring with him the wherewithal for a priori classification, the subjects must be randomly assigned to levels (conditions) or combinations of levels on the manipulated variable(s). There must be independence of subjects from row to row, from column to column, and/or from cell to cell.

There is one definite restriction on the simultaneous choice of two variables (characteristics) as a basis for the classifications: the two independent variables must themselves be independent in the sense of being uncorrelated either linearly or curvilinearly or in the contingency sense. This restriction is met when both variables are manipulable or when one is manipulable; but when both variables are characteristics of the subjects (such as personality traits, abilities, height, weight, or such as socioeconomic level, educational level, residence locale, membership in groups), there is risk that correlation (association) exists between the two "independent" variables. If there is any possibility of association, the interpretations of significant effects become blurred. We will return to this problem later (see under "nonorthogonality" in Index).

The bases of classification, as the independent variables in a study, are sometimes referred to as *factors* (not to be confused with the factors that emerge from a technique known as "factor analysis"). It will have been noted that the foregoing illustrations of factors, as either qualitative or quantitative variables, did not include any for which the selection of "levels" (categories or groupings) depends upon a random process. All involve fixed effects (see p. 299). After giving an example of the required computations and illustrations that should help one understand the meaning of interaction, we will consider in detail a case of two-way classification in which rows stand for a random (not fixed) variable.

## COMPUTATIONS, $m$ SCORES PER CELL

The computations are relatively easy, and arranging the scores into rows and columns will facilitate the task. A $\Sigma X$ and $\Sigma X^2$ is calculated for each cell. Summing the $RC \, \Sigma X^2$ values gives $\Sigma\Sigma X^2$ as the sum of all the $mRC$ squared scores. Summing the $\Sigma X$ values in each row gives $\Sigma X_{rc}$, and summing the $\Sigma X$ values in each column gives $\Sigma X_{rc}$. These become sums along the margins, which marginal values sum down, and across, to the total sum of the $mRC$ scores, $\Sigma\Sigma X_{rc}$. The sum of scores in any particular cell will be symbolized as $\Sigma X_{rc}$. The formulas are:

$$\text{Total sum of squares} = \frac{1}{mRC}[mRC\Sigma\Sigma X^2_{rc} - (\Sigma\Sigma X_{rc})^2] \quad (16.1)$$

$$\text{Between-rows squares} = \frac{1}{mRC}\left[R\Sigma_r\left(\Sigma_c X_{rc}\right)^2 - (\Sigma\Sigma X_{rc})^2\right] \quad (16.2)$$

$$\text{Between-columns squares} = \frac{1}{mRC}\left[C\Sigma_c\left(\Sigma_r X_{rc}\right)^2 - (\Sigma\Sigma X_{rc})^2\right] \quad (16.3)$$

$$\text{Within-cells squares} = \frac{1}{m}[m\Sigma\Sigma X^2_{rc} - \Sigma(\Sigma X_{rc})^2] \quad (16.4)$$

## Table 16.3. Coded learning scores (sum of scores on 29th and 30th trials) for Koerth pursuit rotor*

| Rest Interval | Practice Sessions | | | | | | | |
|---|---|---|---|---|---|---|---|---|
| | 5(M T W Th F) | | | | 3(M W F) | | | |
| 3 minutes | 9 | 14 | 6 | 10 | 8 | 10 | 11 | 14 |
| | 10 | 15 | 10 | 11 | 9 | 7 | 9 | 10 |
| | 14 | 17 | 10 | 11 | 9 | 12 | 13 | 14 |
| | 10 | 7 | 8 | 15 | 12 | 13 | 7 | 17 |
| | 12 | 8 | 14 | 6 | 9 | 12 | 8 | 15 |
| 1 minute | 2 | 6 | 1 | 9 | 11 | 12 | 9 | 7 |
| | 5 | 9 | 2 | 11 | 9 | 6 | 11 | 9 |
| | 14 | 1 | 1 | 8 | 6 | 8 | 11 | 12 |
| | 14 | 4 | 11 | 5 | 9 | 7 | 4 | 10 |
| | 6 | 8 | 2 | 5 | 13 | 6 | 7 | 8 |

\* Data from Renshaw, M. J., *The effects of varied arrangements of practice and rest on proficiency in the acquisition of a motor skill*, Unpublished Doctor's Dissertation, Stanford University, California, 1947.

## Table 16.4. Sums and means for data of Table 16.3

| Rest Interval | Practice Session | | Totals |
|---|---|---|---|
| | 5(M T W Th F) | 3(M W F) | |
| 3 minutes | $\Sigma X_{11} = 217$ <br> $\Sigma X^2_{11} = 2543$ <br> $\bar{X}_{11} = 10.8500$ | $\Sigma X_{12} = 219$ <br> $\Sigma X^2_{12} = 2547$ <br> $\bar{X}_{12} = 10.9500$ | $\Sigma X_{1c} = 436$ <br> $\Sigma X^2_{1c} = 5090$ <br> $\bar{X}_{1.} = 10.9000$ |
| 1 minute | $\Sigma X_{21} = 124$ <br> $\Sigma X^2_{21} = 1102$ <br> $\bar{X}_{21} = 6.2000$ | $\Sigma X_{22} = 175$ <br> $\Sigma X^2_{22} = 1643$ <br> $\bar{X}_{22} = 8.7500$ | $\Sigma X_{2c} = 299$ <br> $\Sigma X^2_{2c} = 2745$ <br> $\bar{X}_{2.} = 7.4750$ |
| Totals | $\Sigma X_{r1} = 341$ <br> $\Sigma X^2_{r1} = 3645$ <br> $\bar{X}_{.1} = 8.5250$ | $\Sigma X_{r2} = 394$ <br> $\Sigma X^2_{r2} = 4190$ <br> $\bar{X}_{.2} = 9.8500$ | $\Sigma\Sigma X_{rc} = 735$ <br> $\Sigma\Sigma X^2_{rc} = 7835$ <br> $\bar{X} = 9.1875$ |

The interaction sum of squares is obtained as the remainder when the numerical values of the last three are subtracted from the total sum of squares.

Table 16.3 contains data on learning with two variations as to practice sessions and two variations as to rest interval between trials. For each combination of conditions there are 20 ($= m$) cases. The scores are recorded in a 2 by 2 or 4-cell table. Table 16.4 is a work-sheet layout in which are recorded sums of scores, sums of squared scores, and means, for cells and for the margins. The lower-right corner contains values for the total group of 80 cases. For the sums of squares (of deviations) we have the following:

Total: $\frac{1}{80}[80(7835) - (735)^2] = 1082.1875$.
Rows: $\frac{1}{80}[2(436^2 + 299^2) - (735)^2] = 234.6125$.
Columns: $\frac{1}{80}[2(341^2 + 394^2) - (735)^2] = 35.1125$.
Within cells: $\frac{1}{20}[20(7835) - (217^2 + 219^2 + 124^2 + 175^2)] = 782.4500$
Interaction: $1082.1875 - (234.6125 + 35.1125 + 782.4500) = 30.0125$.

The interaction sum of squares can also be calculated by direct substitution into the definition formula of Table 16.2, which will involve $RC$ quantities to be squared, summed, and multiplied by $m$. We have

$$(10.85 - 10.90 - 8.525 + 9.1875)^2 = (.6125)^2$$
$$(10.95 - 10.90 - 9.85 + 9.1875)^2 = (-.6125)^2$$
$$(6.20 - 7.475 - 8.525 + 9.1875)^2 = (-.6125)^2$$
$$(8.75 - 7.475 - 9.85 + 9.1875)^2 = (.6125)^2$$

which when added and multiplied by 20 lead to 30.0125, or the value obtained by subtraction.

Any reader who is surprised that the above four values involved in computing the interaction sum of squares directly are numerically equal should ponder the fact that for the given situation the $df$ for the interaction term is $(2 - 1)(2 - 1)$, or 1.

Actually, the easiest way to compute the interaction sum of squares for a 2 by 2 table is to work with the four cell sums of scores. The formula is

$$\frac{1}{4m}(\Sigma X_{11} + \Sigma X_{22} - \Sigma X_{12} - \Sigma X_{21})^2$$

For this problem we have

$$\frac{1}{80}(217 + 175 - 219 - 124)^2 = \frac{1}{80}(49)^2 = 30.0125$$

The sums of squares and resulting variance estimates are brought together in Table 16.5. We have four variance estimates which for the

**Table 16.5. Analysis of variance for pursuit learning**

| Source | Sum of Squares | df | Variance Estimate |
|---|---|---|---|
| Rest interval (rows) | 234.6125 | 1 | 234.6125 |
| Sessions (columns) | 35.1125 | 1 | 35.1125 |
| Interaction | 30.0125 | 1 | 30.0125 |
| Individual differences (within cells) | 782.4500 | 76 | 10.2954 |
| Total | 1082.1875 | 79 | |

given situation are all estimates of the same population variance under the null hypothesis conditions: no row effect, no column effect, and no interaction. It is appropriate for this table to use $s^2_w$ as the denominator of $F$ to test the row, the column, and the interaction effects. We have for interaction, $F_{rc} = 30.0125/10.2954 = 2.92$, which falls short of the $F$ of about 4.0 required for significance at the .05 level. This indicates that the apparent failure of the four cell means to be consistent, in either direction, with the marginal means (or with each other) is attributable to chance fluctuations. For this particular problem the chance fluctuation is the sampling of individuals (plus a relatively small component having to do with errors of measurement).

Next consider the effect on pursuit learning of varying the rest interval and varying the sessions. For sessions we have $F_c = 35.1125/10.2954 = 3.41$, which is not large enough to lead us to reject the null hypothesis; but since nonrejection of the null hypothesis does not prove the hypothesis, we can conclude only that the effect, if it exists, is not large enough to be demonstrated by the number of cases used. The between-rows or rest-interval effect is highly significant as judged by $F_r = 234.6125/10.2954 = 22.79$, which is double the $F$ needed for the .001 level of significance. Now the fact that the interaction is not significant permits us to conclude that the rest-interval effect is similar for five sessions and for three sessions per week. If the interaction had been significant, we would need to qualify our conclusion about the effect of the rest interval.

## INTERACTION: MEANING AND ILLUSTRATIONS

One way to acquire insight into the meaning of interaction is based on the fact, as we learned earlier (p. 328), that the deviations contributing thereto emerge as a remainder after the two main effects (and the within-cells part) are taken out. If we were dealing with the cell means as deviations from $\overline{X}$, i.e., $(\overline{X}_{rc} - \overline{X})$, and wrote these deviations as

$$(\overline{X}_{rc} - \overline{X}) = (\overline{X}_{r.} - \overline{X}) + (\overline{X}_{.c} - \overline{X})$$

we would need to tack on $(\overline{X}_{rc} - \overline{X}_{r.} - \overline{X}_{.c} + \overline{X})$ in order to have an identity. The tacked-on term is obtained as a remainder. It represents what is left over after allowing for $\bar{X}_{r.}$ and $\bar{X}_{.c}$ variations. In a way the remainder is what is left over after adjustments to the $(\bar{X}_{rc} - \bar{X})$ values for the row and the column differences.

We will resort to a numerical example, oversimplified in order to make the calculations easy. Suppose nine cell means, along with row and column means, as in the left-hand part of Table 16.6. The sum of squares of the cell means about the over-all mean, 4, can readily be written as $\sum_{r}\sum_{c}(\bar{X}_{rc} - \bar{X})^2 = 2^2 + 1^2 + 0^2 + 0^2 + 1^2 + 4^2 + 2^2 + 1^2 + 1^2 = 28$. For

**Table 16.6. Illustration of adjustment procedure to obtain interaction sum of squares as a remainder**

| Starting means | | | | cols. adjust | | | | rows adjust | | | |
|---|---|---|---|---|---|---|---|---|---|---|---|
| 2 | 3 | 4 | 3 | 2 | 4 | 3 | 3 | 3 | 5 | 4 | 4 |
| 4 | 3 | 8 | 5 | 4 | 4 | 7 | 5 | 3 | 3 | 6 | 4 |
| 6 | 3 | 3 | 4 | 6 | 4 | 2 | 4 | 6 | 4 | 2 | 4 |
| 4 | 3 | 5 | 4 | 4 | 4 | 4 | 4 | 4 | 4 | 4 | 4 |

the row sum squares we have $C\sum_{r}(\bar{X}_{r.} - \bar{X})^2 = 3(1^2 + 1^2 + 0^2) = 6$; for columns, $R\sum_{c}(\bar{X}_{.c} - \bar{X})^2 = 3(0^2 + 1^2 + 1^2) = 6$. By subtraction we get $28 - 6 - 6 = 16$ as the sum of squares for the remainder, or interaction. The student can check this by substituting directly into

$$\sum_{r}\sum_{c}(\bar{X}_{rc} - \bar{X}_{r.} - \bar{X}_{.c} + \bar{X})^2.$$

Next we consider a simple adjustment to the cell means so as to equalize the column means, or make them all $= 4$, the total mean. Since the first column mean is already 4, no adjustment is needed for the cell means therein. The second column mean is 3; adding 1 to the cell means in the second column will change the column mean to 4; and since the third column mean is 5, we would need to adjust the cell means in this column by subtracting 1 from each. The results are set forth in the middle part of Table 16.6, from which it will be seen that these adjustments did not change the row means. Exactly the same procedure can be used to adjust the means in the row cells in order to equalize the row means (to 4). This is accomplished by working with the middle part of the table, and when done produces the right-hand part of Table 16.6, in which we have nine cell means adjusted for both row and column effects. These nine "adjusted" means still show variation, and $\sum_{r}\sum_{c}(_a\bar{X}_{rc} - \bar{X})^2$, with $_a\bar{X}_{rc}$ as the adjusted values, may again be used to compute a sum of squares. Thus we get $1^2 + 1^2 + 0^2 + 1^2 + 1^2 + 2^2 + 2^2 + 0^2 + 2^2 = 16$, which is exactly the same as was obtained by subtraction. In this sense, the interaction sum of squares may be regarded as what is left over after adjustment for the two main effects; i.e., the $\bar{X}_{rc}$ are twice adjusted in calculating the interaction sum of squares.

A by-product of this adjustment approach emerges when it is noted that the sum of squares for $B \times C$ interaction in the left-hand part of Table 16.6, where main effects are present, is precisely the same as that for the right-hand part, where there are no main effects. You don't need to have a main effect in order to have interaction.

Reference to actual examples of statistically significant interaction may help clarify its meaning. For this purpose we shall use some data on visual acuity from an experiment by Walker.* For visual acuity (low

**Table 16.7.  Visual acuity: interaction of type of measurement with eyes**

|  | Depth | Vernier | Total |
|---|---|---|---|
| Binocular | .08 | 1.07 | .57 |
| Monocular | .24 | 1.50 | .87 |
| Total | .16 | 1.28 | .72 |

score, better acuity) by two methods of measurement (depth and vernier) with binocular and monocular vision, we have means as given in Table 16.7. The marginal means are markedly different, and it is readily seen that the cell means (each based on 108 determinations) are not consistent with the marginal values. The difference, $.24 - .08$, is not of the same order as the difference, $1.50 - 1.07$; or stated in another way, the two differences differ from each other. In other words, the amount of difference between binocular and monocular acuity depends upon the type of measurement.

One variable investigated in the experiment was the distance of the stimulus from the subject. Since distance is an ordered variable, it is possible to picture the interaction by making a graph, with acuity as the ordinate and distance along the $x$ axis. Fig. 16.1 shows the relationship of acuity (average of the two types of measures) and the three distances used. Note the difference between the two curves—the significant interaction for eyes and distance actually means that the two curves are different. This lack of parallel behavior of curves is more striking in Fig. 16.2, which illustrates the interaction of measures with distance, for binocular and monocular combined. In this study there was also a significant

* Walker, E. L., *Factors in vernier acuity and distance discrimination*, Unpublished Doctor's Dissertation, Stanford University, California, 1947.

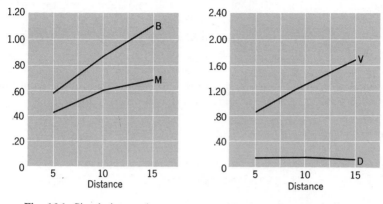

**Fig. 16.1.** Simple interaction: eyes by distance.

**Fig. 16.2.** Simple interaction: measures by distance.

variance for the subjects by distance interaction, from which we conclude that the relationship between acuity and distance varies from person to person (see Fig. 16.3).

Walker also investigated the effect of stimulus rod width and size of aperture. A plot of the results for acuity (ordinate) against rod width (abscissa) for three apertures (*A* large, *B* medium, *C* small) is given in Fig. 16.4 as another possible example of interaction except that this time the apparent interaction is so slight as not to possess statistical significance. This being the case, it can be said that the effect of rod width is independent of aperture (and vice versa). Contrast this with the possible conclusion, based on a highly significant *F*, that distance affects acuity. When we note the interaction effect depicted in Fig. 16.2, we see that such a conclusion does not hold at all for the depth measure. Thus, significant interaction

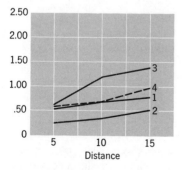

**Fig. 16.3.** Simple interaction: distance by subjects.

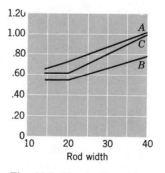

**Fig. 16.4.** Nonsignificant interaction: aperture by stimulus rod width.

always calls for a qualification, sometimes drastic, regarding a main effect. It is entirely possible for an effect to be in opposite directions for different conditions, and the over-all effect need not be significant for this to occur.

If there is no interaction in the population, it follows that each of the $RC$ values of $(\mu_{rc} - \mu_{r.} - \mu_{.c} + \mu)$ would be exactly zero, and from this it follows that $\mu_{rc} = \mu_{r.} + \mu_{.c} - \mu$, or that $\mu_{rc}$ is an additive function of $\mu_{r.}$ and $\mu_{.c}$ as variates. That is, the cell means are an additive function of the two main effects. Accordingly, it is sometimes said that a test of interaction is a test of *additivity* of the main effects.

## ROWS FOR PERSONS AND COLUMNS FOR EXPERIMENTAL CONDITIONS

We next consider the setup in which the $R$ rows stand for $R$ subjects who are measured just once (usually) under each of the $C$ conditions that define the columns. That is, we have $R$ persons for a total of $RC$ scores. This is a repeat-measure design which is frequently used in psychology. Since this is a bit less complicated than the two-way design with $m$ scores per cell, we will give more detail regarding the breakdown of the total sum of squares than was feasible when variation within cells was present.

The deviation of any score, $X_{rc}$, from the total mean can be expressed in terms of the deviation of its row mean from the total mean, $(\bar{X}_{r.} - \bar{X})$, plus the deviation of its column mean from the total mean, $(\bar{X}_{.c} - \bar{X})$, plus a sort of remainder term which represents an individual variation over and above that due to the groups to which the score belongs. To secure an expression for this term, we note that by definition the term must be the part of the score deviation (from the total mean) left over after the sum of the two parts specified above have been subtracted. Accordingly, we have

$$(X_{rc} - \bar{X}) - [(\bar{X}_{r.} - \bar{X}) + (\bar{X}_{.c} - \bar{X})]$$

which simplifies to

$$(X_{rc} - \bar{X}_{r.} - \bar{X}_{.c} + \bar{X})$$

We may therefore write the following identity:

$$(X_{rc} - \bar{X}) = (\bar{X}_{r.} - \bar{X}) + (\bar{X}_{.c} - \bar{X}) + (X_{rc} - \bar{X}_{r.} - \bar{X}_{.c} + \bar{X})$$

With $r$ running from 1 to $R$, and $c$ taking values from 1 to $C$, there will, of course, be $RC$ individual deviations. We need the sum of their squares, which sum will involve the squares of the three parts, plus three cross-product terms that can be shown to vanish when summed. It may be instructive to indicate how the sum of squares for all $RC$ scores can be set up. Suppose we begin by writing the squares of the deviations for scores in the first column. Each of these squares will involve cross-product terms, which we shall here ignore except for a plus sign to indicate their existence. We have for the first-column scores:

$$(X_{11} - \bar{X})^2 = (\bar{X}_{1\cdot} - \bar{X})^2 + (\bar{X}_{\cdot1} - \bar{X})^2 + (X_{11} - \bar{X}_{1\cdot} - \bar{X}_{\cdot1} + \bar{X})^2 +$$

$$(X_{21} - \bar{X})^2 = (\bar{X}_{2\cdot} - \bar{X})^2 + (\bar{X}_{\cdot1} - \bar{X})^2 + (X_{21} - \bar{X}_{2\cdot} - \bar{X}_{\cdot1} + \bar{X})^2 +$$

$$(X_{r1} - \bar{X})^2 = (\bar{X}_{r\cdot} - \bar{X})^2 + (\bar{X}_{\cdot1} - \bar{X})^2 + (X_{r1} - \bar{X}_{r\cdot} - \bar{X}_{\cdot1} + \bar{X})^2 +$$

$$(X_{R1} - \bar{X})^2 = (\bar{X}_{R\cdot} - \bar{X})^2 + (\bar{X}_{\cdot1} - \bar{X})^2 + (X_{R1} - \bar{X}_{R\cdot} - \bar{X}_{\cdot1} + \bar{X})^2 +$$

The summing of these squares of deviations for the scores of column 1 involves $R$ cases, i.e., $r$ runs from 1 to $R$; hence we need a symbol which denotes this fact. Let us use $\sum_r$ for this purpose. Note that the second term on the right is constant for all $R$ scores, which permits us to replace the summation sign by $R$.

The sum of the first column squares, and by analogy the sums for the other columns, can be written as:

1st col.:

$$\sum_r (X_{r1} - \bar{X})^2 = \sum_r (\bar{X}_{r\cdot} - \bar{X})^2 + R(\bar{X}_{\cdot1} - \bar{X})^2 + \sum_r (X_{r1} - \bar{X}_{r\cdot} - \bar{X}_{\cdot1} + \bar{X})^2$$

2nd col.:

$$\sum_r (X_{r2} - \bar{X})^2 = \sum_r (\bar{X}_{r\cdot} - \bar{X})^2 + R(\bar{X}_{\cdot2} - \bar{X})^2 + \sum_r (X_{r2} - \bar{X}_{r\cdot} - \bar{X}_{\cdot2} + \bar{X})^2$$

$c$th col.:

$$\sum_r (X_{rc} - \bar{X})^2 = \sum_r (\bar{X}_{r\cdot} - \bar{X})^2 + R(\bar{X}_{\cdot c} - \bar{X})^2 + \sum_r (X_{rc} - \bar{X}_{r\cdot} - \bar{X}_{\cdot c} + \bar{X})^2$$

$C$th col.:

$$\sum_r (X_{rC} - \bar{X})^2 = \sum_r (\bar{X}_{r\cdot} - \bar{X})^2 + R(\bar{X}_{\cdot C} - \bar{X})^2$$
$$+ \sum_r (X_{rC} - \bar{X}_{r\cdot} - \bar{X}_{\cdot C} + \bar{X})^2$$

We may now sum over the $C$ columns, and for the results we will need double summation signs. Since the first right-hand term does not vary from column to column, its sum is merely $C$ times its value. The second right-hand set of terms involves a constant times a variable; hence the constant $R$ comes from under the summation sign. Finally we have the following expression for the sum of squares for the $RC$ scores:

$$\sum_r \sum_c (X_{rc} - \bar{X})^2 = C\sum_r (\bar{X}_{r\cdot} - \bar{X})^2 + R\sum_c (\bar{X}_{\cdot c} - \bar{X})^2$$
$$+ \sum_r \sum_c (X_{rc} - \bar{X}_{r\cdot} - \bar{X}_{\cdot c} + \bar{X})^2 \quad (16.5)$$

The reader who is worried about whether the cross-product terms really vanish should note that for the $c$th column the product term

$$\sum_r (\overline{X}_{r.} - \overline{X})(\overline{X}_{.c} - \overline{X}) = (\overline{X}_{.c} - \overline{X})\sum_r(\overline{X}_{r.} - \overline{X})$$

vanishes because $\sum_r(\overline{X}_{r.} - \overline{X}) = 0$. The other two cross-product sums have as one factor the remainder, or residual term; we have already had examples of a general principle that product terms involving residuals vanish.

From formula (16.5) we see that the total sum of squares can be broken into three additive components: between row means with $R - 1$ degrees of freedom, between column means with $df$ of $C - 1$, and a remainder, which will have $df = (R - 1)(C - 1)$. Note that the $df$s for the three parts sum to the $df$ for the total sum of squares, $RC - 1$.

Division of the three sums of squares by their $df$s leads to variance estimates, $s^2_r$, $s^2_c$, and $s^2_{rc}$. That for columns is of primary interest; it is tested by $F_c = s^2_c/s^2_{rc}$. The row variance estimate involves the ever-present variation among individuals, which is testable by $F_r = s^2_r/s^2_{rc}$ provided an assumption is tenable, which is unlikely except in one special case to be discussed (p. 355). We will later examine the meaning of the component represented by $\sum\sum(X_{rc} - \overline{X}_{r.} - \overline{X}_{.c} + \overline{X})^2$. Some readers will have anticipated that this may involve something akin to row by column interaction, even though it contains $X_{rc}$ instead of $\overline{X}_{rc}$ when there are $m$ scores per cell. Note that since $s^2_{rc}$, sometimes referred to as the error variance, is based on a remainder, or residual, after the parts of the total sum of squares associated with the between-row and the between-column variation have been subtracted, it follows that we are using as our error term (the denominator of $F_c$) a variance which has been freed of any variation produced by the column conditions and also freed of the main "effect" of subject variation as represented by the $\overline{X}_{r.}$ values. Eventually we will see that $s^2_{rc}$ is not completely free of variation due to individual differences.

## SIGNIFICANCE OF THE DIFFERENCES BETWEEN TWO CORRELATED MEANS

Let us consider the limiting case of $C = 2$. The between-columns sum of squares, $R\sum_c(\overline{X}_{.c} - \overline{X})^2$, may be written as

$$R(\overline{X}_{.1} - \overline{X})^2 + R(\overline{X}_{.2} - \overline{X})^2$$

which we have already shown (p. 304) reduces to $(R/2)(\overline{X}_{.1} - \overline{X}_{.2})^2$, or to a function of the difference between the two means.

Let us next examine the remainder or error term. If we turn back to p. 339, where we summed over columns, we readily see that the remainder sum can be expressed as

$$\sum_r (X_{r1} - \bar{X}_{r\cdot} - \bar{X}_{\cdot1} + \bar{X})^2 + \sum_r (X_{r2} - \bar{X}_{r\cdot} - \bar{X}_{\cdot2} + \bar{X})^2$$

in which the $c$ of formula (16.5) has the explicit values of 1 and 2. Now the mean of any row, say the $r$th, is merely the mean of $C = 2$ scores; i.e., $\bar{X}_{r\cdot} = (X_{r1} + X_{r2})/2$, and the total mean must be the average of the two column means, or $\bar{X} = (\bar{X}_{\cdot1} + \bar{X}_{\cdot2})/2$. Making these substitutions, we have

$$\sum_r \left( X_{r1} - \frac{X_{r1} + X_{r2}}{2} - \bar{X}_{\cdot1} + \frac{\bar{X}_{\cdot1} + \bar{X}_{\cdot2}}{2} \right)^2$$

$$+ \sum_r \left( X_{r2} - \frac{X_{r1} + X_{r2}}{2} - \bar{X}_{\cdot2} + \frac{\bar{X}_{\cdot1} + \bar{X}_{\cdot2}}{2} \right)^2$$

which simplifies to

$$\tfrac{1}{4}\sum_r (X_{r1} - X_{r2} - \bar{X}_{\cdot1} + \bar{X}_{\cdot2})^2 + \tfrac{1}{4}\sum_r (X_{r2} - X_{r1} - \bar{X}_{\cdot2} + \bar{X}_{\cdot1})^2$$

These two terms become indentical when we change the signs within the second parentheses, which change is permissible since the square of a function is the same as the square of its negative, e.g., $(a)^2 = (-a)^2$. Hence we have

$$\tfrac{1}{2}\sum_r [(X_{r1} - X_{r2}) - (\bar{X}_{\cdot1} - \bar{X}_{\cdot2})]^2$$

Now the first parentheses term is the difference between any individual's two scores, say $D_r$, and the second is the difference between the two column means, which difference it will be recalled is the same as the mean of the differences, $\bar{D}$. We have finally the remainder sum of squares as $\tfrac{1}{2}(\sum_r D_r - \bar{D})^2$, or one-half the sum of the squares of the difference scores about the mean difference.

The $F$ for comparing two column means becomes

$$F = \frac{s^2_c}{s^2_{rc}} = \frac{\dfrac{R}{2}(\bar{X}_{\cdot1} - \bar{X}_{\cdot2})^2}{\dfrac{1}{\dfrac{\tfrac{1}{2}\sum_r (D_r - \bar{D})^2}{R - 1}}}$$

with $n_1 = 1$ and $n_2 = R - 1$. This reduces to

$$F = \frac{(\bar{X}_{.1} - \bar{X}_{.2})^2}{\dfrac{\sum_r (D_r - \bar{D})^2}{R(R-1)}}$$

which the reader will recognize as $t^2$ for comparing the difference between means based on sets of correlated scores with the standard error of the mean difference estimated by formula (7.1).

We have seen in Chapter 6 that in testing the difference between the means of correlated scores we can, for the large sample situation, determine the needed sampling error either from the distribution of differences between paired scores or by means of the standard error of the difference formula with the correlational term included. The important thing to note is that the analysis of variance technique provides a method for testing the significance of the difference between two or *more* means based on sets of correlated scores. The scores may be correlated either because they are based on the *same* individuals working under $C$ conditions or having $C$ trials on some stunt, or because siblings or litter mates are involved (each of the $C$ groups containing one case from each of $R$ families), or because we started with $R$ sets of matched individuals, one from each set being assigned to the several $C$ groups.

The $F$ just discussed has to do with column means. What of the row means for the given setup? The means of the $R$ rows represent the mean performance of each of the several individuals, and a test of the significance of the estimate of variance based on the between-row sum of squares becomes a test of the significance of individual differences. Since it is known that individuals do differ on practically all psychological variables, such a test is usually a trivial test of the obvious, and hence it is seldom needed. We may, however, have the situation in which we wonder whether individual variation is significant in the light of known measurement or response errors. To this question we now turn.

## RELIABILITY OF MEASUREMENT

Suppose the scores in each row represent either the performance of an individual on different forms of a scale or $C$ measurements for a given variable. The column means would be the means for the forms or successive sets of measurements, and the test of the significance between column means would be a test of the difference between the several form means or of the difference between the means for the $C$ successive sets of trials. For form means or for trial means, $F = s^2_c/s^2_{rc}$, as outlined previously, provides an over-all test of the significance of these *correlated* means.

In order to understand better the meaning in this situation of $F = s^2_r/s^2_{rc}$, let us again take the limiting case of $C = 2$; e.g., two forms of a test have been administered to $R$ individuals. Previously it was shown that the remainder sum of squares reduces to $\frac{1}{2}\Sigma_r(D_r - \bar{D})^2$ in which $D_r$, a difference score, is expressed as a deviation from the mean of the $R$ differences; hence this term is $\frac{1}{2}\Sigma d^2$ (cf. p. 90). Since $S^2_D = \Sigma d^2/R$, the term becomes $\frac{1}{2}RS^2_D$. When we recall the expression for the variance of difference scores for correlated values (6.6), it is readily seen that the remainder sum of squares can be written as

$$\frac{1}{2}\Sigma_r(D_r - \bar{D})^2 = \frac{R}{2}(S^2_1 + S^2_2 - 2r_{12}S_1S_2)$$

If we make the usual assumption that the two forms have the same variance, we can let $S^2_1 = S^2_2 = S^2_x$. Then noting that $r_{12}$ is the form versus form reliability coefficient, $r_{xx}$, we can substitute and get

$$\frac{R}{2}(2S^2_x - 2r_{xx}S^2_x) = RS^2_x(1 - r_{xx}) = RS^2_e$$

where $S^2_e$ is the error of measurement variance (see p. 165). When we divide this remainder sum of squares, $RS^2_e$, by its $df$, we have the previously labeled $s^2_{rc} = RS^2_e/(R - 1)$. Now let us recall the general relationship between $S^2$ and $s^2$ as biased and unbiased estimates of variance. For $NY$ (say) scores, $S^2_y = \Sigma y^2/N$, and $s^2_y = \Sigma y^2/(N - 1)$. From the definition of $S^2_y$ we have $\Sigma y^2 = NS^2_y$, hence $s^2_y = NS^2_y/(N - 1)$. Accordingly, with $S^2_e$ as the error of measurement variance, it is seen that $s^2_{rc}$ is an unbiased estimate of the error of measurement variance. We label this estimate $s^2_e$.

Thus, under the usual assumption of equal form variances, the remainder sum of squares and the variance estimate based thereon has to do with errors of measurement. The remainder sum of squares as actually computed includes an adjustment for possibly differing form means but does not allow for any difference in form variances. (It will be recalled that $S^2_e$ computed via $r_{xx}$ is also unaffected by a difference in form means.) If we have $C = 3$ or more forms, the remainder term is likewise a base for an unbiased estimate of error of measurement variance. When we test the difference between *row* means, we are actually asking whether the individual differences are significant in light of the variability due to measurement errors.

In our earlier discussion of reliability (pp. 163–70), nothing was said about unbiased estimates. Admitting that $s^2_e$ represents a slight improvement over the biased $S^2_e$, we next ask whether the use of unbiased estimates leads to an improvement in the estimate of $r_{xx}$. Reliability is sometimes defined by way of formula (10.15), i.e., $r_{xx} = 1 - S^2_e/S^2_x$ or in population values as $r_{xx(pop)} = 1 - \sigma^2_e/\sigma^2_x$. This definition formula can become a computational formula provided we have a means of estimating $\sigma^2_e$ and $\sigma^2_x$. If we were to plug in the unbiased estimates, $s^2_e$ and $s^2_x$ (the latter as an estimate of the common form variance but based on the scores from one form only), we would have

$$r_{xx} = 1 - s^2_e/s^2_x = 1 - \frac{RS^2_e/(R-1)}{RS^2_x/(R-1)} = 1 - \frac{S^2_e}{S^2_x}$$

from which we see that, when estimating $r_{xx}$ via variances in the two form situation, it matters not whether we use unbiased or biased estimates of the variances.

In some areas of psychology it is feasible to approach the reliability problem by way of $m$ repeated measurements on each of $N$ individuals. Feasibility depends on absence of practice or fatigue effects, permitting us to ignore the ordering of the measuring—as first, or second, and so on. The $m$ measurements for each person are averaged; we wish to know the reliability of these average scores, the coefficient for which we will designate as $r_{\bar{x}\bar{x}}$. By definition, $r_{\bar{x}\bar{x}} = 1 - \sigma^2_{e(\bar{x})}/\sigma^2_{\bar{x}_i}$ in which $\sigma^2_{e(\bar{x})}$ is the true (population) variance for the measurement errors associated with the "score" as an average of $m$ measures and $\sigma^2_{\bar{x}_i}$ is the variance of the distribution of the scores as averages (the subscript $i$ indicating that the $i$th person has such a mean score). These two variances would need to be estimated from the available data consisting of $m$ measures on each of $N$ individuals. The mean score for a person will have a sampling error since it is based on a sample of $m$ measures. The square of the estimated standard error of the mean score for the $i$th individual would be $s^2_{e(i)} = s^2_{w(i)}/m$ in which $s^2_{w(i)}$ is the unbiased estimate of the score (single, not average) variance within individual $i$. If we assume that this within individual score variance is the same from individual to individual, a better estimate of the within person variance will be obtained by averaging the separate $N$ estimates. But this is exactly the meaning of $s^2_w$ (see p. 291), placed now in the context of within persons instead of within groups. (The $N$ individuals provide $N$ sets, or groups, of $m$ scores each.) We will therefore take as our estimate of the error variance for the scores as averages, $s^2_{e(\bar{x})} = s^2_w/m$.

To secure an estimated $\sigma^2_{\bar{x}_i}$, the variance over persons of their scores as averages, we note that since the between-groups estimate $s^2_b = ms^2_{\bar{x}_b}$, we have $s^2_{\bar{x}_b} = s^2_b/m$. For the between-persons situation, this could be written as $s^2_{\bar{x}_i} = s^2_b/m$, which we will take as an unbiased estimate of $\sigma^2_{\bar{x}_i}$. Substituting in the definition formula, we have for the reliability of the scores as averages,

$$r_{\bar{x}\bar{x}} = 1 - \frac{s^2_w/m}{s^2_b/m} = 1 - \frac{s^2_w}{s^2_b} \tag{16.6}$$

which is based on unbiased estimates, computable by formulas given in Chapter 15.

The foregoing was concerned with the reliability of scores, $\bar{X}_i$, each taken as the average of $m$ scores or $m$ $X$s. What of the reliability of these $X$ scores? Recall that when we have one score per person, it can be expressed as $X = X_t + E$ and that for $N$ infinitely large, we have $\sigma^2_x = \sigma^2_t + \sigma^2_e$. The reliability of the $X$ scores,

$$r_{xx(pop)} = 1 - \frac{\sigma^2_e}{\sigma^2_x} = \frac{\sigma^2_t}{\sigma^2_x}$$

An estimate of $r_{xx(pop)}$ can be obtained by replacing $\sigma^2_t$ and $\sigma^2_x$ by their best available estimates. With $m$ measures on each of $N$ persons, we still have (ignoring order of measurement) a one-way analysis of variance setup, random model. The within-groups estimate, $s^2_w$, as a within-persons estimate would be an estimate of $\sigma^2_e$ and the between-groups estimate, $s^2_b$, would be an estimate of between-individual variation. We could replace

$$s^2_b \rightarrow \sigma^2 + m\sigma^2_{\mu_g}$$

by

$$s^2_{b_i} \rightarrow \sigma^2_e + m\sigma^2_{\mu_i}$$

in which $\mu_i$ as the population mean for the $i$th individual is the average of $m$ measures when $m$ is infinitely large; that is, $\mu_i$ is $X_t$ for the individual, and $\sigma^2_{\mu_i}$ is $\sigma^2_t$.

It will be recalled that when we say the expected value of $s^2_b$ is the sum of the two terms on the right of the arrow, we are saying that on the average, $s^2_{b_i} = \sigma^2_e + m\sigma^2_t$. If we solve for $\sigma^2_t$, and substitute the estimate, $s^2_w$, for $\sigma^2_e$, we have

$$\sigma^2_t = \frac{s^2_{b_i} - s^2_w}{m}$$

as an estimate of the variance of true scores.

To obtain an estimate of $\sigma^2_x$, we substitute the foregoing value of $\sigma^2_t$ and the estimate of $\sigma^2_e$, given by $s^2_w$, into $\sigma^2_x = \sigma^2_t + \sigma^2_e$. These substitutions lead to

$$\sigma^2_x = \frac{s^2_{b_i} - s^2_w}{m} + s^2_w = \frac{s^2_{b_i} - s^2_w + ms^2_w}{m}$$

which can be written as

$$\sigma^2_x = \frac{s^2_{b_i} + (m - 1)s^2_w}{m}$$

Then substituting the above estimates into $r_{xx(pop)} = \sigma^2_t/\sigma^2_x$ we have the estimate

$$r_{xx} = \frac{\dfrac{s^2_{b_i} - s^2_w}{m}}{\dfrac{s^2_{b_i} + (m - 1)s^2_w}{m}} = \frac{s^2_{b_i} - s^2_w}{s^2_{b_i} + (m - 1)s^2_w} \tag{16.7}$$

which the alert student will recognize as $r'$, the intraclass correlation coefficient. We have, therefore, incidentally provided a derivation of the $r'$ defined on p. 322. We see that $r'$, in the situation involving $m$ measures on each of $N$ persons, is a measure of the reliability of the $X$ scores, when taken singly.

Earlier (p. 237) we derived the generalized Brown-Spearman formula as a coefficient for the reliability of a sum (or average) of $m$ scores:

$$r_{\bar{x}\bar{x}} = \frac{m r_{xx}}{1 + (m - 1)r_{xx}}$$

Let us substitute $r_{xx}$ from equation (16.7), with $s^2_b$ standing for $s^2_{b_i}$:

$$r_{\bar{x}\bar{x}} = \frac{m \dfrac{s^2_b - s^2_w}{s^2_b + (m - 1)s^2_w}}{1 + (m - 1)\dfrac{s^2_b - s^2_w}{s^2_b + (m - 1)s^2_w}}$$

$$= \frac{m \dfrac{s^2_b - s^2_w}{s^2_b + (m - 1)s^2_w}}{\dfrac{s^2_b + (m - 1)s^2_w + (m - 1)s^2_b - (m - 1)s^2_w}{s^2_b + (m - 1)s^2_w}}$$

$$= \frac{m s^2_b - m s^2_w}{m s^2_b}$$

$$r_{\bar{x}\bar{x}} = 1 - \frac{s^2_w}{s^2_b}$$

Thus the Brown-Spearman formula leads to the value given by formula (16.6).

Since the reliability coefficient is a function of the error variance relative to the observed trait variance, it follows that a significant between-individuals variance is evidence for statistically significant reliability. But we cannot conclude from this that the test or instrument possesses satisfactory reliability since coefficients as low as .20 or .30 or even .10 can be statistically different from zero if $R$ is sufficiently large. The author does not recommend this approach to the question of the reliability of measurement for the simple reason that it is more important to know *how* reliable a test is or how near its reliability approaches unity than to know only that it *is* reliable in the sense of yielding a coefficient significantly different from zero.

This possible application of the variance technique, however, points up the fact that it is sometimes meaningful to speak of the remainder variance as "error" variance. In a wider sense, the remainder variance can be thought of as the uncontrolled variation which contributes to the variation of the means of the groups being compared. Now a little reflection leads to the conclusion that the sources of error in research are many and varied. Sometimes instrumental and/or measurement errors loom large, sometimes the error associated with the sampling of individuals is paramount, at other times the intraindividual variation is sizable, and frequently if the sources of variation are unknown the term experimental error is used as a catchall. When a particular variance estimate is referred to as *the* error variance to be used as the denominator of the $F$ ratio, the "error" may be any one of or a combination of the many types of error. In this sense, the variance estimate based on the remainder sum of squares may be the error variance even for those situations where we have classifications into $R$ groups rather than as $R$ individuals, but as will presently be seen the term which we are now calling the remainder may not always be the one to utilize as "error." The within-groups variance estimate of Chapter 15 was an "error" variance for testing the significance of the between-groups variation. In more complex setups in the analysis of variance, rules are required for choosing the appropriate error term.

## ILLUSTRATION, C CORRELATED MEANS

The required computations for testing variation between column means will now be set forth. It makes no difference in the computational procedure whether we have $R$ individuals with $C$ scores each or $R$ sets of $C$ individuals matched.

The computation of the required sums of squares involves formulas

(16.1) to (16.3) with $m = 1$:

$$\underset{r\,c}{\Sigma\Sigma}(X_{rc} - \bar{X})^2 = \frac{1}{RC}\left[RC\underset{r\,c}{\Sigma\Sigma}X^2_{rc} - \left(\underset{r\,c}{\Sigma\Sigma}X_{rc}\right)^2\right] \quad \text{for total} \qquad (16.8)$$

$$R\underset{c}{\Sigma}(\bar{X}_{\cdot c} - \bar{X})^2 = \frac{1}{RC}\left[C\underset{c}{\Sigma}\left(\underset{r}{\Sigma}X_{rc}\right)^2 - \left(\underset{r\,c}{\Sigma\Sigma}X_{rc}\right)^2\right] \quad \text{for columns} \quad (16.9)$$

$$C\underset{r}{\Sigma}(\bar{X}_{r\cdot} - \bar{X})^2 = \frac{1}{RC}\left[R\underset{r}{\Sigma}\left(\underset{c}{\Sigma}X_{rc}\right)^2 - \left(\underset{r\,c}{\Sigma\Sigma}X_{rc}\right)^2\right] \quad \text{for rows} \qquad (16.10)$$

The sum of squares for the remainder can be obtained by subtracting the sums for between columns and for between rows from the total sum of

**Table 16.8. Data for visual acuity, 4 individuals, 3 distances (Monocular, vernier method, coded scores)\***

| Subjects | Distance (in Meters) | | | $\underset{c}{\Sigma X_{rc}}$ | $\bar{X}_{r\cdot}$ |
| | 5 | 10 | 15 | | |
|---|---|---|---|---|---|
| 1 | 13 | 29 | 17 | 59 | 19.7 |
| 2 | 4 | 9 | 19 | 32 | 10.7 |
| 3 | 8 | 30 | 37 | 75 | 25.0 |
| 4 | 9 | 27 | 53 | 89 | 29.7 |
| $\underset{r}{\Sigma X_{rc}}$ | 34 | 95 | 126 | 255 | |
| $\bar{X}_{\cdot c}$ | 8.5 | 23.7 | 31.5 | $21.2 = \bar{X}$ | |

$$\underset{r\,c}{\Sigma\Sigma X_{rc}} = 255 \qquad\qquad \underset{r\,c}{\Sigma\Sigma X^2_{rc}} = 7709$$

$$\underset{r\,c}{\Sigma(\Sigma X_{rc})^2} = 18{,}051 \qquad\qquad \underset{c\,r}{\Sigma(\Sigma X_{rc})^2} = 26{,}057$$

\* From Walker, E. L., *Factors in vernier acuity and distance discrimination*, Doctoral Dissertation, Stanford University, California, 1947.

squares. Sum each row, and write the sums on the right-hand margin; sum each column, and write the sums along the bottom margin. Summing down the right-hand margin gives the total sum, and summing across the bottom margin should give the same total sum. Square all scores and sum to get the first sum in (16.8); square all the right-hand margin sums and then sum to get the first part of (16.10); square all the bottom margin sums and then sum to get the first part of (16.9).

The computations are illustrated by the data of Table 16.8. Casual examination of the table indicates that acuity measures are influenced by distance. Do the means for the three distances differ significantly?

The required sums are included in the table. Substituting these in the foregoing formulas gives:

$\frac{1}{12}[12(7709) - (255)^2] = 2290.25$ for the total sum of squares

$\frac{1}{12}[3(26,057) - (255)^2] = 1095.50$ for between-columns sum of squares

$\frac{1}{12}[4(18,051) - (255)^2] = 598.25$ for between-rows sum of squares

Subtracting the sum of the last two from the total gives 596.50 as the remainder sum of squares.

These results are assembled in Table 16.9 along with the *df*s and the

**Table 16.9. Variance table for data of Table 16.8**

| Source | Sum of Squares | df | Variance Estimate |
|---|---|---|---|
| Distance | 1095.50 | 2 | 547.75 |
| Subjects | 598.25 | 3 | 199.42 |
| Remainder | 596.50 | 6 | 99.42 |
| Total | 2290.25 | 11 | |

variance estimates. For the influence of distance we have $F = 547.75/99.42 = 5.51$, which for $n_1 = 2$ and $n_2 = 6$ is significant at slightly better than the $P = .05$ level (additional data in Walker's dissertation leave no doubt—distance does have an effect). This is a situation in which experimentally induced differences are so large that they can be demonstrated with only four cases.

## CHOICE OF ERROR TERM IN TWO-WAY CLASSIFICATION

Now that we have learned something about the meaning of interaction and have had a couple of examples which illustrate the computations and the way hypotheses can be tested, we must specifically consider an as yet unmentioned question: Which variance estimate is the correct one to use as the error term, that is, as the denominator for the $F$ ratio? The answer depends on the mathematical model that is appropriate for a given situation. Three models have been set forth by the mathematical statisticians. These are referred to as the random model, the fixed constants model, and the mixed model. Let us define these for the two-way classification setup.

We have the *random model* when both classifications involve sampling. Such would be the case when rows stand for individuals and columns stand for judges (each of whom has rated each individual). The individuals and the judges are regarded as random samples from normally distributed populations: normal distribution of individuals with respect to the ratings and normal distribution for the rating characteristics of the judges.

We have a *fixed constants model* when no random sampling is involved so far as the bases of the classifications are concerned. Such is the case when the classifications depend on such things as size, distance, time interval, degree of illumination, etc.; or on such unordered things as sense modality, sex, method, diagnostic group, etc. The setup in Table 16.3 involves the fixed constants model; neither the rest intervals nor the sessions were chosen at random.

We have a *mixed model* when one basis of classification involves sampling and the other fixed constants. Table 16.8 illustrates a typical mixed model, typical in that one basis of classification is individuals.

Each of the three models calls for precisely the same breakdown of the sum of squares and of the degrees of freedom, and each leads to three variance estimates plus a within-cells estimate in case we have more than one score per cell. It should be noted that the within-cells scores can stand for two kinds of *replication*. We might have replication in the sense of having carried out the experiment with more than one person in each cell (but with different persons from cell to cell) as in Table 16.3, or we might have a replication of measures on the same person or persons. Thus in Table 16.8 we could have $m$ measures per person under each of the $C$ conditions. (We are not here concerned with replication in the sense of a repetition of the entire experiment by another investigator.)

Actually, for the working statistician the precise formula for the possible mathematical models is not nearly so important as the deductions therefrom regarding the meanings of the several variance estimates. Earlier (pp. 292–301) we attempted to explain the meaning of the variance estimate, $s^2{}_b$, under nonnull conditions. Perhaps the student should turn back and review the argument that led to specifying the expected value of $s^2{}_b$ as either

$$s^2{}_b \rightarrow \sigma^2 + m\sigma^2{}_{\mu_g}$$

or as

$$s^2{}_b \rightarrow \sigma^2 + \frac{m\Sigma(\mu_{.g} - \mu_{..})^2}{G - 1}$$

From now on we will simply write out the expected values as set forth by the mathematical statistician.

The general model for two-way classification may be written as

$$(X_{rck} - \mu) = \alpha_r + \alpha_c + \alpha_r\alpha_c + e_{rck} \qquad (16.11)$$

in which the deviation score from the over-all population mean is thought of in terms of a row contribution, $\alpha_r$; a column contribution, $\alpha_c$; an interactive effect, $\alpha_r\alpha_c$; and a normally distributed random (error) part, $e_{rck}$. The subscript $k$ indicates that we have replication, $m$ scores per cell, with $k$ taking on values $1, \cdots, m$, but the $m$ scores in each cell are independent of the scores in all other cells. The $\alpha_r$, $\alpha_c$, and $\alpha_r\alpha_c$ are all expressed in deviation form, i.e., possess the property that $\sum_r \alpha_r = 0$, $\sum_c \alpha_c = 0$, $\sum_r \sum_c \alpha_r\alpha_c = 0$. Actually, $\alpha_r$ is used here instead of $\mu_r - \mu$ (or instead of the more precise notation, $\mu_{r.} - \mu_{..}$), with a similar meaning for $\alpha_c$. The $\sum_r \sum_c \alpha_r\alpha_c$ stands for $\sum_r \sum_c (\mu_{rc} - \mu_{r.} - \mu_{.c} + \mu_{..})$.

For the fixed constants (sometimes called fixed effects, sometimes fixed factor) model, we replace $\alpha$ by $A$, thus

$$(X_{rck} - \mu) = A_r + A_c + A_rA_c + e_{rck} \qquad (16.12)$$

For the random (sometimes called components of variance) model we replace $\alpha$ by $a$, thus

$$(X_{rck} - \mu) = a_r + a_c + a_ra_c + e_{rck} \qquad (16.13)$$

and the mixed model can be written (with columns standing for fixed constants) as

$$(X_{rck} - \mu) = a_r + A_c + a_rA_c + e_{rck} \qquad (16.14)$$

The $a_r$, $a_c$, $a_ra_c$, and $a_rA_c$ are all assumed to be random variates from normally distributed populations of effects having variances $\sigma^2_r$, $\sigma^2_c$, $\sigma^2_{rc}$, and $\sigma^2_{rC}$. Note that the lower-case subscripts to a $\sigma^2$ refer to random factors whereas the upper-case subscript refers to a fixed factor. (No such distinction is needed for subscripts to $s^2$ or to $X$.) For the fixed values $A_r$, $A_c$, and $A_rA_c$ no assumption as to distribution of effects is required. Indeed, it is difficult to imagine a distribution of, say $A_c$ when $C = 2$. The "population" of effects consists of just two values, $A_1$ and $A_2$, which symbols stand for $(\mu_{.1} - \mu)$ and $(\mu_{.2} - \mu)$. Two values, or for that matter the usual small number of fixed effects, cannot very well be described as to distribution, hence the differences among them or their variation about an over-all $\mu$ cannot aptly be described in terms of a $\sigma^2$. Consequently, in the sequel the variation among them will be specified in terms of $\sum_c A^2_c$. Likewise, for the $A_r$ and the $A_rA_c$ we will have $\sum_r A^2_r$ and $\sum_r \sum_c (A_rA_c)^2$, respectively.

When the $m$ scores per cell represent measurement replication, $s^2_w$ will be taken as an estimate of $\sigma^2_e$; when the $m$ scores per cell involve $m$ individuals (measured once), $s^2_w$ will be regarded as an estimate of individual difference variance, designated $\sigma^2_i$. It is to be understood that $\sigma^2_i$

has two components: true score variance and error of measurement variance.

We are now ready to examine the various possible situations involving two-way classification in order to point out just what is being estimated by $s^2_r$, $s^2_c$, $s^2_{rc}$, and $s^2_w$. Once this is done, we will be in a position to choose an appropriate variance estimate, if such is available, as the denominator, or error, term for $F$. The question of variance homogeneity will be discussed after a consideration of eight setups (cases) involving two-way classification.

*Case I.* Fixed constants model, with $m$ scores ($m$ persons) per cell, a total of $mRC$ individuals:

$$s^2_r \rightarrow \sigma^2_i + \frac{mC}{R-1} \sum_r A^2_r$$

$$s^2_c \rightarrow \sigma^2_i + \frac{mR}{C-1} \sum_c A^2_c$$

$$s^2_{rc} \rightarrow \sigma^2_i + \frac{m}{(R-1)(C-1)} \sum_r \sum_c (A_r A_c)^2$$

$$s^2_w \rightarrow \sigma^2_i$$

The general principle in forming an $F$ ratio is to choose two estimates which differ (in their expected values) by one term only, the term involving the effect being tested. Accordingly, $s^2_w$ is the correct denominator for $F_r$, $F_c$, and $F_{rc}$, for testing row, column, and interaction effects, respectively. Note that interaction, if present, has nothing whatever to do with the main (row or column) effects. This is true because the interaction is a fixed, not a random, effect. If the interaction is significant, we must be on guard in drawing conclusions about the main effects—qualifications will be needed, as we learned in our discussion of the meaning of interaction.

*Case II.* Fixed constants model, $RC$ individuals, one per cell, with each measured $m$ times: the expectations for the first three estimates will be precisely the same as in Case I, but now $s^2_w \rightarrow \sigma^2_e$. We see immediately that this design leads to difficulties. The resulting $s^2_w$ estimate is useless; if we did use $s^2_w$ as the denominator for testing, say, $s^2_r$, a significant $F$ would be meaningless because we would not know whether its significance was attributable to a real row effect or to real individual differences or to a combination of the two.

*Case III.* Fixed constants model, only one person measured $m$ times under each of the $RC$ conditions: if we replace $\sigma^2_i$ by $\sigma^2_e$ in the set of expected values for Case I we will have indicated what each $s^2$ estimates. As for Case I, the appropriate error term for all three $F$s is $s^2_w$, but any conclusion we draw from a significant $F$ must be carefully scrutinized for meaning. It can mean only that the effect holds for the one person used in the experiment, with no assurance whatsoever that a repetition of the experiment with another person—either in the same or in a different laboratory—will lead to a confirmation of the results. In other words, no generalization is possible except the trivial one that the effect holds for a particular individual, a useful generalization if the scientific horizon is limited to one person.

These three cases exhaust the possible situations for two-way analysis of variance involving the fixed constants model. If it has occurred to the reader that each of $m$ cases might be measured under all the $RC$ conditions, he should be apprised that this would involve three-way classification, to be discussed later. The important thing to have noted is that clear-cut results, permitting generalizations to a population of individuals, are possible only by the setup of Case I. We have listed the other two cases because it may be helpful to know what not to do. However, we must point out a possible exception to a sweeping dismissal of Case III: there are some areas (sensory-perceptual) in experimental psychology for which experimentally produced effects are so large relative to individual differences that we can be reasonably sure that similar significant results will hold for other persons; sure, that is, provided some knowledge of the extent of individual differences is available. Rarely will the effects be of the same order of magnitude for two persons—individual by conditions interaction is the rule rather than the exception.

*Case IV.* Random model, rows stand for $R$ individuals and columns stand for $C$ judges with $m$ (ordinarily $m$ will not exceed 2) ratings by each judge on each individual. The ratings, which must be directed toward some trait and involve at least a 10-point scale, might be based on observed, or on a transcribed record of, behavior of the $R$ individuals. (The judges might find it difficult to rule out memory when making two or more ratings for each individual.) Instead of $C$ judges making ratings we might have $C$ examiners or testers, each testing the $R$ individuals twice on, say, the Rorschach. We have a sample of individuals and a sample of judges (or examiners). The expected values of the variance estimates are:

$$s^2_r \rightarrow \sigma^2_e + m\sigma^2_{rc} + mC\sigma^2_r$$
$$s^2_c \rightarrow \sigma^2_e + m\sigma^2_{rc} + mR\sigma^2_c$$
$$s^2_{rc} \rightarrow \sigma^2_e + m\sigma^2_{rc}$$
$$s^2_w \rightarrow \sigma^2_e$$

It is obvious that $s^2_w$ can be used as the error term for testing the interactive effect, but since $s^2_w$ is nothing more than an estimate of error of measurement variance for the ratings, the conclusion from a significant $F$ is that interaction holds for these particular $R$ individuals and $C$ judges—there is no assurance that repetition of the investigation with $R$ other individuals and $C$ other judges would lead to interaction. As for the main effects, it is obvious that $s^2_{rc}$ becomes the appropriate (and only correct) term to use for $F_r$ and $F_c$. A significant $F_r$ would mean a dependable differentiation of individuals over and above the variation due to measurement error and judge by individual interaction, and a significant $F_c$ would indicate real variation from judge to judge in a possible population of judges.

*Case V.* Random model, same as Case IV except that $m = 1$. No estimate of $\sigma^2_e$ is available, but $s^2_{rc}$ would still be the error term for $F_r$ and $F_c$.

Remark: Actually we are hard put to find good illustrations in psychology for the random model. Any student who attempts to find other illustrations should keep in mind that it must be possible to classify a score simultaneously in two different ways, each involving sampling.

*Case VI.* Mixed model, rows stand for individuals (or matched persons), columns involve $C$ fixed constants (fixed conditions having fixed effects), and measurement replication leading to $m$ scores per cell:

$$s^2_r \rightarrow \sigma^2_e + mC\sigma^2_r$$

$$s^2_c \rightarrow \sigma^2_e + m\sigma^2_{rC} + \frac{mR}{C-1}\Sigma_c A^2_c$$

$$s^2_{rc} \rightarrow \sigma^2_e + m\sigma^2_{rC}$$

$$s^2_w \rightarrow \sigma^2_e$$

The reader will need to recall that lower-case and upper-case letters as subscripts to a $\sigma^2$ indicate random and fixed factors, respectively. The interaction term can be tested by $F_{rc} = s^2_{rc}/s^2_w$; if $F_{rc}$ is significant, we can conclude that the differential responses (failure of the individuals to maintain the same rank order under the $C$ conditions) are larger than expected on the basis of errors of measurement. Individual by conditions interactions are usually found to be significant. It will be recalled that in the mixed model the interaction term, $a_r A_c$, is regarded as a random variate, and as such it becomes a source of random variation which, if real, will affect the between-columns term. We see from the foregoing that $s^2_{rc}$ is the proper error term for testing $s^2_c$. To use $s^2_w$ for this purpose is simply not defensible; if, for example, $s^2_c/s^2_w$ is significant, it might be so because of real column differences or because of real interaction or because of a combination of the two. Ordinarily $s^2_r$ in this situation is not tested for significance since it reflects individual differences which are always real unless the measurements are completely unreliable. Indeed, it would be sad to carry out an experiment without first ascertaining that the measurements have reliability.

*Case VII.* Mixed model, same as Case VI except that $m = 1$ (no measurement replication). This does not provide an $s^2_w$, which is non-essential anyway, but $s^2_{rc}$ is again the proper error term for testing $s^2_c$. The setup in Table 16.8, falls under this case. In psychological research, Case VII is used quite frequently; it provides a significance test for the differences among a series of $C$ correlated means, correlated for reasons previously specified.

*Case VIII.* Mixed model, $R$ rows stand for $R$ individuals and columns stand for $C$ forms of a test (the reliability of measurement setup discussed on pp. 342–44):

$$s^2_r \rightarrow \sigma^2_e + C\sigma^2_r$$

$$s^2_c \rightarrow \sigma^2_e + \frac{R}{C-1} \Sigma A^2_c$$

$$s^2_{rc} \rightarrow \sigma^2_e$$

It will be recalled that $s^2_{rc}$ was shown to depend solely on errors of measurement under the assumptions usually made in connection with test reliability. We see now that these assumptions involve the a priori assumption of no interaction, an assumption which implies, among other things, that possible practice effects are not different from person to person. Note that in case interaction is operating, $s^2_c$ will involve an interaction component (as in Case VII); hence $s^2_{rc}$ is the appropriate error term, regardless of whether there is or is not interaction, for $F_c$ as a test of the difference between the $C$ form means or over-all practice effects or both (we would not know which). But a test of the significance of $s^2_r$ requires the assumption of no interaction.

Remark about measurement replication: We have seen that having $s^2_w$ as an estimate of $\sigma^2_e$ does not provide us with a useful error term (for $F$) in the testing of hypotheses about main effects (and sometimes about interaction) under any of the three mathematical models. This illustrates a general principle: when an estimate of error of measurement variance is used as the denominator of $F$, no generalization to a population of persons is possible, and hence no generalization of import to science. This raises the question as to whether measurement replication is worth while. The answer is yes, particularly when it is known that a single measurement is not very reliable. By replicating measurement we will obtain more reliable scores in the form of the average of $m$ values; hence one source of variability in the data will be reduced. The student who has not noticed that the analyses involving measurement replication are, in essence, dealing with average scores for individuals should ponder further.

**Homogeneity of variance assumption.** For Cases I and II it is assumed that individual difference variance is the same from cell to cell. For Cases III through VIII it is assumed that error of measurement variance is homogeneous from cell to cell. The assumption is testable (say, by Bartlett's test) only for Cases I, III, IV, and VI.

An additional assumption, seldom mentioned in textbooks of applied statistics, is required for those cases where, for $C$ greater than 2, $s^2_{rc}$ is used as the error term. This is the assumption of homogeneity of interaction variances, the meaning of which may not be quickly obvious to the reader. Let us consider the mixed model, with rows standing for individuals and with columns for levels on some factor. The interaction sum of squares involves

$$(X_{rc} - \bar{X}_{r.} - \bar{X}_{.c} + \bar{X})$$

which, it will be recalled, was a simplification of the remainder

$$(X_{rc} - \bar{X}) - [(\bar{X}_{r.} - \bar{X}) + (\bar{X}_{.c} - \bar{X})]$$

from which we see that every deviation being squared and summed to get the interaction sum of squares is one in which the deviation $(X_{rc} - \bar{X})$ is twice "adjusted," once for the row effect and once for the column effect. Suppose that the adjustment has been made for the column effect and that we examine what is left, which is $(X_{rc} - \bar{X}) - (\bar{X}_{r.} - \bar{X})$, or simply $(X_{rc} - \bar{X}_{r.})$.

Thus, after adjustment for column effect, the interaction sum of squares is represented by $\underset{r\ c}{\Sigma\Sigma}(X_{rc} - \bar{X}_{r.})^2$, which in turn can be written as

$$\underset{r}{\Sigma}(X_{r1} - \bar{X}_{r.})^2 + \underset{r}{\Sigma}(X_{r2} - \bar{X}_{r.})^2 + \cdots$$
$$+ \underset{r}{\Sigma}(X_{rc} - \bar{X}_{r.})^2 + \cdots + \underset{r}{\Sigma}(X_{rC} - \bar{X}_{r.})^2$$

each term of which is based on $R - 1$ degrees of freedom and, as we already know, the sum of these terms is based on $(R - 1)(C - 1)$ degrees of freedom. The important thing to note is that the interaction sum of squares is made up of $C$ components, every one of which when divided by $R - 1$ provides a variance estimate. In effect, when we combine these $C$ sums of squares as a basis for estimating the interaction variance, we are averaging $C$ separate estimates (with due allowance for the number of degrees of freedom). Homogeneity of interaction variances means that the $C$ possible estimates are all estimates of a common population variance.

The ramifications of this interaction business might be better understood by pursuing further the foregoing breakdown into $C$ parts. Consider the contribution of the first column, $\sum_r (X_{r1} - \bar{X}_{r.})^2$. The $X_{r1}$ are variates with mean $\bar{X}_{.1}$, and the $\bar{X}_{r.}$ are also variates with mean $\bar{X}$, hence the differences $(X_{r1} - \bar{X}_{r.}) = (x_{r1} - \bar{x}_{r.})$ in deviation units. Then

$$\sum_r (x_{r1} - \bar{x}_{r.})^2 = \sum_r x^2_{r1} + \sum_r \bar{x}^2_{r.} - 2\sum_r x_{r1}\bar{x}_{r.}$$

$$= RS^2_{x_{r1}} + RS^2_{\bar{x}_{r.}} - 2Rr_{x_{r1}\bar{x}_{r.}} S_{x_{r1}} S_{\bar{x}_{r.}}.$$

in which $S^2_{x_{r1}}$ is the variance (not unbiased estimate) of the scores in the first column and $S^2_{\bar{x}_{r.}}$ is the variance of the distribution of the row means, and $r_{x_{r1}\bar{x}_{r.}}$ is the correlation between these two variates. Similarly for the second column and the $c$th column we have

$$RS^2_{x_{r2}} + RS^2_{\bar{x}_{r.}} - 2Rr_{x_{r2}\bar{x}_{r.}} S_{x_{r2}} S_{\bar{x}_{r.}}.$$

$$RS^2_{x_{rc}} + RS^2_{\bar{x}_{r.}} - 2Rr_{x_{rc}\bar{x}_{r.}} S_{x_{rc}} S_{\bar{x}_{r.}}.$$

and so on to the $C$th column.

We can now tease out two possible sources for the heterogeneity of variance for the subparts entering into the row by column interaction. The $C$ components can differ either because the variances in the several columns differ and/or because the degree of correlation between the row means and the column scores varies from column to column. When the latter variation occurs, it implies that the $C(C - 1)/2$ intercorrelations among the columns are heterogeneous, a heterogeneity that can readily be a by-product of the $C$ experimental conditions or the $C$ levels.

It might be guessed that violation of the assumption of homogeneity of variance for the components of interaction would have effects on significance levels similar to the effect of heterogeneous variances in the one-way setup. However, we are here dealing with a more complex situation. Heterogeneity of interaction components springs from two sources: heterogeneity of column variances and differing degrees of correlation between the columns. A recent study (R. O. Collier et al., *Psychometrika*, 1967, **32**, 339–353) suggests that with homogeneous column variances and heterogeneous correlations the effect on the $F$ test is such as to produce 7 or 8 per cent at the .05 level and 2 or 3 per cent at the .01 level. These results are similar to those of the Norton study referred to earlier. But when the situation involves heterogeneity both of correlations and of column variances, the effect on significance levels becomes marked: 10 per cent of $F$s reaching the .05 level, and from 3 or 4 even up to 5 per cent reaching the .01 level. It therefore appears that when using a repeat-measure design, the researcher would be wise to require $F$ to reach the .01

level in order to have assurance that significance can be claimed at the .05 level. It is also of interest to note that in the Collier et al. empirical study, the increase of sample size ($R$) from 5 to 15 did not improve the validity of the $F$ test. This is in contrast to the findings of Boneau (cited in Chapter 7) and Norton, which indicate that increase in sample size(s) does improve the $t$ and $F$ tests if assumptions are violated when $s^2_w$ is the error term.

Incidentally, since the foregoing development permits us to write

$$R\Sigma_c S^2_{x_{rc}} + CRS^2_{\bar{x}_{r.}} - 2R\Sigma_c r_{x_{rc}\bar{x}_{r.}}.S_{x_{rc}}S_{\bar{x}_{r.}}.$$

as an expression for the interaction sum of squares after adjustment for differences in column means, we have the basis for additional insight into the meaning of interaction for the mixed two-way model. Suppose, for argument's sake, that the variance in each column equals the variance of the distribution of row means and that there is perfect correlation between the scores in each column and the row means. Under these conditions the interaction sum of squares is seen to be zero. Next note that as soon as these correlations cease to be perfect, the interaction sum of squares ceases to be zero. The lower the correlations, the greater the interaction. Now since the correlation of the scores in, say, the first column with the row means will equal the correlation with the row sums (because the means are merely the sums divided by the constant, $C$), we can write this $r$, in simpler notation, as

$$r_{x_1\bar{x}_r} = \frac{\Sigma x_1(x_1 + x_2 + \cdots + x_c + \cdots + x_C)}{RS_1 S_{sum}}$$

the numerator of which tells us that this correlation is a function of the extent to which the scores in the first column correlate with the scores in the other columns; similarly, for the correlation of any column against the marginal means. Thus the interaction is a function of the intercorrelations among the columns—the lower these are, the greater the interaction. When it is recalled that errors of measurement tend to lower correlations, it is readily seen that the computed interaction depends in part on measurement errors, as was specified in the expectations given under Case VI.

The foregoing argument holds, of course, when the variances within the columns are unequal. As an exercise the student might consider the situation where the variance in the first column is, say, 4 while that for all other columns is 1 and where the correlations are all unity, and thereby demonstrate that any differences in column variances also contribute to the interaction sum of squares. Real interaction can produce differences in variances.

## THREE-WAY CLASSIFICATION

Suppose that we wish to arrange an investigation so as to let one set of data serve to determine whether the variation of a dependent variable is due to or associated with variation on three independent variables. Again, the term independent variable is being used in its broad sense. It might be a "real" variable like illumination, temperature, amount of food, length of rest interval; or it might be a variable having to do with qualitative differences, such as kind of food, type of motivation or incentive, various psychological sets. Correlated variables cannot be used (see p. 330–31).

It is necessary that we be able to assign individuals or scores to each combination of groupings made possible by whatever classifications we have on the three independent variables. Let us suppose that there are $C$ categories on one variable, $R$ on another, and $B$ on a third. For purposes of exposition and as a systematic way of arranging the data, let the $C$ categories define $C$ columns, the $R$ categories $R$ rows, and the $B$ categories $B$ blocks. Let $X_{rbc}$ represent a score in the $r$th row, $b$th block, and $c$th column. Thus $X_{324}$ would be any score in the third row, second block, and fourth column. The scores may be arranged in some such systematic order as that in Table 16.10, which should be studied carefully by the reader. Each $X$ stands for $m$ scores.

Note in particular how the various sums are specified and their location in the table. The first two subscripts in $\sum_c X_{11c}$ indicate that this sum has to do with scores in the first row and first block, and that in the summing process $c$ takes on values running from 1 to $C$. The general expression for all such sums is $\sum_c X_{rbc}$. The symbol $\sum_r X_{r11}$ stands for the sum of scores in the first column and first block; $r$ takes on values of 1 to $R$. The corresponding general symbol is $\sum_r X_{rbc}$. In next to the bottom section of the table will be found $\sum_b X_{1b1}$ as the sum for all the cases in row 1 and column 1, the summing being through blocks; i.e., $b$ takes on values from 1 to $B$. The general expression for such sums is $\sum_b X_{rbc}$. The sum of all the scores in the first block is symbolized as $\sum_r \sum_c X_{r1c}$, and in the $b$th block as $\sum_r \sum_c X_{rbc}$. For the sum of all the scores in the first column, irrespective of row and block, we have $\sum_r \sum_b X_{rb1}$, and the general expression is $\sum_r \sum_b X_{rbc}$. The symbol $\sum_b \sum_c X_{1bc}$ stands for the sum of all scores in the first row, and $\sum_b \sum_c X_{rbc}$ is the corresponding general expression. Note also how the "dot" notation is used to specify the several means. The subscript which has been replaced by a dot indicates the direction of the addition required to obtain the sum for the given mean. Thus in $\bar{X}_{.24}$ the dot replaces $r$; this mean is based on $mR$ scores, with $r$ running from 1 to $R$ when we sum. The subscripts which

Table 16.10. Score and sum schema for three-way classification

| | | Column 1 | Column c | Column C | Sum | Mean |
|---|---|---|---|---|---|---|
| Block 1 | Row 1 | $X_{111}$ | $X_{11c}$ | $X_{11C}$ | $\sum_c X_{11c}$ | $\bar{X}_{11\cdot}$ |
| | $r$ | $X_{r11}$ | $X_{r1c}$ | $X_{r1C}$ | $\sum_c X_{r1c}$ | $\bar{X}_{r1\cdot}$ |
| | $R$ | $X_{R11}$ | $X_{R1c}$ | $X_{R1C}$ | $\sum_c X_{R1c}$ | $\bar{X}_{R1\cdot}$ |
| | Sum | $\sum_r X_{r11}$ | $\sum_r X_{r1c}$ | $\sum_r X_{r1C}$ | $\sum_r\sum_c X_{r1c}$ | $\bar{X}_{\cdot1\cdot}$ |
| | Mean | $\bar{X}_{\cdot11}$ | $\bar{X}_{\cdot1c}$ | $\bar{X}_{\cdot1C}$ | $\bar{X}_{\cdot1\cdot}$ | Mean block 1 |
| Block $b$ | 1 | $X_{1b1}$ | $X_{1bc}$ | $X_{1bC}$ | $\sum_c X_{1bc}$ | $\bar{X}_{1b\cdot}$ |
| | $r$ | $X_{rb1}$ | $X_{rbc}$ | $X_{rbC}$ | $\sum_c X_{rbc}$ | $\bar{X}_{rb\cdot}$ |
| | $R$ | $X_{Rb1}$ | $X_{Rbc}$ | $X_{RbC}$ | $\sum_c X_{Rbc}$ | $\bar{X}_{Rb\cdot}$ |
| | Sum | $\sum_r X_{rb1}$ | $\sum_r X_{rbc}$ | $\sum_r X_{rbC}$ | $\sum_r\sum_c X_{rbc}$ | $\bar{X}_{\cdot b\cdot}$ |
| | Mean | $\bar{X}_{\cdot b1}$ | $\bar{X}_{\cdot bc}$ | $\bar{X}_{\cdot bC}$ | $\bar{X}_{\cdot b\cdot}$ | Mean block $b$ |
| Block $B$ | 1 | $X_{1B1}$ | $X_{1Bc}$ | $X_{1BC}$ | $\sum_c X_{1Bc}$ | $\bar{X}_{1B\cdot}$ |
| | $r$ | $X_{rB1}$ | $X_{rBc}$ | $X_{rBC}$ | $\sum_c X_{rBc}$ | $\bar{X}_{rB\cdot}$ |
| | $R$ | $X_{RB1}$ | $X_{RBc}$ | $X_{RBC}$ | $\sum_c X_{RBc}$ | $\bar{X}_{RB\cdot}$ |
| | Sum | $\sum_r X_{rB1}$ | $\sum_r X_{rBc}$ | $\sum_r X_{rBC}$ | $\sum_r\sum_c X_{rBc}$ | $\bar{X}_{\cdot B\cdot}$ |
| | Mean | $\bar{X}_{\cdot B1}$ | $\bar{X}_{\cdot Bc}$ | $\bar{X}_{\cdot BC}$ | $\bar{X}_{\cdot B\cdot}$ | Mean block $B$ |
| Sums through blocks | 1 | $\sum_b X_{1b1}$ | $\sum_b X_{1bc}$ | $\sum_b X_{1bC}$ | $\sum_b\sum_c X_{1bc}$ | $\bar{X}_{1\cdot\cdot}$ |
| | $r$ | $\sum_b X_{rb1}$ | $\sum_b X_{rbc}$ | $\sum_b X_{rbC}$ | $\sum_b\sum_c X_{rbc}$ | $\bar{X}_{r\cdot\cdot}$ |
| | $R$ | $\sum_b X_{Rb1}$ | $\sum_b X_{Rbc}$ | $\sum_b X_{RbC}$ | $\sum_b\sum_c X_{Rbc}$ | $\bar{X}_{R\cdot\cdot}$ |
| | Sum | $\sum_r\sum_b X_{rb1}$ | $\sum_r\sum_b X_{rbc}$ | $\sum_r\sum_b X_{rbC}$ | $\sum_r\sum_b\sum_c X_{rbc}$ | $\bar{X}_{\cdots}$ |
| Means for rows by columns | 1 | $\bar{X}_{1\cdot1}$ | $\bar{X}_{1\cdot c}$ | $\bar{X}_{1\cdot C}$ | $\bar{X}_{1\cdot\cdot}$ | Means for |
| | $r$ | $\bar{X}_{r\cdot1}$ | $\bar{X}_{r\cdot c}$ | $\bar{X}_{r\cdot C}$ | $\bar{X}_{r\cdot\cdot}$ | rows |
| | $R$ | $\bar{X}_{R\cdot1}$ | $\bar{X}_{R\cdot c}$ | $\bar{X}_{R\cdot C}$ | $\bar{X}_{R\cdot\cdot}$ | |
| Column means | | $\bar{X}_{\cdot\cdot1}$ | $\bar{X}_{\cdot\cdot c}$ | $\bar{X}_{\cdot\cdot C}$ | $\bar{X}_{\cdots}$ | $=\bar{X}$ |

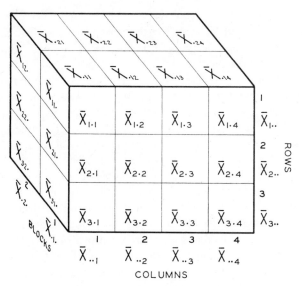

**Fig. 16.5.** Geometric picture of three-way classification.

are left denote that the mean is for scores in the second block and fourth column. The total number of means will be as follows:

$RBC$ means of the form $\bar{X}_{rbc}$ = means for cubicles

$\quad RB$ means of the form $\bar{X}_{rb\cdot}$

$\quad RC$ means of the form $\bar{X}_{r\cdot c}$

$\quad BC$ means of the form $\bar{X}_{\cdot bc}$

$\quad\ \ R$ means of the form $\bar{X}_{r\cdot\cdot}$

$\quad\ \ B$ means of the form $\bar{X}_{\cdot b\cdot}$

$\quad\ \ C$ means of the form $\bar{X}_{\cdot\cdot c}$

One mean of the form $\bar{X}_{\cdots}$ = total mean = $\bar{X}$

Perhaps a better appreciation of the meaning of all these means can be obtained by a study of Fig. 16.5, which pictures geometrically the situation for two blocks, three rows, and four columns. The individual scores can be thought of as in the cubicles of a 2 by 3 by 4 box. Summing through the box in the vertical direction leads to the 8 means on the top; summing in the forward-backward direction leads to the 12 means on the front surface; and summing through right-leftward leads to the 6 means on the side. Summing the means (or summing sums) across the front leads to the means placed along the vertical axis for the groups defined by the rows;

summing the means (or sums) downward on the front leads to the means placed along the right-left axis for the groups defined by the columns; summing down on the side leads to the means along the third axis for the groups defined by the blocks. To get any of these means it is, of course, assumed that the sum involved is divided by the proper number.

Of primary interest is the question: Is the variation among the means along the edges, considered separately, larger than expected on the basis of chance? To answer this we need to break down the sum of squares of deviations from the total mean into appropriate components. A score, $X_{rbc}$, in the cubicle defined by the $r$th row, $b$th block, and $c$th column will vary more or less from $\bar{X}$, and three possible sources of variation for $X_{rbc}$ are obvious: the deviation of its row mean, its column mean, and its block mean from $\bar{X}$. Now, if we recall the situation for double classification, it is fairly obvious that, when the score $X_{rbc}$ is considered as belonging in row $r$ and column $c$, one source of variation becomes the remainder or interaction for rows and columns; considered next as also falling in row $r$ and block $b$, another source of variation is the possible interaction of rows and blocks; and then thought of as belonging to column $c$ and block $b$, the score also involves the interaction of columns and blocks.

When we write the deviation $(X_{rbc} - \bar{X})$ as a function of these components plus a within-cubicles part we have

$$(X_{rbc} - \bar{X}) = (\bar{X}_{r..} - \bar{X}) + (\bar{X}_{.b.} - \bar{X}) + (\bar{X}_{..c} - \bar{X})$$
$$+ (\bar{X}_{rb.} - \bar{X}_{r..} - \bar{X}_{.b.} + \bar{X}) + (\bar{X}_{r.c} - \bar{X}_{r..} - \bar{X}_{..c} + \bar{X})$$
$$+ (\bar{X}_{.bc} - \bar{X}_{.b.} - \bar{X}_{..c} + \bar{X}) + (X_{rbc} - \bar{X}_{rbc})$$
$$+ (\bar{X}_{rbc} - \bar{X}_{rb.} - \bar{X}_{r.c} - \bar{X}_{.bc} + \bar{X}_{r..} + \bar{X}_{.b.} + \bar{X}_{..c} - \bar{X})$$

the last term of which has been included to establish an identity, or equation. This last term is a remainder after the other seven terms on the right (of the equality sign) have been subtracted from $(X_{rbc} - \bar{X})$. This "remainder" has to do with three-way interaction, and it obviously involves rows, blocks, and columns. The reader, having in mind the idea that the simple row by column (two-way) interaction has to do with possible failure of cell entries (means) to be consistent with the marginal means, must now try imagining that the $RBC$ means in the cubical cells of our box may not be entirely consistent with the three sets of means on the surfaces. Stated differently, significant $R \times B \times C$ interaction indicates that two-way interactions within levels on any one of the factors are more or less different from level to level. The two-way interactions are not alike for the several levels.

The total sum of squares breaks down into eight components, with all cross-product terms vanishing. Table 16.11 gives the essentials, in symbols,

**Table 16.11. Variance table for three-way classification into $R$ rows, $B$ blocks, and $C$ columns**

| Source | Sum of Squares | $df$ | Variance Estimate |
|---|---|---|---|
| Rows | $mBC\sum_{r}(\bar{X}_{r..} - \bar{X})^2$ | $R - 1$ | $s^2_r$ |
| Blocks | $mRC\sum_{b}(\bar{X}_{.b.} - \bar{X})^2$ | $B - 1$ | $s^2_b$ |
| Columns | $mRB\sum_{c}(\bar{X}_{..c} - \bar{X})^2$ | $C - 1$ | $s^2_c$ |
| $R \times B$ interaction | $mC\sum_{r}\sum_{b}(\bar{X}_{rb.} - \bar{X}_{r..} - \bar{X}_{.b.} + \bar{X})^2$ | $(R - 1)(B - 1)$ | $s^2_{rb}$ |
| $R \times C$ interaction | $mB\sum_{r}\sum_{c}(\bar{X}_{r.c} - \bar{X}_{r..} - \bar{X}_{..c} + \bar{X})^2$ | $(R - 1)(C - 1)$ | $s^2_{rc}$ |
| $B \times C$ interaction | $mR\sum_{b}\sum_{c}(\bar{X}_{.bc} - \bar{X}_{.b.} - \bar{X}_{..c} + \bar{X})^2$ | $(B - 1)(C - 1)$ | $s^2_{bc}$ |
| $R \times B \times C$ interaction | $m\sum_{r}\sum_{b}\sum_{c}(\bar{X}_{rbc} - \bar{X}_{rb.} - \bar{X}_{r.c} - \bar{X}_{.bc}$ $+ \bar{X}_{r..} + \bar{X}_{.b.} + \bar{X}_{..c} - \bar{X})^2$ | $(R - 1)(B - 1)(C - 1)$ | $s^2_{rbc}$ |
| Within | $\sum_{r}\sum_{b}\sum_{c}(X_{rbc} - \bar{X}_{rbc})^2$ | $mRBC - RBC$ | $s^2_w$ |
| Total | $\sum_{r}\sum_{b}\sum_{c}(X_{rbc} - \bar{X})^2$ | $mRBC - 1$ | |

for the analysis of variance for the triple-classification setup. In order to specify the interactions, we here adopt the abbreviation scheme generally used. Thus $R \times B$, read $R$ by $B$, indicates the row and block interaction, and $R \times B \times C$ stands for the row by block by column or three-way interaction. In a given investigation, the rows, blocks, and columns refer to particular independent or classificatory variables.

It will be noted in Table 16.11 that the $df$ for the three-way interaction term is given as $(R - 1)(B - 1)(C - 1)$. The student may be helped in understanding the reasoning which leads to this $df$ by referring again to Fig. 16.5. The surface means tend to restrict the deviation of the means within the box. How many cubical cell means can we assign before these restrictions operate? The general rule-of-thumb procedure for determining the $df$ for interaction sums of squares is to take the product of the $df$s of the variables involved in the given interaction. This holds for two-way, three-way, and higher-order interactions.

## SPECIAL CASE WHERE THE ROWS STAND FOR PERSONS OR MATCHED INDIVIDUALS, $m = 1$ PER CUBICLE

Suppose the purpose of a study is to ascertain whether variation on a dependent variable is influenced by or associated with variation on two independent variables. This, of course, involves the double classification idea previously discussed, but we are now in a position to accomplish, by means of three-way classification, two closely related things which could not be done by the simpler two-way classification scheme.

1. If transfer, practice, fatigue, etc., effects are such that it is permissible to make observations on an individual under each of the $RC$ combinations of conditions, we may increase the precision of an experiment by using only $m$ individuals instead of $mRC$ individuals, as in the illustration involving pursuit learning. Or we may make observations on $mRC$ cases so as to have in each of the $RC$ cells $m$ scores which are based on $m$ sets of matched individuals, thereby reducing error.

2. If we are dealing with a situation in which it is required that observations be made on the same individual in each of the $RC$ conditions, and if more than one case is used either to reduce errors or to provide a basis for generalizing to a population, it is necessary that we make statistical allowance for the fact that the $RC$ observations on the $m$ cases are nonindependent, or correlated. This allowance was not possible by the two-way classification scheme, for which it was assumed that the $m$ scores in one cell were independent of the observations in the other cells.

It will be recalled that in the two-way classification setup, by letting one classification refer to $R$ individuals or sets of matched cases, we were provided with an over-all test of significance for several correlated means for groups classified on a *single* independent variable. Triple classification permits a similar test of correlated means for groups involved in *double* classification.

Since the assigning of the bases of classification to rows, blocks, and columns is arbitrary, we shall let the $R$ rows stand for $R$ individuals (or $R$ matched persons), with the blocks and columns representing the independent variables to be investigated.

The task of computing the required sums of squares (see Table 16.11) is tedious. The first step is to arrange the data in some such systematic order as that depicted in Table 16.10 and do the necessary adding to secure the various sums indicated in that table. The total sum of squares for all $RBC$ cases is obtained as usual: sum all the scores, sum all the squared scores, and substitute in the general formula $(1/RBC)[RBC\Sigma X^2 - (\Sigma X)^2]$.

To secure the three between-groups and the three simple interaction sums of squares, we form three subtables involving sums taken in various directions. For the first of these subtables we take row by column sums obtained by adding cell entries from block to block, i.e., through the $B$ blocks. The next to the bottom section of Table 16.10 contains these row by column sums, which we reproduce here as Table 16.12a. The reader will note that the values for Table 16.12b are the right-hand margin sums of Table 16.10 and that the values for Table 16.12c are found as the sums in Table 16.10 along the bottom of each block.

With these auxiliary tables in mind, we can write the required computational formulas. The simple interaction terms are secured by computing

**Table 16.12a. Required sums for row by column analysis**

|     | 1 | $c$ | $C$ | Sum |
|-----|---|-----|-----|-----|
| 1   | $\sum_b X_{1b1}$ | $\sum_b X_{1bc}$ | $\sum_b X_{1bC}$ | $\sum_b \sum_c X_{1bc}$ |
| $r$ | $\sum_b X_{rb1}$ | $\sum_b X_{rbc}$ | $\sum_b X_{rbC}$ | $\sum_b \sum_c X_{rbc}$ |
| $R$ | $\sum_b X_{Rb1}$ | $\sum_b X_{Rbc}$ | $\sum_b X_{RbC}$ | $\sum_b \sum_c X_{Rbc}$ |
| Sum | $\sum_r \sum_b X_{rb1}$ | $\sum_r \sum_b X_{rbc}$ | $\sum_r \sum_b X_{rbC}$ | $\sum_r \sum_b \sum_c X_{rbc}$ |

**Table 16.12b. Required sums for row by block analysis**

|     | 1 | $b$ | $B$ | Sum |
|-----|---|-----|-----|-----|
| 1   | $\sum_c X_{11c}$ | $\sum_c X_{1bc}$ | $\sum_c X_{1Bc}$ | $\sum_b \sum_c X_{1bc}$ |
| $r$ | $\sum_c X_{r1c}$ | $\sum_c X_{rbc}$ | $\sum_c X_{rBc}$ | $\sum_b \sum_c X_{rbc}$ |
| $R$ | $\sum_c X_{R1c}$ | $\sum_c X_{Rbc}$ | $\sum_c X_{RBc}$ | $\sum_b \sum_c X_{Rbc}$ |
| Sum | $\sum_r \sum_c X_{r1c}$ | $\sum_r \sum_c X_{rbc}$ | $\sum_r \sum_c X_{rBc}$ | $\sum_r \sum_b \sum_c X_{rbc}$ |

**Table 16.12c. Required sums for block by column analysis**

|     | 1 | $c$ | $C$ | Sum |
|-----|---|-----|-----|-----|
| 1   | $\sum_r X_{r11}$ | $\sum_r X_{r1c}$ | $\sum_r X_{r1C}$ | $\sum_r \sum_c X_{r1c}$ |
| $b$ | $\sum_r X_{rb1}$ | $\sum_r X_{rbc}$ | $\sum_r X_{rbC}$ | $\sum_r \sum_c X_{rbc}$ |
| $B$ | $\sum_r X_{rB1}$ | $\sum_r X_{rBc}$ | $\sum_r X_{rBC}$ | $\sum_r \sum_c X_{rBc}$ |
| Sum | $\sum_r \sum_b X_{rb1}$ | $\sum_r \sum_b X_{rbc}$ | $\sum_r \sum_b X_{rbC}$ | $\sum_r \sum_b \sum_c X_{rbc}$ |

a subtotal sum of squares for each table and then subtracting therefrom the two appropriate "between" sums of squares. These subtotal sums of squares will not be the same as the total sum of squares obtained for double classification by formula (16.8) because we are now dealing with cell entries which are the sums of scores rather than single scores. Due allowance for this can be made by a slight change in formula (16.8). The amended formula, with notation appropriate for and specific to the three auxiliary tables, may be written as follows:

Subtotal: row by column

$$\frac{1}{RBC}\left[ RC\sum_r\sum_c\left(\sum_b X_{rbc}\right)^2 - \left(\sum_r\sum_b\sum_c X_{rbc}\right)^2\right] \qquad (16.15a)$$

Subtotal: row by block

$$\frac{1}{RBC}\left[ RB\sum_r\sum_b\left(\sum_c X_{rbc}\right)^2 - \left(\sum_r\sum_b\sum_c X_{rbc}\right)^2\right] \qquad (16.15b)$$

Subtotal: block by column

$$\frac{1}{RBC}\left[ BC\sum_b\sum_c\left(\sum_r X_{rbc}\right)^2 - \left(\sum_r\sum_b\sum_c X_{rbc}\right)^2\right] \qquad (16.15c)$$

From the right-hand margin of either Table 16.12a or 16.12b we can compute the sum of squares for

Between rows: $$\frac{1}{RBC}\left[ R\sum_r\left(\sum_b\sum_c X_{rbc}\right)^2 - \left(\sum_r\sum_b\sum_c X_{rbc}\right)^2\right] \qquad (16.15d)$$

From the bottom of either Table 16.12a or 16.12c we can obtain the sum of squares for

Between columns: $$\frac{1}{RBC}\left[ C\sum_c\left(\sum_r\sum_b X_{rbc}\right)^2 - \left(\sum_r\sum_b\sum_c X_{rbc}\right)^2\right] \qquad (16.15e)$$

From the bottom of Table 16.12b or from the right-hand margin of Table 16.12c we can calculate the sum of squares for

Between blocks: $$\frac{1}{RBC}\left[ B\sum_b\left(\sum_r\sum_c X_{rbc}\right)^2 - \left(\sum_r\sum_b\sum_c X_{rbc}\right)^2\right] \qquad (16.15f)$$

Then from the above six sums of squares the simple interaction sums of squares may be secured by the following subtractions:

Row by column interaction:   (16.15a) − (16.15d) − (16.15e)   (16.16a)
Row by block interaction:   (16.15b) − (16.15d) − (16.15f)   (16.16b)
Block by column interaction:   (16.15c) − (16.15e) − (16.15f)   (16.16c)

And finally, again by subtraction, we have the sum of squares for the row by column by block, or

Three-way interaction: Total sum of squares minus (16.15*def*)
minus (16.16*abc*).

We will illustrate the procedure by using the data of Table 16.13, in which the blocks represent two levels of illumination, the columns three

**Table 16.13. Data used in illustrating computations for three-way classification: 2 levels of illumination (blocks), 3 albedos (columns), and 4 observers (rows)\***

| Illumination | Observer | Albedo .07 | .14 | .26 | Sum | Mean |
|---|---|---|---|---|---|---|
| | 1 | 11 | 24 | 60 | 95 | 31.67 |
| | 2 | 22 | 26 | 44 | 92 | 30.67 |
| | 3 | 16 | 22 | 55 | 93 | 31.00 |
| 1.20 | 4 | 20 | 32 | 82 | 134 | 44.67 |
| | Sum | 69 | 104 | 241 | 414 | 34.50 |
| | Mean | 17.25 | 26.00 | 60.25 | 34.50 | |
| | 1 | 14 | 24 | 65 | 103 | 34.33 |
| | 2 | 27 | 36 | 47 | 110 | 36.67 |
| | 3 | 18 | 24 | 62 | 104 | 34.67 |
| 2.00 | 4 | 24 | 59 | 84 | 167 | 55.67 |
| | Sum | 83 | 143 | 258 | 484 | 40.33 |
| | Mean | 20.75 | 35.75 | 64.50 | 40.33 | |
| | 1 | 25 | 48 | 125 | 198 | 33.00 |
| | 2 | 49 | 62 | 91 | 202 | 33.67 |
| Sums through | 3 | 34 | 46 | 117 | 197 | 32.83 |
| blocks | 4 | 44 | 91 | 166 | 301 | 50.17 |
| | Sum | 152 | 247 | 499 | 898 | 37.42 |
| | 1 | 12.50 | 24.00 | 62.50 | 33.00 | |
| Means for rows | 2 | 24.50 | 31.00 | 45.50 | 33.67 | |
| by columns | 3 | 17.00 | 23.00 | 58.50 | 32.83 | |
| | 4 | 22.00 | 45.50 | 83.00 | 50.17 | |
| Column means | | 19.00 | 30.87 | 62.38 | 37.42 | |

\* Data from R. E. Taubman, *J. Exp. Psychol.*, 1945, **35**, 235–241.

degrees of albedo, and the rows four individuals, and the scores are judged whiteness. Notice that each subject made judgments under all six of the combinations of conditions. The sums given in Table 16.13 become the entries for the auxiliary computational Tables 16.14*abc*. The needed value of $\Sigma\Sigma\Sigma X_{rbc}$ is 898, and the sum of all the squared scores, $\Sigma\Sigma\Sigma X^2_{rbc}$, is 44,394. From these figures we have

$$\tfrac{1}{24}[24(44,394) - (898)^2] = 10,793.83 = \text{total sum of squares}$$

The various "between" sums can readily be obtained by adding the squares of the appropriate marginal sums of auxiliary Tables 16.14*abc*, and substituting in formulas (16.15*def*).

**Table 16.14*a*. Required sums for block by column analysis**

|  | Albedo | | | |
|---|---|---|---|---|
| Illumination | .07 | .14 | .26 | Sum |
| 1.20 | 69 | 104 | 241 | 414 |
| 2.00 | 83 | 143 | 258 | 484 |
| Sum | 152 | 247 | 499 | 898 |

**Table 16.14*b*. Required sums for row by block analysis**

|  | Individuals | | | | |
|---|---|---|---|---|---|
| Illumination | 1 | 2 | 3 | 4 | Sum |
| 1.20 | 95 | 92 | 93 | 134 | 414 |
| 2.00 | 103 | 110 | 104 | 167 | 484 |
| Sum | 198 | 202 | 197 | 301 | 898 |

**Table 16.14*c*. Required sums for row by column analysis**

|  | Albedo | | | |
|---|---|---|---|---|
| Individual | .07 | .14 | .26 | Sum |
| 1 | 25 | 48 | 125 | 198 |
| 2 | 49 | 62 | 91 | 202 |
| 3 | 34 | 46 | 117 | 197 |
| 4 | 44 | 91 | 166 | 301 |
| Sum | 152 | 247 | 499 | 898 |

For between blocks we need $(414)^2 + (484)^2 = 405,652$;
For between columns we need $(152)^2 + (247)^2 + (499)^2 = 333,114$;
For between rows we need $(198)^2 + (202)^2 + (197)^2 + (301)^2 = 209,418$.

Then we have

$\frac{1}{24}[2(405,652) - (898)^2] = 204.17$ for between-blocks sum of squares
$\frac{1}{24}[3(333,114) - (898)^2] = 8039.08$ for between-columns sum of squares
$\frac{1}{24}[4(209,418) - (898)^2] = 1302.83$ for between-rows sum of squares

In order to secure the subtotal sums of squares we add the squares of the cell entries in the auxiliary tables. For the block by column subtotal we have from Table 16.14a:

$$(69)^2 + (83)^2 + (104)^2 + (143)^2 + (241)^2 + (258)^2 = 167,560$$

Similarly for the row by block subtotal we have from Table 16.14b:

$$(95)^2 + (103)^2 + \cdots + (167)^2 = 105,508$$

and for the row by column subtotal we have from Table 16.14c:

$$(25)^2 + \cdots + (44)^2 + \cdots + (166)^2 = 87,814$$

These three sums can now be substituted into formulas (16.15abc):

$\frac{1}{24}[6(167,560) - (898)^2] = 8289.83 =$ block by column subtotal sum of squares
$\frac{1}{24}[8(105,508) - (898)^2] = 1569.17 =$ row by block subtotal sum of squares
$\frac{1}{24}[12(87,814) - (898)^2] = 10,306.83 =$ row by column subtotal sum of squares

Next we get the simple interaction sum of squares by the subtractions indicated in formulas (16.16abc):

$8289.83 - 204.17 - 8039.08 = 46.58 =$ block by column interaction
$1569.17 - 204.17 - 1302.83 = 62.17 =$ row by block interaction
$10,306.83 - 8039.08 - 1302.83 = 964.92 =$ row by column interaction

Then for the three-way interaction sum of squares we have

$10,793.83 - 204.17 - 8039.08 - 1302.83$
$- 46.58 - 62.17 - 964.92 = 174.08$

The several sums of squares, their *dfs*, and the resulting variance estimates are brought together in Table 16.15.

**Table 16.15. Analysis of variance for judged whiteness by 4 observers for 3 degrees of albedo and 2 levels of illumination**

| Source | Sum of Squares | df | Variance Estimate |
|---|---|---|---|
| Illumination | 204.17 | 1 | 204.17 |
| Albedo | 8,039.08 | 2 | 4,019.54 |
| Subjects (individual differences) | 1,302.83 | 3 | 434.28 |
| Interaction: $I \times A$ | 46.58 | 2 | 23.29 |
| Interaction: $I \times S$ | 62.17 | 3 | 20.72 |
| Interaction: $A \times S$ | 964.92 | 6 | 160.82 |
| Interaction: $I \times A \times S$ | 174.08 | 6 | 29.01 |
| Total | 10,793.83 | 23 | |

We are not yet ready to discuss the principles controlling the choice of the error term appropriate for the possible $F$s. When the models have been presented, the student may check back to see whether we have used, in the next paragraph, the correct denominator for the $F$ ratios.

For this situation, the primary interest is testing for the significance of the effects of the two manipulated variables and their interaction. For the influence of illumination level on judged whiteness we have $F = 204.17/20.72 = 9.85$, which falls near the 10.13 required for $P = .05$, and is therefore only suggestive of a real difference due to illumination. For albedo, we have $F = 4019.54/160.82 = 24.99$, which is highly significant. For the illumination by albedo interaction, $F = 23.29/29.01 = .80$, which we immediately know is not significant. The reader will have noted that the denominators for these $F$s were all different—an explanation will come later, at which time the reader will also learn why no $F$s were calculated for the row (subject) effect or for the subject by illumination and subject by albedo interactions.

Actually, the foregoing results are not to be regarded as conclusive. The data which we have used to illustrate the computations are only a part of more complete data which involved additional degrees of albedo and other levels of illumination. Partly because of space limitations and partly because it is easier to illustrate the computations when only a few rows, columns, and blocks are involved, we have ignored a part of the available data.

It should be kept in mind that this illustration is an example of the use of the three-way classification scheme as a method for making allowance for the use of correlated observations in a problem of double classification involving the influence of two variables on a third. In this special use of three-way classification, in which the rows correspond to individuals, the objective is identical with that in the earlier analysis of pursuit rotor learning (Table 16.5). The two situations are similar in that there are $m$ (or $R$) scores in each cell; they are different in that the $m$ scores in any one cell for the pursuit learning problem are independent of the $m$ scores in other cells, whereas the $R$ scores in each of the albedo-illumination cells are correlated—each person contributes a score to each cell. Both schemes permit a check on the interaction effect of the two independent variables used to classify the observations. The use of $BC$ observations on each of $R$ cases (if feasible) will yield more precise information than obtainable by having scores for $m$ individuals in each of the $BC$ cells. This is analogous to the well-known principle that experimentation in which individuals serve as their own controls tends to be more precise than that in which an independent control group is set up. A word of caution is, however, in order: we are dealing with a repeat-measure design for which possible heterogeneity of variances for an interaction seriously disrupts the $F$ test.

## $B = 2$,  $C = 2$,  ROWS $= R$  PERSONS

It is instructive to consider in some detail the situation with just two levels on each of two factors with each of $R$ subjects measured under the four combinations of conditions. With $b$ and $c$ each taking on values of 1 and 2, we may write the $B \times C$ interaction sum of squares as

$$R\underset{b\ c}{\Sigma\Sigma}(\bar{X}_{.bc} - \bar{X}_{.b.} - \bar{X}_{..c} + \bar{X})^2 = R(\bar{X}_{.11} - \bar{X}_{.1.} - \bar{X}_{..1} + \bar{X})^2$$
$$+ R(\bar{X}_{.12} - \bar{X}_{.1.} - \bar{X}_{..2} + \bar{X})^2$$
$$+ R(\bar{X}_{.21} - \bar{X}_{.2.} - \bar{X}_{..1} + \bar{X})^2$$
$$+ R(\bar{X}_{.22} - \bar{X}_{.2.} - \bar{X}_{..2} + \bar{X})^2$$

Now all of the means with less than two subscripts (more than one dot) are simple averages of other means; therefore we may rewrite each part on the right side of the equality sign in terms of the means of appropriately chosen means. If we do this for the first right-hand part, we have

$$R\left(\bar{X}_{.11} - \frac{\bar{X}_{.11} + \bar{X}_{.12}}{2} - \frac{\bar{X}_{.11} + \bar{X}_{.21}}{2} + \frac{\bar{X}_{.11} + \bar{X}_{.12} + \bar{X}_{.21} + \bar{X}_{.22}}{4}\right)^2$$

which simplifies to

$$\frac{R}{16}[(\bar{X}_{.11} - \bar{X}_{.12}) - (\bar{X}_{.21} - \bar{X}_{.22})]^2$$

By similar substitution it can easily be shown that the other three parts reduce to the same expression except for signs within the brackets, which are irrelevant because of the squaring process. (Any connection with the fact that the $df = 1$ ?) Thus the sum of the four equivalent parts becomes

$$\frac{R}{4}\left[(\bar{X}_{\cdot 11} - \bar{X}_{\cdot 12}) - (\bar{X}_{\cdot 21} - \bar{X}_{\cdot 22})\right]^2$$

which is, of course, the sum of squares for the $B \times C$ interaction. It will be noted that the interaction is a function of the difference between differences between means. The first difference holds for block 1, the second for block 2. We can, of course, reshuffle the means (yet maintain their signs) so that within the brackets we have $(\bar{X}_{\cdot 11} - \bar{X}_{\cdot 21}) - (\bar{X}_{\cdot 12} - \bar{X}_{\cdot 22})$. Now the first difference holds for column 1, the second for column 2. Regardless of which direction we take differences, we see that the interaction with one $df$ is a difference between differences.

Let us return to the $R \times B \times C$ sum of squares to see what it means for the situation involving $B = 2$ and $C = 2$, with the same $R$ subjects measured under all conditions. It will be recalled that an interaction term emerges as a sort of remainder; what is left in the deviation $(X_{rc} - \bar{X})$ in the two-way setup or $(X_{rbc} - \bar{X})$ in the three-way layout after certain specified deviations have been subtracted. On p. 335 this was shown to be the equivalent of what remains after adjusting for row and column means in the two-way situation. It will be recalled that the $R \times C$ interaction term for two-way classification did not require cell means, or $\bar{X}_{rc}$ values; a potential interactive effect can be operative even with one score per cell. Similarly, an $R \times B \times C$ interaction can be present with one score per cubicle.

We now propose to examine the $R \times B \times C$ interaction sum of squares by adjusting it for block differences, for column differences, and for the $B \times C$ interaction part. The adjustment for block differences involves adding or subtracting so as to make $\bar{X}_{\cdot 1 \cdot} = \bar{X}_{\cdot 2 \cdot} = \bar{X}$; the adjustment for column differences leads to $\bar{X}_{\cdot \cdot 1} = \bar{X}_{\cdot \cdot 2} = \bar{X}$; and the adjustment for $B \times C$ interaction implies that we have corrected the $BC$ cell means so that $\bar{X}_{\cdot 11} = \bar{X}_{\cdot 12} = \bar{X}_{\cdot 21} = \bar{X}_{\cdot 22} = \bar{X}$, or that all $\bar{X}_{\cdot bc}$ have been adjusted to the same value. Now we examine the effect of these adjustments on the interaction sum of squares,

$$\underset{r\ b\ c}{\Sigma\Sigma\Sigma}(X_{rbc} - \bar{X}_{rb\cdot} - \bar{X}_{r\cdot c} - \bar{X}_{\cdot bc} + \bar{X}_{r\cdot\cdot} + \bar{X}_{\cdot b\cdot} + \bar{X}_{\cdot\cdot c} - \bar{X})^2$$

The adjustments lead to $\bar{X}_{\cdot bc} = \bar{X}_{\cdot b\cdot} = \bar{X}_{\cdot\cdot c} = \bar{X}$, and since two of these appear with positive signs and two with negative signs, all drop out. Thus the sum of squares for the $R \times B \times C$ interaction is somewhat simplified to

$$\Sigma\Sigma\Sigma_{r\ b\ c}(X_{rbc} - \bar{X}_{rb\cdot} - \bar{X}_{r\cdot c} + \bar{X}_{r\cdot\cdot})^2 = \Sigma_r(X_{r11} - \bar{X}_{r1\cdot} - \bar{X}_{r\cdot 1} + \bar{X}_{r\cdot\cdot})^2$$

$$+ \Sigma_r(X_{r12} - \bar{X}_{r1\cdot} - \bar{X}_{r\cdot 2} + \bar{X}_{r\cdot\cdot})^2$$

$$+ \Sigma_r(X_{r21} - \bar{X}_{r2\cdot} - \bar{X}_{r\cdot 1} + \bar{X}_{r\cdot\cdot})^2$$

$$+ \Sigma_r(X_{r22} - \bar{X}_{r2\cdot} - \bar{X}_{r\cdot 2} + \bar{X}_{r\cdot\cdot})^2$$

This time the means in the four parts can be replaced by the simple averages of appropriately chosen *scores*. The first term may be written as

$$\Sigma_r\left(X_{r11} - \frac{X_{r11} + X_{r12}}{2} - \frac{X_{r11} + X_{r21}}{2} + \frac{X_{r11} + X_{r12} + X_{r21} + X_{r22}}{4}\right)^2$$

which simplifies to

$$\frac{1}{16}\Sigma_r\left[(X_{r11} - X_{r12}) - (X_{r21} - X_{r22})\right]^2$$

Similar replacing of means by scores shows that the other three terms reduce to the same value, except for signs within the brackets. Accordingly we have

$$\frac{1}{4}\Sigma_r[(X_{r11} - X_{r12}) - (X_{r21} - X_{r22})]^2$$

as the sum of squares for the $R \times B \times C$ interaction when $B = C = 2$, with each of $R$ subjects measured under all four conditions.

Obviously, $(X_{r11} - X_{r12}) = D_{r1\cdot} = $ difference score for subject $r$ under block 1 condition, and $(X_{r21} - X_{r22}) = D_{r2\cdot} = $ difference score for the $r$th subject for the block 2 condition. Therefore we have the $R \times B \times C$ interaction sum of squares as

$$\frac{1}{4}\Sigma_r(D_{r1\cdot} - D_{r2\cdot})^2 = \frac{1}{4}\Sigma_r D^2{}_{D_r} = \frac{1}{4}\Sigma_r d^2{}_{D_r}$$

or as a function of the difference between difference scores. Note that replacing $D_{D_r}$ by $d_{D_r}$ implies that $\bar{D}_{D_r} = 0$. This is true because the adjustments set $\bar{X}_{\cdot 11} = \bar{X}_{\cdot 12} = \bar{X}_{\cdot 21} = \bar{X}_{\cdot 22}$, which leads to $\bar{D}_{\cdot 1} = \bar{D}_{\cdot 2\cdot} = 0$, hence $\bar{D}_{D_r} = 0$ since a mean of differences is always equal to the difference between means.

When we divide the foregoing expression for the $R \times B \times C$ interaction sum of squares by its degrees of freedom, $R - 1$, we have $s^2{}_{rbc} = \frac{1}{4}\Sigma_r d^2{}_{D_r}/$ $(R - 1)$, which may be regarded as $1/4$ of the unbiased estimate of the variance of the difference between the difference scores, $s^2{}_{D_D}$.

Recall that a few paragraphs earlier it was shown that the $B \times C$ interaction was reducible to a function of the difference between the differences between means. Division by its $df = 1$ will yield $s^2_{bc}$. Thus we have

$$F_{bc} = \frac{s^2_{bc}}{s^2_{rbc}} = \frac{\dfrac{R}{4}\left[(\bar{X}_{.11} - \bar{X}_{.12}) - (\bar{X}_{.21} - \bar{X}_{.22})\right]^2}{\dfrac{1}{4}\sum_r d^2_{D_r}/(R-1)}$$

$$= \frac{\left[(\bar{X}_{.11} - \bar{X}_{.12}) - (\bar{X}_{.21} - \bar{X}_{.22})\right]^2}{s^2_{D_D}/R}$$

With $n_1 = 1$, this $F = t^2$. When we have $s^2_x/N$ we take it as an unbiased estimate of the sampling variance of $\bar{X}$. Hence $s^2_{D_D}$ divided by its $N$ (i.e., $R$) must also be an unbiased estimate of the sampling variance of a mean. What mean? Answer: $\bar{D}_D$, which is the mean difference of differences. We have for the $r$th individual

$$D_{D_r} = D_{r1.} - D_{r2.} = (X_{r11} - X_{r12}) - (X_{r21} - X_{r22})$$

which when summed over $r$ and divided by $R$ gives

$$\bar{D}_D = \bar{D}_{.1.} - \bar{D}_{.2.} = (\bar{X}_{.11} - \bar{X}_{.12}) - (\bar{X}_{.21} - \bar{X}_{.22})$$

Hence the above $F_{bc} = t^2$ is the equivalent of $(\bar{D}_D)^2/s^2_{\bar{D}_D}$, which tells us that the $F$ test for the $B \times C$ interaction with $s^2_{rbc}$ as denominator is a test of whether there is a significant difference between two mean differences, $\bar{D}_{.1.} - \bar{D}_{.2.}$, each of which in turn can be expressed as the difference between two means. This development should help the student see that it makes sense to use $s^2_{rbc}$ as the denominator for $F_{bc}$. This usage of an interaction to test an interaction will emerge when we discuss the expected values of the variance estimates derivable from three-way classification, mixed model.

Let us next consider further the case in which we have in each cubicle $m$ scores, which are independent of the $m$ scores in other cubicles. The total number of scores will, of course, be $mRBC$, and the breakdown of the total sum of squares will include the components specified in Table 16.11. Since each cubicle defines a group, the within-cubicles sum of squares does not differ from previously discussed "within" sums of squares. The formula in this case is

$$\frac{1}{m}[m\Sigma\Sigma\Sigma X^2_{rbc} - \Sigma(\Sigma X_{rbc})^2]$$

in which it is understood that the $\Sigma X^2$ term contains $mRBC$ squares and that the subtractive term indicates that we first sum the $m$ scores separately for each cubicle, then square each of these sums, and finally sum all these $RBC$ squared sums. The $df$ for this term will be $mRBC - RBC$ because we are dealing with the deviations of $mRBC$ scores about $RBC$ different means.

With $m$ independent scores per cubicle, the six computational formulas (16.15) need only be modified by the use of $1/mRBC$ instead of $1/RBC$ as the factor outside the brackets. It must be understood, however, that the sums within the parentheses of formulas (16.15) will involve $m$ times as many scores as for the simpler situation with one case per cubicle. The computation is again accomplished by auxiliary tables, the main cell entries of which will, of course, also involve sums with $m$ times as many scores. If we think of the orderly arrangement of the original data, as exemplified in Table 16.10, it will be seen that each cell in the separate block designations will consist of $m$ score entries; i.e., we will have $m$ scores of the type $X_{111}$ or $X_{324}$. A more precise notation would be to let $X_{irbc}$ stand for the score of the $i$th person in the $r$th row and $c$th column of the $b$th block, with $i$ taking on values of $1, 2, \cdots m$.

Except for the use of $1/mRBC$ in place of $1/RBC$ in formulas (16.15), the computation of the between and simple interaction sums of squares follows exactly the steps outlined for a single score per cubicle. The three-way interaction sum of squares is again obtained by subtraction, *but* now we must also deduct the within-cubicles sum of squares.

## CHOICE OF ERROR TERM IN THREE-WAY CLASSIFICATION

The general mathematical model for the breakdown of a score in the three-way classification setup may be written as

$$(X_{rbck} - \mu) = \alpha_r + \alpha_b + \alpha_c + \alpha_r\alpha_b + \alpha_r\alpha_c + \alpha_b\alpha_c + \alpha_r\alpha_b\alpha_c + e_{rbck}$$

in which the subscripts, $r$, $b$, and $c$ refer to rows, blocks and columns, and $k$ takes on values $1 \cdots m$, there being $m$ independent replications (either of measurement or of individuals) in each cell. The mean value of each term on the right of the equality sign is zero; that is, all values are expressed in deviation units. Note the manner in which the interactive effects are designated—$\alpha_r\alpha_b$ is to be read as row by block interaction. Using notation like that employed in specifying equations (16.12–16.14) from equation (16.11) for two-way classification, we may replace the alphas by $A$s to represent fixed values (fixed constants model) and by $a$s for classifications involving samplings (random model). The mixed model would, of course, contain one lower-case and two capital letters or two lower-case and one capital.

Rather than rewrite the model equation with Latin letters specifying the particular models, we can indicate the models by the following symbols:

$[A_r A_b A_c]$ for fixed constants model
$[a_r a_b a_c]$ for the random model
$[a_r A_b A_c]$ and $[a_r a_b A_c]$ for mixed models

It is assumed that the $a_r$, $a_b$, $a_c$, $a_r a_b$, $a_r a_c$, $a_b a_c$, $a_r A_b$, $a_r A_c$, $a_b A_c$, $a_r a_b a_c$, $a_r a_b A_c$, $a_r A_b A_c$, and $e_{rbck}$ are random variates from normally distributed populations of effects having the respective variances: $\sigma^2_r$, $\sigma^2_b$, $\sigma^2_c$, $\sigma^2_{rb}$ $\sigma^2_{rc}$, $\sigma^2_{bc}$, $\sigma^2_{rB}$, $\sigma^2_{rC}$, $\sigma^2_{bC}$, $\sigma^2_{rbc}$, $\sigma^2_{rbC}$, $\sigma^2_{rBC}$, and $\sigma^2_e$ when $k = 1 \cdots m$ represents measurement replication or $\sigma^2_i$ when $k = 1 \cdots m$ involves replication of individuals. There is seldom, if ever, an opportunity to check on the normality of the several interactive effects—a fact which may be disturbing to the reader. No such assumptions are made regarding the effects $A_r$, $A_b$, $A_c$, $A_r A_b$, $A_r A_c$, $A_b A_c$, and $A_r A_b A_c$, which are associated with the fixed constants. Since all effects are expressed in terms of deviation units, the sum of each particular set of effects, such as $a_r$ or $A_r$ or $a_r A_b$ or $A_b A_c$, is zero; that is, e.g., $\sum_r \sum_b a_r A_b = 0$.

In order to choose the appropriate variance estimate for the denominator of $F$ for a given significance test, we again need to indicate just what each possible variance estimate ($s^2$) estimates under nonnull conditions. A summary statement will be given later regarding the assumption of homogeneity of variance for the several cases involving three-way classification.

*Case IX.* Fixed constants model $[A_r A_b A_c]$, with $m$ different individuals in each of the *RBC* cubicles. This is a simple, straightforward case in which $s^2_w \rightarrow \sigma^2_i$, and *all* the other seven $s^2$ values are estimates of $\sigma^2_i$ *plus* a single (possible) effect, the one to be tested. Examples:

$$s^2_r \rightarrow \sigma^2_i + \frac{mBC}{R-1} \Sigma_r A^2_r$$

and

$$s^2_{rb} \rightarrow \sigma^2_i + \frac{mC}{(R-1)(B-1)} \Sigma\Sigma_{rb} (A_r A_b)^2$$

Thus $s^2_w$ is the proper error term for testing all three main effects, all three two-way interactions, and the three-way interaction. Generalizations are to the population(s) from which the *mRBC* persons were drawn, but conclusions regarding main effects of the factors will need to be qualified in case a given factor is involved in a significant interaction.

A subcase under Case IX in which $m = 1$ will not provide the needed $s^2_w$ as the error term, hence is not a fruitful plan unless an estimate of $\sigma^2_i$ is possible under the assumption that an interaction is zero, but such an assumption in psychological research is hazardous.

*Case X.* Fixed constants model, one person per cubicle but each person measured $m$ times. This leads to an $s^2_w$ which is an estimate of $\sigma^2_e$ rather than the needed estimate of $\sigma^2_i$. Now it might be thought that this $s^2_w$ could be used to test $s^2_{rbc}$ for the presence of three-way interaction, but note that since

$$s^2_{rbc} \rightarrow \sigma^2_i + \frac{m}{(R-1)(B-1)(C-1)} \sum_r \sum_b \sum_c (A_r A_b A_c)^2$$

and $s^2_w \rightarrow \sigma^2_e$, the division of $s^2_{rbc}$ by $s^2_w$ leads to a noninterpretable $F$ (if significant) because there is no way of knowing whether the significance is due to individual differences (remember that $\sigma^2_i$ contains an error of measurement part) or to three-way interaction. Stated differently, the $s^2_{rbc}$ is an estimate in which error of measurement variance, true individual difference variance, and possible three-way interaction effects are all *confounded*, a term used to indicate that a given setup does not allow a disentangling of the sources of variation which enter into a particular estimate.

*Case XI.* Fixed constants model, with only one person supplying all scores, i.e., a score (or scores) under each of the $RBC$ combinations of conditions. If we have $m$ measures on the one person under each of the $RBC$ conditions, $s^2_w \rightarrow \sigma^2_e$ and each of the other seven variance estimates has an expected value including $\sigma^2_e$ plus an effect. A significant $F$ with $s^2_w$ as the error term permits only the conclusion that repetition of the experiment on this same person would be expected to yield similar results—a "generalization" which has no generality, and hence is worthless.

*Case XII.* Mixed model $[a_r A_b A_c]$. Typically, this will involve $R$ individuals assigned to the rows with each measured at least once under the $BC$ conditions. We have (with no measurement replication):

$$s^2_r \rightarrow \sigma^2_e + BC\sigma^2_r$$

$$s^2_b \rightarrow \sigma^2_e + C\sigma^2_{rB} + \frac{RC}{B-1} \sum_b A^2_b$$

$$s^2_c \rightarrow \sigma^2_e + B\sigma^2_{rC} + \frac{RB}{C-1} \sum_c A^2_c$$

$$s^2_{rb} \rightarrow \sigma^2_e + C\sigma^2_{rB}$$

$$s^2_{rc} \rightarrow \sigma^2_e + B\sigma^2_{rC}$$

$$s^2_{bc} \rightarrow \sigma^2_e + \sigma^2_{rBC} + \frac{R}{(B-1)(C-1)} \sum_b \sum_c (A_b A_c)^2$$

$$s^2_{rbc} \rightarrow \sigma^2_e + \sigma^2_{rBC}$$

Scrutiny of the foregoing expected values indicates that $s^2_{rbc}$ is appropriate for testing the $B \times C$ interaction, that $s^2_c$ should be tested against

$s^2_{rc}$ and $s^2_b$ against $s^2_{rb}$. No test for $s^2_r$ is possible, but this is not serious since it would only be a test of the significance of individual differences. Nor is there a test for $s^2_{rb}$ and $s^2_{rc}$, the two interaction terms having to do with individual differences in reaction to the defined experimental conditions. We would need an estimate of $\sigma^2_e$ for this purpose; ordinarily, such individual by condition interactions are real.

*Case XIII.* Mixed model $[a_r a_b A_c]$, with one score per cell. Researches calling for this model in psychology are not plentiful. Suppose $R$ children are observed under $C$ different social conditions by $B$ observers, each of whom rates (on a 10-point scale) each child in each of the situations for a particular aspect of behavior, e.g., social participation. Primary interest would be in the effect of the conditions (the $A_c$ effects) with secondary interest in observer bias (the raters being regarded as a sample of observers having $a_b$ "effects") and possible interest in two-way interaction effects. For model $[a_r a_b A_c]$ the meaning of the several variance estimates is as follows:

$$s^2_r \to \sigma^2_e + C\sigma^2_{rb} + BC\sigma^2_r$$

$$s^2_b \to \sigma^2_e + C\sigma^2_{rb} + RC\sigma^2_b$$

$$s^2_c \to \sigma^2_e + \sigma^2_{rbC} + B\sigma^2_{rC} + R\sigma^2_{bC} + \frac{RB}{C-1}\sum_c A^2_c$$

$$s^2_{rb} \to \sigma^2_e + C\sigma^2_{rb}$$

$$s^2_{rc} \to \sigma^2_e + \sigma^2_{rbC} + B\sigma^2_{rC}$$

$$s^2_{bc} \to \sigma^2_e + \sigma^2_{rbC} + R\sigma^2_{bC}$$

$$s^2_{rbc} \to \sigma^2_e + \sigma^2_{rbC}$$

When we examine the foregoing expected values, we see that both $s^2_{rc}$ and $s^2_{bc}$ are testable against $s^2_{rbc}$ as the denominator for the $F$s and that $s^2_r$ and $s^2_b$ can be tested against $s^2_{rb}$, but $s^2_{rb}$ itself is not testable. The great difficulty is that the main effect of primary interest, the $A_c$ effect, is not amenable to test unless we can assume either $\sigma^2_{rC}$ or $\sigma^2_{bC}$ (or both) to be zero. If $\sigma^2_{rC}$ were zero, $s^2_c$ could be tested against $s^2_{bc}$; if $\sigma^2_{bC}$ were zero, we could use $s^2_{rc}$ to test $s^2_c$; if both were zero, we could use $s^2_{rbc}$ for testing the main effect variance, $s^2_c$. We can scarcely make a priori the assumption that either of these two two-way interactions is zero; in fact, the safest presumption is that neither is zero. It is frequently asserted that the failure of a two-way interaction to be significant when tested against *its* appropriate error term can be used to justify the assumption of zero interaction, but failure to be significant means only that it could be zero. Furthermore, if $R$ and $B$ are small, a sizable interaction can go undetected. This issue, along with a similar one, will be discussed later under the heading "Preliminary tests and pooling." Suffice it to say that model $[a_r a_b A_c]$ is not recommended.

(For the situation involving $R$ children and $B$ observers with main interest in the effect of the $C$ conditions, we can simply sum or average the $B$ possible ratings for each child for each of the $C$ conditions, then use these sums or averages as the scores in a two-way mixed model setup with $R$ rows and $C$ columns, which leads to a straightforward test of the effect of the $C$ experimental conditions.)

*Case XIV.* Random model $[a_r a_b a_c]$. If anyone should find a situation in which all three bases of classification involve sampling, he will need to know that the two-way interactions can be tested against $s^2_{rbc}$ but that there is no way of testing the main effects without making untenable assumptions regarding two-way interactions. This sad state of affairs is not too sad simply because experimentation involving the random model, three-way classification, is hard to come by.

*Case XV.* Mixed model $[a_r A_b A_c]$, *but* a pseudo three-way classification. Suppose a sample of $R$ individuals in block 1, a sample of $R$ different individuals in block 2, and so on. The $B$ blocks represent $B$ experimental conditions, or $B$ levels for a factor, the effects of which are to be determined, and at the same time the $C$ columns stand for another factor which is also to be evaluated. The $B$ sets of $R$ individuals are used because it is not feasible to use each person under each block condition. Or suppose the blocks stand for different groups (say, diagnostic) from each of which $R$ cases are drawn at random. We wish to compare the groups and also the $C$ conditions and perhaps the $B \times C$ interaction. This setup is often referred to as the "split-plot design," the plot concept coming from agricultural experimentation. More recently, this design is said to involve "nesting"—one group of $R$ persons are nested in one block, another set of $R$ persons are nested in a second block, and so on, with never a move from nest to nest.

Let us re-examine Table 16.10 in order to determine how to set up the model for this situation. We first note that for Case XII the variation among the row means $(\bar{X}_r..)$ contributes to $s^2_r$ as an estimate of individual difference variation, whereas for Case XV each of these row means is an average for $B$ different individuals; hence row means do not hold for individuals. We do, however, have individual difference variation within each block, as represented by means of the type $\bar{X}_{rb}.$ (right-hand part of Table 16.10). Accordingly, we can anticipate a sum of squares for individual differences which will involve combining the sums of squares within each block; i.e., $C\sum_r\sum_b(\bar{X}_{rb}. - \bar{X}._b.)^2$, with $RB - B$ degrees of freedom. The resulting variance estimate may be labeled $s^2_i$, for individual differences.

In ordinary three-way classification (Case XII) the $B$ sets of means of the type $\bar{X}_{rb.}$ have to do with row (individual) by block interaction, an interaction which reflects the failure of the individuals to maintain similar score positions from block to block. But with independent cases in each block, no block by row interaction is possible; a person cannot react differently from one block to another unless he has been measured under more than one block condition. Consider next the $\bar{X}_{r.c}$ type of mean at the bottom of Table 16.10. These means ordinarily enter into row by column interaction, but in the present case each of these means is the average for $B$ different individuals who just happened to have been assigned the same row number. Therefore, there can be no row by column interaction in the usual sense. We have, nevertheless, $RB$ independent individuals in a total of $RB$ (instead of $R$) rows; hence there could be a meaningful individual by column interactive effect (not testable with one score per cell, but present as a source of variation).

What of a possible three-way interaction involving rows, blocks, and columns? This does not make sense since an individual can in no way react inconsistently from one block condition to another without having been subjected to different block conditions.

With the foregoing in mind, we may write the following specific model for Case XV:

$$(X_{rbc} - \mu) = a_i + A_b + A_c + A_bA_c + h_{rbc}$$

in which $a_i$ indicates individual difference effects and $h_{rbc}$ is the remainder after the first four parts have been subtracted from $(X_{rbc} - \mu)$. The several sums of squares and their $dfs$ are given in Table 16.16. Note how the first line differs from the first line of Table 16.11; note also the similarity of the remainder sum of squares to the remainder (or last) term in equation (16.5) and to the row by column interaction term in Table 16.2. Actually, the remainder in Table 16.16 involves possible individual by column interaction, composed of ordinary row by column interaction within each block, then summed over blocks.

The expected values of the several variance estimates are as follows (recall that $\sigma^2_i$ contains $\sigma^2_e$ as a component):

$$s^2_i \rightarrow \sigma^2_i$$

$$s^2_b \rightarrow \sigma^2_i + \frac{RC}{B-1} \sum_b A^2_b$$

$$s^2_c \rightarrow \sigma^2_e + \sigma^2_{rC} + \frac{RB}{C-1} \sum_c A^2_c$$

$$s^2_{bc} \rightarrow \sigma^2_e + \sigma^2_{rC} + \frac{R}{(B-1)(C-1)} \sum_b \sum_c (A_bA_c)^2$$

$$s^2_h \rightarrow \sigma^2_e + \sigma^2_{rC}$$

**Table 16.16. Modification of variance Table 16.11 for case XV: $R$ different and independent individuals in each block**

| Source | Sum of Squares | $df$ | Variance Estimate |
|---|---|---|---|
| Individuals* | $C \sum_{r} \sum_{b} (\overline{X}_{rb.} - \overline{X}_{.b.})^2$ | $RB - B$ | $s^2_i$ |
| Blocks | $RC \sum_{b} (\overline{X}_{.b.} - \overline{X})^2$ | $B - 1$ | $s^2_b$ |
| Columns | $RB \sum_{c} (\overline{X}_{..c} - \overline{X})^2$ | $C - 1$ | $s^2_c$ |
| $B \times C$ interaction | $R \sum_{b} \sum_{c} (\overline{X}_{.bc} - \overline{X}_{.b.} - \overline{X}_{..c} + \overline{X})^2$ | $(B - 1)(C - 1)$ | $s^2_{bc}$ |
| Remainder | $\sum_{r} \sum_{b} \sum_{c} (X_{rbc} - \overline{X}_{rb.} - \overline{X}_{.bc} + \overline{X}_{.b.})^2$ | $B(R - 1)(C - 1)$ | $s^2_h$ |
| Total | $\sum_{r} \sum_{b} \sum_{c} (X_{rbc} - \overline{X})^2$ | $RBC - 1$ | |

* The sum of squares for individuals is computed by substituting in

$$\frac{1}{RC} [R \sum_{r} \sum_{b} (\sum_{c} X_{rbc})^2 - \sum_{b} (\sum_{r} \sum_{c} X_{rbc})^2].$$

From these values we see at a glance that $s^2_i$ is the error term for testing $s^2_b$, a test which is analogous to $s^2_b / s^2_w$ in the one-way classification setup for the difference between means of independent groups. For testing $s^2_c$ the remainder estimate, $s^2_h$, is appropriate. Since $s^2_h$ is, in part, an estimate of individual by column interaction, we find an analogue in the two-factor setup (Case VI) for which the row by column interaction provides the correct variance estimate for testing column effects when the column means are correlated (based on the same or related or matched individuals).

The remainder variance estimate is also appropriate for testing the $B \times C$ interaction. Note that this interaction involves $C$ means in each block that are independent of the $C$ means in every other block but at the same time the $C$ means within each block are not independent of each other. This interaction has a special meaning when $B$ stands for different groups and $C$ stands for $C$ tests all scored in comparable standard score form. The column means for each block are the basis for a given group's profile; hence a test of the $B \times C$ interaction tells us whether there are significant differences among the profiles for the $B$ groups.

Caution: Case XV as here outlined calls for the same number of individuals per block (or group).

**Assumption of homogeneity of variance.** Cases IX, X, and XI require similar variances for all cubicles, but only Case IX permits a test of the assumption. For Cases XI, XII, XIII, and XIV it is assumed that error of measurement variance is the same from cubicle to cubicle. The assumption for these cases is not testable unless we have measurement replication with $m$ scores per cubicle. Case XV assumes that the row variance within blocks is homogeneous from block to block when $s^2_i$ is used to test $s^2_b$, and, again, we must recall that for repeat-measure designs the $F$ test may be seriously disturbed (see p. 357). For Case XV, the disruption affects the $F$ test for columns and $R$ by $C$ interaction, not for blocks. For the test of differences between group profiles by way of $B \times C$ interaction, one source for heterogenity of variances for the person by test interaction is not present because the use of standard scores tends to equalize the variances within the columns in each block (group). Since profiles based on personality scales (such as the MMPI) involve tests with markedly diverse intercorrelations (far more so than in the Collier et al. study), the $F$ test for differences in group profiles will lead to too many "significant" values. To guard against the type I error, it is definitely advisable to require $F$ to reach the .01 level in order to safely claim significance at the .05 level.

## PRELIMINARY TESTS AND POOLING

When we discussed Cases XIII and XIV, we found that certain effects could not be tested without assuming that an interaction is zero. The temptation is to assume an interaction is zero if it fails to be significant when tested against an appropriate error term. The writers of textbooks on mathematical statistics are remarkably mum on this point, presumably because the situation gets too "iffy": a main effect is significant *if* it reaches, say, the .05 level, and *if* a certain interaction was not significant at a specified level. Under such circumstances a $P$ for an effect ceases to have the same meaning as when unencumbered by conditional probabilities.

Note that preliminary tests may have to do with the assumption of zero interaction in the numerator term of $F$ (as for Case XIII) or in the denominator term (as for Cases II and X). Failure to satisfy the assumption of a zero interaction in the numerator will lead to too many "significant" $F$s. Stated differently, significance for a main effect cannot be safely claimed because the numerator involves a possible confounding of interactive and main effects. Failure to satisfy an assumption of zero interaction in the denominator term will lead to too few significant $F$s, which means that an obtained $F$ possesses greater significance than its $P$ indicates.

Preliminary tests are also used in connection with the "pooling" of sums of squares and of their $df$s. To understand the meaning of pooling, let us consider Case IX in which all effects are testable against $s^2_w$. The advocated steps are: first, $s^2_{rbc}$ is tested against $s^2_w$. If this $F$ is not significant at, say, the .05 level, the sum of squares for the three-way interaction term is combined with that of $s^2_w$, with the $df$s also being summed. Dividing the pooled sum by the pooled $df$ gives another estimate of variance for the error term. This estimate is next used to test the two-way interactions, which if insignificant provide additional sums of squares and $df$s for adding to the pool already made up.

The claimed advantage of pooling is that the number of degrees of freedom for the denominator, or error, term of $F$ is thereby increased, with a resultant more stable estimate of variance. But whether this procedure provides an improved or better estimate depends, of course, on whether the interactions judged to be insignificant are really zero in the sampled population. Actually, the $F$ based on the pooled values may be either larger or smaller than the $F$ based on the appropriate variance estimate obtained without pooling. When we examine the $F$ table, we see that the gain in $df$ does not have an appreciable effect, in the sense that a smaller $F$ is required for significance, except when $n_2$ is very small, say less than 8 or 10. It should be clearly noted that the gain in $df$ by pooling does not lead to a reduction in the sampling errors of the means being tested.

The use of preliminary tests as a basis for pooling is not nearly so defensible as textbooks written prior to 1951 would have us believe. The work of Paull† indicates that the usually advocated rule (that when $F$ is less than the value required for the .05 level, pooling is permissible and advisable) is far from satisfactory. He sets up an elaborate set of rules leading to the decision "never pool" or "sometimes pool" or "always pool." Space does not permit an exposition of his rules here. A simple rule to follow when the $df$s are equal, or when unequal provided both are greater than 6, is to pool only when $F$ is less than 2. Even when we follow the rules, $F$s based on pooling do not lead to $P$s of precisely the same meaning as $P$s obtained from $F$s which do not involve pooling.

## FACTORIAL AND LATIN SQUARE DESIGNS

The student who encounters the term "factorial design" will need to know that it is difficult to make a distinction between factorial design and the analysis of variance setups discussed in this chapter. The bases for classification are referred to as factors; the categories within a classification are termed "levels." Perhaps the term factorial design is inappropriate when one basis for classification is persons.

The Latin square design had its origins in agricultural experimentation. If $T$ different treatments (fertilizers) are to be evaluated, a plot of land is laid off into $T$ rows and $T$ columns and the treatments are so assigned that

† Paull, A. E., On a preliminary test for pooling mean squares in the analysis of variance, *Annals math. Stat.*, 1950, **21**, 539–556.

each treatment occurs only once in each row and only once in each column. With Latin letters standing for the treatments, there might be the accompanying square, an examination of which reveals that this is a scheme for

Columns

|      |     | 1 | 2 | 3 | 4 |
|------|-----|---|---|---|---|
|      | I   | A | D | B | C |
|      | II  | B | A | C | D |
| Rows | III | C | B | D | A |
|      | IV  | D | C | A | B |

balancing out the effects of possible fertility differentials from row to row and also from column to column.

Some researchers in psychology have used the Latin square principle as a way of balancing the effect of individual differences and order of testing. That is, with $T$ conditions to be evaluated, the rows stand for $T$ individuals and the columns for $T$ orders of testing, with Latin letters representing the $T$ conditions. The design also can be and has been used in lieu of a complete three-way factorial design when all three factors involve the same number of levels. For example, sixteen properly arranged observations may be used instead of the sixty-four observations required for a complete three-way classification plan with four levels per classification. This second use of the Latin square principle is not for the purpose of balancing out the effect of a factor but rather for evaluating the effect of factors which are deliberately varied.

Thus, it would seem that the Latin square design might be very useful in psychology, but before we accept it uncritically (as some advocates have), we need to examine the underlying mathematical model, which may be written as

$$(X_{rct} - \mu) = \alpha_r + \alpha_c + \alpha_t + f_{rct}$$

The $\alpha$s refer to row, column, and treatment effects, and $f_{rct}$ is a remainder, or residual. It follows from the model that the breakdown of the total sum of squares and degrees of freedom will lead to sums of squares for rows, for columns, and for treatments, each with $T - 1$ degrees of freedom. These sums of squares will use up $3T - 3$ of the total $df$, $T^2 - 1$; hence there remain $T^2 - 3T + 2$ degrees of freedom for the residual sum of squares, which provides the error term for testing $s^2_r$, $s^2_c$, and $s^2_t$.

When the foregoing model is compared with that for the complete three-way classification, we see a marked difference: the absence of interaction terms. *For the Latin square design it is assumed that all interactions are zero.* This assumption is necessary for three (not necessarily independent) reasons: (1) there are not enough degrees of freedom available for taking out possible interactions, (2) the main effects are confounded with interactions, and (3) the residual simply does not provide an error term appropriate for testing any of the three main effects. These considerations can be made more explicit by examining the expected values of the several variance estimates.

For the fixed effects model $[A_r A_t A_c]$, which may more aptly be specified as $[A_r A_b A_c]$ since there are $T$ levels for each of three factors rather than just $T$ "treatments" for the one factor designated by blocks, the expected mean squares are (aside from a common $\sigma^2_e$):

$$s^2_r \rightarrow \sigma^2_i + \left(1 - \frac{2}{T}\right)\frac{\underset{r\,b\,c}{\Sigma\Sigma\Sigma}(A_r A_b A_c)^2}{(T-1)^3} + \frac{\underset{b\,c}{\Sigma\Sigma}(A_b A_c)^2}{(T-1)^2} + \frac{T}{T-1}\underset{r}{\Sigma}A^2_r$$

$$s^2_b \rightarrow \sigma^2_i + \left(1 - \frac{2}{T}\right)\frac{\underset{r\,b\,c}{\Sigma\Sigma\Sigma}(A_r A_b A_c)^2}{(T-1)^3} + \frac{\underset{r\,c}{\Sigma\Sigma}(A_r A_c)^2}{(T-1)^2} + \frac{T}{T-1}\underset{b}{\Sigma}A^2_b$$

$$s^2_c \rightarrow \sigma^2_i + \left(1 - \frac{2}{T}\right)\frac{\underset{r\,b\,c}{\Sigma\Sigma\Sigma}(A_r A_b A_c)^2}{(T-1)^3} + \frac{\underset{r\,b}{\Sigma\Sigma}(A_r A_b)^2}{(T-1)^2} + \frac{T}{T-1}\underset{c}{\Sigma}A^2_c$$

$$s^2_{res} \rightarrow \sigma^2_i + \left(1 - \frac{3}{T}\right)\frac{\underset{r\,b\,c}{\Sigma\Sigma\Sigma}(A_r A_b A_c)^2}{(T-1)^3}$$
$$+ \frac{\underset{r\,b}{\Sigma\Sigma}(A_r A_b)^2}{(T-1)^2} + \frac{\underset{r\,c}{\Sigma\Sigma}(A_r A_c)^2}{(T-1)^2} + \frac{\underset{b\,c}{\Sigma\Sigma}(A_b A_c)^2}{(T-1)^2}$$

These expected values are for the Latin square design used in lieu of a complete three-way layout, fixed effects model, where the interest is in testing all the main effects, that is, the effect of the factor assigned to rows, of that assigned to columns, and of that assigned to blocks. We see immediately that the possible presence of interactions snarls the obtaining of a valid $F$ ratio for any of the three main effects. This sad state of affairs can, of course, be avoided by using the regular three-way classification design—more work to be sure, but the rewards are twofold: main effects are readily testable and interactions also can be extracted and tested.

Table 16.17 has been prepared for the reader who is puzzled by the manner in which main effects are confounded with interaction effects in the foregoing expected values and who is also curious as to how to proceed to set up a Latin square in lieu of a complete three-way fixed constants design. In this table we presume, for purposes of illustration, that all scores in blocks $A$, $B$, and $C$ are population means. All row means, all column means, and all block means are equal to 4; that is, there are no main effects at all. In each block there is the same row by column

### Table 16.17. A Latin square generated from a three-way layout

| Blocks | A | | | B | | | C | | | Square | | |
|---|---|---|---|---|---|---|---|---|---|---|---|---|
| | 1 | 2 | 3 | 1 | 2 | 3 | 1 | 2 | 3 | 1 | 2 | 3 |
| I | 4 | 4 | 4 | 4 | 4 | 4 | 4 | 4 | 4 | A4 | B4 | C4 |
| II | 2 | 4 | 6 | 2 | 4 | 6 | 2 | 4 | 6 | B2 | C4 | A6 |
| III | 6 | 4 | 2 | 6 | 4 | 2 | 6 | 4 | 2 | C6 | A4 | B2 |

interaction; thus summing through blocks and dividing by 3 will yield means that will show row by column interaction, but since this interaction is exactly the same within each block there is no three-way interaction.

The boldface numerals in the three blocks are the "scores" for the 3 × 3 Latin square to the right. Each of these boldface values enters the Latin square with its row and column designation intact and with its block source designated by $A$ or $B$ or $C$. For the Latin square so generated, it will be seen that the row means are all 4; ditto, the column means. But for the block effect we have from the Latin square the following means:

$$\bar{X}_A = (4 + 6 + 4)/3 = 4.67$$
$$\bar{X}_B = (4 + 2 + 2)/3 = 2.67$$
$$\bar{X}_C = (4 + 4 + 6)/3 = 4.67$$

which are illusory as indications of a main effect because the effect was produced by the row by column interaction—no block differences held for the starting three-way situation. Compare this outcome with the expected value for $s^2_b$ and note that the row by column interaction is not involved in the expected values for $s^2_r$ and $s^2_c$.

For the second, more common use in psychology of the Latin square, with rows standing for persons (animals), columns for order or sequence in testing, and Latin letters for experimental conditions (treatments), we have a mixed model $[a_r A_t A_c]$ with rows as random variates. The expected mean squares are (again omitting the common $\sigma^2_e$):

$$s^2_r \rightarrow \left(1 - \frac{1}{T}\right)\sigma^2_{rTC} + \frac{\underset{t\,c}{\Sigma\Sigma}(A_t A_c)^2}{(T-1)^2} + T\sigma^2_r$$

$$s^2_t \rightarrow \left(1 - \frac{2}{T}\right)\sigma^2_{rTC} + \sigma^2_{rT} + \sigma^2_{rC} + \frac{T}{T-1}\underset{t}{\Sigma}A^2_t$$

$$s^2_c \rightarrow \left(1 - \frac{2}{T}\right)\sigma^2_{rTC} + \sigma^2_{rT} + \sigma^2_{rC} + \frac{T}{T-1}\underset{c}{\Sigma}A^2_c$$

$$s^2_{res} \rightarrow \left(1 - \frac{2}{T}\right)\sigma^2_{rTC} + \sigma^2_{rT} + \sigma^2_{rC} + \frac{\underset{t\,c}{\Sigma\Sigma}(A_t A_c)^2}{(T-1)^2}$$

The primary interest is in testing $s^2{}_t$, but we see no suitable error term unless it can be assumed that the order by treatment interaction is zero. Such an assumption is equivalent to saying that the influence of the order $A$, $D$, $B$, $C$ (see Latin square on p. 384) is the same as the influence of the order $B$, $A$, $C$, $D$; and so on. Whatever the order effect, whether it be practice, fatigue, boredom, something physiological, change in mental set, etc., it must be assumed that any such effects or combination thereof are independent of particular treatments. If, for example, treatments were various drugs, differences in residual effects would lead to order by treatment interaction.

The reader will have noted that when $F$ is taken as $s^2{}_t/s^2{}_{res}$, the presence of order by treatment interaction will mitigate against getting a significant $F$, and that if $F$ reaches the $\alpha$ level of significance he can claim significance at better than the $\alpha$ level, though how much better remains unknown. The reader will have also noted that for a (typically) small number of treatments, a single Latin square design uses so few cases ($T$ in number) that sampling errors will tend to be very large. The advantages of larger $N$ can be attained by replication—additional sets of $T$ persons provide additional Latin squares, for a discussion of which the reader is referred to Cochran and Cox.‡ And, finally, the reader may *not* have noted that the presence, in the expectations for both $s^2{}_t$ and $s^2{}_{res}$, of the three interactions involving rows indicates that the $A^2{}_t$ component must be relatively sizable in order to lead to an appreciable $F$.

Concerning the merits of the Latin square design in psychological research, there has been a difference of opinion attributable in part to the until recently unsettled question regarding the expected values of the variance estimates when interactions are present. Now that the expectations are known, it is seen that the substitution of a Latin square design in lieu of a fixed constants factorial design has disadvantages that far outweigh its only advantage, i.e., the making of observations on fewer individuals. But the use of the Latin square design as a method of balancing sequence effects and also as a method for using repeated observations on the same individuals (individual differences are extracted as a row effect, which is the gain from having correlated treatment means) has an appeal that must be evaluated against the worth of less "iffy" designs such as (1) random assignments of $m$ individuals to each of the $T$ treatment (or experimental) conditions or (2) the use of matched cases with matching on the basis of some relevant variable(s) or on the basis of pretest measures of the dependent variable under consideration.

‡ Cochran, W. G. and Cox, G. M. *Experimental designs*, 2nd ed., New York: John Wiley, 1957.

## SELECTED COMPARISONS

As in one-way classification, when $F$ indicates that a main effect is significant, we may proceed to test specific contrasts among the means. The reasons for doing this and the general procedures are the same as those set forth in the last section of Chapter 15, which the reader should review at this time. We limit the discussion here to the two-way classification setup, fixed effects model with equal $ms$ in the cells and the mixed model. For the column means (or the row means), we could have a $D$ or a $D'$ computed exactly as before for the case of equal $ms$.

In the fixed-effects situation, the needed standard error for a contrast of column means is given by the square root of

$$s^2{}_{D'} = \frac{s^2{}_w}{mR}\left(\frac{1}{a} + \frac{1}{b}\right)$$

in which $s^2{}_w$ is the within-cells variance estimate and $a$ and $b$ are the number of means being averaged for a contrast. Again, when $a = b = 1$, we have the error for a $D$-type contrast. The significance of a contrast springing from an a priori hypothesis can be ascertained from $t = D/s_D$ (or $t = D'/s_{D'}$), with $df = mRC - RC$. A contrast of the data-snooping variety will be judged significant at the $\alpha$ level if $D/s_D$ (or $D'/s_{D'}$) reaches $K$ where $K$ is now defined as the square root of the product of $(C - 1)$ times the $F$ required for the $\alpha$ level of significance for $n_1 = C - 1$ and $n_2 = mRC - RC$ degrees of freedom. For comparisons involving row means, $R$ and $C$ are simply interchanged.

For the mixed model with $C$ means based on the same $R$ persons (or $R$ sets of matched individuals), a contrast of the $D$ type will have $s_D = s_{rc}\sqrt{2/R}$. Given an a priori hypothesis, we have $t = D/s_D$ with $(R - 1)(C - 1)$ degrees of freedom whereas for a contrast suggested by an examination of the data, $D/s_D$ must reach $K$ which this time is the square root of the product of $(C - 1)$ times the $F$ required for $\alpha$ with $n_1 = C - 1$ and $n_2 = (R - 1)(C - 1)$ degrees of freedom.

It should be noted that for the mixed model situation neither the procedure involving $t$ nor that involving $K$ makes any allowance for the possibility that the correlation between the scores in the columns involved in a particular contrast may differ from the average of the $C(C - 1)/2$ intercorrelations entering into $s^2{}_{rc}$. The value of $t$ could, of course, be calculated independently of the over-all row by column interaction, but it is not clear whether the Scheffé method permits this alteration.

Apparently neither the $t$ approach nor the Scheffé method is applicable for contrast of the $D'$ type in the mixed model, but there appears to be little need for $D'$ comparisons in the mixed model situation.

# Chapter 17

# TRENDS AND DIFFERENCES IN TRENDS

So-called trend analysis is, in essence, a part of the larger problem of the relationship between variates when we have an independent-dependent variable situation. Correlational analysis is appropriate for specifying relationships between individual difference variables regardless of whether or not one variable can be characterized as dependent on the other as an independent variable. When it can be argued that one variable is dependent (consequent) and the other independent (antecedent), there may be some interest in the regression of the dependent on the independent variable, both variates being individual difference variables. We have already given methods for testing the significance of regression coefficients (p. 160), for the equivalent testing of the significance of linear regression (p. 308), for testing linearity (p. 311), and for testing the difference between regression coefficients based on independent samples (p. 161).

Although our discussion of the analysis of variance has been mainly concerned with the significance of the differences between means, the perceptive reader will have noted that when a basis of classification involves an ordered variable, or factor, such as distance, degree of illumination, size, etc., which is manipulable as an independent variable, the $F$ test for a main effect is really concerned with whether or not some dependent variable, $X$, is being affected by the factor. That is, is $X$ as a dependent variable influenced by or related to the manipulated variable? This may be regarded as a question of regression (most mathematical statisticians subsume all analysis of variance under regression analysis) or more simply a question of trend and its form. For this situation the correlation coefficient ceases to be a useful descriptive term, but the presence of linear

389

trend and the slope thereof is of interest as will also be the possible curvilinearity of the relationship. Or differences among trends may be of primary interest.

Some of the techniques to be presented in this chapter are frequently subsumed under the topic "Orthogonal Polynomials."

**Review and recast.** When we have $G$ levels on a factor, or independent variable, with $m$ different individuals randomly assigned into each of the $G$ groups, we have a one-way classification design (possible analyses suggested on pp. 308–17) with $Y$ as the dependent and $X$ as the independent variable. The $m$s per group need not be the same, although equal $m$s are preferable.

When we have $C$ levels on the ordered factor and $R$ levels on a second ordered factor (or $R$ conditions not orderable), with $m$ independent cases assigned to each of the $RC$ cells, we have a two-way design. A plot of the $X$ means against the $C$ values (or levels), done separately for each of the $R$ levels (or conditions), will permit the drawing of $R$ trend lines (as in Figs. 16.1–16.4, p. 337). Or a plot of the appropriate means for $X$ against the $R$ values (or levels for the factor identified with the rows), this time separately for the $C$ levels, will permit the drawing of $C$ trend lines. The test of the $R \times C$ interaction provides a test of the differences between the $R$ trend lines (or the $C$ trend lines when the row factor is ordered).

When we have $C$ levels for one ordered variable and $B$ levels on a second factor (quantitative or qualitative) with each of $R$ individuals measured under all the $BC$ combinations of conditions (a three-way, mixed model), a test of $s^2_{bc}$ against $s^2_{rbc}$ is a test of the difference between the $B$ trends plotted with appropriate $X$ means against the column factor (or between the $C$ trends when $X$ means are plotted against the $B$ levels when blocks stand for an ordered variable). Note that the $C$ means entering into the trend for each of the $B$ levels are correlated (based on the same individuals); ditto, the $B$ means for $C$ trends. The use of $s^2_{rbc}$ as the error term allows for the correlation.

If for the $B$ levels we used $B$ sets of different persons, $R$ persons per set, the $C$ means for the trend of $X$ against the $C$ levels would again be correlated but the $B$ trend lines would be uncorrelated. The test of the $B \times C$ interaction, as specified in Case XV, p. 380, is appropriate for testing the differences among the $B$ trends.

A significant interaction in any of the foregoing types of situations simply means that the trends or curves are not parallel, regardless of their general shape, or the form of the relationships. A presentation of the trend lines or a description thereof is necessary for an interpretation of any claimed statistically significant differences among the trends.

## LINEAR TRENDS

The specification and testing of linear trends in psychology is of special interest for two reasons: (1) many relationships are linear in form, sometimes predictably so from theory, and (2) the question as to whether a relationship is nonlinear is readily approached via a test of departure from linearity. Although we have already set forth a method (pp. 308–17) for testing linear trend (linear regression) and for testing nonlinearity, there is a somewhat shorter approach which is applicable only when the levels on the factor being varied experimentally are evenly spaced and there are $m$ scores (measures) at each of the $G$ levels. We will need to distinguish between two situations: (1) when the $m$ scores are uncorrelated from level to level (i.e., $m$ independent cases assigned randomly to the $G$ groups) and (2) when the $m$ scores are correlated (i.e., based on just $m$ persons or $m$ sets of matched persons). The first of these will involve one-way, the other two-way, analysis of variance but some computational methods developed for the first will also be applicable to the second.

**Linear trend: uncorrelated $Y$ observations.** First, a little algebra. It will be recalled that the regression sum of squares was shown to be equal to $Nr^2S^2_y$, where $Y$ is the dependent variate and $X$ the independent variate, the variable for which we now choose the $G$ levels. We have

$$\Sigma(Y' - \overline{Y})^2 = Nr^2S^2_y = N\left(\frac{\Sigma xy}{NS_xS_y}\right)^2 S^2_y = \frac{(\Sigma xy)^2}{NS^2_x} = \frac{(\Sigma xy)^2}{\Sigma x^2}$$

But

$$\Sigma xy = \Sigma(X - \overline{X})(Y - \overline{Y})$$
$$= \Sigma XY - \overline{Y}\Sigma X - \overline{X}\Sigma Y + \Sigma \overline{X}\overline{Y}$$
$$= \Sigma XY - \overline{Y}N\overline{X} - \overline{X}\Sigma Y + N\overline{X}\overline{Y}$$
$$\Sigma xy = \Sigma XY - \overline{X}\Sigma Y$$

Now consider the sum, $\Sigma xY$, with $x$ in deviation units and $Y$ in original $Y$ units.

$$\Sigma xY = \Sigma(X - \overline{X})Y = \Sigma XY - \overline{X}\Sigma Y$$

Thus $\Sigma xy = \Sigma xY$.

To simplify computations we may code the $X$ variates into numerically small values, with a mean of zero so as to possess one property of deviation scores. Let us use $v$ for the coded values of the $G$ $X$ values, or points, used to define the $G$ levels. If $G$, the number of levels, is an odd number, we can assign a $v$ of 0 to the middle level and have coded values of

$$\cdots -4, -3, -2, -1, 0, 1, 2, 3, 4 \cdots$$

and if $G$ is an even number we can assign $-1$ and $+1$ to the two middle levels and have coded values of

$$\cdots -7,\ -5,\ -3,\ -1,\ 1,\ 3,\ 5,\ 7,\ \cdots$$

Let $v_g$, with $g = 1 \cdots G$, be the coded value for the $g$th group (level) and let $Y_g$ be the $Y$ scores for those in the $g$th group. Then

$$\Sigma vy = \Sigma vY = \Sigma v_1 Y_1 + \cdots + \Sigma v_g Y_g + \cdots + \Sigma v_G Y_G$$
$$= v_1 \Sigma Y_1 + \cdots + v_g \Sigma Y_g + \cdots + v_G \Sigma Y_G$$

Simply sum the $m$ $Y$ scores for each group (level), multiply by the $v$ for the level, and sum over groups, thus obtaining what we will designate as $\Sigma v Y$ instead of $\Sigma v_g \Sigma Y_g$, a more exact symbolization. (The $G$ separate sums of $Y$ scores will already have been obtained when computing the total, between-groups, and within-groups sums of squares.)

The regression sum of squares, $(\Sigma xy)^2/\Sigma x^2$, will be $(\Sigma vy)^2/\Sigma v^2$ $= (\Sigma v Y)^2/\Sigma v^2$ in terms of the $v$s, or coded $X$s. With $m$ cases per level, we have $\Sigma v^2 = \Sigma m v^2_g = m \Sigma v^2_g$. Simply square the (numerically small) $v$ values, sum, and multiply by $m$. Thus, we have for the regression sum of squares,

$$\Sigma\Sigma(Y' - \overline{Y})^2 = m\Sigma(Y'_g - \overline{Y})^2 = (\Sigma v Y)^2/m\Sigma v^2 \qquad (17.1)$$

which, since it has 1 degree of freedom, corresponds to the $s^2_p$ of p. 311. This is sometimes called the variance estimate for the linear component. It must not be forgotten that the foregoing computationally simple approach holds only for equal spacings for the levels on the independent variable, $X$, and for equal $m$s in the $G$ groups (at the $G$ levels of $X$).

By computational methods already given (formulas 15.6–15.8) we can obtain the total, the within-levels, and the between-levels sums of squares for the $Y$s. Recall from p. 307 that

$$\Sigma\Sigma(Y - \overline{Y})^2 = \Sigma\Sigma(Y - \overline{Y}_g)^2 + m\Sigma(\overline{Y}_g - \overline{Y})^2$$

and from p. 313 that

$$m\Sigma(\overline{Y}_g - \overline{Y})^2 = m\Sigma(\overline{Y}_g - Y'_g)^2 + m\Sigma(Y'_g - \overline{Y})^2$$

From the last equation we see that

$$m\Sigma(\overline{Y}_g - Y'_g)^2 = m\Sigma(\overline{Y}_g - \overline{Y})^2 - m\Sigma(Y'_g - \overline{Y})^2$$

provides a way of calculating the sum of squares for the deviation of the array, or group, means from linear form.

The breakdown of the total sum of squares along with $df$s and variance estimates may be assembled, as in Table 17.1. It will be noted that this table does *not* contain the "residual from line" component of Table 15.4, which was used there as the error term for testing the significance of $s^2_p$. Actually, we do not need this as an error term; we may take $F = s^2_p/s^2_w$. The $n_2$ for the $F$ table will be $mG - G$, a value that will be somewhat (usually slightly) smaller than the $n_2 = mG - 2$ when the variance estimate based on the residuals about the line is used as error. This slight loss in $df$ will, in the long run, be compensated for by the fact that $s^2_w$ tends always to be smaller than $s^2_{res}$.

**Table 17.1. Analysis of variance for linear trend, $Y$ as dependent on $G$ levels of $X$**

| Source | Sum of Squares | df | Variance Estimate |
|---|---|---|---|
| Between levels | $m\Sigma_g(\bar{Y}_g - \bar{Y})^2$ | $G - 1$ | $s^2_b$ |
| Linear trend | $m\Sigma_g(Y'_g - \bar{Y})^2$ | $1$ | $s^2_p$ |
| Deviation of means from line | $m\Sigma_g(\bar{Y}_g - Y'_g)^2$ | $G - 2$ | $s^2_d$ |
| Within levels | $\Sigma\Sigma_g(Y - \bar{Y}_g)^2$ | $mG - G$ | $s^2_w$ |
| Total | $\Sigma\Sigma(Y - \bar{Y})^2$ | $mG - 1$ | |

A significant $F = s^2_p/s^2_w$ has various connotations for various people: a significant linear relationship, a significant linear trend, a significant linear correlation, a significant linear regression, a significant linear slope ($B_{yx}$ significantly different from zero), a significant linear rate of change, a significant linear component of trend.

The departure from linear trend is testable by way of $F = s^2_d/s^2_w$ and the main effect of $X$ (differences among the $Y$ means for the $G$ groups) is tested by $F = s^2_b/s^2_w$. Ordinarily, if $s^2_b$ is significant, we would expect either $s^2_p$ or $s^2_d$ to be significant. It is possible for $s^2_b$ to be insignificant while $s^2_p$ is significant simply because the latter takes into consideration a progressive increase (or decrease) in the $G$ means. For example, if for five successive levels of the $X$ variable the $Y$ means were 16, 19, 21, 23, and 26, we would intuitively regard $X$ as having a greater effect than if the means were 19, 26, 16, 23, and 21. For both of these sets of means the values of $s^2_b$ are, of course, identical. Suppose $s^2_w$ is such that $F = s^2_b/s^2_w$ is significant only at the .10 level and that we test the $X$ effect for both sets via the significance of the linear trend. So doing yields significance at less than the .001 level for the first set and no significance whatever for the second set. The important point is that greater significance than that reached by $s^2_b$ should emerge if there is a systematic (linear) trend of the means. This is somewhat analogous to the advantage accruing from a one-tail test when the direction of the difference between two means has been predicted from theory. Certainly, if we have predicted a systematic increase (or decrease) of $Y$ means for the $G$ levels of $X$, the extent to which observations do show the predicted trend should somehow emerge in the statistical analysis.

**Linear trend: correlated observations.** So far our treatment of a linear trend has been confined to the setup where the $m$ scores for each of the $G$ groups, or levels, are independent from group to group. Suppose each person is measured at each level, and that the levels are again chosen to be equally spaced on the factor, or independent variable. This becomes a two-way analysis of variance setup, mixed model, with $R$ rows for $R$ persons and $C$ columns for $C$ levels on the factor. The differences among the resulting correlated column means, it will be recalled, are tested by $F = s^2_c/s^2_{rc}$. The means for the $C$ columns when plotted against the $C$ values of the independent variable may show a trend the linear component of which we may wish to test. With equal spacings for the levels on the independent variable, we may again set up coded, or $v$, scores for the points on the independent variable and proceed to compute the sum of squares for the linear component of the trend, $R\Sigma_c(X'._c - \bar{X})^2$, as $(\Sigma vX)^2/R\Sigma v^2$ in exactly the manner indicated earlier for $G$ independent groups. (Note: we use $X$ here as the dependent variable since the entire discussion of two- and three-way analysis of variance has been in terms of $X$s.) The sum of squares has 1 degree of freedom, hence is equal to an $s^2_p$.

What we have done here is to break up the between-column sum of squares into two component parts, a linearly predicted part and deviations of the means from linearity:

$$R\Sigma_c(\bar{X}._c - \bar{X})^2 = R\Sigma_c(X'._c - \bar{X})^2 + R\Sigma_c(\bar{X}._c - X'._c)^2$$

which again allows us to obtain the sum of squares for deviations from linearity by subtraction. This sum divided by $C - 2$ will give an $s^2_d$. Thus $F_c = s^2_c/s^2_{rc}$ is a test of the main effect of the factor; $F_p = s^2_p/s^2_{rc}$ tests the linear trend or linear regression; and $F_d = s^2_d/s^2_{rc}$ provides a test of the departure from linearity.

The use of $s^2_{rc}$ instead of $s^2_w$ as the error term distinguishes between two situations involving the relationship of a dependent and a manipulable variable (factor): $s^2_w$ is used when the dependent variate scores are independent from level to level of the factor, and $s^2_{rc}$ is used when the dependent variate scores are themselves correlated from level to level either because each of $R$ individuals is measured at each level or because we have $R$ sets of matched individuals, $C$ per set with random assignment from within each set to the $C$ levels.

## DIFFERENCES AMONG SLOPES

Earlier (p. 161) a method was given for testing the difference between two regression coefficients (linear slopes) based on independent samples. It will be recalled that this difference was tested against an estimated standard error that depended on the within groups residuals about the two regression lines.

**Slope differences for independent groups.** A test of the difference among three or more regression coefficients likewise depends on the within

**Table 17.2. Calculations for testing differences among $G$ slopes**

| | 1 | | $g$ | | $G$ | | Sum | | Within |
|---|---|---|---|---|---|---|---|---|---|
| $df$: | $m_1-1$ | $+$ | $m_g-1$ | $+$ | $m_G-1$ | $=$ | $\sum_g m_g - G$ | $=$ | $\sum_g m_g - G$ |
| $\sum xy$: | $(XY)_1$ | $+$ | $(XY)_g$ | $+$ | $(XY)_G$ | $=$ | $\sum_g (XY)_g$ | $=$ | $(XY)_w$ |
| $\sum x^2$: | $X_1$ | $+$ | $X_g$ | $+$ | $X_G$ | $=$ | $\sum_g X_g$ | $=$ | $X_w$ |
| $\sum y^2$: | $Y_1$ | $+$ | $Y_g$ | $+$ | $Y_G$ | $=$ | $\sum_g Y_g$ | $=$ | $Y_w$ |
| $B_{yx}$: | $(XY)_1/X_1$ | | $(XY)_g/X_g$ | | $(XY)_G/X_G$ | | | | $(XY)_w/X_w$ |
| $\sum(Y'-\bar Y)^2$: | $(XY)^2_1/X_1$ | | $(XY)^2_g/X_g$ | | $(XY)^2_G/X_G$ | | | | $(XY)^2_w/X_w$ |
| $\sum(Y-Y')^2$: | $Y_1-(XY)^2_1/X_1$ | $+$ | $Y_g-(XY)^2_g/X_g$ | $+$ | $Y_G-(XY)^2_G/X_G$ | $=$ | $\sum_g[Y_g-(XY)^2_g/X_g]$ | $\neq$ | $Y_w-(XY)^2_w/X_w$ |
| $df$ for residual: | $m_1-2$ | $+$ | $m_g-2$ | $+$ | $m_G-2$ | $=$ | $\sum_g m_g - 2G$ | $\neq$ | $\sum_g m_g - G - 1$ |

groups residuals. The procedure entails the calculation of the sum of squares, $\sum x^2$ for $X$, $\sum y^2$ for $Y$, and $\sum xy$, all three separately for each of the $G$ groups. The two sums of squares are computed by (3.6) and the cross-product sum by $\sum XY - \sum X \sum Y/N$, with $N$ replaced by $m_g$.

A tabular arrangement (Table 17.2) of these sums, along with additional indicated calculations, will facilitate the exposition. In this table the boldface **X**s, **Y**s, and **(XY)**s represent $\sum x^2$, $\sum y^2$, and $\sum xy$, respectively. The slope, or $B_{yx}$, is calculated as $\sum xy/\sum x^2$; the regression sum of squares, $\sum(Y' - \bar Y)^2$, is given by $(\sum xy)^2/\sum x^2$; and the residual sum of squares, $\sum(Y - Y')^2$, is obtained by subtracting the regression sum of squares from $\sum y^2$. The first four and the last two rows are summed across to get the "sum" column, and the first four of these sums are entered as the first four values in the "within" column. The next three entries under "within" are obtained from the $\mathbf{X}_w$, $\mathbf{Y}_w$, and $(\mathbf{XY})_w$ of the "within" column, not by summing across. Note that $\mathbf{X}_w$, $\mathbf{Y}_w$, and $(\mathbf{XY})_w$, being $\sum_g \mathbf{X}_g$, $\sum_g \mathbf{Y}_g$, and $\sum_g (\mathbf{XY})_g$, are nothing more than the familiar within-group sums, obtained by first summing within groups, then summing over groups; hence, the subscript $w$. The student should convince himself that $(\mathbf{XY})_w/\mathbf{X}_w$ does not correspond to the regression coefficient that would hold in case all the groups were combined, or thrown together in one scattergram, with sums of squares and the sum of products computed about the grand total means. It is also true that the value of $(\mathbf{XY})_w/\mathbf{X}_w$ is not a simple average of the $(\mathbf{XY})_g/\mathbf{X}_g$ values.

Under the null hypothesis that the population slopes for $G$ groups do not differ, we are in effect saying that a common slope holds for the $G$ populations. The $(\mathbf{XY})_w/\mathbf{X}_w$ is taken as the best estimate of this common slope, an estimate that in no way depends on possible group differences in the $X$ and $Y$ means—we need not assume equality of means. The residual, $\mathbf{Y}_w - (\mathbf{XY})^2_w/\mathbf{X}_w$, about the regression line with slope $(\mathbf{XY})_w/\mathbf{X}_w$ will have $(\Sigma_g m_g - G - 1)$ degrees of freedom; $G$ degrees of freedom are lost in the calculation of $\mathbf{Y}_w$ and an additional $df$ is used up in calculating the one slope, $(\mathbf{XY})_w/\mathbf{X}_w$. The $df$ for $\sum_g [\mathbf{Y}_g - (\mathbf{XY})^2_g/\mathbf{X}_g]$ is simply the sum of the $df$s for the parts being summed; i.e., $\Sigma_g m_g - 2G$.

If all $G$ slopes were exactly the same, each would equal $(\mathbf{XY})_w/\mathbf{X}_w$, and the sum of the $G$ residual sums of squares would be exactly the same as the residual sum of squares in the "within" column. That is, $\sum_g [\mathbf{Y}_g - (\mathbf{XY})^2_g/\mathbf{X}_g]$ would equal $\mathbf{Y}_w - (\mathbf{XY})^2_w/\mathbf{X}_w$ exactly. But in practice the $G$ slopes will not be the same, even when the population slopes are identical, simply because of sampling errors. If it is recalled that for any sample the slope $B_{yx}$ taken as $rS_y/S_x$, or the exact equivalent $\Sigma xy/\Sigma x^2 = (\mathbf{XY})/\mathbf{X}$, is that value of the slope (of the regression line) which minimizes the residual sum of squares, it is readily seen that the residual sum of squares for, say, group $g$ will be larger about the line with slope $(\mathbf{XY})_w/\mathbf{X}_w$ than about the line with slope $(\mathbf{XY})_g/\mathbf{X}_g$ (unless the two slopes happen to be equal). The same will hold for all $G$ groups simply because $(\mathbf{XY})_w/\mathbf{X}_w$ is not the optimum value for the separate groups. The greater the divergence of the separate $G$ slopes from $(\mathbf{XY})_w/\mathbf{X}_w$, the larger the residual sum of squares in the "within" column compared to the sum of the $G$ residual sums of squares. That is, $\mathbf{Y}_w - (\mathbf{XY})^2_w/\mathbf{X}_w$ will be larger than $\sum_g [\mathbf{Y}_g - (\mathbf{XY})^2_g/\mathbf{X}_g]$. This means that $\mathbf{Y}_w - (\mathbf{XY})^2_w/\mathbf{X}_w$ as a sum of squares may have a source of variation which does not affect the sum of the $G$ separate residual sums of squares. That source is the possible differences among the $G$ regression coefficients, or slopes.

Accordingly, we may break down the residual sum of squares in the "within" column into two parts: a within-groups residual about the *separate* regression lines, or $\sum_g [\mathbf{Y}_g - (\mathbf{XY})^2_g/\mathbf{X}_g]$, plus differences among slopes. The sum of squares for slopes is obtained by subtraction:

$$[\mathbf{Y}_w - (\mathbf{XY})^2_w/\mathbf{X}_w] - \sum_g [\mathbf{Y}_g - (\mathbf{XY})^2_g/\mathbf{X}_g]$$

Likewise, the $df$ for the slopes part is obtained by subtraction:

$$(\sum_g m_g - G - 1) - (\sum_g m_g - 2G) = G - 1$$

Division of the sum of squares for slopes by $G - 1$ will yield a variance estimate, $s^2_{sl}$, and division of $\sum_g [\mathbf{Y}_g - (\mathbf{XY})^2_g/\mathbf{X}_g]$ by its $df$ will yield a within-groups residual variance estimate, $s^2_{res(w)}$. Then $F = s^2_{sl}/s^2_{res(w)}$ provides a test of the differences among the $G$ slopes, or a test for the homogeneity of regressions.

This test for the differences among slopes is general in that it is applicable (a) when $Y$ and $X$ are both individual difference variables and the $G$ groups are independent or (b) when $Y$ is regarded as dependent on $X$ as a manipulated variable and the $G$ groups are independent and there is also independence from level to level on $X$. (See next section for a special case of the latter situation.) It is preferable, although not required, to have equal group $N$s, i.e., equal $m_g$. When both the $m_g$ and the spacings of $X$ are equal, $(\mathbf{XY})_w/\mathbf{X}_w$ will be the simple average of the $(\mathbf{XY})_g/\mathbf{X}_g$. Otherwise, it is a weighted average, $m_g$ being the weight for the $g$th group.

**Slope differences, independent variable manipulable and no repeat measures.** When we have $G$ groups of equal size and the subjects are assigned randomly to, say, $C$ levels on a manipulable variable, $X$, with $m$ cases per level and the levels on $X$ equally spaced, we can simplify the computations by using the coded values for $X$ (i.e., the $v$s discussed earlier) in order to calculate the $\Sigma xy$ and the $\Sigma x^2$ terms. We can also reorient the problem so as to tie in more closely with analysis of variance, with a shift to an ordinary $s^2_w$ instead of $s^2_{res(w)}$ as the appropriate error term for $F$. To simplify transition from the preceding general test for slope differences, we will continue to designate $Y$ as the dependent variable. We will have a two-way analysis of variance layout with $C$ columns standing for $C$ levels on $X$ and $R$ rows standing for $G$ groups, with $m$ scores in each of the $RC$ cells. Again for transitional purposes we will continue using $g$ and $G$ as group designations (instead of $r$ and $R$). Incidentally, the classification into $G$ groups could be on the basis of either a qualitative or a quantitative factor.

With equal $N$s for the $G$ groups and equal spacings of the levels on $X$, all $\mathbf{X}_g$ ($= m\Sigma v^2$) will have the same value, i.e., be a constant. Now by definition (see Table 17.2), $(\mathbf{XY})_w = \sum_g (\mathbf{XY})_g$ and $\mathbf{X}_w = \sum_g \mathbf{X}_g$, but with $\mathbf{X}_g$

a constant we have $\mathbf{X}_w = G\mathbf{X}_g$. With these in mind we may write an expression for the average of the $G$ slopes, or the $B_{yx}$ values, as

$$\bar{B}_{yx} = \frac{\underset{g}{\Sigma}(\mathbf{XY})_g/\mathbf{X}_g}{G} = \frac{\underset{g}{\Sigma}(\mathbf{XY})_g}{G\mathbf{X}_g} = \frac{(\mathbf{XY})_w}{\mathbf{X}_w}$$

Thus the "within" slope is, for the given situation, a simple average of the $G$ separate slopes.

From the previous development we saw that in general the sum of squares for between slopes could be obtained by subtraction:

$$SS_{sl(b)} = [\mathbf{Y}_w - (\mathbf{XY})^2_w/\mathbf{X}_w] - \underset{g}{\Sigma}[\mathbf{Y}_g - (\mathbf{XY})^2_g/\mathbf{X}_g] \qquad (17.2)$$

$$= \mathbf{Y}_w - (\mathbf{XY})^2_w/\mathbf{X}_w - \underset{g}{\Sigma}\mathbf{Y}_g + \underset{g}{\Sigma}(\mathbf{XY})^2_g/\mathbf{X}_g$$

$$= \underset{g}{\Sigma}(\mathbf{XY})^2/\mathbf{X}_g - (\mathbf{XY})^2_w/\mathbf{X}_w$$

since by definition $\mathbf{Y}_w = \underset{g}{\Sigma}\mathbf{Y}_g$. Now recalling that $\mathbf{X}_g$ is a constant, that $\mathbf{X}_w = G\mathbf{X}_g$, and that $(\mathbf{XY})_w = \underset{g}{\Sigma}(\mathbf{XY})_g$, we get by substitution

$$SS_{sl(b)} = \frac{\underset{g}{\Sigma}(\mathbf{XY})^2_g}{\mathbf{X}_g} - \frac{\left[\underset{g}{\Sigma}(\mathbf{XY})_g\right]^2}{G\mathbf{X}_g}$$

which, by factoring out an established common denominator and replacing $(\mathbf{XY})_g$ and $\mathbf{X}_g$ by their equivalents in terms of the coded $v$s, becomes

$$SS_{sl(b)} = \frac{1}{Gm\Sigma v^2}\left[G\underset{g}{\Sigma}\left(\underset{i}{\Sigma}vY_{ig}\right)^2 - \left(\underset{g}{\Sigma}\underset{i}{\Sigma}vY_{ig}\right)^2\right] \qquad (17.3)$$

as a simple computation formula for the sum of squares for slope differences. A $\underset{i}{\Sigma}vY_{ig}$ term is computed for each group by multiplying the $i$th individual's $Y$ scores by the appropriate $v$ (the coded $X$ value for his $X$ level), then summing the products. This is made easier by first summing the $mY$ scores separately for each of the $C$ levels in group $g$ (required sums for analysis of variance—a line like the $\Sigma X$ in Table 15.2). Each of these $C$ sums is then multiplied by the appropriate $v$ and the $C$ products summed to get $\underset{i}{\Sigma}vY_{ig}$. There will be $G$ such sums, which are next summed to get the double sum that is squared for the second term, and the $G$ sums are each squared with the $G$ squared values summed to get the first term. This process will be far shorter than the general method of getting the sum of squares for slope differences as a difference by formula (17.2).

The formulation in (17.3) may help us understand this difference in slopes business. Notice that a part of formula (17.3) is precisely of the form $(1/N)[N\Sigma X^2 - (\Sigma X)^2]$ used in computing a sum of squares. The correspondence may be carried a bit farther. The $\underset{i}{\Sigma} v Y_{ig}$ are the numerator terms for group slopes. With the denominators all having the same constant value, $m\Sigma v^2$, the variation among the slopes is solely a function of the differences among the $\underset{i}{\Sigma} v Y_{ig}$, hence a familiar formula for calculating a sum of squares by which to estimate a variance, the variance of the slopes.

When $SS_{sl(b)}$ is divided by its $df = G - 1$, we will have the needed unbiased estimate, $s^2_{sl(b)}$, for a significance test. What do we use as the error term? Here we depart from the scheme of Table 17.2 and say that $s^2_w$ will be used instead of $s^2_{res(w)}$. This is exactly what we did in moving from Table 15.4 to Table 17.1 when testing for linear trend. It is here presumed that the computations for a two-way analysis of variance are available. And this brings us to a conceptual tie-up with analysis of variance. Recall that a significance test for interaction has to do with over-all differences in trend—do the sets of $C$ means for the $R$ (or $G$) groups behave similarly? Now each group's trend may show a linear component, specifiable as slope, and these slopes may differ from group to group. What the foregoing test for the differences between slopes tells us is whether or not the linear components of the interaction are significantly different from one another. This idea that interaction is being scrutinized for the linear aspect of its form holds for the next two situations involving slope differences. We should, perhaps, add that when the $G$ groups stand for equally spaced levels on a second factor, the whole process can be repeated to examine the differences between slopes when appropriate $\overline{Y}$ values are plotted against the $G$ levels. There will be $C$ such trends, each with its own slope, or linear component. It will be recalled that an interaction involving two quantitative factors can be viewed in either of two ways—two sets of trend lines.

**Slope differences, independent groups but correlated observations within the groups.** The scores on the dependent variable $X$ (note shift back to $X$ as a symbol for the dependent variable) may be arranged as for a three-way analysis of variance, with blocks for $B$ independent groups that are measured under the $B$ conditions (either qualitatively different or as $B$ levels on a quantitative factor), with columns for $C$ levels on a quantitative factor or as $C$ trials in a learning task, and the rows for $R$ individuals. The observations from column to column are (likely) correlated because we have repeated measures on each individual. In ordinary analysis of variance this setup is Case XV (p. 379) for which the test of the $B \times C$ interaction provides a test of the differences among the $B$ trends (p. 381).

Our present concern is the differences among the linear trends, or slopes, shown by the $B$ groups.

When this is of interest to the experimenter he should, for sake of computational simplicity, have equal spacings for the $C$ levels with exactly the same levels for all $B$ groups. (In the learning setup, it is usually tacitly, perhaps gratuitously, assumed that trials constitute equal spacings.) The method to be given here presumes equal spacings for the $C$ levels. The linear part of any possible trend for the $b$th group can be specified in terms of the best fitting line to the successive $C$ means (the $\bar{X}_{.b1}, \cdots, \bar{X}_{.bc}, \cdots, \bar{X}_{.bC}$) for the dependent variable, here designated as $X$. But the $r$th individual in the $b$th group has $C$ scores which permit the plotting of an *individual* trend line which, in turn, may be described in part by a straight line the slope of which will represent the linear component for the individual's trend. These individual slopes will always show variation from person to person, hence are variates—we may regard the slope for an individual as a sort of "score." The average of these scores (slopes) for individuals in the $b$th group will correspond to the slope for the $b$th group, thus permitting us to regard the group slope as a mean. We will have $B$ such means.

If we code the $C$ levels as $v$s, we can readily calculate a slope (for the linear regression of $X$ on the coded scores of the factor) for each individual. It is simply $\Sigma vX/\Sigma v^2$ (no $m$ here since each person has just one $X$ score at each of the $C$ levels). The calculation of $RB$ (or $N_t$ where $N_t = \Sigma N_b$) different values of $\Sigma vX$ is greatly facilitated by having the $v$ values on a strip that can be placed just under the $X$s in a row. The $\Sigma v^2$ is a constant—the same for all individuals. For the present purpose the sizes of the $B$ groups need not be equal.

With the individual slopes calculated, the test of the significance of the differences among the group slopes is not only easy to carry out but also easy to conceptualize. When we regard the individual slopes as "scores," we have a simple one-way analysis of variance setup with a breakdown of the total sum of squares (for the slopes) into between-group and within-group sums of squares with $B - 1$ and $RB - B$ (or $N_t - B$) degrees of freedom, respectively. The computations for this part are by formulas (15.6–15.8) or (15.9–15.11), with the individual slopes taken as the $X$s for *those* formulas. Actually, the calculations for the test of significance can be made on the individual $\Sigma vX$ values as the "scores" for the one-way analysis of variance since the $\Sigma v^2$ part of the individual slopes is a constant. $F = s^2_b/s^2_w$ is the desired test for judging whether the group slopes (or linear regressions) are heterogeneous.

If the blocks represent $B$ equally spaced levels on a factor, we may wish to consider the $B \times C$ interaction in terms of the possible $C$ trends of $X$

on the block factor. There will be a mean for the $c$th column for each block; a plot of these will yield a trend. The differences between the $C$ slopes can be handled by way of the $v$ codings provided the number of subjects in each of the $B$ groups is the same; i.e., same value for $R$, or number of subjects per block. Each mean for plotting the $C$ trends of $X$ on the block factor will be based on $R$ scores, so $R$ will replace $m$ in formula (17.3) and $G$ will be replaced by $C$ for computing the between slopes sum of squares, or

$$SS_{sl(b)} = \frac{1}{CR\Sigma v^2}\left[C\sum_c\left(\sum_i vX_{ic}\right)^2 - \left(\sum_c\sum_i vX_{ic}\right)^2\right]$$

with $C - 1$ degrees of freedom. But since these $C$ trends are based on correlated scores, the appropriate error term for $F$ will not be an $s^2_w$; it will be $s^2_h$ as specified in Table 16.16 for Case XV.

**Slope differences, observations correlated two ways.** The scores can be arranged as in a complete three-way analysis of variance into $B$ blocks, $C$ columns, and $R$ rows, there being a total of just $R$ individuals. Each person has a score in every column and in every block—there will be intercolumn and also interblock correlation. Provided the $C$ levels are equally spaced, we can again code in order to compute $\Sigma vX/\Sigma v^2$ as an individual's slope for $X$ on the column factor (independent variable) coded as $v$s, but now each individual has $B$ slopes since he has a set of $C$ $X$ scores in each block. The total number of individual slopes will be $RB$, and either these slopes or the $\Sigma vX$ values ($\Sigma v^2$ again being a constant) can be arranged into a new table with $R$ rows and $C$ columns (each of the block conditions is now assigned to a column position for this new table). Thus we have a two-way analysis of variance setup, mixed model, for which $F = s^2_c/s^2_{rc}$ with the usual $df$s will provide a test of the differences between the $B$ slopes (one for each block) since the means of the columns in the new table correspond to the slopes for the $B$ blocks, each block slope being the mean of individual slopes. Due allowance has been made for any possible (and likely) correlation between the blocks. The foregoing analysis is possible because for equal spacings on the $C$ factor the slope for the $b$th block of scores is the same as the mean of the $R$ individual slopes in the block. The $B$ blocks may stand for $B$ qualitatively different conditions or for $B$ levels on a quantitative factor, with no requirement that the $B$ levels be equally spaced.

Now suppose that the $B$ blocks represent $B$ equally spaced levels on a factor and that we wish also to consider the linear parts of the $C$ trends which are exhibited when we plot the appropriate means (the $\overline{X}_{.1c}, \cdots,$ $\overline{X}_{.bc}, \cdots, \overline{X}_{.Bc}$), separately for each of the $C$ sets, against the $B$ levels of the block factor. (If the student is confused about how to pick out these sets of means, in contrast with the sets used a couple of pages earlier, he should refer to Table 16.10. For the earlier discussed trends for $X$ against the $C$ factor, the means along the bottom of each block are used, whereas for the present trends of $X$ against the $B$ factor, a mean is picked from each block as one goes down a column.) It will be recalled that, so far as differences in general trend are of interest, the testing of $s^2_{bc}$ against $s^2_{rbc}$ provides a basis for saying whether there are significant differences among the $B$ trends for $X$ against the $C$ levels *and* also for saying whether there are significant differences among the $C$ trends for $X$ against the $B$ levels (see p. 390). However, for the linear components of the trends the test of the differences among the $B$ slopes does not simultaneously provide a test of the differences among the $C$ slopes. For the latter test we must calculate $C$ slopes, for $X$ against the $B$ levels, for each of the $R$ individuals. The $B$ levels are now coded as $v$s. For the computation of the required $\Sigma vX$, $RC$ in number, the tedium of picking in turn each of the $B$ appropriate scores from the blocks may be avoided by first rearranging the entire table so as to have the $B$ scores for each person under the $c$th condition in a row. That is, the original block and column designations are interchanged; if, for example, the original blocks and columns stood for distances and illumination levels, respectively, blocks and columns in the rearranged table would stand for illumination levels and distances, respectively. Regardless of the arrangement for computing the $\Sigma vX$, once we have them they are the values which become the "scores" for a two-way analysis of variance, as before.

## HYPOTHESES ABOUT CURVATURES

The previously given test for curvilinearity merely permits us to accept or reject the hypothesis that the fit of the observations (the means) to a straight line is sufficiently close to be regarded either as within the limits of chance or as showing discrepancies too large to be attributed to chance. Rejection of the linearity hypothesis implies nothing about the possible form of the curving relationship. Now the form of a relationship may at times be predicted from theory, thus permitting us to go beyond the general statement that $Y$ depends on $X$, or $Y = f(X)$, to an equation involving a specified (predicted) form for the relationship, such as $Y' = B \log X + A$ or $Y' = Ae^{BX}$ or $Y' = X/(A + BX)$, and so on. Or on the basis of a plot of $Y$ (or the $Y$ means) against $X$ we may proceed empirically. With knowledge concerning the shapes of various mathematical curves, we select the form of the curve that might fit the observations. Whether the form is arrived at from theory or empirically, we determine the numerical values

TRENDS AND DIFFERENCES IN TRENDS

for the constants called for in the mathematical equation of the chosen form. Since the general problem of curve fitting is far beyond the scope of this book, the author refers the reader to the excellent discussion of the topic in Don Lewis' *Quantitative methods in psychology* (New York: McGraw-Hill, 1960).

Here we shall only be concerned with going a step beyond the question of whether a significant departure from linearity is haphazard or shows sufficient regularity to suggest that some type of systematic curvature is present. This need not be empirical in that theory might predict that a relationship involves an increasing, leveling off, then decreasing function (or a decrease, leveling, then increase) or a rapid rise followed by a leveling off. The theory might not be sufficiently well developed to permit a prediction of a more specific form for the relationship, particularly for parts of the curve beyond (above or below) certain chosen levels for the independent variable. In other words, we may merely predict that a segment (that for the chosen levels of $X$) of the relationship between $Y$ and $X$ should show curvature.

The argument in favor of proceeding to a curvature component of a trend is similar to that given earlier (p. 393) for going from the ordinary $F$ test for between levels to the testing of the significance of possible linear trend. We may here use the earlier illustration in which we presumed that the means for the dependent variable were 19, 26, 16, 23, 21 for five consecutive levels on the independent variable in contrast to 16, 19, 21, 23, 26 for successive levels. Although the identical $F$s for between levels might fail to reach significance, a significant effect might be claimed for the second set via linear trend. Now suppose the means are 16, 19, 26, 23, 21 for the successive levels. A plot of these will show apparent systematic curvature, but very little linear trend (near zero slope). Would such an observed curvature prove to be a nonchance affair if tested by a method that gives some consideration to the systematic curving trend? If so, could $X$ be claimed as having a significant effect on $Y$?

As a first approximation, we may regard a segment of a quadratic curve, defined by the equation $Y' = A + BX + CX^2$, as "fitting" (maybe) the segment of the "curve" based on available data. The quadratic component resides in the $CX^2$ term, so in effect we have the question of whether $C$ differs from zero. It must be understood that rarely in psychological research will we find a logical reason for predicting a quadratic form of relationship between a dependent variable and an independent variable over a wide range of values for the latter. The quadratic form is here used merely as a basis for testing the hypothesis that some curvature exists which, if taken into account, would explain a significant portion of the between means (for levels) variance. Since the method does take into consideration the apparent systematic curvature, an effect is more apt to be detected than by the ordinary $F$ test for between levels.

When the levels on the independent variable are equally spaced with an equal number of observations per level, there is a relatively simple way for testing the quadratic component of the trend. The method involves the use of a type of coded score, which we symbolize as $u$. The set of $u$s to be employed depends on the number of levels, and possesses the property that $\Sigma u = 0$. Table 17.3 gives the values of $u$ for 3 up to 10 levels. Once the $u$s

**Table 17.3. Coded values, $u$, for quadratic component of trends for 3 to 10 levels on an independent variable**

| | | | | | | | | | | $\Sigma u^2$ |
|---|---|---|---|---|---|---|---|---|---|---|
| Level | | | 1 | 2 | 3 | | | | | |
| $u$ | | | +1 | −2 | +1 | | | | | 6 |
| Level | | 1 | 2 | 3 | 4 | | | | | |
| $u$ | | +1 | −1 | −1 | +1 | | | | | 4 |
| Level | | 1 | 2 | 3 | 4 | 5 | | | | |
| $u$ | | +2 | −1 | −2 | −1 | +2 | | | | 14 |
| Level | 1 | 2 | 3 | 4 | 5 | 6 | | | | |
| $u$ | +5 | −1 | −4 | −4 | −1 | +5 | | | | 84 |
| Level | 1 | 2 | 3 | 4 | 5 | 6 | 7 | | | |
| $u$ | +5 | 0 | −3 | −4 | −3 | 0 | +5 | | | 84 |
| Level | 1 | 2 | 3 | 4 | 5 | 6 | 7 | 8 | | |
| $u$ | +7 | +1 | −3 | −5 | −5 | −3 | +1 | +7 | | 168 |
| Level | 1 | 2 | 3 | 4 | 5 | 6 | 7 | 8 | 9 | |
| $u$ | +28 | +7 | −8 | −17 | −20 | −17 | −8 | +7 | +28 | 2,772 |
| Level | 1 | 2 | 3 | 4 | 5 | 6 | 7 | 8 | 9 | 10 |
| $u$ | +6 | +2 | −1 | −3 | −4 | −4 | −3 | −1 | +2 | +6 |  132 |

appropriate for $G$ (or $C$) levels have been arranged alongside the sum of the dependent variable scores for the successive levels, we simply multiply the $\Sigma Y$ for the $g$th group (or level) by $u_g$, thus finding $G$ products, which are then summed over groups to obtain $\sum_g u_g \Sigma Y_g$, or $\Sigma u Y$. (Substitute $X$ for $Y$ if $X$ has been used to designate the dependent variable.) The sum of squares for the quadratic component is given by $(\Sigma u Y)^2/m\Sigma u^2$. Note the similarity to the earlier given sum of squares for the linear component— the only difference is in the coded values used. This new sum of squares has $df = 1$; this $df$ has to do with the one constant $C$ in $Y' = A + BX + CX^2$ which controls the curvature, just as the linear component was concerned with the one constant $B$ in $Y' = A + BX$. With $df = 1$, the value of $(\Sigma u Y)^2/m\Sigma u^2$ is automatically a variance estimate, which we will symbolize by $s^2_q$.

To test $s^2_q$ for significance, we need an error term appropriate to the situation. If the observations are independent from level to level, the error term is the ordinary $s^2_w$ of one-way analysis of variance. If the observations are correlated (same persons measured at each level), we have a two-way analysis of variance layout with $C$ columns for $C$ levels and $R$ rows for persons ($m = R$ for foregoing indicated computations), and the error term is $s^2_{rc}$.

No attempt will be made here to explain the derivation of the sets of $u$s in Table 17.3. Aside from the property that $\Sigma u = 0$ always, an examination of any one set may help us understand more fully their use in testing for a curvature component of trend. If the reader will plot any one set of $u$s, say for $G = 5$, against any imagined five equally spaced values for a quantitative factor, $X$, he will have five points that follow a curve. Now suppose the $\Sigma Y$ values for all five levels are identical—a plot of the five corresponding $Y$ means against the five $X$ values will, of course, be a horizontal line. With $\Sigma Y$ a constant from level to level, the *linear* component, using the earlier defined $v$s ($-2, -1, 0, +1, +2$), will give $\Sigma v Y = \Sigma_g v_g \Sigma Y_g$ $= \Sigma Y_g \Sigma_g v_g = 0$ since $\Sigma_g v_g = 0$. The linear component is zero as it should be for a zero slope. When we proceed to compute $\Sigma u Y$ (equivalent to $\Sigma u_g \Sigma Y_g$) with $u$s of $+2, -1, -2, -1, +2$ (see Table 17.3), we have $\Sigma Y_g \Sigma_g u_g$, which is also zero because $\Sigma_g u_g = 0$. No curvature when all five $Y$ means are identical! Now, just as the departure of $\Sigma v Y$ from zero indicates the presence of a linear component in the trend, the departure of $\Sigma u Y$ from zero indicates a quadratic component. If the five means were such that the successive $\Sigma Y$ values were 10, 20, 30, 40, 50, we would have $\Sigma u Y = 2(10) - 1(20) - 2(30) - 1(40) + 2(50) = 0$. An obvious linear component, but no curvature. (In this last example there was no constant $\Sigma Y$ which could be taken from under the indicated summation over groups.)

Let us again consider the three sets of five means mentioned earlier (p. 403):

| Set $A$ | 19 | 26 | 16 | 23 | 21 |
| Set $B$ | 16 | 19 | 21 | 23 | 26 |
| Set $C$ | 16 | 19 | 26 | 23 | 21 |

With each mean based on $m$ cases, the required $\Sigma Y$ for group $g$ will be $m\overline{Y}_g$; hence the $\Sigma v Y$ for a set will be $\Sigma vm\overline{Y}_g = m\Sigma v \overline{Y}_g$ and the $\Sigma u Y$ for a set would be $\Sigma um\overline{Y}_g = m\Sigma u \overline{Y}_g$.

For Set $A$ we have for the linear component

$$\Sigma v Y = m[-2(19) - 1(26) + 0(16) + 1(23) + 2(21)] = +1(m)$$

and for the quadratic component we have

$$\Sigma u Y = m[+2(19) - 1(26) - 2(16) - 1(23) + 2(21)] = -1(m)$$

from which we see that both the linear and the quadratic components are perhaps negligible ("perhaps" because no significance test has been applied).

For Set $B$ we have

$$\Sigma v\, Y = m[-2(16) - 1(19) + 0(21) + 1(23) + 2(26)] = 24(m)$$

and

$$\Sigma u\, Y = m[+2(16) - 1(19) - 2(21) - 1(23) + 2(26)] = 0(m)$$

for which we see a sizable linear component with no quadratic component.

For Set $C$ we have

$$\Sigma v\, Y = m[-2(16) - 1(19) + 0(26) + 1(23) + 2(21)] = 14(m)$$

and

$$\Sigma u\, Y = m[+2(16) - 1(19) - 2(26) - 1(23) + 2(21)] = -20(m)$$

which indicates the possible presence of both components.

For sake of illustration, let us suppose that $s^2_w = 71$ and that $m = 10$. The between-groups sum of squares becomes 580, which with $df$ of 4 leads to an $s^2_b$ of 145. The simple $F$ test for the differences among the five means yields $F = s^2_b/s^2_w = 145/71 = 2.04$ which for 4 and 45 degrees of freedom does not quite reach the .10 level of significance. This $F$ holds, of course, for all three sets.

For Set $A$ the linear component sum of squares, $(\Sigma v\, Y)^2/m\Sigma v^2 = 100/100 = 1$; hence $s^2_p = 1$, and $F = 1/71$, which is far from significant. The quadratic component sum of squares, $(\Sigma u\, Y)^2/m\Sigma u^2 = 100/140$ leading to $s^2_q$ of .7, thence $F = .7/71$, which is likewise insignificant.

For Set $B$ in which there appears to be a definite linear trend, we have the sum of squares for the linear component, $(\Sigma v\, Y)^2/m\Sigma v^2 = 57,600/100 = 576$, which yields $F = 576/71 = 8.11$, which is significant beyond the .001 level. The quadratic sum of squares is zero.

For Set $C$ the sum of squares for the linear component is 19,600/100, leading to $F = 196/71 = 2.76$, which for $dfs$ of 1 and 45 does not reach the .10 level of significance. For the quadratic component sum of squares we have 40,000/140 = 285.71, from which we get $F = 285.71/71 = 4.02$ which is significant at the .05 level.

Admittedly, the foregoing starting means, with $ms$ of 10 and $s^2_w$ of 71, were contrived for a purpose: to show that the ordinary $F$ ($s^2_b/s^2_w$ or $s^2_c/s^2_{rc}$) test is not sensitive to possible systematic trends and that the sensitivity of the statistical analysis for an effect can be improved by a method that takes into consideration the systematic trend shown by the data. This bonus in sensitivity is particularly deserved by the experimenter who has made an a priori prediction either that the trend will involve a linear component or that it will have simple curvature (or both).

## DIFFERENCES AMONG CURVATURES

We saw earlier that testing for slope differences had to do with the linear components of interaction. A further examination of the nature of the interaction can be achieved by way of possible differences in curvatures for the trends. Here we will consider only the quadratic components of the trends. Ordinarily, a plot of the means entering into the trends will reveal whether any or all of the trend lines show evidence of quadratic-type of curvature—if differences in curvature are not apparent, the calculating of a significance test for the differences may be a waste of time. But if there are any noticeable differences in curvatures, we may wish to have a test for judging whether they are significantly different. Again, we will need to distinguish between the possible experimental setups, as was done on p. 391 for slope differences.

**Curvature differences, independent variable manipulable, no repeat measures.** The starting setup is exactly as described in the previous section labeled "Slope differences, independent variable manipulable and no repeat measures." With the dependent variable designated as $X$, the appropriate formula for calculating the sum of squares for differences between the quadratic components is of the form (17.3) with $Y$ replaced by $X$ and $v$ replaced by $u$, the $u$ being the coded values in Table 17.3. Thus

$$SS_{q(b)} = \frac{1}{Gm\Sigma u^2}\left[ G\Sigma_g\left(\sum_i uX_{ig}\right)^2 - \left(\sum_g\sum_i uX_{ig}\right)^2\right] \qquad (17.4)$$

Again, the computations are, as for (17.3), made easier by utilizing the $\Sigma X$ values as in Table 15.2. Each of $C$ such sums (for a given group) is now multiplied by the appropriate $u$, instead of by the $v$ used in studying slope differences. Then these products are summed to get the $\sum_i uX_{ig}$.

The $df$ for the sum of squares, $SS_{q(b)}$, is $G - 1$ and $F_{q(b)} = s^2_{q(b)}/s^2_w$ provides the significance test. If this $F$ is significant, one concludes that a part of the over-all interaction is descriptively attributable to differences in quadratic components of curvature for the trends of $X$ on the factor represented by the $C$ levels. If the $G$ groups also stand for equally spaced levels on the second factor, there will be $C$ possible trend lines for $X$ on the second factor. By exactly the same procedure (with different sets of $u$ values if $C$ does not equal $G$), these trends may also be tested for differences in their quadratic components. For formula (17.4), $C$ will play the role of

$G$ and $c$ the role of $g$. The $df$ for curvature differences becomes $C - 1$, and $s^2_w$ is again the denominator for $F$.

**Differences in quadratic curvature, independent groups but correlated observations within groups.** The layout corresponds to Case XV, with repeated measures on the column factor. There will be $B$ trend lines for $X$ on the column factor. Each subject will also have his own individual trend line, the quadratic component of which will be $\Sigma uX/\Sigma u^2$, where $u$ is the coded value for the equally spaced $C$ levels. As with slopes, we may regard these individual quadratic values as scores for an ordinary one-way analysis of variance. The scores so defined will lead to an $s^2_b$ which is the desired $s^2_{q(b)}$, and these scores will also provide an $s^2_w$ based on $RB - B$ degrees of freedom or on $N_t - B$ $df$ in case of unequal $R$s. This test does not, of course, use any of the computations utilized in connection with Case XV although the test is concerned with the $B \times C$ interaction.

In case the $B$ groups are subjects assigned to $B$ equally spaced levels on a quantitative factor, the $B \times C$ interaction can be looked at in terms of $C$ trend lines for $X$ against the $B$ levels. If all $R$ are equal for the $B$ groups, the means entering into the $C$ trend lines will each be based on $R$ cases, hence $R$ will play the role of $m$ in formula (17.4). For column $c$ we will have a sum of $R$ scores for each of the $B$ groups. When each of these sums is multiplied by $u$, we will have a value corresponding to the $\underset{i}{\Sigma}uX_{ig}$, which we can now label as $\underset{r}{\Sigma}uX_{rb}$. Then these sums are summed over $c$, and $C$ replaces $G$ in formula (17.4). The resulting sum of squares, $SS_{q(b)}$, for between quadratic components will have $C - 1$ degrees of freedom. The error term for $F$ will be the $s^2_h$, as defined in Table 16.16. Its use allows for the fact that the means entering into the $C$ trend curves are not independent from one value of $C$ to the next.

**Curvature differences, observations correlated two ways.** Here we have a three-way design with $R$ subjects with repeat measures on the column factor and on the row factor. If both the block and the column factors involve equally spaced levels, we will, as with slopes (refer back to the discussion thereof for this situation), have $B$ trend lines for $X$ on the column factor and $C$ trends for $X$ on the block factor. This time we can calculate two sets of quadratic components for each of the $R$ subjects. Each person will have $B$ quadratic "scores" when we are dealing with the trend of $X$ on the column factor, and each person will have $C$ quadratic "scores" for the trend of $X$ on the block factor. The subject's $B$ quadratic scores will be of the form $\Sigma uX/\Sigma u^2$ with the $u$ values appropriate to the $C$ levels, and the subject's $C$ quadratic scores will have a similar form with the $u$ now appropriate to the $B$ levels. We will have, respectively, a total of $RB$ scores and a total of $RC$ scores, and each of these two sets of scores can be arranged as a two-way layout, with the rows for the $R$ subjects and the columns standing for the $B$ levels in the first instance and for the original $C$ levels in the second instance. Each of these two-way layouts is the basis for a mixed model analysis of variance. One $F_c = s^2_c/s^2_{rc}$ will test the differences between the $B$ quadratic components of the trends for $X$ on the original column factor; the second $F_c = s^2_c/s^2_{rc}$ provides a test for the differences between the $C$ quadratic components of the trends for $X$ on the original block factor. Obviously, we will have two different $s^2_c$ values; ditto, $s^2_{rc}$. The whole procedure parallels that for differences between the slopes (linear components) for the given situation. Also, instead of the quadratic scores as $\Sigma uX/\Sigma u^2$, we can simply use $\Sigma uX$ since $\Sigma u^2$ is a constant.

## TRENDS, INTERACTION, AND MAIN EFFECTS

It should be obvious that the presence of a significant interaction is sufficient statistical justification for further tests to learn whether the general, or over-all, trend differences represent differences in either linear or in quadratic components. But is it necessary to have significant interaction before proceeding to the extraction of these components? Stated differently, with insignificant interaction is it possible to find significant differences in either slopes or quadratic curvatures as aspects of interaction? To this question we will give only a tentative or suggestive answer. When the interaction is nearly significant, it would seem reasonable to say that there could be significant linear and/or quadratic differences in trend merely because each of these components has to do with ordered, systematic trends, whereas the over-all test for interaction does not take ordering into account. Recall that we earlier saw that tests for the linear and the quadratic parts of a single trend could be significant even though the over-all $F$ for the differences between the means was insignificant. In case the interaction is far from significant it seems unlikely that breakdown into components will lead to significance. One can easily think of two non-statistical considerations that might bear on the decision to extract components of trend differences: (a) Did an a priori, theoretic based hypothesis specify differences in trend form or (b) in the absence of a hypothesis does an examination of a plot of the separate trend lines suggest either linear or curvature differences?

Another rather obvious observation may be made. Just as significant interaction forces one to qualify the interpretation of the main effects for the factor(s) involved, so also must significant differences among slopes and/or significant differences between curvatures restrict what is said about the linear and/or curving aspects of the trend (or lack thereof) shown by a plot of the main effect means against the factor. This leads us to the question of testing for the linear and the quadratic portions of a main effects trend. Such a test or tests can be carried out by easy extensions of the calculations for a single trend as in the one-way analysis of variance setup, discussed previously under the headings Linear Trends and Hypotheses about Curvatures. We will limit ourselves to the situations specified in subheadings (i.e., in boldface type) under the topics of differences among slopes and among curvatures, and for equal spacings and an equal number of scores entering into the main effect means involved in a trend.

For the two-way design with $m$ independent subjects in the $RC$ cells, the means for the main effect of the column factor will each be based on $mR$ scores; thus to get the required sum of squares for the single slope we would start with formula (17.1), now written with $X$ as the dependent variable as $(\Sigma vX)^2/m\Sigma v^2$, and replace its $m$ by the $mR$ of the two-factor design. The $\Sigma vX$ would, of course, involve multiplying each $\Sigma X$, obtained by adding *all* the $X$s in a column, by its $v$ value, then summing the $C$ products so obtained. For the sum of squares for the quadratic component we simply use the appropriate $u$s in place of the $v$s. The $df$ for each sum of squares will be 1, and the error term for both $F$s is $s^2_w$, the error term that is used to test the main effect of the column factor. If the row factor involves equal spacings, exactly the same procedure would be followed with choice of appropriate $v$ and $u$ values as multipliers for the $\Sigma X$ values that enter into the main effect means for the row factor. The $m$ in the denominator becomes $mC$. Again, $s^2_w$ is the error term for the two $F$s.

For the two-factor design with repeat measures for one factor (say that assigned to the columns) and independent subjects from block to block, which is a Case XV layout, the sum of squares for the over-all slope, and also the quadratic component, is computed by again using formula (17.1), with $v$ and $u$, respectively, and the $m$ replaced by $RB$, which is the number of scores for each of the $C$ main effect means. But this time the error term for the two $F$s must be calculated by way of the $RB$ individual slopes and the $RB$ individual quadratic "scores" (which values will have been computed for the previously given tests for differences between the $B$ slopes and among the $B$ quadratic components). These two sets of $RB$ "scores" become the $X$s for one-way analysis of variance. Each set will yield an $s^2_w$ with $RB - B$ degrees of freedom as an error term for an $F$. It will be noted that the two error terms are both free of any variation in slopes and in quadratic curvatures that might have been produced by the block factor. Actually, both of these $F$s correspond to testing whether an over-all mean is significantly different from zero. For example, the slope shown by the $C$ means is simply the mean of the $RB$ individual slope "scores," and departure of the mean from zero can be tested by $t\ (= \sqrt{F})$ with $RB - B$ degrees of freedom. If for the Case XV layout we have $B$ equally spaced levels on the block factor, we can test for the linear and the quadratic components of the trend of $X$ on the block variable. The required sums of squares for these components are obtained by using formula (17.1). This time $m$ is taken as $RC$. The sum of the $X$s in each block is multiplied by the appropriate $v$ and $u$ values to get the numerator parts, $\Sigma vX$ and $\Sigma uX$. Then when the $B$ values of $(\Sigma vX)^2/RC\Sigma v^2$, also those of $(\Sigma uX)^2/RC\Sigma u^2$, are added we have the needed sums of squares, each with $df = 1$. The error term for both $F$s is the $s^2_i$ of Table 16.16.

For the two-factor situation with repeat measures on both factors (a three-way mixed model, rows as subjects, hence random) for which we have equal spacing for the $C$ levels, also for the $B$ levels, the slope (linear) and quadratic components for the column and for the block factors, as main effects, can be tested by way of individual "scores" for linear and for quadratic trend; each subject will have two linear "scores," one against the column factor, the other against the block factor. Likewise, there are two quadratic "scores." The scores become the bases for four one-way analysis of variance layouts from which emerge $s^2_w$ values as appropriate denominators for four $F$s. The procedure is precisely the same as outlined in the previous paragraph for a factor involving repeat measures. The $m$ for formula (17.1) is $RB$ for the trend on the column factor, and $RC$ for the trend on the block factor; the $df$s are $RB - B$ and $RC - C$, respectively. Again, the individual "scores" for linear and for quadratic trend may have already been ascertained for trend difference analysis.

## POSSIBLE EXTENSIONS

The foregoing methods of analysis can be expanded in two directions.

1. When three or more factors are in the design, the computations can fairly easily be extended to test for possible linear and quadratic components of the trend on a given factor and for the differences among the linear and quadratic aspects of two-way interactions.

2. A trend can be broken down into additional components—cubic, quartic, quintic, etc.—each with $df$ of 1. Interactions can also be examined for differences corresponding to these equations of higher degree. Since these polynomial forms of relationship are scarce in the empirical data of psychology and even more scarce in the minds of psychological theorists, there would seem to be no good reason for going beyond the second degree polynomial (the quadratic) in this business of extracting components, with its implication that lawful

relationships among psychological variables somehow involve a cubic or higher-order polynomial curve. Admittedly, the quadratic form of relationship may rarely hold for psychological variables, but the testing of the quadratic component does provide us with a more sensitive statistical test of an effect than is possible by $F = s^2_b/s^2_w$ when systematic curvature is present and/or has been predicted. It would be easy to devote a small book to the near-silly use of trend analysis during the past dozen years in psychology.

# Chapter 18

## ANALYSIS OF VARIANCE: COVARIANCE METHOD

It is usually possible in experimentation to choose, either by random methods or by pairing or matching, groups that are comparable on variables judged relevant to the comparisons to be made. There are times, however, when it is more practicable to use intact groups which may differ in important respects, and occasionally we may wish to make an unanticipated comparison which does not seem justifiable in light of known differences between groups. Experimental control is the ideal, but, if this cannot be attained, we may resort to statistical allowances and thereby arrive at valid conclusions.

Suppose that two intact groups are being used to evaluate the relative merits of two methods of memorizing and that the mean IQ is 105 for group A and 111 for group B. Now, if there is an appreciable correlation between the particular memorizing ability involved and intelligence, the results will need qualifying because of the difference in intelligence of the two groups. It would seem logical to use the regression equation, for estimating memory score from intelligence, as a basis for predicting how much of a difference in memorizing would arise because of the group difference in IQs. Let us suppose that the mean memory performance is 60 for group A and 70 for group B, and that substituting 105 and 111 in the regression equation yields a predicted value of 62 for group A and of 68 for group B. Thus our prediction would lead us to expect a difference of 6 points, and accordingly it would be said that 6 of the obtained difference of 10 could be attributed to lack of comparability of the two groups with respect to intelligence.

The next question concerns the proper sampling error to use in evaluating the adjusted difference. It should be obvious that the ordinary procedure is inapplicable for the simple reason that we have tampered with the obtained means and in so doing have interfered somewhat with the operation of chance.

It is the purpose of this chapter to give a precise method for making allowance for an uncontrolled variable and to set forth the sampling error adjustment which is needed in testing the statistical significance of the difference between "corrected" means. The method is applicable whenever it seems desirable to correct a difference on a dependent variable for a known difference on another variable which for some reason could not

**Table 18.1. Schema of scores for covariance**

Group

| 1 | | 2 | | $g$ | | $G$ | |
|---|---|---|---|---|---|---|---|
| $X_{11}$ | $Y_{11}$ | $X_{12}$ | $Y_{12}$ | $X_{1g}$ | $Y_{1g}$ | $X_{1G}$ | $Y_{1G}$ |
| $X_{21}$ | $Y_{21}$ | $X_{22}$ | $Y_{22}$ | $X_{2g}$ | $Y_{2g}$ | $X_{2G}$ | $Y_{2G}$ |
| $X_{i1}$ | $Y_{i1}$ | $X_{i2}$ | $Y_{i2}$ | $X_{ig}$ | $Y_{ig}$ | $X_{iG}$ | $Y_{iG}$ |
| $X_{m1}$ | $Y_{m1}$ | $X_{m2}$ | $Y_{m2}$ | $X_{mg}$ | $Y_{mg}$ | $X_{mG}$ | $Y_{mG}$ |

be controlled by matching or by random sampling procedures. Since the scheme about to be proposed has an analysis of variance setting, the reader can readily guess that it will provide an adjustment for, and a test of significance of, the differences between two or more groups, and that it will be usable for either large or small samples.

In order to present the required adjustments, we need first to consider *covariance*, which is defined as $\Sigma xy/N$ or $\Sigma(X - \bar{X})(Y - \bar{Y})/N$. The sum of products of deviations can be broken down into components in a manner similar to that used with a sum of squares. In the simplest situation we can have $m$ pairs of $X$ and $Y$ scores in each of $G$ groups. These pairs of scores can be recorded in some such fashion as that depicted in Table 18.1. Note that $X_{ig}$ and $Y_{ig}$ stand for the $X$ and $Y$ values of the $i$th individual in the $g$th group. Note also that in allowing $i$ to take on values running from 1 to $m$ we do not imply any order for the individual, and that the $i$th individual in one group is in no sense paired with the $i$th case in another group. The product of the deviation scores for the $i$th individual in the $g$th group would be $(X_{ig} - \bar{X})(Y_{ig} - \bar{Y})$, in which $\bar{X}$ and $\bar{Y}$ are the means for all $mG$ cases. The total sum of products would be $\Sigma\Sigma(X_{ig} - \bar{X})(Y_{ig} - \bar{Y})$.
Now each deviation can be expressed in terms of two components in

**Table 18.2. Setup for analysis of variance by covariance adjustments; sums indicated by boldface X, Y, and XY**

|  | Total | Within | Between |
|---|---|---|---|
| 1. Sum of products | $(\mathbf{XY})_t$ | $(\mathbf{XY})_w$ | $(\mathbf{XY})_b$ |
| 2. Sum of squares: $X$s | $\mathbf{X}_t$ | $\mathbf{X}_w$ | $\mathbf{X}_b$ |
| 3. Sum of squares: $Y$s | $\mathbf{Y}_t$ | $\mathbf{Y}_w$ | $\mathbf{Y}_b$ |
| 4. $df$ | $N-1$ | $N-G$ | $G-1$ |
| 5. Correlation coefficients | $\dfrac{(\mathbf{XY})_t}{\sqrt{\mathbf{X}_t}\,\sqrt{\mathbf{Y}_t}}$ | $\dfrac{(\mathbf{XY})_w}{\sqrt{\mathbf{X}_w}\,\sqrt{\mathbf{Y}_w}}$ | $\dfrac{(\mathbf{XY})_b}{\sqrt{\mathbf{X}_b}\,\sqrt{\mathbf{Y}_b}}$ |
| 5a. $df$ for $r$ | $N-2$ | $N-G-1$ | $G-2$ |
| 6. $b_{xy}$ | $(\mathbf{XY})_t/\mathbf{Y}_t$ | $(\mathbf{XY})_w/\mathbf{Y}_w$ | $[(\mathbf{XY})_b/\mathbf{Y}_b]$ |
| 7. Adjusted sum of squares: $X$s | $[\mathbf{X}_t - (\mathbf{XY})^2_t/\mathbf{Y}_t]$ | $-\ [\mathbf{X}_w - (\mathbf{XY})^2_w/\mathbf{Y}_w] = $ adj. $\mathbf{X}_b$ |  |
| 8. $df$ | $N-2$ | $N-G-1$ | $G-1$ |

exactly the same way as in Chapter 15; i.e., one part is the deviation of the score from the mean of the group to which it belongs, and the other part is the deviation of the group mean from the total mean. Thus we have

$$(X_{ig} - \bar{X}) = (X_{ig} - \bar{X}_g) + (\bar{X}_g - \bar{X})$$

and

$$(Y_{ig} - \bar{Y}) = (Y_{ig} - \bar{Y}_g) + (\bar{Y}_g - \bar{Y})$$

Then the foregoing sum of the products becomes

$$\sum_i\sum_g [(X_{ig} - \bar{X}_g) + (\bar{X}_g - \bar{X})][(Y_{ig} - \bar{Y}_g) + (\bar{Y}_g - \bar{Y})]$$

When the bracketed expressions are multiplied together, four terms result, and, since two of these vanish, we have left that the total sum of products is equal to

$$\sum_i\sum_g (X_{ig} - \bar{X}_g)(Y_{ig} - \bar{Y}_g) + m\sum_g(\bar{X}_g - \bar{X})(\bar{Y}_g - \bar{Y})$$

The first of these terms involves a *within*-groups sum of products, whereas the second is for *between* groups. If there happens to be an unequal number of cases per group, the $m$ of the second term goes under the summation sign as $m_g$. The degrees of freedom for the total sum of products is $mG - 1$, or $N - 1$, where $N$ is the sum of the $m_g$s; the $df$s for the within and between terms are $mG - G$ (or $N - G$) and $G - 1$ respectively.

It will be of convenience to assemble in a table the sums of products, along with the sums of squares, for both the $X$ and $Y$ variables. These will be found in the first three lines of Table 18.2.

Although we are here presenting the covariance technique as a method for making such adjustments as discussed in introducing this chapter, it is of interest to link covariance with the problem of correlation. The product moment correlation coefficient is usually defined as

$$r = \frac{\Sigma xy}{N S_x S_y}$$

which may be written as

$$r = \frac{\Sigma xy}{N \sqrt{\frac{\Sigma x^2}{N}} \sqrt{\frac{\Sigma y^2}{N}}} = \frac{\Sigma xy}{\sqrt{\Sigma x^2}\sqrt{\Sigma y^2}} = \frac{\Sigma(X - \bar{X})(Y - \bar{Y})}{\sqrt{\Sigma(X - \bar{X})^2}\sqrt{\Sigma(Y - \bar{Y})^2}}$$

or as a function of a sum of products and two sums of squares. Using the sums of Table 18.2, we may specify three correlations: one based on the total sums, one based on the within sums, and one based on the between sums. These three correlations are indicated in line 5 in terms of appropriate sums. Line 5a gives the dfs for the rs.

Note that the between-groups r is actually the correlation between the X means and the Y means for the groups. If this r is significant, it follows that one source of the correlation for the total group is the heterogeneity resulting from the throwing together of groups with unlike means. (This between-groups correlation is meaningless when only two groups are involved. Why?) Stated differently, an appreciable between-groups r indicates that the total r is spurious; this spuriousness is eliminated when r is computed from the within sums. The similarity of the within-groups r to the partial correlation coefficient will be recognized by the discerning student, especially if he recalls the derivation of the latter.

We now turn to the use of covariance as a basis for allowing for the influence of an uncontrolled variable on the differences between group means. The question here is not what the result would be if the uncontrolled variable were held constant, as in partial correlation, but rather what the result would be if the groups were made comparable with respect to the uncontrolled variable. Let X represent the dependent variable, and Y the uncontrolled variable. It is presumed that the $\bar{Y}_g$ values differ, and that X is correlated with Y in a linear fashion. For purposes of exposition we shall refer to Table 18.2, which will serve as an outline of the required computations. Line 6 of this table gives the regression coefficients, $b_{xy}$, for predicting X from Y. Since no use is made of this regression coefficient based on between sums, it need not be computed.

That these $(XY)/Y$ values are regression coefficients can readily be demonstrated. In Chapter 9 the regression of $X$ on $Y$ was given as

$$b_{xy} = r \frac{S_x}{S_y}$$

Since, as we have seen previously,

$$r = \frac{\Sigma xy}{\sqrt{\Sigma x^2}\sqrt{\Sigma y^2}}, \qquad S_x = \sqrt{\Sigma x^2/N}, \quad \text{and} \quad S_y = \sqrt{\Sigma y^2/N}$$

we have

$$b_{xy} = \frac{\Sigma xy}{\sqrt{\Sigma x^2}\sqrt{\Sigma y^2}} \cdot \frac{\sqrt{\Sigma x^2/N}}{\sqrt{\Sigma y^2/N}}$$

$$= \frac{\Sigma xy}{\Sigma y^2} = \frac{\mathbf{XY}}{\mathbf{Y}}$$

In order to make allowance for the uncontrolled differences in $\bar{Y}_g$, we need not only to adjust the $\bar{X}_g$ values but also to make an adjustment to the error term, which is used as the denominator of the $F$ ratio in testing the difference between the adjusted $X$ means. As in the simpler situation of Chapter 15, $F$ will involve the ratio of between-groups to a within-groups variance estimate.

First, let us consider the method of making the adjustment to the total and to the within-groups variance estimates. The problem here is that of specifying how much of the variation in $X$ can be predicted from variation in $Y$ and then of subtracting this to secure the left-over variation as an adjusted value. But this left-over variance is nothing more than the residual variance, or square of the standard error of estimate, obtainable from formula (9.6):

$$S^2_{x \cdot y} = S^2_x - r^2 S^2_x$$

Actually the adjustment is to be made to the sum of squares. In order to state the residual variance in terms of sums, we may substitute for $S^2_x$ and $r^2$. Thus,

$$S^2_{x \cdot y} = \frac{\Sigma x^2}{N} - \frac{(\Sigma xy)^2}{(\Sigma x^2)(\Sigma y^2)} \cdot \frac{\Sigma x^2}{N}$$

hence,

$$NS^2_{x \cdot y} = \Sigma x^2 - \frac{(\Sigma xy)^2}{\Sigma y^2}$$

Since $NS^2$ always equals a sum of squares, the value of $NS^2_{x \cdot y}$ is obviously the sum of squares for the residuals. In the notation of this chapter,

$$NS^2_{x \cdot y} = \sum_i \sum_g (X_{ig} - \bar{X})^2 - \frac{\left[ \sum_i \sum_g (X_{ig} - \bar{X})(Y_{ig} - \bar{Y}) \right]^2}{\sum_i \sum_g (Y_{ig} - \bar{Y})^2}$$

would be the residual sum of squares after the regression adjustment. This sum can be written as

$$NS^2_{x \cdot y} = \mathbf{X}_t - (\mathbf{XY})^2_t / \mathbf{Y}_t$$

which is the entry for the *total* group in line 7 of Table 18.2. Similarly, the corresponding residual, or adjusted sum of squares for *within* groups is $\mathbf{X}_w - (\mathbf{XY})^2_w / \mathbf{Y}_w$.

At first thought it would seem logical to adjust $\mathbf{X}_b$ by the use of $(\mathbf{XY})_b$ and $\mathbf{Y}_b$, but the between-groups regression (and $r$) is affected by the differences between the $X$ means, which are the differences to be adjusted and then tested for statistical significance. Our adjustment should be one that is independent of the differences to be tested. This suggests that the regression for within groups, or $(\mathbf{XY})_w / \mathbf{Y}_w$, should be used since the regression for total is also affected by the differences which we are out to test. Insofar as we are concerned solely with the adjustment of the between-group $X$ means, the best adjustment would be by use of the within-groups regression. This could take the form of either an adjustment to the between-groups sum of squares for $X$ or a direct adjustment to the several $\bar{X}_g$ values.

Although the latter would be the best way of ascertaining how much of an effect the noncomparability of the groups with respect to $Y$ had on the $X$ means, there is another consideration as to whether the within regression is appropriate for adjusting the between-groups sum of squares. It will be recalled that $F$ is to be taken as the ratio of a variance estimate based on the between sum of squares to that based on within groups, and that the two variance estimates being so compared must be independent estimates. Now, if we adjust both the within and the between sum of squares by means of the same regression coefficient (say, that based on within groups), any sampling error in this regression coefficient would have a similar effect on both adjustments; hence it could not be argued that the resulting adjusted sums of squares possess the requisite independence. Therefore variance estimates based thereon would not be strictly independent.

This difficulty is overcome by taking the adjusted sum of squares for between groups as the difference between the adjusted total sum and the

adjusted within sum of squares. Thus, for the purpose of testing significance,

$$[\mathbf{X}_t - (\mathbf{XY})^2{}_t/\mathbf{Y}_t] - [\mathbf{X}_w - (\mathbf{XY})^2{}_w/\mathbf{Y}_w]$$

leads to the proper adjustment for the between sum of squares for $X$.

Perhaps the reader has anticipated that the $df$s may change as a result of
these manipulations. The new $df$s are recorded in line 8 of Table 18.2.
Note that the $df$ for the between sum has not changed since the adjustment
was not made by using the between-groups regression.

Aside from the usual methods for calculating sums of squares, we need
formulas for computing sums of products in terms of raw scores. The
following formulas are written for unequal $m_g$ values, but are of course
applicable for equal $m$s.

$$\sum_i\sum_g (X_{ig} - \bar{X})(Y_{ig} - \bar{Y}) = \sum_i\sum_g X_{ig}Y_{ig} - \frac{\sum_i\sum_g X_{ig}\sum_i\sum_g Y_{ig}}{N} \quad \text{for total} \quad (18.1)$$

$$\sum_i\sum_g (X_{ig} - \bar{X}_g)(Y_{ig} - \bar{Y}_g) = \sum_i\sum_g X_{ig}Y_{ig} - \sum_g \frac{\sum_i X_{ig}\sum_i Y_{ig}}{m_g} \quad \text{for within} \quad (18.2)$$

$$\sum_g m_g(\bar{X}_g - \bar{X})(\bar{Y}_g - \bar{Y}) = \sum_g \frac{\sum_i X_{ig}\sum_i Y_{ig}}{m_g} - \frac{\sum_i\sum_g X_{ig}\sum_i\sum_g Y_{ig}}{N} \quad \text{for between}$$

$$(18.3)$$

Thus to compute the sums of products of deviations, we need the sum of
all $N$ raw score products or $\sum_i\sum_g X_{ig}Y_{ig}$, the sum of all the $X$s or $\sum_i\sum_g X_{ig}$,
the sum of all the $Y$s or $\sum_i\sum_g Y_{ig}$, the sum of the $X$s separately for each group
or $\sum_i X_{ig}$, and the sum of the $Y$s for each separate group or $\sum_i Y_{ig}$. Adding
the several $X$ sums gives the sum of all the $X$s; likewise for $Y$s. Note that
to get the second term of (18.2), or the first term of (18.3), we must divide
the product of the two sums for a group by its $m$ and then sum such
quotients over all $G$ groups. The reader may find some interest in comparing formulas (18.1–18.3) with formulas (15.9–15.11) and it should be
apparent that in the case of equal $m$s formulas (18.1–18.3) can be written
in the simpler way of formulas (15.6–15.8).

The required computations are illustrated by using the data (fictitious)
of Table 18.3, which contains $Y$ and $X$ scores for ten cases in each of
three groups. The scores in each of the six columns are separately summed
to yield 109, 73, etc. The scores are squared and summed to yield 1213,
557, etc. Summing the products of the $X$ and $Y$ values gives 810, 571, and
307 for the three groups. Summing over groups yields the double summa-
tions 173, 268, etc. Certain of these sums are then substituted into

**Table 18.3. Score data and sums based on raw scores for analysis of variance by covariance adjustments**

| | | | | | | | |
|---|---|---|---|---|---|---|---|
| | **Group** | | | | | | |
| | **1** | | **2** | | **3** | | |
| | $Y$ | $X$ | $Y$ | $X$ | $Y$ | $X$ | |
| | 14 | 10 | 11 | 5 | 7 | 5 | |
| | 9 | 6 | 9 | 2 | 6 | 4 | $\Sigma\Sigma X = 173$ |
| | 11 | 8 | 8 | 6 | 2 | 1 | $\Sigma\Sigma Y = 268$ |
| | 12 | 6 | 10 | 5 | 10 | 7 | |
| | 10 | 9 | 10 | 4 | 7 | 9 | $\Sigma\Sigma X^2 = 1161$ |
| | 11 | 7 | 10 | 8 | 7 | 4 | $\Sigma\Sigma Y^2 = 2642$ |
| | 11 | 9 | 12 | 10 | 6 | 5 | |
| | 8 | 5 | 9 | 6 | 3 | 2 | $\Sigma\Sigma XY = 1688$ |
| | 11 | 6 | 10 | 4 | 2 | 2 | |
| | 12 | 7 | 11 | 6 | 9 | 5 | $\Sigma(\Sigma X)^2 = 10,401$ |
| | | | | | | | $\Sigma(\Sigma Y)^2 = 25,362$ |
| Sum | 109 | 73 | 100 | 56 | 59 | 44 | $\bar{X} = 5.77$ |
| Mean | 10.9 | 7.3 | 10.0 | 5.6 | 5.9 | 4.4 | $\bar{Y} = 8.93$ |
| $\Sigma Y^2$ or $\Sigma X^2$ | 1213 | 557 | 1012 | 358 | 417 | 246 | |
| $\Sigma XY$ | 810 | | 571 | | 307 | | |

formulas (18.1–18.3) to secure the total, within, and between sums of products of deviations. By substituting the proper sums into formulas (15.6–15.8), we get the required sums of squares for the $X$s and for the $Y$s. Then these three sets of sums are entered as the first three rows of Table 18.4, which follows the pattern set forth in Table 18.2.

Before proceeding to the covariance adjustment, let us consider the means given in Table 18.3. It will be noticed that the groups differ considerably on $X$, or the dependent variable, and that they also differ on $Y$, the relevant but not controlled variable. An analysis of variance based on the sum of squares for the $X$s leads to a between-groups variance estimate of 42.47/2, or 21.26, and a within-groups estimate of 120.90/27, or 4.48. The $F$ for testing the significance of the between-groups variance becomes 21.26/4.48, or 4.75, which for the given $df$s is significant at about the .02 or .03 level of significance. This analysis does not, of course, allow for the fact that the groups differ on $Y$. If there is correlation between $X$ and $Y$, the observed differences on $X$ may be mainly a reflection of the group differences on $Y$. As previously stated, the purpose of the covariance adjustment is to make statistical allowance for such uncontrolled differences.

**Table 18.4. Analysis of variance for $X$ variable of Table 18.3 by covariance adjustments for uncontrolled $Y$**

|  | Total |  | Within |  | Between |
|---|---|---|---|---|---|
| 1. Sum of products | 142.53 |  | 72.70 |  | 69.83 |
| 2. Sum of squares: $X$ | 163.37 |  | 120.90 |  | 42.47 |
| 3. Sum of squares: $Y$ | 247.87 |  | 105.80 |  | 142.07 |
| 4. df | 29 |  | 27 |  | 2 |
| 5. Correlation | .709 |  | .643 |  | .912 |
| 5a. df for r | 28 |  | 26 |  | 1 |
| 6. $b_{xy}$ value | .5750 |  | .6871 |  | . . . . . . |
| 7. Adjusted $\Sigma x^2$ | 81.42 | minus | 70.95 | equals | 10.47 |
| 8. df | 28 |  | 26 |  | 2 |

By following the steps indicated in Table 18.2, we determine the values in lines 5 to 7 of Table 18.4. Note that the adjusted $\Sigma x^2$ for between groups, 10.47, is secured by subtracting 70.95 from 81.42. The analysis of variance based on the adjusted sums of squares (for the $X$s) gives a between-groups variance estimate of 10.47/2, or 5.23, and a within-groups estimate of 70.95/26, or 2.73. Then $F = 5.23/2.73 = 1.92$, which for 2 and 26 degrees of freedom yields a $P$ of about .20. Accordingly, it cannot be concluded that there are significant group differences on $X$ over and above those which would be expected because of the differences on $Y$.

It should be obvious that the use of the covariance adjustment method must be justified by logical and experimental considerations. When it is logical to control a variable by pairing or matching, the covariance adjustment is defensible as a way of making proper allowance for a failure, because of infeasibility, to control the variable. The use of the covariance adjustment is not predicated on the degree of correlation between the dependent and the uncontrolled variable. If the correlation is relatively low, the adjusted values will differ but little from the unadjusted values; if high, both the total and within adjusted variances will differ considerably from the unadjusted variances, but, as we shall presently see, the extent to which the adjusted and unadjusted between-groups variances differ is not solely a function of the correlation.

It is of interest to make an actual adjustment of the $X$ means of Table 18.3 for the group differences on $Y$. The adjustments can be made by

$$\overline{X}_{ga} = \overline{X}_g - b_{xy}(\overline{Y}_g - \overline{Y})$$

in which $\overline{X}_{ga}$ is the adjusted value for the $g$th group, and $b_{xy}$ is the *within*-groups regression coefficient. For the data of Table 18.3 we have

$$\bar{X}_{1a} = 7.30 - .687(10.90 - 8.93) = 5.95$$

$$\bar{X}_{2a} = 5.60 - .687(10.00 - 8.93) = 4.86$$

$$\bar{X}_{3a} = 4.40 - .687(5.90 - 8.93) = 6.48$$

Should the reader be surprised that the adjustment puts group three ahead, he should ponder the fact that, relative to the *within*-groups $X$ and $Y$ variances, the third group's $\bar{X}$ of 4.40 was not as far below the means of the other two groups as was its $\bar{Y}$ of 5.90.

From a careful consideration of the foregoing, it will be seen that the covariance adjustment method will not necessarily reduce the differences between the means on the dependent variable. Situations arise in which groups that show marked differences on some correlated but uncontrolled variable may yield similar means on the variable being studied. Suppose that we are using two intact groups to investigate the relative merits of two learning methods, and that the intial means of the two groups are markedly different. We would, accordingly, expect a difference on final standing even though the two methods were equally efficacious. If this expected difference is not found, it follows that the method used by the group with the lower initial score was more effective in that this group overtook the other group. With groups differing on an uncontrolled variable, it is not only as proper, but also as necessary, to use the covariance technique when the groups are nearly the same on the dependent variable as when they are different. For such situations the adjustment will *increase* the between-groups variance.

The extent to which the adjusted variances lead to a level of significance different from that based on an analysis of the unadjusted values will obviously depend on three things: the degree of correlation between the dependent and uncontrolled variable, the size of the differences between the groups on the uncontrolled variable, and the found differences on the dependent variable. The applicability of the covariance technique does not depend on a minimum degree of correlation or on a definite amount of group differences on the uncontrolled variable. But, if the within-groups correlation is low and/or there is only a small, chance difference between the groups on the uncontrolled variable, the use of the covariance adjustment may not be worth the effort. Obviously, if a variable correlates near zero with the dependent variable, it need not be controlled experimentally or statistically.

The covariance method can be extended to make adjustments for group differences on more than one uncontrolled variable. This involves the use of multiple regression, but computationally it is perhaps simpler to handle the adjustments in terms of multiple $r$s. We need two multiple correlation coefficients, one obtained by way of correlations based on within-groups sums of squares and of products, and the other by way of correlations based on total sums of squares and of products.

If, for example, allowance is to be made for three uncontrolled variables, $Y_1$, $Y_2$, and $Y_3$, we will need six (one for each pair of variables—$X$ is the fourth or dependent variable) auxiliary tables consisting of entries like those in lines 1, 2, and 3 under the "total" and the "within" columns of Table 18.2 (or Table 18.4). We can then calculate two sets of intercorrelations (each auxiliary table will lead to two $r$s when the substitutions called for in line 5 of Table 18.2 are made) among the four variables, and from these we compute, by the methods set forth in Chapter 11, two $r^2_{x \cdot y_1 y_2 y_3}$ values. Let us designate the multiple based on the total sums as $R_t$ and that based on the within sums as $R_w$.

With these two multiple $r$s available, we may rewrite line 7 of Table 18.2 as

$$\mathbf{X}_t(1 - R^2_t) - \mathbf{X}_w(1 - R^2_w) = \text{adjusted } \mathbf{X}_b$$

with respective $df$s of

$$N - n, \qquad N - G - (n - 1), \qquad G - 1$$

for the $n$ variable problem (one dependent, plus the number of uncontrolled variables included in the adjustments).

The covariance technique can also be extended to adjust for uncontrolled differences in the two-way and higher-order designs. Here we will indicate the extension for the two-way, fixed effects design with $m$ independent subjects in the $RC$ cells. The breakdown of the total sum of squares for $X$ and $Y$, and the breakdown of the total sum of products, will of course involve exactly the same sources as for two-way analysis of variance: row, column, $R \times C$ interaction, and within cells. In order to explicate, we will again adopt boldface $\mathbf{X}$, $\mathbf{Y}$, and $(\mathbf{XY})$ as symbols for the sum of squares for $X$ and for $Y$ and for the sum of cross-products, respectively, along with appropriate subscripts: $t$, $r$, $c$, $rc$, and $w$. Table 18.5 gives, in symbols, the necessary sums. Each row in the table implies an equation: total sum = sum of four parts.

**Table 18.5. Sums required for covariance adjustments for two-factor setup**

| Sum | Total | Row | Column | $R \times C$ | Within |
|---|---|---|---|---|---|
| $\Sigma xy$ | $(\mathbf{XY})_t$ | $(\mathbf{XY})_r$ | $(\mathbf{XY})_c$ | $(\mathbf{XY})_{rc}$ | $(\mathbf{XY})_w$ |
| $\Sigma x^2$ | $\mathbf{X}_t$ | $\mathbf{X}_r$ | $\mathbf{X}_c$ | $\mathbf{X}_{rc}$ | $\mathbf{X}_w$ |
| $\Sigma y^2$ | $\mathbf{Y}_t$ | $\mathbf{Y}_r$ | $\mathbf{Y}_c$ | $\mathbf{Y}_{rc}$ | $\mathbf{Y}_w$ |
| $df$ | $mRC - 1$ | $R - 1$ | $C - 1$ | $(R-1)(C-1)$ | $RC(m-1)$ |

Assuming for the time being that all of the sums in Table 18.5 have been computed, we proceed to the method for making the "corrections" for uncontrolled group differences on the covariate, or concomitant variable, $Y$. Obviously, if the $RC$ groups in the cells have differing $Y$ means, the effect of $Y$ as an uncontrolled variable could be such as to produce differences among the $X$ means for the cells, which in turn could disrupt the means for the rows and for the columns and also affect the interaction. If we had the $\bar{X}_{rc}$ adjusted for the differing $\bar{Y}_{rc}$, an average of these across and downward would provide adjusted row and column means. The problem, however, is to adjust the four sums of squares in such a way as to obtain independent variance estimates for $F$ ratios.

The trick here is to note that we can, in effect, reduce a two-way analysis to a one-way analysis, and thereby eventually obtain values for a layout like that in Table 18.2. If we had not included, say, the column factor we would not have it and the $R \times C$ interaction as sources of variation. We would simply have between and within sources with the groups defined in terms of the levels on the factor assigned to rows. What we need is an expression for a total sum of squares, $X_t$, that reflects only these two sources so as to have $X_t = X_r + X_w$. But the $X_t$ in Table 18.5 includes two components having to do with columns. If from the $X_t$ in the table we subtracted those two components we would have a new or reduced $X_t$ that would contain only $X_w$ and $X_r$. Let $X_{t \cdot c(rc)}$ be the symbol for this left over (residual type) sum of squares. The subscripts and the dot indicate that the original $X_t$ has been reduced by taking out the column and the $R \times C$ interaction parts. Thus

$$X_{t \cdot c(rc)} = X_t - X_c - X_{rc}$$

but since $X_t = X_w + X_r + X_c + X_{rc}$, we have

$$X_{t \cdot c(rc)} = X_w + X_r \qquad (18.4)$$

Also

$$Y_{t \cdot c(rc)} = Y_w + Y_r \qquad (18.5)$$

and

$$(XY)_{t \cdot c(rc)} = (XY)_w + (XY)_r \qquad (18.6)$$

These nine sums, in (18.4) to (18.6), become the basic entries in the first three lines of the layout depicted in Table 18.2, or subsequently as the nine numerical values for a table like Table 18.4. From here on the procedure is exactly the same as that indicated in Table 18.2 and carried out in Table 18.4. The number of degrees of freedom is $R - 1$ for the adjusted row sum of squares and that for error, $N - G - 1$ becomes $N - RC - 1 = mRC - RC - 1$. Although the number of groups for the between term is $G = R$, the number of groups for the within terms is actually $RC$, the number of cells.

In order to have an $F$ test for the *column* effect with allowance for the uncontrolled $Y$, we need to get new sums for total by reducing the original total sums for the variation produced by the row and by the $R \times C$ interaction effects. The reduced sums will have just two sources: within cells and between columns when the row and interaction parts are subtracted from the original total sums. If in formulas (18.4) to (18.6) we switch subscripts $c$ to $r$ and $r$ to $c$, with $t$, $w$, and $(rc)$ unchanged, we will have the needed expressions for another set of nine values for a Table 18.2 layout. For example,

$$\mathbf{X}_{t \cdot r(rc)} = \mathbf{X}_t - \mathbf{X}_r - \mathbf{X}_{rc} = \mathbf{X}_w + \mathbf{X}_c$$

The *interaction* sum of squares can also be adjusted to allow for the uncontrolled $Y$ variable. Again, we will get a reduced total sum that is composed of two parts, this time a within cells and the $R \times C$ interaction. Consider the sums for the $X$ variable. If from

$$\mathbf{X}_t = \mathbf{X}_w + \mathbf{X}_r + \mathbf{X}_c + \mathbf{X}_{rc}$$

we subtract $\mathbf{X}_r + \mathbf{X}_c$ we get

$$\mathbf{X}_{t \cdot rc} = \mathbf{X}_w + \mathbf{X}_{rc}; \quad \text{also} \quad \mathbf{Y}_{t \cdot rc} = \mathbf{Y}_w + \mathbf{Y}_{rc};$$
$$\text{and} \quad (\mathbf{XY})_{t \cdot rc} = (\mathbf{XY})_w + (\mathbf{XY})_{rc}$$

as the nine entries for another Table 18.2 layout. Line 7 of that table becomes

$$\left[ \mathbf{X}_{t \cdot rc} - \frac{(\mathbf{XY})^2_{t \cdot rc}}{\mathbf{Y}_{t \cdot rc}} \right] - \left[ \mathbf{X}_w - \frac{(\mathbf{XY})^2_w}{\mathbf{Y}_w} \right] = \text{adjusted } \mathbf{X}_{rc}$$

The adjusted $\mathbf{X}_{rc}$ divided by its $df = (R - 1)(C - 1)$ provides a variance estimate that reflects the interactive effect after adjustment for the uncontrolled $Y$ variable. The error term for the $F$ test is, of course,

$$[\mathbf{X}_w - (\mathbf{XY})^2_w / \mathbf{Y}_w]$$

divided by its $df = mRC - RC - 1$. This within-cells variance estimate is, as a residual variance adjusted for $Y$, exactly the same as that used for testing both the row and column variances after they have been adjusted for the covariate $Y$. Perhaps it is not necessary to say that the reduced total sums are not the same for the three adjustment situations. For example, $\mathbf{X}_{t \cdot c(rc)} \neq \mathbf{X}_{t \cdot r(rc)} \neq \mathbf{X}_{t \cdot rc}$.

There remains the tedium of computing the basic starting sums of squares and of cross-products. Let $X_{irc}$ and $Y_{irc}$ stand for the pair of scores for the $i$th individual in the $r$th row and $c$th column. The sums of squares for $X$ and for $Y$ are computable by way of formulas (16.1) to (16.4), which were written without the subscript $i$. The directions, however, had summings over individuals as a part of the procedure. Things become a bit more complicated when we turn to the sums for products. The scores, as deviations, are expressed as

$$(X_{irc} - \bar{X}) = (\bar{X}_{r\cdot} - \bar{X}) + (\bar{X}_{\cdot c} - \bar{X})$$
$$+ (\bar{X}_{rc} - \bar{X}_{r\cdot} - \bar{X}_{\cdot c} + \bar{X}) + (X_{irc} - \bar{X}_{rc})$$
$$(Y_{irc} - \bar{Y}) = (\bar{Y}_{r\cdot} - \bar{Y}) + (\bar{Y}_{\cdot c} - \bar{Y})$$
$$+ (\bar{Y}_{rc} - \bar{Y}_{r\cdot} - \bar{Y}_{\cdot c} + \bar{Y}) + (Y_{irc} - \bar{Y}_{rc})$$

The cross-product, $(X_{irc} - \bar{X})(Y_{irc} - \bar{Y})$, is to be written in terms of the sources indicated to the right of the equality signs in the foregoing. There will be 16 such cross-products among the right-hand terms, but when we get the 16 products for every one of the $N = mRC$ subjects and then sum over each set of products separately, it will be found that only four are nonvanishing. That is,

$$\sum_{i}\sum_{r}\sum_{c}(X_{irc} - \bar{X})(Y_{irc} - \bar{Y})$$

$$= mC\sum_{r}(\bar{X}_{r\cdot} - \bar{X})(\bar{Y}_{r\cdot} - \bar{Y}) + mR\sum_{c}(\bar{X}_{\cdot c} - \bar{X})(\bar{Y}_{\cdot c} - \bar{Y})$$
$$+ m\sum_{r}\sum_{c}(\bar{X}_{rc} - \bar{X}_{r\cdot} - \bar{X}_{\cdot c} + \bar{X})(\bar{Y}_{rc} - \bar{Y}_{r\cdot} - \bar{Y}_{\cdot c} + \bar{Y})$$
$$+ \sum_{i}\sum_{r}\sum_{c}(X_{irc} - \bar{X}_{rc})(Y_{irc} - \bar{Y}_{rc})$$

The computational formulas are as follows.

For the total sum of products:

$$(\mathbf{XY})_t = \frac{1}{mRC}\left[mRC\sum_r\sum_c\sum_i X_{irc}Y_{irc} - \left(\sum_r\sum_c\sum_i X_{irc}\right)\left(\sum_r\sum_c\sum_i Y_{irc}\right)\right] \quad (18.7)$$

in which we have shifted the ordering of summing—$\sum_r\sum_c\sum_i$ directs you to sum over $i$ first, then over $c$, then over $r$. The $X$ and $Y$ sums and $XY$ sums for each cell should be recorded for future use.

For the row sum of products:

$$(\mathbf{XY})_r = \frac{1}{mRC}\left[R\sum_r\left(\sum_c\sum_i X_{irc}\right)\left(\sum_c\sum_i Y_{irc}\right) - \left(\sum_r\sum_c\sum_i X_{irc}\right)\left(\sum_r\sum_c\sum_i Y_{irc}\right)\right] \quad (18.8)$$

which states explicitly that you first sum over $i$ to get a sum for $X$, also for $Y$, in the $rc$th cell; then these sums are summed over $c$, thus producing a sum for $X$, and for $Y$, for each row; next, these $R$ pairs of sums are multiplied together to get $R$ products which are then summed over $r$.

For the column sum of products:

$$(\mathbf{XY})_c = \frac{1}{mRC}\left[ C\sum_c\left(\sum_r\sum_i X_{irc}\right)\left(\sum_r\sum_i Y_{irc}\right) - \left(\sum_r\sum_c\sum_i X_{irc}\right)\left(\sum_r\sum_c\sum_i Y_{irc}\right)\right] \quad (18.9)$$

which is similar to the procedure for the row part except that $C$ products of sums are being summed.

For the within-cell sum of products:

$$(\mathbf{XY})_w = \frac{1}{m}\left[ m\sum_i\sum_r\sum_c X_{irc}Y_{irc} - \sum_r\sum_c\left(\sum_i X_{irc}\right)\left(\sum_i Y_{irc}\right)\right] \quad (18.10)$$

The sum of products for the interaction part is obtained by subtraction:

$$(\mathbf{XY})_{rc} = (\mathbf{XY})_t - (\mathbf{XY})_r - (\mathbf{XY})_c - (\mathbf{XY})_w \quad (18.11)$$

The reader may wish to compare and contrast these formulas with those for the sums of squares, (16.1) to (16.4). For every term involving a square in those formulas we have a corresponding term in (18.7) to (18.10) that involves a product. For example, consider (16.2) rewritten with the subscript $i$ with summing over $i$ explicitly stated. We have

$$\frac{1}{mRC}\left[ R\sum_r\left(\sum_c\sum_i X_{irc}\right)^2 - \left(\sum_r\sum_c\sum_i X_{irc}\right)^2\right] \quad (16.2)$$

which would, of course, also hold for $Y$ substituted for $X$ to get the between-row sum of squares for $Y$. Thus the following will have been calculated:

$$\sum_c\sum_i X_{irc}, \qquad \sum_c\sum_i Y_{irc}, \qquad \sum_r\sum_c\sum_i X_{irc}, \qquad \text{and} \quad \sum_r\sum_c\sum_i Y_{irc}$$

When the two involving $X$ are used in (16.2), we get the between-rows sum of squares for $X$; the two involving $Y$ give the between-rows sum of squares for $Y$; and, more importantly, these four, already used to get two sums of squares, are inserted in (18.8) to get a sum of products. The same parallelism holds for (16.3) and (18.9). Unfortunately, the sum for the first part of (18.7), which appears also in (18.10), is not computable from separate sums for $X$ and $Y$. The last part of (18.10) requires the sum of $X$s and $Y$s for each of the $RC$ cells, then the products of the $RC$ sums are summed.

Remark: The use of the covariance adjustment technique is far superior to attempts at pairing individuals from the intact groups on the basis of one or more uncontrolled variables, a procedure which inevitably leads to a reduction of sample size and also runs astride a regression difficulty (see p. 179).

**Evaluation of changes.** In Chapter 6 we discussed the usually advocated method for comparing changes shown by experimental and control groups (applicable also for two experimental groups). We have, with $i$ and $f$ standing for the pretest and posttest measures and $E$ and $C$ standing for experimental and control groups,

$$D = \bar{D}_E - \bar{D}_C = (\bar{X}_{fE} - \bar{X}_{iE}) - (\bar{X}_{fC} - \bar{X}_{iC})$$

as the net change, the change shown by the experimentals corrected for that shown by the controls. We may rearrange the $\bar{X}$s, yet maintain the numerical value of $D = \bar{D}_E - \bar{D}_C$, as follows:

$$D = (\bar{X}_{fE} - \bar{X}_{fC}) - (\bar{X}_{iE} - \bar{X}_{iC})$$

from which it is seen that the net change may also be thought of as the final difference between the two groups corrected for their initial difference. Such a correction involves the assumption that each unit of difference in initial standing will produce a unit of difference in final standing. In other words, this type of adjustment implies a 1-to-1 relationship between initial and final scores. Since a perfect correlation is never found or approached in practice, one may question whether the usual procedure of comparing changes is really defensible.

It is, of course, entirely logical that group differences on final scores, which we may here call the dependent variable, should be corrected for group differences on initial standing as an uncontrolled variable. The covariance adjustment technique provides a way of correcting final means for initial differences, with due allowance for the *degree* of correlation between initial and final scores. The ordinary and the covariance method differ not only in the correction but also in the resultant sampling error. The ordinary technique uses a standard error which definitely includes, either explicitly or implicitly, the variance for both initial and final scores and the correlation of initial with final, whereas the error term used in the covariance method is a direct function of the degree of correlation and of the variance for the final scores only. In other words, the net differences being tested are not the same, and neither are the error terms the same. The covariance method will, in general, be more sensitive. The student should read Professor R. A. Fisher's discussion on this point.*

**Cautions.** The covariance adjustment technique is based on certain assumptions, which we here subsume under three categories.

1. Since the $F$ test is used, it is not surprising that a homogeneity of variance assumption is involved. What variances? The residual variances for the dependent variable about the $G$ (or $RC$) regression lines, which are

_____

* Chapter IX in Fisher, R. A., *Design of experiments*, London: Oliver and Boyd.

pooled to get a within-groups variance estimate as the error term for $F$. Presumably violations of this assumption will not have serious consequences.

2. It is assumed that the group regressions for the dependent variable on the control variable are linear and, more importantly, that the slopes are homogeneous. Obviously, the use of linear regression adjustment will be inadequate in a curvilinear situation. The fact that one slope (that based on within sums) is involved in the adjustments implies that a single slope holds for the $G$ (or $RC$) populations represented by the $G$ (or $RC$) samples. This assumption, which can easily fail to hold, may or may not be crucial—a guess is that minor violations are tolerable.

3. Aside from variance and slope homogeneity, it does not seem reasonable to believe that the covariance adjustment method can be valid when the groups represent populations that differ markedly in average level on the control variable. In other words the (intact) groups must be regarded as belonging to one population or to populations that are essentially similar on the control variable.

# Chapter 19
## DISTRIBUTION-FREE METHODS

The $F$, $t$, and $z$ techniques are sometimes referred to as parametric because they involve the estimation of at least one parameter (population value); e.g., a population variance estimate is needed as the error term for $F$. These techniques also involve, at the derivation stage, an assumption of a normal distribution (of some variate). The techniques to be presented in this brief chapter are sometimes called nonparametric because they do not involve the estimation of parameters and they are sometimes referred to as distribution-free methods because no assumptions are made about the distribution of variates.

Advocates of nonparametric tests usually do not stress their nonparametric character as much as their distribution-free property. It is said that distribution-free methods should be used because the assumption of normality, on which parametric tests are based, may not hold. But in light of Norton's study (p. 288) and Boneau's results (p. 118), the worry about violating this assumption seems ill-founded.

Another argument advanced in favor of nonparametric methods springs from the level of measurement achieved by measuring instruments. These levels are referred to as nominal, ordinal, interval, and ratio scales of measurement. The so-called nominal scale is not a scale at all since it involves nothing more than classifying individuals into categories that are qualitatively different, with no ordering implied. Numbers may be used for coding the categories, and frequencies may be expressed as proportions or percentages. The ordinal scale connotes an ordering with rank-order positions usually specified by numbers. An interval scale is one for which equal units can be claimed; e.g., when the interval 140–150 represents exactly the same amount as the interval 110–120. Such is the case if we are measuring length in terms of inches or weighing objects in terms of pounds. Without giving any reasons, we will make the dogmatic-sounding assertion that very little "measurement" in psychology involves equal units—the scales and tests provide a basis for ordering which may be regarded as much better than subjective rank-ordering. For a scale to be called a ratio scale it must have a true zero point, in addition to qualifying as an interval scale.

Now it is claimed by some that the level of measurement definitely restricts possible statistical treatment of data. It is easy to see that adding numbers, used to code qualitatively different categories, in order to compute a mean is nonsensical. It is easy to see that a ratio scale is required for a meaningful coefficient of variation, $(S/M)$. It is easy to see that rank positions as "scores" will lead to absurd standard deviations—if ten persons are ranked the ranks range from 1 to 10 whereas if twenty-five persons are ranked the rank scores range from 1 to 25—the amount of variation is a direct function of $N$, hence a $\sigma$ or an $S$ is not descriptive of group variation. And it is easy to see that the adding of scores that do not qualify as being on an interval scale may make the mathematical purist dubious about the precise descriptive property of the mean, variance, and product moment $r$, all of which call for addition.

The crucial question, however, is whether or not the $F$, $t$, and $z$ tests can, in view of their dependence on means and variances, be safely used when the scale of measurement is, as is the rule in psychology, somewhere between the ordinal and the interval scales. The question boils down to this: Will $F$s, $t$s, and $z$s follow their respective theoretical sampling distributions when the underlying scores are not on an interval scale? The answer to this is a firm yes provided the score distributions do not markedly depart from the normal form. Nowhere in the derivations purporting to show that various ratios will have sampling distributions which follow either the $F$ or the $t$ or the normal distribution does one find any reference to a requirement of equal units. The attaining of an interval scale of measurement, though desirable for some reasons, will not alter the risks of type I and type II errors when statistical inferences are made.

There is, of course, no denying the fact that the type of data available does dictate the type of statistical technique that can be used. We have already discussed methods for handling nominal data—either by the binomial or by $\chi^2$, both of which may be called distribution-free methods because no assumptions are made about the distribution of the variable or variables underlying the categories. Data in the form of ranks may force one to use Spearman's rho or Kendall's tau or the tests to be presented later in this chapter.

In general, distribution-free methods, when applied for comparative purposes to data which are normal or nearly normal, are not as sensitive (that is, as powerful for avoiding type II errors) as the appropriate $z$,

$t$, or $F$ technique. Consequently, in using a nonparametric method as a short-cut, we are throwing away dollars in order to save pennies.

**The sign test.** Perhaps the simplest of all distribution-free methods is the "sign" test, which is applicable for testing the difference between two correlated sets of scores. The procedure is to consider the $N$ pairs of differences, $X_1 - X_2$, some of which will be plus, some minus (with an occasional zero). If there is no difference between the two sets of scores we would expect the plus and minus signs to be equally divided. To test whether there are more plus signs than reasonable on a chance basis, the binomial, $(p + q)^N$ with $p = .50$, is used ($N$ is for the pair differences having a sign; it is the sample size less the number of zero differences) in the manner discussed earlier (pp. 48–50). For effective $N$ larger than 10 we may use either the normal curve approximation to the binomial (pp. 46–48) or the $\chi^2$ approximation (pp. 245–48). Whether we use the binomial itself or one of the approximations, we must take care to secure a $P$ that represents whichever—a one-tailed or a two-tailed—test is appropriate for the hypothesis being tested.

**The "median" test.** A procedure for testing the difference between two sets of *independent* scores is to use the median for the two groups combined as a basis for dichotomizing. This leads to a fourfold table: above vs. below (the median) on one axis, group vs. group on the other. Then the $\chi^2$ test for the fourfold table may be employed, with Yates' correction if necessary. With very small $N$s the exact probability method (pp. 272–75) would be used. The idea back of the median test is simply that two samples drawn from two populations having the same median should yield equal splits. In practice, difficulties are sometimes encountered in attempting to dichotomize exactly at the median. When the median in an integer and several scores are equal to the median, the dichotomy can be taken as those scores which *exceed* the median vs. those which do not exceed the median.

**Median test for more than two independent groups.** This is a straightforward extension of the median test to provide an over-all test of the differences between, say, $C$ independently drawn groups. On the basis of the median of the distribution of the $C$ groups combined, the scores are dichotomized (as near the median as possible). This will lead to a 2 by $C$ table from which one may obtain a $\chi^2$ with $C - 1$ degrees of freedom.

Whether we are dealing with two groups or with $C$ groups, the $N$s for the groups need not be equal for use of the median test.

**Mean and variance for rank scores.** Since the next four tests are based on ranks, we now digress to consider the mean and variance of the distribution of ranks, a rectangular distribution running from 1 to $N$ when

$N$ persons have been ranked. Let $X$ stand for any rank score. A little thought leads to the conclusion that $\bar{X} = (N + 1)/2$, hence $\Sigma X = N\bar{X} = N(N + 1)/2$ as the sum of the rank scores. In some college algebra textbooks there is proof that the sum of the squares of the first $N$ natural numbers is given by $N(N + 1)(2N + 1)/6$. This gives us the value of $\Sigma X^2$. When the foregoing given values for $\Sigma X$ and $\Sigma X^2$ are substituted into the general formula (3.6) for the sum of squares of deviations about the mean, we find after simplification that $\Sigma x^2 = (N^3 - N)/12$, hence the variance,

$$\sigma^2 = \frac{\Sigma x^2}{N} = \frac{(N^2 - 1)}{12} \tag{19.1}$$

It is no accident that 6, or $\frac{1}{2}$ of 12, appears in the formula for rank-difference correlation, and that 12 appears in Sheppard's correction to $S$ for the grouping error. Why in the latter?

**Mann-Whitney $U$ test.** This test, which is applicable only to results based on two independent groups, involves rank ordering the scores, for the two groups combined, from *greatest* (rank 1) to *least* (for which the rank will be $N = N_1 + N_2$ unless there are ties for the bottom position). When ties occur, each person involved is assigned the average of the ranks that would be assigned in case the tied persons could be differentiated (see p. 233). Then the ranks so assigned are summed separately for each group. Let $T_1$ and $T_2$ represent these two sums. (As a check on the arithmetic, $T_1 + T_2$ should equal $\dfrac{N(N + 1)}{2}$, the sum of the first $N$ natural numbers.)

When both $N_1$ and $N_2$ are 8 or greater, the statistic

$$U_1 = N_1 N_2 + \frac{N_1(N_1 + 1)}{2} - T_1 \tag{19.2}$$

is distributed normally about a chance expected value, or mean, given by $N_1 N_2/2$, and with variance of $N_1 N_2(N_1 + N_2 + 1)/12$. We then have

$$\frac{x}{\sigma} = \frac{U_1 - N_1 N_2/2}{\sqrt{\dfrac{N_1 N_2(N_1 + N_2 + 1)}{12}}}$$

as a unit normal deviate by which the significance of $U$ as a deviation from the null hypothesis expected value is determined. If, as an alternate, we define $U$ by replacing $T_1$ with $T_2$ and $N_1$ with $N_2$ (in the second term), we will have $U_2$. Now $U_1$ and $U_2$ will deviate to the same extent, but in opposite directions, from $N_1 N_2/2$.

When $U_1$ is larger than $N_1 N_2/2$, the direction of the difference between the two sets of scores is such that group 1 is superior to group 2. (If ranks are assigned with the least score as rank 1, and so on, the value of $U_1$ will be smaller than $N_1N_2/2$ when group 1 is superior.) For $N_1$ and $N_2$ less than 8, special tables are required for judging the significance of $U$.

**Kruskal-Wallis one-way analysis of variance by ranks.** This test is applicable for testing the difference between $G$ independent groups, with varying numbers, $m_g$, of cases per group. All $N$ ($N = \Sigma m_g$) scores are ranked with a rank of 1 assigned to the lowest score and a rank of $N$ to the highest, but the group identity of the cases is maintained so that we can sum the rank scores within each group, which sum we will designate as $T_g$ for the $g$th group. Then the quantity

$$H = \frac{12}{N(N+1)} \Sigma \frac{T^2_g}{m_g} - 3(N+1) \tag{19.3}$$

is computed. Under null conditions (no difference in averages for the populations), and for all $m_g$ greater than 5, the sampling distribution of $H$ follows closely the $\chi^2$ distribution with $df = G - 1$.

When there are sets of ties, with $t_s$ the number of cases tied in the $s$th set, it is necessary to apply a correction to $H$:

$$H_c = \frac{H}{1 - \Sigma_s (t^3_s - t_s)/(N^3 - N)}$$

The corrected value will be higher than the uncorrected value and will therefore tend to help us reject the null hypothesis. If $H$ is significant, we would not bother to compute $H_c$.

**Friedman two-way analysis of variance by ranks.** This test is for the mixed model situation where the columns stand for $C$ experimental conditions (or levels on a factor) and the rows stand either for $R$ individuals, each of whom is measured under all $C$ conditions, or for $R$ sets of matched persons, with random assignment within a set to the $C$ conditions. The $X$ scores in each *row* are assigned ranks, 1 to $C$, then the rank scores within each column are added to obtain a total for each column, $T_c$ for the $c$th column. The quantity

$$\chi^2_r = \frac{12}{RC(C+1)} \Sigma_c T^2_c - 3R(C+1) \tag{19.4}$$

is computed. The designation of this as a chi square implies that in the sampling sense the quantity is a chi square variate. When $C > 3$ and $R > 9$ or $C > 4$ and $R > 4$, the random sampling values of $\chi^2_r$ tend to follow approximately the chi square distribution, with $df = C - 1$.

The rationale for the Friedman test is, briefly, as follows. The assigning of ranks within rows as 1 to $C$ reduces all row means to the same value,

$(C + 1)/2$, thus "taking out" row differences. (Recall that for the ordinary two-way $F$ test a row sum of squares was extracted.) Under null conditions that the original $X$ scores all belong to the same population, i.e., that there is no column effect, we would expect the rank scores, $RC$ in number, to be distributed evenly over the columns in such a way that the distributions within columns would be the same. Therefore, under null conditions the column means for the rank scores would tend to be the same (that is, we would expect equal $T_c$), and the within-column variances would also tend to be the same.

Next we note that with all the row means equal to $(C + 1)/2$, the mean of means, or total mean, would also be $(C + 1)/2$. When we consider the general expression for the sum of squares for the $R \times C$ interaction, $\sum_r\sum_c(X_{rc} - \bar{X}_r. - \bar{X}_{\cdot c} + \bar{X}..)^2$, we see that when $\bar{X}_r. = \bar{X}..$, this sum of squares reduces to $\sum_r\sum_c(X_{rc} - \bar{X}_{\cdot c})^2$, which is nothing more than the sum of squares within columns. But under null conditions the distribution of ranks within any one column will, of course, be rectangular (the rank scores will run from 1 to $C$ within a column) with variance that can be specified theoretically as $(C^2 - 1)/12$. Strictly speaking the theoretically specified, under null conditions, distribution of ranks within all columns cannot be exactly the same unless $R$ is a multiple of $C$.

The $\chi^2_r$ of the Friedman test is an $F = s^2_c/\sigma^2_{rc}$, in which $s^2_c$ has $C - 1$ degrees of freedom and $s^2_{rc}$ has been replaced by the theoretical population variance, or a variance with $df = \infty$. As usual, when $n_2 = \infty$, $F$ becomes a $\chi^2/n_1$, hence $\chi^2_r$ is $n_1$ times an $F$. Ties of ranks within rows do not disturb the Friedman test, and it is claimed that the Friedman test agrees very closely with an $F$ test applied to the original $X$ scores.

**Kendall's coefficient of concordance, $W$.** Suppose $C$ judges each rank order $R$ individuals, and we wish a measure of the agreement among the judges. Arrange the rankings into a table of $R$ rows and $C$ columns. The rank scores in the $c$th column will run from 1 to $R$ (except when ties occur for either the top or bottom position—unlikely in practice). Sum across columns and enter the several sums along the right-hand margin. We might regard these sums as "scores" for the $R$ individuals. If there were perfect agreement among the judges, these sums would range from $C$ to $RC$, with in between values of $2C, 3C, \cdots, (R - 1)C$, though not necessarily in that order. Consider the variance of these sum scores. Since each sum is simply $C$ times a rank, with the ranks running from 1 to $R$, the variance of these sum scores will be $C^2$ times the variance of the first $R$ natural numbers. The maximum variance possible will be

$$\sigma^2_{max} = \frac{C^2(R^2 - 1)}{12}$$

The same value can be obtained by considering the variance of the sums in terms of the variance theorem. The variance of the sum scores will be equal to the sum of the separate $C$ variances, one for each column and all identical, i.e., all will equal $(R^2 - 1)/12$, plus a sum of $C(C - 1)/2$ correlational terms of the form $2r_{12}\sigma_1\sigma_2$. Under the condition of perfect agreement among the judges, all of these correlational terms will become $2(1)\sigma_1\sigma_2$, but since all columns have the same variance, $\sigma_1$ will equal $\sigma_2$ and each correlational term will become $2(1)\sigma^2$ in which $\sigma^2 = (R^2 - 1)/12$. Summing $C(C - 1)/2$ such terms yields $C(C - 1)\sigma^2$, which when added to the sum of the $C$ variances (i.e., added to $C\sigma^2$) gives

$$\sigma^2{}_{max} = C\sigma^2 + C(C - 1)\sigma^2 = C^2\sigma^2 = C^2(R^2 - 1)/12$$

In practice, perfect agreement will rarely if ever occur. As a measure of the extent of agreement, Kendall proposed that the variance of the *obtained* sums be taken relative to the maximum possible variance. Accordingly, the coefficient of concordance is defined as

$$W = \frac{S^2{}_{sum}}{\sigma^2{}_{max}} \tag{19.5}$$

in which

$$S^2{}_{sum} = \frac{1}{R^2}\left[ R\sum_r T^2{}_r - \left(\sum_r T_r\right)^2 \right]$$

with $T_r$ as the total (or sum) for the $r$th row. Obviously, $W$ can never be negative; it will be 1 when the agreement is perfect. The value of $W$ tends to be higher than the average of all possible Spearman rank-difference correlations between the judges. When ties occur, the denominator term for $W$ becomes $\sigma^2{}_{max} - \frac{C}{12R}\sum_s (t^3{}_s - t_s)$ in which $t_s$, the number of cases tied in the $s$th set of ties, will take on values $2, 3, 4 \cdots$, and all the sets irrespective of their column location are included.

For $R > 7$, $W$ may be tested for significance by

$$\chi^2{}_W = \frac{12RS^2{}_{sum}}{CR(R + 1)} \tag{19.6}$$

which follows approximately the $\chi^2$ distribution with $R - 1$ degrees of freedom. A significant $W$ may be interpreted in two ways: either as indicative of better than chance agreement among the $C$ judges or as a significant (reliable) difference among the $R$ sums (or the $R$ possible means) for the $R$ individuals. But mere statistical significance may not be as crucial as the knowledge that $W$ is fairly high.

There is a direct connection between the Friedman test and the significance test for Kendall's $W$, although as significance tests they differ as to purpose. Friedman's test is concerned with the significance of the differences among the $C$ column averages, the number of which is usually very small, whereas the test of $W$ is concerned with the significance among $R$ row averages, the number of which is usually not very small. Friedman's $\chi^2_r$, is, typically, used to test for the effect of a fixed constants factor, whereas typically $\chi^2_W$ tests for the significance of a random factor (individual differences) although applicable to the ranking of $C$ objects by judges. Kendall's $W$ provides a useful descriptive measure of agreement among judges, but such a measure is not a relevant part of the Friedman technique.

It can be shown by simple algebra that the test for $W$ can be written in the alternate form:

$$\chi^2_W = \frac{12}{CR(R+1)} \Sigma_r T^2_r - 3C(R+1) \tag{19.7}$$

which bears a marked resemblance to the expression for $\chi^2_r$. Simply transpose the roles assigned to the rows and columns and also interchange the $R$ and $C$ designations in $\chi^2_W$ and you have $\chi^2_r$.

# *Chapter 20*

## SUPPLEMENTARY TOPICS

It is the purpose of this chapter to (*a*) provide an outline of the extensions of analysis of variance to four-way classification, with injection of a frequently used notational scheme, (*b*) discuss a limitation of the use of analysis of variance and develop certain connections with multiple regression analysis, (*c*) introduce a sometimes useful estimation procedure known as "variance components estimation," and (*d*) summarize and integrate some of the principles and/or methods of error reduction, including therewith an analysis of variance design as an alternative to control by pairing, or matching, subjects on relevant variables.

Analysis of variance, can, of course, be extended to higher-order classification designs involving more than three factors. Each additional factor doubles the number of parts in the partitioning of the sum of squares and degrees of freedom. With *m* scores per group (cell, cubicle) we have two parts for one-way classification, four parts for the two-way setup, eight parts for a three-factor design. One general rule for specifying the number of parts when there is only one score per cell (hypercubicle) is to utilize the coefficients in the binomial expansion, with the first coefficient, 1, standing for the over-all mean. Thus if we have a five-way design, we will have 1 over-all mean, 5 main effects, 10 two-way interactions, 10 three-way interactions, 5 four-way interactions, and 1 five-way interaction for a total of 31 sums of squares. With more than one score per cubicle, 32 sums of squares would be needed for the breakdown. The author once examined a planned study which proposed to use an eleven-way design, thus leading to 2048 sums of squares, a computationally staggering colossus that might stump even a sophisticated electronic computer. There were to be two levels for eight and three levels for three of the factors, thus generating a design with $2 \times 2 \times 2 \times 2 \times 2 \times 2 \times 2 \times 2$ $\times 3 \times 3 \times 3 = 6912$ cells. To have the needed $s^2_w$, a minimum of two scores per cell would be required, or a total of 13,824 subjects. Such an $N$ would force the experimenter to sample beyond Psych 1 students! Suppose the study were carried out and one of the 11 possible ten-way interactions is significant—how would it be comprehended? Or suppose 100 of the possible 2047 $F$s were significant at the .05 level, with 20 of them reaching the .01 level? One can very well imagine extending this sort of thing until the $n$th factor is the presence or absence of the proverbial kitchen sink.

It is beyond the scope of this book to present the computational procedures for four-way and higher-order designs. The researcher who does not have access to computer facilities might turn to B. J. Winer's *Statistical principles in experimental design* (McGraw-Hill, 1962), a book that contains illustrations and computations for a variety of simple and complicated designs. Here we will attempt to sketch the required breakdown and variance estimates and $F$s for certain designs involving four factors. In doing so we will introduce another notation which may help the student understand what he may encounter in other textbooks and in the research literature. It was hoped that the translation of factors into row, column, and block designation, with the resultant arrangement of scores as an aid to computation and to discussion, would have its heuristic value. But to carry this farther would call for "levels" (already used in another sense) or "strata" as a basis for designating the fourth factor. A frequently used scheme that avoids this search for geographic-space concepts is the use of capital letters, $A, B, C, D, \cdots$, to identify the factors, with $n_a, n_b, n_c, n_d, \cdots$, as the respective number of levels on the factors. In this system, one might sensibly use $S$ instead of $D$ or $E$ or $F$ if the fourth or fifth or sixth factor was subjects (persons). This we will do for repeat-measure designs because of the need for easy identification of which letter stands for subjects.

## FOUR-WAY CLASSIFICATION

The typical breakdown for the four-way layout leads to sums of squares for 4 main effects, 6 two-way interactions, 4 three-way interactions, 1 four-way interaction, plus a not always possible within-cubicles term. The $df$s for the main effects would be $n_a - 1, \cdots, n_d - 1$ (or $n_s - 1$) and the $df$s for all interactions are typically the products of the $df$s for the factors entering into the given interactions.

**Fixed effects model.** The several variance estimates would be labeled as $s^2_a, s^2_b, s^2_c, s^2_d, s^2_{ab}, s^2_{ac}, s^2_{ad}, s^2_{bc}, s^2_{bd}, s^2_{cd}, s^2_{abc}, s^2_{abd}, s^2_{acd}, s^2_{bcd}, s^2_{abcd},$

and $s^2_w$ if $m$ scores per cubicle. The denominator, or error term, for the $F$ ratios for testing all 15 possible effects is $s^2_w$. Very simple, once the sums of squares have been computed except for the task of interpreting the four-way interaction in case it is significant. When the investigator has hypothesized that an interaction will occur, he will be pleased at finding it to be significant. Otherwise, significant interactions are somewhat of a nuisance in that interpretation may not be easy, and the presence of a significant interaction should lead to a qualification of the results for lower-order interactions and main effects.

**Mixed model.** With $A$, $B$, and $C$ as the factors to be studied, and subjects ($S$) as the fourth (random) factor, we have the typical repeat-measure design in which each of the $n_s$ subjects is measured once under

**Table 20.1. Variance estimates for mixed model; factors, $A$, $B$, and $C$, with $S$ random**

| Main | $s^2_a$ | $s^2_b$ | $s^2_c$ | $s^2_s$ | | | |
|---|---|---|---|---|---|---|---|
| Two-way I | $s^2_{as}$ | $s^2_{bs}$ | $s^2_{cs}$ | | $s^2_{ab}$ | $s^2_{ac}$ | $s^2_{bc}$ |
| Three-way I | | | $s^2_{abc}$ | | $s^2_{abs}$ | $s^2_{acs}$ | $s^2_{bcs}$ |
| Four-way I | | | $s^2_{abcs}$ | | | | |

all of the $n_a n_b n_c$ combinations of conditions. The variance estimates may be specified as above with the subscript $d$ replaced by $s$. As for the three-way, mixed model, there is no one single error term to use as the denominator for the $F$ ratios. In Table 20.1 will be found the several variance estimates so arranged that if a variance is testable, the appropriate denominator for $F$ is in the line immediately beneath it. Note that none of the variances involving the subscript $s$ for subjects as a random factor is testable. This is not serious since variances involving subjects are, typically, of nuisance rather than of scientific value. The reader should note the pattern of subscripts for the possible $F$ tests: for each $F$ the denominator variance estimate contains the same subscript(s) as the numerator plus $s$.

Anyone who incubates a research design calling for the mixed model with two factors fixed and two factors random should bear in mind that the variances of particular interest (the main effects of, and interaction between, the fixed factors) are not testable in any exact way.

Repeat-measure, four-way designs in which each subject is not measured under each of the $n_a n_b n_c$ conditions are, in effect, extensions of the three-way setup discussed in Chapter 16 as Case XV. To convert to the notation of this section we might let $B$ be the factor involving repeat measures (the column factor in Table 16.16), let $A$ be the factor the levels of which involve different and independently assigned subjects (the block factor in Table 16.16), and let $S$ stand for subjects, $n_s$ in number for each level on $A$. The sources of variation for this type of design are frequently subsumed under two general headings: "between subjects" and "within subjects." If we think of summing over the levels on $B$ (over columns) we can obtain a "score," thence a mean for each of the $n_s n_a$ subjects. These scores differentiate between subjects and the total between-subject variation has two parts, between-groups (levels on $A$) and a residual, within-groups. Now the "within subjects" heading indicates the factor(s) over which we have repeat measures, thus providing a within-subject variation—the

**Table 20.2. Factor $A$, independent subjects; factor $B$, repeat measures; factor $S$, subjects**

| Source | $df$ | Variance Estimate | $F$ |
|--------|------|-------------------|-----|
| Between | | | |
| $A$ | $n_a - 1$ | $s^2_a$ | $s^2_a / s^2_s$ |
| $S(A)$ | $n_a(n_s - 1)$ | $s^2_s$ | |
| Within | | | |
| $B$ | $n_b - 1$ | $s^2_b$ | $s^2_b / s^2_{sb(a)}$ |
| $AB$ | $(n_a - 1)(n_b - 1)$ | $s^2_{ab}$ | $s^2_{ab} / s^2_{sb(a)}$ |
| $SB(A)$ | $n_a(n_b - 1)(n_s - 1)$ | $s^2_{sb(a)}$ | |
| Total | $n_s n_a n_b - 1$ | | |

variation of a subject's $n_b$ scores about his own mean. This variation will spring from two possible sources: differences produced by the $B$ factor and those that result from the interaction of $A$ and $B$.

Accordingly, we could reset Table 16.16 in modified notation as shown in Table 20.2. The symbol $S(A)$, to be read as subject within levels on $A$, stands for a source which is frequently referred to as "error $b$," and $SB(A)$ is referred to as "error $w$." Instead of $S \times B$ as an abbreviation for interaction we have simply $SB$, and the symbol $SB(A)$ is to be read as subject by factor $B$ interaction within levels on $A$, averaged over the $n_a$ levels. The subscripts in $s^2_{sb(a)}$ indicate the variance estimate of this interaction. Computationally, it is based on calculation of a separate sum of squares for the $SB$ interaction in each level of $A$, which sums are then summed over all the $n_a$ levels. Table 20.2 also indicates the ratios of variances for the permissible $F$ tests.

To extend this sort of thing to a four-way design with three factors fixed and subjects, $S$, a random factor, we will need to specify which of the factors involve independent, or different, subjects for levels thereon and which involve repeat, or correlated, measures—same subjects measured at each level (or combination of levels). If there are repeat measures under all combinations of conditions, we have the already discussed Table 20.1; all three factors involve correlated scores.

**Four-way, repeat measures on two factors.** Let $A$ stand for the factor with independent subjects per level, a total of $n_a n_s$ subjects, with $B$ and $C$ the factors for which repeat measures are available; the means for the levels on $B$ and for the levels on $C$ will be based on correlated scores. For

**Table 20.3. Factor $A$, independent subjects; factors $B$ and $C$, repeat measures; factor $S$, subjects**

| Source | df | Variance Estimate | F |
|---|---|---|---|
| Between | | | |
| $A$ | $n_a - 1$ | $s^2_a$ | $s^2_a/s^2_s$ |
| $S(A)$ | $n_a(n_s - 1)$ | $s^2_s$ | |
| Within | | | |
| $B$ | $n_b - 1$ | $s^2_b$ | $s^2_b/s^2_{sb(a)}$ |
| $C$ | $n_c - 1$ | $s^2_c$ | $s^2_c/s^2_{sc(a)}$ |
| $AB$ | $(n_a - 1)(n_b - 1)$ | $s^2_{ab}$ | $s^2_{ab}/s^2_{sb(a)}$ |
| $AC$ | $(n_a - 1)(n_c - 1)$ | $s^2_{ac}$ | $s^2_{ac}/s^2_{sc(a)}$ |
| $BC$ | $(n_b - 1)(n_c - 1)$ | $s^2_{bc}$ | $s^2_{bc}/s^2_{sbc(a)}$ |
| $ABC$ | $(n_a - 1)(n_b - 1)(n_c - 1)$ | $s^2_{abc}$ | $s^2_{abc}/s^2_{sbc(a)}$ |
| $SB(A)$ | $n_a(n_s - 1)(n_b - 1)$ | $s^2_{sb(a)}$ | |
| $SC(A)$ | $n_a(n_s - 1)(n_c - 1)$ | $s^2_{sc(a)}$ | |
| $SBC(A)$ | $n_a(n_s - 1)(n_b - 1)(n_c - 1)$ | $s^2_{sbc(a)}$ | |

each level on $A$ we have the equivalent of a block by column by subject layout which corresponds to Case XII, p. 377, with $S$ playing the role of $R$. The sources, $df$s, variance estimates, and possible $F$s are given in Table 20.3. It will be seen that again we have an interaction that results from calculating an interaction sum of squares at each level of $A$, then summing these over the $n_a$ levels of $A$. This idea is, of course, reflected in the number of degrees of freedom. Consider the interaction represented by $SBC(A)$. For each level on $A$ one could have an $SBC$ interaction sum of squares with $(n_s - 1)(n_b - 1)(n_c - 1)$ degrees of freedom. When these are summed over the $n_a$ levels of factor $A$, we also sum the $df$s which is the equivalent of multiplying the three-way $df$ by $n_a$.

Notice also that in order to have a factor appear in an interaction involving subjects ($S$), it must be one on which there are repeat measures.

The pattern of subscripts for the variances for the several $F$ ratios should be studied sufficiently to see the similarity to $F$ ratios discussed in connection with Table 20.1. And again we see that interactions involving $S$ are not testable.

**Four-way, repeat measures on one factor.** Here we will say that for both factors $A$ and $B$ there is no repeating of measures. This is another way of saying that for $n_a$ and $n_b$ levels, leading to $n_a n_b$ cells in a two-way layout, the subjects, $n_s$ in number in each cell, have been randomly (or independently) assigned to the cells. The total number of subjects would be $n_a n_b n_s$. This is, of course, the row by column design with $m$ scores per

**Table 20.4. Factors $A$ and $B$, independent subjects; factor $C$, repeat measures; factor $S$, subjects**

| Source | $df$ | Variance Estimate | $F$ |
|---|---|---|---|
| Between | | | |
| $A$ | $n_a - 1$ | $s^2_a$ | $s^2_a / s^2_{s(ab)}$ |
| $B$ | $n_b - 1$ | $s^2_b$ | $s^2_b / s^2_{s(ab)}$ |
| $AB$ | $(n_a - 1)(n_b - 1)$ | $s^2_{ab}$ | $s^2_{ab} / s^2_{s(ab)}$ |
| $S(AB)$ | $n_a n_b (n_s - 1)$ | $s^2_{s(ab)}$ | |
| Within | | | |
| $C$ | $n_c - 1$ | $s^2_c$ | $s^2_c / s^2_{sc(ab)}$ |
| $AC$ | $(n_a - 1)(n_c - 1)$ | $s^2_{ac}$ | $s^2_{ac} / s^2_{sc(ab)}$ |
| $BC$ | $(n_b - 1)(n_c - 1)$ | $s^2_{bc}$ | $s^2_{bc} / s^2_{sc(ab)}$ |
| $ABC$ | $(n_a - 1)(n_b - 1)(n_c - 1)$ | $s^2_{abc}$ | $s^2_{abc} / s^2_{sc(ab)}$ |
| $SC(AB)$ | $n_a n_b (n_c - 1)(n_s - 1)$ | $s^2_{sc(ab)}$ | |

cell as illustrated by Table 16.6. Each subject within the $n_a n_b$ groups is measured at each of the $n_c$ levels on factor $C$; hence factor $C$ involves correlated scores. Table 20.4 indicates the variance estimates provided by this design. This time there are only two error terms for $F$ ratios, one for testing effects based on groups that are independent, the other for effects that are based on nonindependent, or correlated, scores. The latter, as usual, is an interactive effect with subjects as one variable: subject interaction with factor $C$ as averaged over the $n_a n_b$ groups.

**Summary remarks.** The extension of Table 20.1 to situations with additional fixed effects factors is relatively easy conceptually if not computationally. The inclusion of additional factors so as to set up tables like 20.3 and 20.4 for various combinations of four or more fixed effects factors with repeat measures for some is not a particularly easy task; nor do the required computations appeal to anything more animate than an electronic computer. The reader can turn to E. F. Lindquist's *Design and analysis of experiments* (Houghton Mifflin, 1951) or to B. J. Winer's *Statistical principles in experimental design* (McGraw-Hill, 1962) for information on these, and other, designs. We should not forget that lack of homogeneous correlations and variances in repeat-measure designs may be such that an $F$ at the .01 level means significance at the .04 or .05 level. The nominal .05 level may be reached .10 of the time by chance.

## NONORTHOGONALITY OF FACTORS

Earlier, when the bases (factors, or variables) for classification were discussed, a caution was injected concerning the requirement that if two or more bases are involved, the variables, or factors, must be independent of one another. The implication of the $F$ tests, say, for a row and a column factor, fixed effects model, is that whatever the row effect it was such that its statistical significance did not depend on the presence or absence of a column effect in any way whatsoever. Nor did it depend on a possible $R \times C$ interaction. The presence of an interactive effect meant that in a sense the row and column effects were not additive (see p. 338), but lack of additivity did not mean that the row and column variables were somehow correlated. When the two factors have been manipulated in the laboratory, it is obvious that they are uncorrelated. (As an apparent exception, consider distance and illumination as factors to be varied in a perception study: if the greater distance also meant farther away from the light source, the illumination level would be related to distance. But no well-designed experiment would ever permit this to happen.) In contrast, when the variables used for classification into "levels" have to do with a person belonging to subgroups or with a subject's score for some measured trait, then one must face questions as to: (*a*) whether there may be association between belonging to categories, or subgroups, on any two bases of classification; (*b*) whether any two measurable factors are correlated; and (*c*) whether scores on a measured trait are associated with subgroup membership.

As an example of nonorthogonality, let us suppose that a newcomer to the school of education arrives equipped with enthusiasm to apply his undergraduate-acquired knowledge of statistical inference to educational problems. Being amazingly unsophisticated, he hits upon that old chestnut, the "dependence" of achievement in college on high-school record and aptitude. Since his instructor in statistics had a disdain for correlational techniques, our newcomer does not see the problem in terms of correlation but rather as one that would be amenable to analysis of variance. Accordingly, on the basis of records readily obtainable from the registrar of his alma mater, he secures the necessary data on 400 freshmen who had entered the previous year. He decides to use the quartile points on both aptitude and high-school grade-point ratio as a basis for classifying. Thus he has a 4 × 4 "levels" design, with college grade-point average as the dependent variable. Despite some difficulties, which we shall not discuss here, he manages to do the required computations for analysis of variance. He finds highly significant $F$s for both high-school record and aptitude, with insignificant interaction, and concludes that each of these two variables has an "effect" on success in college.

Now it would indeed be trifling to pursue this trite example if it did not involve more or less (to some) obvious consequences. Anyone who is well-versed in the problems of correlational analysis would not make the mistake of presuming that because he had used the $F$ test, which is usually associated with experimental manipulation of variables, he is therefore justified in interpreting a significant $F$ as indicative of "effect." An effect must, by definition, be the resultant of some causative factor. The beauty of carefully controlled experimental manipulation is that any found difference on a so-called dependent variable permits the conclusion that the manipulated factor has had an effect. The $F$ test is a test of whether the manipulation *produces* variation. When the levels on a factor do not depend on manipulation, the researcher does not have a situation in which variation is produced; he can only say that a part of the ever-present variance in his "dependent" variable is associated with some other variable. Whether or not the "dependent" variable is really dependent on one or more factors is an exercise in logic. A significant $F$ for a main effect is no more indicative of causation than a significant correlation coefficient. Does high-school record have an "effect" on college achievement? Or do both depend on aptitude?

Incidentally, the question of causation is not confined to situations involving two or more factors of the nonmanipulable type. It is also pertinent when only one variable is of this type, a fact that seems to have been overlooked in a lot of experimentation wherein one factor is represented by levels, typically groups chosen as high and low, on some measured variable, such as anxiety, aggressive tendency, ability.

Aside from the question as to when an "effect" is an effect, the use of two or more factors that involve characteristics of individuals and/or group membership poses a couple of complications in the analysis of variance paradigm which hold when the "levels" are based on factors that are correlated. One of these has to do with interpreting a significant $F$ for a main effect. The point can be illustrated by the previous example: Does a significant $F$ for high-school record indicate the "effect" of this factor independently of the operation of the other factor, aptitude? Could the high-school record effect represent in part the effect of aptitude operating through high-school performance? What of aptitude as a main effect when it is argued that aptitude score differences depend in part on achievement in high school? Does either main effect represent an effect which is independent of the other? These are the types of questions of interpretation that must be faced when the $F$ test is used with non-orthogonal factors.

The second complication arises from the fact that classification on the basis of nonorthogonal factors leads not only to unequal cell frequencies but also to empty cells, particularly when more than two levels are used for each factor. It is, of course, desirable to have three or more levels on an ordered factor so as to obtain some information regarding the form of the relationship between the dependent variable and the factor. Consider again the college achievement problem, for which the use of four levels on each factor could very well lead to at least two empty cells if, as is likely, the correlation between aptitude and high-school record is .50 or higher. Unequal cell frequencies and empty cells could easily arise when only two levels are used for each factor if the levels are defined in terms of extremes. Obviously, empty cells will prevent any very clear-cut test or statement concerning interaction. It is beyond the scope of this book to consider the adjustments that can be safely made when unequal cell frequencies have resulted from fortuitous circumstances or when the cell frequencies are proportional from column to column. Knowledge of these adjustments would not help in the situation being discussed simply because the requisite conditions do not hold. There seems to be no definite way to adjust for unequal cell frequencies and empty cells that result from using factors which are correlated (nonorthogonal). In short, analysis of variance models call for the decomposition of a total sum of squares into independent components in such a way as to obtain $F$ tests for the independent effects of the components.

Is there any way for ascertaining the "effect" of factors that are correlated? If the factors are measurable characteristics, the answer is an easy yes. Multiple regression will do the trick. If we can assume causation, then a significant beta weight is interpretable as evidence for the effect of the factor. A beta is testable in terms of the difference between two multiple $r$s when three or more factors are involved or as the difference between a multiple $r$ and an ordinary bivariate $r$ when only two factors are being studied (see p. 321). Anyone who has been led to believe that factorial designs and the analysis of variance constitute the sole approach to statistical inference should be reminded that problems involving measured factors and measured dependent variables may need to be attacked by correlational techniques.

**Factorial design vs. multiple regression, factors correlated.** At this point it may be instructive to compare and contrast the factorial design and multiple regression when the factors are correlated individual difference variables. For simplicity, we will (a) continue using the two-way design which has as its counterpart a three-variable multiple regression setup and (b) develop the argument by way of population parameters. We suppose measures for $N_p$ (the population $N$) subjects on all three variables. A simplification, without loss of generality, will result from converting all three sets of scores to standard score form, $\mu = 0$, $\sigma = 1.00$. Let $Y$ be the dependent variable and $X$ and $Z$ the factors (or independent variables, although correlated with each other). Let $y$, $x$, and $z$ be the standard scores, and let $X$ be the row factor with $R$ levels, and $Z$ be the column factor with $C$ levels. ($R$ and $C$, as number of grouping intervals, should be 10 or more for meaningful correlations.)

The scattergram for $x$ and $z$ will involve unequal cell frequencies for the $RC$ cells, with the number of empty cells a function of the degree of correlation between $X$ and $Z$. For the nonempty cells we would have $\mu_{rc}$ as the cell means for the dependent, or $Y$ variable. When $m_{rc} = 1$, we would simply say that the one score is the cell mean. Obviously, $\mu_{r.}$ would be the mean $y$ for all cases in the $r$th row and $\mu_{.c}$ the mean $y$ for the $c$th column.

Now let us consider two ways for estimating $y$ scores: by way of cell means and by way of multiple regression. When the cell means are used for estimation we are simply saying that for a person in the $r$th row and $c$th column the best possible estimate of his $y$ score will be $\mu_{rc}$. It will greatly simplify matters, and also clinch an important point, if we assume no $R \times C$ interaction. This means (see p. 338) that the row and column effects are additive, so $\mu_{rc} = \mu_{r.} + \mu_{.c}$. Now any score $y_{rc}$ can be regarded as made up of two parts, a predictable component and a residual component. Thus

$$y_{rc} = \mu_{rc} + (y_{rc} - \mu_{rc})$$

or

$$y_{rc} = \mu_{r.} + \mu_{.c} + (y_{rc} - \mu_{rc}) \tag{20.1}$$

Prediction by way of a multiple regression equation would take the form $y'_{rc} = \beta_x x + \beta_z z$. From this, any score can likewise be regarded as made up of a predictable and a residual part. Thus

$$y_{rc} = y'_{rc} + (y_{rc} - y'_{rc})$$

or

$$y_{rc} = \beta_x x + \beta_z z + (y_{rc} - \mu_{rc}) \tag{20.2}$$

in which we have replaced $y'_{rc}$ in the parentheses by $\mu_{rc}$ on the assumption that exact linearity holds for the population regression plane. Later we will consider what ensues if this assumption does not hold—in practice it will never hold exactly for a sample.

From (20.1) and (20.2) we can write two expressions for the breakdown of the total sum of squares for $y$:

$$\underset{r\ c}{\Sigma\Sigma} y^2_{rc} = \underset{r}{\Sigma} m_r \mu^2_{r.} + \underset{c}{\Sigma} m_c \mu^2_{.c} + \underset{r\ c}{\Sigma\Sigma}(y_{rc} - \mu_{rc})^2 \qquad (20.3)$$

$$\underset{r\ c}{\Sigma\Sigma} y^2_{rc} = N_p \beta^2_x + N_p \beta^2_z + 2N_p \beta_x \beta_z r_{xz} + \underset{r\ c}{\Sigma\Sigma}(y_{rc} - \mu_{rc})^2 \quad (20.4)$$

Both of these expressions reflect simplifications that result from standard scores: no over-all mean (since $\mu = 0$) in (20.3); in writing (20.4) we utilize the fact that for standard scores, $\Sigma x^2/N_p = \Sigma z^2/N_p = 1$, hence $\Sigma x^2 = \Sigma z^2 = N_p$; and also since $\Sigma xz/N_p = r_{xz}$ we have $\Sigma xz = N_p r_{xz}$. The $r$ is a population value. As usual, the cross-product terms involving the residuals, $(y_{rc} - \mu_{rc})$, vanish. Under the assumption of linearity, it follows that

$$\mu_{r.} = r_{yx} x_r \quad \text{and} \quad \mu_{.c} = r_{yz} z_c$$

so that

$$\underset{r}{\Sigma} m_r \mu^2_{r.} = r^2_{yx} \Sigma x^2 \quad \text{and} \quad \underset{c}{\Sigma} m_c \mu^2_{.c} = r^2_{yz} \Sigma z^2$$

We may therefore rewrite (20.3) as

$$\underset{r\ c}{\Sigma\Sigma} y^2_{rc} = N_p r^2_{yx} + N_p r^2_{yz} + \underset{r\ c}{\Sigma\Sigma}(y_{rc} - \mu_{rc})^2 \qquad (20.5)$$

in which we have again utilized the fact that for standard scores $\Sigma x^2 = \Sigma z^2 = N$, in general.

With the sum of squares for, say, columns equal to $N_p r^2_{yz}$ in the population, it is obvious that the column effect is a function of the correlation between $Y$ and the column variable $Z$. (No great surprise!) Therefore a sample $F$ test (if a valid $F$ test were available) of the significance of the column factor would seem to be exactly the same as a test of the significance of a sample $r_{yz}$, but despite the apparent equivalence this is not the case because an analysis of variance $F_c = s^2_c/s^2_w$ does not reflect a (linear) trend as does a test of the significance of a correlation coefficient (see p. 317). Regardless of the relative power of these two significance tests, the major issue here is a question of interpretation. If $F_c$ is significant, what interpretation do we place on the "effect" of the column factor, or the $Z$ variable? To say that $Z$ has an effect on $Y$ not only requires an assumption that $Z$ is a causal factor but also that none of the $Z$ "effect" reflects the "effect" of $X$ operating through $Z$. Stated differently, an investigator has not escaped from the logical question of direction, and independence, of causation by merely imposing an analysis of variance design on a situation involving interrelated variables.

In contrast, when we test the significance of $\beta_z$ (and $\beta_x$) for significance by using the method described on p. 321, we are in a sense asking whether the teaming of $Z$ with $X$ in the regression equation for predicting $Y$ has significantly increased the predictable variance of $Y$ over and above that predicted by using $X$ alone. In this connection, it is necessary to recall that the predicted variance by way of multiple regression is decomposable into parts: $\beta^2_x + \beta^2_z + 2r_{xz}\beta_x\beta_z$. In effect, when we say that $\beta_z$ is significant, we are including the joint "effect." This does not, of course, get us off the cause-effect hook, but it does indicate that the significance test by way of multiple regression makes allowance for the presence of correlation between the two factors, whereas an attempted significance test of the factors by way of a factorial design does not make any allowance whatsoever for the correlation between the factors. A naive use of the $F$ test along with the additional step of specifying the proportion of variance explained by each of several factors could conceivably lead to a paradox: suppose that each of three individual difference variables, as factors, correlated .60 with the dependent variable. Each would account for 36 per cent of the variance; if regarded as uncorrelated, the three together would account for 108 per cent of the variance! The multiple regression approach will never yield such an absurdity in that redundancy represented by overlapping (correlated) measures is taken into consideration.

**Multiple regression vs. factorial design, factors uncorrelated.** It is of some interest to examine and compare the analysis of variance tests for main effects and the tests via the significance of the betas when in fact $X$ and $Z$ are uncorrelated in the population. Such would be the case if $X$ and $Z$ were manipulable factors and could be the case if $X$ and $Z$ were uncorrelated individual difference variables. Under this uncorrelated condition, the population value of $\beta_x$ is $r_{yx}$ and that for $\beta_z$ is $r_{yz}$ (see the formulas for the betas on p. 192), and from (11.6) we have $r^2_{y \cdot xz} = r^2_{yx} + r^2_{yz}$. Although these relationships hold, under the given assumption, for the population, it does not follow that such would be the case for a sample if $X$ and $Z$ were individual difference variables. The test for the significance of $\beta_z$ will, of course, allow for the chance, likely small, joint contribution of $X$ and $Z$. If we used analysis of variance for testing $Z$ as a factor, we would not have any allowance for the joint contribution exhibited in the sample because of random sampling error.

Under the condition that $X$ and $Z$ are manipulable variables, $r_{xz} = 0$ and the test for $\beta_z$ will be given by (15.18):

$$F_{\beta_z} = \frac{(r^2_{y \cdot xz} - r^2_{yx})/1}{(1 - r^2_{y \cdot xz})/(N - 2 - 1)} = \frac{r^2_{yz}}{(1 - r^2_{y \cdot xz})/(N - 2 - 1)} \quad (20.6)$$

which resembles the $t^2 = F$ test formula (15.15) for the significance of $r_{yz}$. The difference in degrees of freedom for (20.6) and (15.15) is indeed trivial unless $N$ is small, but the difference between $r_{y \cdot xz}$ and $r_{yz}$, as the appropriate $r$ in (15.15), can be appreciable when $r_{yx}$ is sizable. Thus it would seem that $\beta_z$ can be significant even though $r_{yz}$ is insignificant. How to explain this apparent paradox? With $r_{xz} = 0$ we may replace $r^2_{y \cdot xz}$ by $r^2_{yx} + r^2_{yz}$. Thus we have in the denominator $1 - r^2_{yz} - r^2_{yx}$, which indicates clearly that the amount of variance *produced* by manipulating $X$ is being removed from the error term when testing $r_{yz}$. If we had varied $Z$ and had not introduced the $X$ factor, the residual variance would be a function of $1 - r^2_{yz}$, which would, of course, be the appropriate value for the denominator of (15.15). Thus the significance of $\beta_z$ is identical to that for $r_{yz}$ for $X$ constant. The error term in (20.6) is similar to the $s^2_w$ of the fixed effects factorial design in that both are freed of any variance produced by the factors.

The author sees no particular reason for recommending the multiple regression approach in lieu of analysis of variance for situations involving manipulable factors. Two facts do emerge from the test via the betas: (*a*) any linear trend for $Y$ on $X$ and for $Y$ on $Z$ is reflected in the significance test, and (*b*) it becomes clear that since nonzero linear correlations are increased by range on the independent variable, a significant effect is more apt to emerge if the levels on the factors are chosen to be as widespread as feasible. The advantage of the linear trend can, of course, be had by resort to trend analysis, which is computationally easier than correlations if the factor levels are equally spaced with an equal number of cases per level.

Let us now look further at the difficulties of attempting to use a factorial design when the factors are uncorrelated (in the population) *individual difference* variables. As stated earlier, if linearity holds exactly for the population, we will have $\sum_r m_r \mu^2_r = N_p r^2_{yx}$ and $\sum_c m_c \mu^2_{\cdot c} = N_p r^2_{yz}$. We would also have that the residual variance about the assumed linear regression plane would exactly equal the within-cells variance; i.e., $\sigma^2_w = 1 - r^2_{y \cdot xz}$ for the population. These three relationships would not hold for a sample of $N$ cases, and hence if we were to write an expression for $F_c = s^2_c/s^2_w$ in terms of sample correlational equivalents we would need to keep in mind that we would have an approximation. Therefore we will use $\doteq$ instead of $=$ in the following:

$$F_c = \frac{\sum_c m_c \bar{y}^2_{\cdot c}/(C - 1)}{s^2_w} \doteq \frac{NS^2_y r^2_{yz}/(C - 1)}{NS^2_y(1 - r^2_{y \cdot xz})/(N - 2 - 1)} \quad (20.7)*$$

* Although $\sigma^2_y = 1$ for population standard scores, we now need to include $S^2_y$ as a sample variance.

Note that multiplying $S^2_y(1 - r^2_{y \cdot xz})$ by $N$, then dividing by $N - 2 - 1$ produces an unbiased estimate of the population residual variance about the regression plane (the $rs$ are now sample values). The $df$ for this estimate is, of course, $N - 2 - 1$, whereas the $df$ for $s^2_w$ would be $N -$ (number of cells in which at least two cases fall). We have under null conditions two estimates of the same population variance. Actually, in terms of biased estimators of variance, i.e., $S^2$ instead of $s^2$, the variance of the residuals, $S^2_{y \cdot xz} = S^2_y(1 - r^2_{y \cdot xz})$, about the regression plane will be larger than $S^2_w$, the within-cells variance, simply because the cell means from which the latter is computed will not, for a sample, fall exactly on the regression plane. The difference in the $dfs$ tends to compensate for the difference in the variances (or sums of squares) when we move to unbiased estimates.

Thus when we contrast the two ratios in (20.7) we first note that, as regards the denominators, the essential difference is the $dfs$; or the $n_2$s, for entering Table F, will differ. Surely, the reader must have by now noted that for $n_2$ in excess of 30 or 40 there is little gain with increased $n_2$ values in the sense of a smaller $F$ for significance. Exactly how much the two $n_2$ values will differ cannot be gauged in the abstract because we cannot specify a priori how many of the $RC$ cells will contain at least two cases, the minimum number required in order for a cell to contribute to $s^2_w$. It does not follow that there will be no empty and one-case cells simply because $X$ and $Z$ are uncorrelated. In general, the number of such cells will be a function of at least three things: the number of levels for the two factors, the form of the distributions of scores for the two variables (factors), and the over-all $N$. To carry along our contrast of the factorial design and correlational approaches, we will need a sizable number of levels. If both variables are nearly normal in distribution and $N$ is 400, the use of nine levels for each factor may very well lead to 24 empty and 12 one-case cells, hence $n_2$ for $s^2_w$ could be near $400 - 12 - (81 - 36) = 333$. If $N$ is 100, the number of empty and one-case cells could easily be 36 and 24, respectively; hence $n_2$ could be near $100 - 24 - (81 - 60) = 55$ as compared with $N - 2 - 1 = 97$ for the multiple regression residual. (The reader may be puzzled about how these $n_2$ values for $s^2_w$ are estimated. First, the effective $N$ is reduced by the number of unusable cases in one-case cells; the part in parentheses is the number of cell means from which we would have deviations.)

A second consideration in the contrast of the two ratios in (20.7) is that $\Sigma_c m_c \bar{y}^2_{\cdot c}$ will exceed $N S^2_y r^2_{yz}$ somewhat because of the chance departure of the $\bar{y}_{\cdot c}$ from linearity (the excess of $\eta_{yz}$ over $r_{yz}$ is involved here). Thus the first ratio would tend to be larger than the second, a seeming

disadvantage of substituting a correlational equivalent for the column sum
of squares, but under the assumption of linearity this disadvantage would
not be sizable. The difference in numerators would balance in part the
slight advantage accruing to the second ratio because of the larger $n_2$ as
the $df$ for the denominator.

Therefore it seems safe to conclude that computation for $F_c$ by way of
correlational equivalents would yield a reasonably good approximation.
At this stage we will do what the reader may have thought should have
been done earlier, namely, cancel $N$ and $S^2{}_y$ in the second (approximate)
ratio. Doing so leads to

$$\frac{r^2{}_{yz}/(C-1)}{(1-r^2{}_{y\cdot xz})/(N-2-1)} \tag{20.8}$$

which, aside from the divisor $(C-1)$ instead of a divisor of 1, is the
same as the $F$ of (20.6) for testing the significance of $\beta_z$. Both (20.6) and
(20.8) were arrived at on the assumption of zero correlation between $X$
and $Z$, and both are concerned with the "effect" of variable $Z$. Obviously,
the numerator of (20.6) will be larger than the numerator of (20.8), hence
a larger $F$ by (20.6). If $C = 9$, $F_{\beta_z}$ will be 8 times larger than $F_c$, a gain
that far eclipses the approximation errors discussed above. But before we
make too much of this, we should turn to Table F to see what effect a
decrease in $n_1$ from 8 to 1 has on the magnitude of $F$ required for significance—a mere twofold increase, or less. The test based on $\beta$ allows for
the trend (if any). This same advantage would go with the simple $t$ or
$F$ test for $r_{yz}$. Further, the use of (20.6) allows for nonzero chance sample
values of $r_{xz}$, the effect of which is not allowed for by $F_c$ computed by the
first ratio in (20.7).

**Question of interaction.** Before closing this discussion, we should
revert to a simplifying assumption made in order to compare testing the
significance of the variables $Z$ and $X$ by the betas and by $F_c$ and $F_r$. For
$F$ tests we presumed no interaction in the population between the factors,
$X$ and $Z$. This permitted the writing of a score on $Y$ as a linear additive
function of row and column effects, as in (20.1). When we assumed
planarity for the regression of $Y$ on $X$ and $Z$ we were in effect saying the
same thing: $Y$ is a linear function of $X$ and $Z$, as in (20.2). This means
that zero interaction and planarity (cell means falling exactly on the
regression plane) are, as assumptions, identical. Does it follow that
interaction and curvilinearity of the regression surface are identical concepts? It is easy to see, and say, that the presence of interaction indicates
that the cell means will not fall on a plane, but it does not follow that

curvilinearity (a curved surface instead of a plane) is indicative of inter-action. To determine why, consider population cell and marginal means as specified in Table 20.5. Obviously, these cell means will not fall on a fitted plane. Equally obvious is the fact that the simple linear relationship of the dependent variable on the column factor is the same for each level of the row factor and the simple curved relationship on the row factor is identical from column to column, hence zero interaction.

**Summary.** When two or more bases for classification are somehow associated or correlated, we have the nonorthogonal case. This means that a breakdown of the sum of squares into components that are func-tionally independent is not possible, hence interpretation that may be

Table 20.5. Means that show curvilinearity with zero inter-action

|   | 1 | 2 | 3 | 4 | |
|---|---|---|---|---|---|
| 1 | 8 | 9 | 10 | 11 | 9.5 |
| 2 | 5 | 6 | 7 | 8 | 6.5 |
| 3 | 3 | 4 | 5 | 6 | 4.5 |
| 4 | 2 | 3 | 4 | 5 | 3.5 |

placed on a factor found to be significant is not unencumbered even when it is safe to say that the dependent variable really depends on the factor as a causative agent. The arguments in this section, based on (a) the as yet insurmountable difficulties of nonorthogonal factors plus (b) comparison and contrast of the factorial design and multiple regression approaches when orthogonality of individual difference factors is assumed (rarely true in practice for a sample) to hold, lead definitely to the conclusion that the factorial design approach is inferior to the multiple regression technique as a method for testing the statistical significance of factors that are characteristics of individuals. The analysis of variance of the data obtained by factorial experiments provides tests as to whether factors have *produced* variation. Multiple regression, in contrast, has traditionally been associated with analyzing *natural* (not laboratory produced) variation into sources with no requirement that the sources be uncorrelated with one another. Although mathematically all analysis of variance can be subsumed under "regression analysis," it is not herein recommended that we lose sight of the uniquely differing situations which provide a basis for distinguishing between problems involving manipulation of factors versus problems of correlation. Each of these approaches has a rightful place, and as of now any data analyzer who replaces multiple regression analysis by analysis of variance via factorial design involving correlated variables as the factors is doing the indefensible.

## VARIANCE COMPONENTS ESTIMATION

In correlational analysis we had for the bivariate situation that $S^2_y = S^2_{y'} + S^2_{y \cdot x}$ and for multiple regression we had, in standard score units, that $S^2_{z_1} = S^2_{z'_1} + S^2_{z_{1 \cdot 23 \cdots n}}$. It will be recalled that for the latter we had a breakdown of the predictable variance into parts. Thus in the three-variable case, we can write

$$S^2_{z_1} = \beta^2_2 + \beta^2_3 + 2r_{23}\beta_2\beta_3 + S^2_{z_{1 \cdot 23}}$$

Since the idea of breaking down variance into component parts is attractive as a descriptive device and since all analysis of variance models involve partitioning sums of squares, a logical next step beyond significance testing in analysis of variance would seem to be a specification of the variance attributable to the factors. This would permit descriptive comparison and contrast of the relative importance of sources of variation. But before we succumb to the temptation to take the ratio of, say, $s^2_c$ to $s^2_r$ as a basis for concluding that the column source of variance is so many times that for the row factor, we should carefully scrutinize the meaning of such a ratio. (Note: $s^2_c/s^2_r$ is not being taken as an $F$ for judging statistical significance.)

We will illustrate a difficulty by considering an example that exists in the recent literature of psychology, the results and conclusions of which have been disseminated beyond technical journal presentation. The research involved a three-factor design with 11 situations as one factor, 14 modes of response as a second factor, and $N$ subjects as the third factor—a repeat-measure design. It was claimed that all three factors were random, hence we have a Case XIV layout. The dependent variable was "anxiousness." Let $i$ stand for individuals (subjects), $s$ for situations, and $r$ for response mode. Then the expected values of the variance estimates for the "main" effects are as follows:

$$s^2_r \rightarrow 11N\sigma^2_r + N\sigma^2_{rs} + 11\sigma^2_{ri} + \sigma^2_{rsi} + \sigma^2_e$$

$$s^2_s \rightarrow 14N\sigma^2_s + N\sigma^2_{rs} + 14\sigma^2_{si} + \sigma^2_{rsi} + \sigma^2_e$$

$$s^2_i \rightarrow (11)(14)\sigma^2_i + 11\sigma^2_{ri} + 14\sigma^2_{si} + \sigma^2_{rsi} + \sigma^2_e$$

Among other things it was concluded on the basis of the ratio $s^2_s/s^2_i = 3.8$ that "situations" as a factor is far more important than persons as a source of variation. In commendable scientific tradition the study was replicated, with sample size, or $N$, increased from 67 to 169. This time the ratio

$s^2_s/s^2_i$ exceeded 11. Apparently the jubilation ensuing from a striking confirmation by replication of a finding dulled the critical powers of the investigators: the question should have been raised as to why the situations factor was, loosely speaking, nearly three times (ratio of 11+ to 3.8) as powerful a source of variation in the second study as in the original.

The reason for this finding is plainly visible in the foregoing expectations: the ratio $s^2_s/s^2_i$ is a function of sample size. $N$ comes in twice in $s^2_s$ and not at all in $s^2_i$; therefore to "prove" beyond a shadow of anybody's incredulity that the situations factor outstrips individuals as a source of variance, one need only beef up $N$. Recall that the variance ratio being considered has nothing whatsoever to do with statistical significance, for which an increased $N$ in no way vitiates results.

If we wish to compare situations and individuals as to their relative importance as sources of variation, we need estimates of $\sigma^2_s$ and $\sigma^2_i$ which are obviously not estimated by $s^2_s$ and $s^2_i$. Even under the untenable assumption that all interactions are zero for the population, we still could not regard $s^2_s$ and $s^2_i$ as appropriate estimates.

To explicate further the problem of variance component estimation, we will confine our attention to two-way and three-way factorial designs in which one factor, two factors, then three factors are random. The last is a bit of unrealism in that designs with three random factors are hard to come by. Parenthetically, the fact that for the mentioned research the investigators claimed that the three factors were random does not make it so. In the first place, neither the situations nor the response modes were chosen at random, a requisite by definition for a random factor. Second, the fact that the replication, carried out in a different university setting, included exactly the same situations and exactly the same response modes is conclusive evidence that the only factor that was random was the persons factor. Perhaps the investigators had been informed (correctly) that variance component estimation holds only for random factors, hence their attempt to invoke by fiat the random model. Given that factors are really random, how can one proceed to make valid estimation of their variances?

Suppose the row factor is random (typically, persons) and that the column factor is fixed. Now turn back to p. 354, Case VI, where you find that $s^2_r$ estimates $\sigma^2_e + C\sigma^2_r$, by which is meant that on the average $s^2_r = \sigma^2_e + C\sigma^2_r$. If we solved that equation for $\sigma^2_r$ we would have

$$\hat{\sigma}^2_r = \frac{s^2_r - \sigma^2_e}{C}$$

in which the "hat" indicates that this gives an estimate of $\sigma^2_r$. Although the starting equation holds on the average, it is not an exact equation for a particular value of $s^2_r$, hence we must distinguish between $\sigma^2_r$ and the estimate thereof. Next we note that we also need an estimate of $\sigma^2_e$. This would be given by $s^2_w$ in case we had measurement replication in each row by column cell, but lacking this replication—the usual situation—we have no available estimate of $\sigma^2_e$. The interaction estimate, $s^2_{rc}$, would be usable as an estimate of $\sigma^2_e$ if it could be assumed that no real interaction existed, but such an assumption is highly unrealistic except when the columns stand for parallel forms of a test. If you wish to estimate $\sigma^2_r$ when columns stand for a manipulated factor, you will need to incorporate measurement replication. Two measures per cell will provide an estimate to replace $\sigma^2_e$ in the above equation.

Suppose next that both the row and the column factors are random. Again one starts with expected values of the variance estimates (p. 353), saying that in the long run

$$s^2_r = \sigma^2_e + m\sigma^2_{rc} + mC\sigma^2_r$$

which is solved for the desired estimate:

$$\hat{\sigma}^2_r = \frac{s^2_r - (\sigma^2_e + m\sigma^2_{rc})}{mC}$$

Since $s^2_{rc}$ is an estimate of the part in parentheses, we may write

$$\hat{\sigma}^2_r = \frac{s^2_r - s^2_{rc}}{mC}$$

which holds regardless of whether $m$ is 1 or more. This time we do not need measurement replication. By a similar line of reasoning we get

$$\hat{\sigma}^2_c = \frac{s^2_c - s^2_{rc}}{mR}$$

as the estimated variance component for the column factor.

Given the two estimates, can one safely take their ratio, say $\hat{\sigma}^2_c/\hat{\sigma}^2_r$, as a basis for assessing the relative importance of the two sources of variation? Or will such a ratio depend upon $R$ and/or $C$? Obviously, $m$ cancels. Your first thought might be that the ratio

$$\frac{\hat{\sigma}^2_c}{\hat{\sigma}^2_r} = \frac{(s^2_c - s^2_{rc})/R}{(s^2_r - s^2_{rc})/C}$$

would not be independent of $R$ and $C$, the sample sizes, simply because an increase in, for example, $C$ would seem to mean a decrease in the denominator, hence an increase in the ratio for a fixed value of $R$. Since the foregoing ratio holds for all values of $m$, it must hold for $m = 1$, and for $m = 1$ we have on the average that $s^2_r = \sigma^2_e + \sigma^2_{rc} + C\sigma^2_r$ and that $s^2_{rc} = \sigma^2_e + \sigma^2_{rc}$. Hence on the average $s^2_r - s^2_{rc} = C\sigma^2_{rc}$. Therefore as $C$ is increased, the numerator of the denominator term tends to increase $C$ fold, which means that the denominator term does not decrease with increase in $C$. Likewise, the numerator of the ratio is not a function of $R$. Accordingly, the ratio of the estimated components of variance is not a function of sample size(s)—a valid deduction can be made concerning the relative importance of sources of variation.

The extension of this sort of variance component estimation to the three-factor setup, mixed model with the row and block factors as random, offers no difficulty. From the expectations given for Case XIII, p. 378, we can quickly write

$$\hat{\sigma}^2_r = \frac{s^2_r - s^2_{rb}}{BC} \quad \text{and} \quad \hat{\sigma}^2_b = \frac{s^2_b - s^2_{rb}}{RC}$$

for the realistic $m = 1$ per cubicle. From these estimates, we see that the fixed factor enters only as a constant, $C$, affecting alike both estimates.

In order to establish the procedure for three factors random, we will need first to write the general expectations for the various variance estimates involved in significance testing:

$$s^2_r \rightarrow \sigma^2_{rbc} + C\sigma^2_{rb} + B\sigma^2_{rc} + BC\sigma^2_r + \sigma^2_e$$

$$s^2_b \rightarrow \sigma^2_{rbc} + R\sigma^2_{bc} + C\sigma^2_{rb} + RC\sigma^2_b + \sigma^2_e$$

$$s^2_c \rightarrow \sigma^2_{rbc} + B\sigma^2_{rc} + R\sigma^2_{bc} + RB\sigma^2_c + \sigma^2_e$$

$$s^2_{rb} \rightarrow \sigma^2_{rbc} + C\sigma^2_{rb} + \sigma^2_e$$

$$s^2_{rc} \rightarrow \sigma^2_{rbc} + B\sigma^2_{rc} + \sigma^2_e$$

$$s^2_{bc} \rightarrow \sigma^2_{rbc} + R\sigma^2_{bc} + \sigma^2_e$$

$$s^2_{rbc} \rightarrow \sigma^2_{rbc} + \sigma^2_e$$

To obtain an estimate of $\sigma^2_r$, we regard the first line as an equation that holds on the average, and solve explicitly for the estimate:

$$\hat{\sigma}^2_r = \frac{s^2_r - \sigma^2_{rbc} - C\sigma^2_{rb} - B\sigma^2_{rc} - \sigma^2_e}{BC}$$

which, to be useful, must have the four $\sigma^2$ values replaced by estimates. From the foregoing expectations, we have the following estimates:

$$\hat{\sigma}^2_{rbc} = s^2_{rbc} - \sigma^2_e$$

$$C\hat{\sigma}^2_{rb} = s^2_{rb} - \sigma^2_{rbc} - \sigma^2_e = s^2_{rb} - (s^2_{rbc} - \sigma^2_e) - \sigma^2_e = s^2_{rb} - s^2_{rbc}$$

$$B\hat{\sigma}^2_{rc} = s^2_{rc} - \sigma^2_{rbc} - \sigma^2_e = s^2_{rc} - (s^2_{rbc} - \sigma^2_e) - \sigma^2_e = s^2_{rc} - s^2_{rbc}$$

The last two equations were obtained by replacing $\sigma^2_{rbc}$ by the estimate $\hat{\sigma}^2_{rbc}$. Substituting, we have

$$\hat{\sigma}^2_r = \frac{s^2_r - (s^2_{rbc} - \sigma^2_e) - (s^2_{rb} - s^2_{rbc}) - (s^2_{rc} - s^2_{rbc}) - \sigma^2_e}{BC}$$

which simplifies to

$$\hat{\sigma}^2_r = \frac{s^2_r - s^2_{rb} - s^2_{rc} + s^2_{rbc}}{BC}$$

Similarly, we may obtain

$$\hat{\sigma}^2_b = \frac{s^2_b - s^2_{rb} - s^2_{bc} + s^2_{rbc}}{RC}$$

and

$$\hat{\sigma}^2_c = \frac{s^2_c - s^2_{rc} - s^2_{bc} + s^2_{rbc}}{RB}$$

It will have been noted that these formulations do not call for an estimate of $\sigma^2_e$, hence are usable with $m = 1$ score per cubicle.

## THE PROBLEM OF ERROR REDUCTION

All research is bedeviled by various types of error. The careful investigator attempts to plan his research to avoid any possible fluke that might bias his results. By tight control of experimental procedures, it becomes less likely that produced effects are attributable to anything other than the experimentally manipulated variables. Even so, he cannot get rid of so-called random variable errors, but he can strive to reduce their effects. We shall here discuss random error reduction under three headings: measurement errors; errors in inferring population parameters in field, or survey, or normative studies; errors in experimental testing of hypotheses.

**Measurement errors.** By now the reader should be fully cognizant of the fact that variable measurement errors tend to increase variances, hence to increase estimated standard errors of means and of differences between means, with a resultant lowering of $t$s or $z$s. Also, in analysis of variance, fixed constants model, these errors will increase $s^2_w$, hence decrease $F$; in mixed models, these errors increase the interaction variance estimates that are used as denominators of $F$s, hence again a decrease in $F$s as a result of measurement errors. In correlational work the attenuating effects of these errors may make it difficult to obtain $r$s that are statistically significant.

How can one reduce measurement errors? When measurement is by way of the typical psychological test, it is merely a question of increasing reliability, which can be achieved either by more careful test construction (item selection) or by increasing the length of the test by the addition of items that are as good as or better than those already selected. If the measurement is by way of a trial or trials with an instrument, errors can be reduced by averaging scores from more trials provided practice and/or fatigue effects are nonoperative.

**Errors in survey, normative, studies.** Surveys for the purpose of gauging opinion, and studies designed to establish normative data, require large scale sampling. The aim is to secure a sample which is unbiased, that is, representative of a defined population, with chance sampling errors as small as possible. We shall limit ourselves to three sampling methods: random, stratified, and area.

**Random sampling.** The conditions of random sampling have been specified earlier (p. 53). By the method of random sampling it is fairly easy to arrive at a representative sample, provided the universe has been catalogued. Thus, if we wish a representative sample of school children of a certain grade in a city, we can secure it by a purely mechanical scheme, such as taking every $n$th card from the files. Although this type of *systematic* sampling does not exactly satisfy the conditions of random sampling, it will assure a random sample unless the cards have been systematically arranged (in a somewhat peculiar order). The use of the random method for sampling an uncatalogued population involves so many difficulties in psychological research that no schemes are to be found in the literature.

Increasing sample size is the only way by which we can reduce chance errors when the random method is being employed. That sheer sample size is not enough to reduce nonrandom errors is evidenced by the *Literary Digest* straw polls, which rested on the assumption that the population of telephone subscribers and car owners was not different in its voting preference from the entire population of potential voters. This happened to hold before 1936, so that replies to ballots mailed at random to telephone subscribers and car owners forecasted fairly accurately the election results. Despite a very large sample, the *Digest* poll failed miserably in 1936; this failure is attributed to the alignment of voting to income levels, an alignment that did not exist in prior years.

**Stratified sampling.** In the stratified method, one or more individuals are pulled at random from each of several strata, the number in the sample from each stratum being proportional to the universe number in the stratum, and the strata are predetermined by knowledge of some control variable or variables. Psychologists who sample so as to secure proportionate representation from the several occupational levels are, in effect, using the principle of stratification. It should be obvious that the method can be used only when information is available on some variable or variables which permits their use in setting up the strata, and when cases within the strata can be drawn randomly.

When the sampling is for attributes by the stratified method, the standard error of an obtained proportion, $P$, is given, in terms of information yielded by the sample, approximately by

$$S_P = \sqrt{\frac{PQ}{N} - \frac{S^2_p}{N}} \qquad (20.9)$$

where $P$ equals the proportion in the total sample, $N$, who possess the attribute, $Q = 1 - P$, and $S^2_p$ is the weighted variance of the several strata proportions about the sample value $P$. A casual examination of formula (20.9) shows that the magnitude of the error is less for a stratified sample than for a random sample, and that the increase in precision depends on our ability to stratify the universe in such a way as to secure strata which are really different with regard to the attribute being studied.

For stratified sampling, the variance of the mean may be written as

$$S^2_{\bar{X}} = \frac{1}{N}(S^2 - S^2_{\bar{X}_s}) \qquad (20.10)$$

where $\bar{X}$ = the sample mean, $S^2$ = the sample variance, and $S^2_{\bar{X}_s}$ = the weighted variance of the means of the several strata about the total sample mean. If stratification has been accomplished by use of a variable, $Y$, which is linearly related to the variable being studied, the formula can be written in the form

$$S^2_{\bar{X}} = \frac{1}{N}(S^2_x - S^2_x r^2_{xy}) \qquad (20.11)$$

It will be noticed that stratified sampling does lead to greater precision in the sense of smaller chance error, but only when the control or stratifying variable is related to the variable being studied.

The *quota* method involves the use of strata, but selection within the strata is not done on a random basis—the field worker merely fills a quota by securing the correct proportion per strata; selective factors leading to bias can easily operate.

**Area sampling.** There is evidence that area or "pin point" sampling is the best method yet devised for drawing samples in survey studies. Its use, however, depends on the availability of extensive facilities. The student who is interested in this, or the stratified, method will wish to turn to detailed treatments of the subject.*

* Yates, F., *Sampling methods for censuses and surveys*, New York: Hafner, 1949; Deming, W. E., *Some theory of sampling*, New York: John Wiley and Sons, 1950.

**Sampling errors in experimentation.** The formation of groups for experimental purpose can be accomplished (1) by random assignment of subjects to the groups to be used for the experimental conditions, (2) by pairing (for two conditions) or using matched sets of subjects for more than two conditions (one of each set assigned randomly to a condition), (3) by using sibs or litter mates, (4) by matching distributions on some control variable. Or (5) we could have just one group, with each person measured under all of the experimental conditions—a repeat-measure design which controls individual differences as a source of random error.

For methods 2, 3, and 5 the statistical analysis is by way of the analysis of variance (mixed model) with rows standing for the matched persons or litters or individuals, respectively, for the three methods. The $F$ test of the significance of the differences among the correlated means (the means for conditions) involves an error term which is freed of the row variation; stated differently, the error term (an estimate of a two-way interaction variance) tends to be small if the correlations between the matched persons or between sibs or between scores on the same persons are large. The foregoing argument holds, of course, for just two experimental groups (or an experimental and a control group) as well as for three or more groups.

Thus, compared to method 1 (random assignment), greater precision is attainable by using method 2 or 3 or 5. Before discussing method 4, let us again consider the situation where groups are needed for just two conditions. If the groups are formed by pairing individuals, the sampling variance of the difference between the two means is, as we learned in Chapter 6, given by

$$S^2{}_D = S^2{}_{\bar{X}_1} + S^2{}_{\bar{X}_2} - 2r_{12}S_{\bar{X}_1}S_{\bar{X}_2} \tag{6.8}$$

The gain in pairing, over random assignment, depends on the magnitude of $r_{12}$. It can be shown that if the pairing is done on the basis of variable $Y$, the value of $r_{12}$ will be $r^2{}_{xy}$, and in case two or more variables are controlled by pairing, $r_{12}$ will be the square of the multiple correlation between the dependent variable, $X$, and the control variables. The reason for pairing, it will be recalled, is to make the groups comparable on certain variables which might affect the outcome of the experiment. We now see explicitly that the advantage of pairing depends definitely on how highly the variables, so controlled, are correlated with the dependent variable. No correlation, no gain; low correlation, little gain.

Method 4 is another way of making groups comparable on pertinent variables. Instead of pairing persons, distributions are matched for the $Y$ variable, to be controlled, in such a manner that the two groups contain the same proportions of cases in the several intervals as hold for a supply distribution on $Y$. The sampling variance of the difference between the two $X$ means is given by

$$S^2{}_D = S^2{}_{\bar{X}_1}(1 - r^2{}_{xy}) + S^2{}_{\bar{X}_2}(1 - r^2{}_{xy}) \qquad (20.12)$$

If the matching has been made on the basis of several control variables, the two correlations (one for each group) become the multiple $r$s between $X$ and the control variables.

From (20.12) we may deduce the following fact. Where two groups have been separately matched as to distribution on the same control variable(s), the standard error of the difference can be obtained without the restriction of the ordinary pairing procedure, which requires that there be an equal number of cases in the two groups. The reader will note that either term in formula (20.12) is, as might be expected, identical to formula (20.11) for the sampling variance of a mean when the stratified method is used. The method of matching distributions is particularly useful when the cost per case is much greater in the experimental group than in the control group. Precision can be increased by taking a larger control group—a possibility also when the groups are chosen by randomization.

The essential advantages of using method 4 to control on a variable can be achieved by way of a two-way analysis of variance design in which the $C$ columns stand for $C$ experimental conditions and the $R$ rows stand for $R$ levels (groups) on the control variable, with $m$ (greater than 1) subjects in each of the $RC$ cells. We here suppose that scores on the control variable are available or can be obtained for a total group of $N$ potential subjects. It is presumed that the reason for using the control variable is that it is known or suspected to have appreciable correlation with the dependent variable; otherwise, there would be little need for, or gain in, using the variable as a control.

How many levels should we plan for the control variable? This will depend in part on the size of $N$ for the potential supply and the value of $C$. The $N$ subjects are to be divided into $RC$ groups with $m$ cases per group; $C$ is fixed, whereas $R$ and $m$ may apparently be chosen at will. There are, however, two relevant considerations, both of which have to do with $s^2{}_w$ as the error term for $F$s. First, if we take $R$ larger and larger we will have a smaller $df$ for the error term—when using $N = mRC$ subjects the $mRC$ part of the $df = mRC - RC$ is fixed, but the $RC$ part increases with $R$, hence the loss in $df$. Whether this loss is consequential may or may not be the case; it is only when we enter the $F$ table with relatively small $n_2$ values that their differences are crucial.

Second, since a level on the control variable is, under the circumstances, to be defined in terms of a group of subjects having the same or nearly the same scores on the control variable, we are in effect led to the use of grouping intervals such as those utilized in making a frequency distribution. But the ordinary equal-sized intervals will, of course, produce markedly differing frequencies, thus making it difficult if not impossible to assign an equal number, $m$, to each of the $RC$ cells. By using such intervals one could attain equal $m$s in a row, with varying $m$s from row to row. This is a feasible design since the cell frequencies would be proportional from column to column, but we have not herein indicated the computations for the unequal $m$ situation. The goal of equal $m$s throughout can be achieved by grouping so as to have (approximately) equal percentages of cases for each level. Unless $N$ happens to be an exact multiple of a potential $R$ times $C$ and if there are tied scores at the division points, it is inevitable that this process will not lead to the use of all $N$ cases of the potential supply. The dropped cases should be eliminated randomly to make the frequency in an interval an exact multiple of $C$. (Incidentally, the use of equal-sized intervals will also lead to dropping cases if proportionality is adhered to.) Now the choice of the number of levels, $R$, allows considerable flexibility. Perhaps a choice of $R = 5$ (20 per cent per group) or $R = 6$ (approximately 17 per cent per group) or $R = 7$ (near 14 per cent per group) or $R = 8$ (12 or 13 per cent) or $R = 9$ with 11 per cent or $R = 10$ with 10 per cent will facilitate the division into $RC$ groups with $m$ per group. Why not $R$ less than 5? This, at last, brings us to the second factor that has to do with the choice of $R$. The error term for $F$s is to be $s^2_w$. As $R$ is taken smaller and smaller the variability on the control variable within a level will tend to increase and since the dependent variable is (supposedly) appreciably correlated with the control variable, the use of broad groupings (small $R$) will lead to a larger $s^2_w$ than will hold for narrower grouping (a larger $R$). The reader who does not immediately see why this is true should do some additional thinking. Obviously, for this sort of thing there is no sharp demarcation between adjacent $R$ values—we somewhat arbitrarily suggest $R = 5$ as a minimum.

With the decision made as to $R$, and its consequent $m$, there remains the task of assigning the subjects within each level (or grouping on the control variable) to the $C$ experimental conditions. This should be done on a strictly random basis. The whole process will, of course, lead to the formation of $C$ groups that have the same distributions on the control variable, the means for which will be more nearly the same than would be the case had the subjects been randomly assigned to the $C$ conditions irrespective of their scores on the control variable. The reader's second thoughts on the desirability of having $R$ not small should also help him see that the narrower the range of control scores per level, the better the matching of the $C$ distributions.

When the measurements on the dependent variable become available, the analysis of variance is straightforward: the usual breakdown for a two-way design with $m$ scores per cell. Then $F_c = s^2_c/s^2_w$ provides a test of the effect of the column factor. With this design the investigator not only has the satisfaction of knowing that he has controlled a relevant variable, he also knows that his experiment has attained greater precision in that $s^2_w$ will tend to be smaller than would hold for $C$ groups chosen randomly. The $s^2_w$ for randomly chosen groups versus that based on matched groups will differ simply because for the latter the within sum of squares represents what is left over after taking out not only the column sum of squares but also a sum of squares for rows and for $R \times C$ interaction. And here we have a bonus: the "levels" by experimental conditions design permits a test of interaction; a significant $F_{rc} = s^2_{rc}/s^2_w$ tells us that the experimental effect is not the same at all levels on the control variable. This type of test is not possible in the situation where the rows stand for pairs (when $C = 2$) or when the "pairing" is extended to triads, quartets, quintets, etc. Obviously, this matching by way of levels on a control variable is not intended for testing the "effect" of the control variable: $s^2_r$ is not tested. Its significance would only verify what we knew at the start; i.e., that the dependent and control variables show correlation. The greater the correlation, the larger the $s^2_r$ value and the greater the statistical gain in using the design.

This way of matching for control purposes, sometimes called "levels by treatments" design, can be extended to control for more than one variable. The optimum gain in precision would be attained by basing the levels on the composite scores that are obtainable by way of a multiple regression equation for predicting dependent variable scores from the two or more control variables. The problem here is that rarely will data be available for establishing the needed regression equation. We may know only, from fragmentary sources, that the possible control variables $Y_1$, $Y_2$, etc., show correlation with $X$, the dependent variable. By pooling information, we may be able to specify roughly the degree of correlation of each control variable with $X$; we may have little or no information regarding the intercorrelations among the control variables. Under the circumstances the best that can be done might be a composite in which the weights for the control variables are taken proportional to their estimated correlations with $X$. Each of these weights should incorporate $1/S_j$, where $S_j$ is the standard deviation for the $j$th $Y$ variable, in order to equalize differences in variation. For the purpose at hand we need not worry about the constant $A$ in the regression equation. Why not? And the numerical values of the weights are arbitrary, the only restriction being that they be proportional to $r_{xj}/S_j$. This will permit the use of integers instead of fractions. The resulting composite scores are then used for groupings according to levels.

The use of pairing or its extension to triads, etc., and the use of the matched distribution procedure by way of levels has long been recognized as a sound procedure. We might argue, however, that the advantages of pairing have been overstressed. The gain in error reduction may not be appreciable. The advocates of pairing say that they are not willing to risk randomization as a method for setting up groups, but it should be noted that there are always numerous variables which might affect the outcome of an experiment that are never controlled except by randomization. Thus we can seldom, if ever, completely avoid placing faith in the randomization process. Random differences between groups never have more than a random effect on the results; the error formulas always include all random effects. When pairing leads to only a slight reduction in error, we have evidence that the pairing procedure may not have been worth the effort involved.

It should be noted that an original group which is split into experimental groups either by the random method or by pairing must be regarded as representative of some defined universe, and that such conclusions as are drawn from the experiment cannot be generalized unless it can be shown that the defined universe is representative of the generality of mankind with respect to the variables being studied. In other words, those who use the college sophomore as a laboratory representative of mankind have not avoided, by showing that selective factors did not render their experimental groups noncomparable, the necessity of bridging the gap between the sophomore's behavior and that of the typical human being.

At this point, we remind the student that the covariance adjustment method (Chapter 18) is an entirely legitimate technique for allowing for uncontrolled variables and at the same time reducing error variance.

It is appropriate to end this discussion (and the text) with an example of an experiment in which error reduction might have been achieved by judicious planning. The Lanarkshire milk experiment in England involved the daily feeding of three-fourths of a pint of raw milk to 5000 children and of an equal amount of pasteurized milk to another group of 5000 over a period of 4 months. These 10,000, plus a control group of 10,000, were measured for height and weight at the beginning and the end of the 4-month period. Since the purpose of the experiment was to check on the relative merits of raw vs. pasteurized milk, the control group was non-essential. (It is an interesting commentary on the magic of the word "control" that very frequently a control group is used when not needed.) Despite large numbers, the feeder and control groups were not comparable as regards initial height and weight, the operating selective factor being the benevolent attitude of school teachers who apparently thought the research would not be harmed if preference was given frail, undernourished children in choosing individuals for the feeder groups. Either a carefully

# EXERCISES AND QUESTIONS*

## CHAPTER 2

**2.1.** *a.* Make separate frequency distributions for the marks of the two groups of students in Table I. Use intervals of size 5.

  *b.* Determine also the cumulative frequencies for each group.

### Table I. Final examination marks for a class in statistics

| Students with No Calculus ($N = 36$) | | | | | | Students with Some Calculus ($N = 22$) | | | |
|---|---|---|---|---|---|---|---|---|---|
| 103 | 150 | 139 | 79 | 150 | 134 | 137 | 139 | 112 | 139 |
| 98 | 79 | 94 | 137 | 118 | 113 | 151 | 124 | 80 | 153 |
| 106 | 93 | 106 | 137 | 91 | 109 | 131 | 94 | 96 | 77 |
| 71 | 101 | 92 | 74 | 106 | 87 | 133 | 123 | 101 | 115 |
| 108 | ˙113 | 103 | 108 | 114 | 105 | 115 | 90 | 154 | 122 |
| 120 | 95 | 83 | 93 | 109 | 97 | 111 | 135 | | |

**2.2.** *a.* Make separate frequency distributions for the two groups of scores in Table II. Use intervals of size 3.

  *b.* Determine also the cumulative frequencies for each group.

**2.3.** *a.* Draw a frequency polygon for the distribution in Table III, part A.

  *b.* Draw an ogive for the data of Table III, part A.

**2.4.** *a.* Draw a frequency polygon for the distribution of Table III, part B.

  *b.* Draw an ogive for the data of Table III, part B.

### Table II. Scores on final examination for a course on psychological tests

| Undergraduates ($N = 32$) | | | | | | Graduate Students ($N = 23$) | | | |
|---|---|---|---|---|---|---|---|---|---|
| 70 | 72 | 76 | 66 | 76 | 80 | 84 | 80 | 90 | 82 | 84 |
| 67 | 69 | 90 | 50 | 76 | 47 | 79 | 62 | 77 | 89 | 70 |
| 51 | 58 | 71 | 88 | 65 | 54 | 73 | 74 | 87 | 76 | 79 |
| 89 | 64 | 80 | 67 | 71 | 90 | 85 | 95 | 78 | 69 | 97 |
| 91 | 71 | 63 | 81 | 87 | 81 | 78 | 86 | 92 | | |
| 79 | 79 | | | | | | | | | |

---

\* These are so arranged that frequently an even-numbered exercise is of the same type as its immediately preceding odd-numbered exercise. "Thought" questions are intended as thinking, not thumbing, exercises.

## CHAPTER 3

**3.1.** For the scores of Table I, compute separately for the two groups:

*a.* the medians, using the undistributed scores.

*b.* the medians, using the frequency distributions.

**3.2.** Repeat exercise 3.1 with the data of Table II.

**3.3.** Compute the mean for each group in Table I by

*a.* the definition formula for the mean.

*b.* the arbitrary origin method.

**3.4.** Repeat exercise 3.3 with the data of Table II.

**3.5.** Combine the two distributions for the data of Table I, compute the mean by the arbitrary origin method, and check by using the formula for securing the mean for a combined group (use the means obtained by the arbitrary origin method for this check).

**3.6.** Repeat exercise 3.5 with the data of Table II.

**3.7.** Compute the median, $Q_1$, and $Q_3$ for the distribution in Table III, part *A*.

**3.8.** Compute the median, the 20th and the 80th percentile points for the distribution in Table III, part *B*.

**3.9.** Using the results of exercise 3.7, locate the three points, $Q_1$, the median, and $Q_3$, on the base line of your *ogive* curve for the distribution of Table III, part *A*. Divide the ordinate on the right-hand side (the ordinate at IQ = 170) into approximate fourths. Draw a line from each of the three base-line points up to the ogive, then horizontally to the right. Notice where these horizontal lines hit the ordinate on the right-hand side.

**3.10.** Using the results of exercise 3.8, locate the three points, the median, $P_{20}$, and $P_{80}$, on the base line of your *ogive* curve for the distribution of Table III, part *B*. Divide the ordinate on the right-hand side (the ordinate at IQ = 180) into approximate fifths. Draw a line from each of the three base-line points up to the ogive, then horizontally to the right. Note where these horizontals hit the ordinate on the right-hand side.

**3.11.** Compute the standard deviations for the two groups in Table I (use arbitrary origin method).

**3.12.** Repeat exercise 3.11 with the data of Table II.

**3.13.** For the distribution of IQs in Table III, part *A*, the mean is 106.68 and the standard deviation is 17.41.

*a.* Determine the two points defined by $M \pm S$.

*b.* Determine the two points defined by $M \pm 2S$.

*c.* Locate these four points, also the mean, on the base line of your frequency polygon for the data of Table III, part *A*. Erect ordinates from each of these five base-line points to the polygon, and study the resulting picture.

*d.* Determine approximately the percentage of cases between $M \pm S$; also between $M \pm 2S$.

# EXERCISES AND QUESTIONS*

## CHAPTER 2

**2.1.** *a.* Make separate frequency distributions for the marks of the two groups of students in Table I. Use intervals of size 5.

  *b.* Determine also the cumulative frequencies for each group.

### Table I. Final examination marks for a class in statistics

| Students with No Calculus (N = 36) | | | | | | Students with Some Calculus (N = 22) | | | |
|---|---|---|---|---|---|---|---|---|---|
| 103 | 150 | 139 | 79 | 150 | 134 | 137 | 139 | 112 | 139 |
| 98 | 79 | 94 | 137 | 118 | 113 | 151 | 124 | 80 | 153 |
| 106 | 93 | 106 | 137 | 91 | 109 | 131 | 94 | 96 | 77 |
| 71 | 101 | 92 | 74 | 106 | 87 | 133 | 123 | 101 | 115 |
| 108 | ˙113 | 103 | 108 | 114 | 105 | 115 | 90 | 154 | 122 |
| 120 | 95 | 83 | 93 | 109 | 97 | 111 | 135 | | |

**2.2.** *a.* Make separate frequency distributions for the two groups of scores in Table II. Use intervals of size 3.

  *b.* Determine also the cumulative frequencies for each group.

**2.3.** *a.* Draw a frequency polygon for the distribution in Table III, part A.

  *b.* Draw an ogive for the data of Table III, part A.

**2.4.** *a.* Draw a frequency polygon for the distribution of Table III, part B.

  *b.* Draw an ogive for the data of Table III, part B.

### Table II. Scores on final examination for a course on psychological tests

| Undergraduates (N = 32) | | | | | | Graduate Students (N = 23) | | | | |
|---|---|---|---|---|---|---|---|---|---|---|
| 70 | 72 | 76 | 66 | 76 | 80 | 84 | 80 | 90 | 82 | 84 |
| 67 | 69 | 90 | 50 | 76 | 47 | 79 | 62 | 77 | 89 | 70 |
| 51 | 58 | 71 | 88 | 65 | 54 | 73 | 74 | 87 | 76 | 79 |
| 89 | 64 | 80 | 67 | 71 | 90 | 85 | 95 | 78 | 69 | 97 |
| 91 | 71 | 63 | 81 | 87 | 81 | 78 | 86 | 92 | | |
| 79 | 79 | | | | | | | | | |

* These are so arranged that frequently an even-numbered exercise is of the same type as its immediately preceding odd-numbered exercise. "Thought" questions are intended as thinking, not thumbing, exercises.

## CHAPTER 3

**3.1.** For the scores of Table I, compute separately for the two groups:
   *a.* the medians, using the undistributed scores.
   *b.* the medians, using the frequency distributions.

**3.2.** Repeat exercise 3.1 with the data of Table II.

**3.3.** Compute the mean for each group in Table I by
   *a.* the definition formula for the mean.
   *b.* the arbitrary origin method.

**3.4.** Repeat exercise 3.3 with the data of Table II.

**3.5.** Combine the two distributions for the data of Table I, compute the mean by the arbitrary origin method, and check by using the formula for securing the mean for a combined group (use the means obtained by the arbitrary origin method for this check).

**3.6.** Repeat exercise 3.5 with the data of Table II.

**3.7.** Compute the median, $Q_1$, and $Q_3$ for the distribution in Table III, part $A$.

**3.8.** Compute the median, the 20th and the 80th percentile points for the distribution in Table III, part $B$.

**3.9.** Using the results of exercise 3.7, locate the three points, $Q_1$, the median, and $Q_3$, on the base line of your *ogive* curve for the distribution of Table III, part $A$. Divide the ordinate on the right-hand side (the ordinate at IQ = 170) into approximate fourths. Draw a line from each of the three base-line points up to the ogive, then horizontally to the right. Notice where these horizontal lines hit the ordinate on the right-hand side.

**3.10.** Using the results of exercise 3.8, locate the three points, the median, $P_{20}$, and $P_{80}$, on the base line of your *ogive* curve for the distribution of Table III, part $B$. Divide the ordinate on the right-hand side (the ordinate at IQ = 180) into approximate fifths. Draw a line from each of the three base-line points up to the ogive, then horizontally to the right. Note where these horizontals hit the ordinate on the right-hand side.

**3.11.** Compute the standard deviations for the two groups in Table I (use arbitrary origin method).

**3.12.** Repeat exercise 3.11 with the data of Table II.

**3.13.** For the distribution of IQs in Table III, part $A$, the mean is 106.68 and the standard deviation is 17.41.
   *a.* Determine the two points defined by $M \pm S$.
   *b.* Determine the two points defined by $M \pm 2S$.
   *c.* Locate these four points, also the mean, on the base line of your frequency polygon for the data of Table III, part $A$. Erect ordinates from each of these five base-line points to the polygon, and study the resulting picture.
   *d.* Determine approximately the percentage of cases between $M \pm S$; also between $M \pm 2S$.

**Table III. Distribution of IQs, form L of 1937 Stanford-Binet scale**

| IQ | A. Ages $2\frac{1}{2}$–$5\frac{1}{2}$ | | B. Ages 6–13 | |
|---|---|---|---|---|
| | $f$ | $cu\,f$ | $f$ | $cu\,f$ |
| 170–179 | | | 1 | 1623 |
| 160–169 | 4 | 728 | 1 | 1622 |
| 150–159 | 4 | 724 | 3 | 1621 |
| 140–149 | 11 | 720 | 29 | 1618 |
| 130–139 | 41 | 709 | 73 | 1589 |
| 120–129 | 82 | 668 | 140 | 1516 |
| 110–119 | 175 | 586 | 308 | 1376 |
| 100–109 | 193 | 411 | 407 | 1068 |
| 90–99 | 107 | 218 | 335 | 661 |
| 80–89 | 76 | 111 | 215 | 326 |
| 70–79 | 20 | 35 | 76 | 111 |
| 60–69 | 7 | 15 | 30 | 35 |
| 50–59 | 5 | 8 | 4 | 5 |
| 40–49 | 2 | 3 | 1 | 1 |
| 30–39 | 1 | 1 | | |
| | $N = 728$ | | $N = 1623$ | |

**3.14.** The distribution of IQs in Table III, part $B$, has a mean of 103.34 and an $S$ of 16.88. Repeat exercise 3.13, using the values and polygon for the data of Table III, part $B$.

**3.15.** Suppose the mean score on a statistics quiz is 35, the median is 36, the $S$ is 6, and the quartile deviation is 4.

    *a.* If to each person's score we added 50 points, what values would we then get for the mean, the median, the $S$, and the quartile deviation?

    *b.* If we doubled each person's score, what would be the values of the new mean and $S$?

**3.16.** Given that the distribution of scores on a quiz leads to a mean of 40, a median of 38, an $S$ of 9, and a quartile deviation of 6.

    *a.* If we added 10 points to the scores of each student, what would be the values for $M$, $Mdn$, $S$, and $Q$?

    *b.* If all scores were halved, what would be the values of the mean and the $S$?

**3.17.** If you were told that the mean final score for the 50 students was 80 and the mean for the 30 men in the class was 82.3, what would you figure as the mean for the women?

**3.18.** Given that the mean weekly pay of the 7 working members of the Jones family is $55 and the median is $50 (both after deductions).

    *a.* What is the weekly "take home" of the family?

*b.* Suppose that Daddy Jones, already the best paid, receives an increase which after deductions amounts to $6 a week. What is the new mean? What is the new median?

**3.19.** If an $S$ is 9 when computed from a frequency distribution with intervals of size 6, what would you expect it to be if computed by using the definition formula for $S$?

**3.20.** How large is the grouping error in an $S$ of 13 computed from a distribution with intervals of size 12?

**3.21.** Why would we usually expect the difference between the 10th and 20th percentile points to exceed the difference between the 40th and 50th percentile points?

**3.22.** Suppose that $A$ knows only that the $Q$ of a distribution is 20, whereas $B$ knows that the 75th percentile is 30 units from the median and the 25th percentile is 10 units from the median. What can $B$ tell about the distribution that $A$ cannot?

## CHAPTER 4

**4.1.** Assume that the IQs for a large number of unselected elementary school children are distributed as a normal curve with a mean of 100 and an $S$ of 17.

*a.* The first quartile point will be near what value?

*b.* The percentage with IQs above 130 will be?

*c.* The middle 80 per cent will fall between what values?

*d.* The 99th percentile will be near what IQ value?

*e.* The percentage with IQs below 70 will be?

**4.2.** Let us presume that the Army General Classification Test yields a normal distribution of scores, with mean of 100 and $S$ of 20.

*a.* The value of the third quartile will be near what score?

*b.* The first percentile point will be at what score?

*c.* Between a score of 70 and a score of 130 will be found what percentage of the cases?

*d.* The middle 60 per cent of scores will fall between what score values?

*e.* The value of the quartile deviation will be what?

**4.3.** One way to comprehend the meaning of either sizable or small differences between groups is to consider the extent to which the distributions overlap. Given the following data for weights of college students:

Men: $M = 142, S = 15$;      Women: $M = 120, S = 12$

Assuming normality for both distributions, how many men per thousand are lighter than the average woman? Determine the number of women per thousand who are heavier than the average man.

**4.4.** If the mean height for college men is 68.5 inches and the $S$ is 2.8, and if the mean height for college women is 64.5 and the $S$ is 2.5, what proportion of women exceed the average man in height? What proportion of men fall below the average height for women?

**4.5.** Suppose that the distribution of numerical grades in a course is normal with a mean of 60 and an $S$ of 10. The instructor wishes to assign letter grades as follows: 15 per cent As, 35 per cent Bs, 35 per cent Cs, and 15 per cent Ds. Determine to the nearest score the dividing line between the As and Bs, between the Bs and Cs, and between the Cs and Ds.

**4.6.** Suppose that it has been decided to use a five-letter grading system, A, B, C, D, and E, and that it is required that the letters shall correspond to "equal" distances on the base line, the whole of which is taken to be 6 $S$s. Assuming normality, what percentage would be assigned As; Bs; Cs?

**4.7.** Determine the height of the unit normal curve at the point which is 1.2 $z$ units below the median; at the third quartile point.

**4.8.** What is the height of the ordinate of the unit normal curve corresponding to the $x/\sigma$ value that cuts off the upper 10 per cent of the curve? The lower 25 per cent?

**4.9.** Frequently, we must be able to translate percentile scores to standard scores and vice versa (assume normality).
   a. What are the standard scores (to the nearest tenth) which correspond to the following percentiles: 44th, 99th?
   b. What are the percentile equivalents (to nearest value) of the following standard scores: $-1.34$, $+2.06$?

**4.10.** Suppose a typical bell-shaped distribution. What is the approximate percentile value of the following points: the mean, $Q_3$, the point which is one $S$ above the mean, and the first decile point?

**4.11.** What is the $x/\sigma$ distance between the following (assume normality):
   a. the 10th and the 90th percentile points?
   b. the 25th and the 75th percentile points?

**4.12.** If a distribution of scores is normal, what is the $x/\sigma$ distance between the 10th and 20th percentile points? Between the 40th and 50th percentile points?

**4.13.** Given that a reading test for unselected 10-year-olds yields a mean of 50 and an $S$ of 10, whereas an arithmetic test gives a mean of 48 and an $S$ of 8. If Joe Bloke scores 52 on reading and 50 on arithmetic, is he better in reading than in arithmetic? Why?

**4.14.** If a student's reading rate score falls at the 20th percentile, and his standard score on reading comprehension is $-1.4$, would you conclude that his comprehension was superior to his rate? Why?

**4.15.** $M \pm 2$ quartile deviations will give two points within which, for normal distributions, one would expect about what per cent of the cases?

**4.16.** For Test $A$ the scores are transformed to $Z$ scores with $M = 50$ and $S = 10$ and for Test $B$ the scores are transformed to $T$ scores with $M = 50$, $S = 10$. Why may a score of 60 on Test $A$ be not comparable to a score of 60 on Test $B$?

**4.17.** Suppose we have a distribution with skewness, $g_1 = .60$. If we transformed the scores into standard scores; also to $T$ scores; and also to percentiles; what can you say regarding the shape of the distribution of

  a. the standard scores?

  b. the $T$ scores?

  c. the percentile scores?

**4.18.** Given that the distributions for two groups, with $N$s equal, are normal in form. Now consider the distribution for the two groups combined. Under what condition would you expect the shape of the combined distribution to be: Platykurtic? Leptokurtic? Normal?

## CHAPTER 5

**5.1.** If you tossed 4 unbiased pennies 160 times, how often would you expect to have 2 heads and 2 tails?

**5.2.** Suppose you roll a pair of fair dice once. What is the probability that exactly 11 spots will turn up?

**5.3.** Suppose that you are rolling 2 fair dice, one red and the other white. What is the probability of obtaining a 3 spot on the red die and a 4 spot on the white one?

**5.4.** In that back-alley game known as "crap shooting," the obtaining of spots on the 2 dice totaling 7 seems to be of paramount importance at certain times. What is the probability of rolling a 7 (assume gentlemen's dice)?

**5.5.** Suppose that we have 3 pyramidal objects (perfectly homogeneous) which can be rolled like dice. The sides of each are numbered 1, 2, 3, 4; and success is defined as the getting of 4s on the down sides. Determine the probability for obtaining exactly three 4s; exactly two 4s; exactly one 4; and no 4s. What is the probability of securing at least two 4s?

**5.6.** If you were dealt 1 card from each of 5 well-shuffled decks, what is the probability of all 5 cards being spades?

**5.7.** The probability of drawing a red card from an ordinary (and well-shuffled) deck is $\frac{1}{2}$ and the probability of drawing a heart is $\frac{1}{4}$. Why isn't $\frac{1}{2}$ plus $\frac{1}{4}$ the probability of drawing either a heart or a red card, or is it?

**5.8.** Suppose that for a class of 100 the number of As given on the first quiz is 15 and that the number of As on the second quiz is also 15. Suppose further that the names of the students are placed on slips which are then well mixed in a hat. We might say that the probability is .15 that a name drawn from the hat will be that of a student who received an A on the first quiz; likewise, the second quiz. Why might it be erroneous to say that the probability is .15 times .15 that the drawn name belongs to a student who made As on both quizzes?

**5.9.** A student takes a four-alternative, 12 question multiple-choice test. If he merely guesses, what is the probability that he will get all 12 questions correct?

**5.10.** The typical ESP deck consists of 25 cards, with 5 cards for each of 5 symbols. The person who claims extra-sensory perception (ESP) ability attempts to name the symbol on the cards as they are exposed (one at a time, after shuffling) by the experimenter in a room remote from the person. The "score" is the number of correct calls in a run through the pack.

  a. What are the numerical values for $p$, $q$, and $n$ for the binomial distribution?

  b. Would "scores" of 3 and 7 be equally likely on a chance basis? Why?

**5.11.**  a. Toss 6 coins 64 times; for each toss tally the number of heads that turn up, thereby obtaining a frequency distribution with an $N$ of 64. Label this Series $A$. Toss the coins 64 more times, and label the resulting distribution as Series $B$. Then combine the two distributions.

  b. Using the binomial expansion, ascertain the expected distribution when 6 coins are tossed 64 times; 128 times.

  c. Compute the mean and standard deviation for each of your three distributions; also for the expected distribution (round to 2 decimals).

  d. Determine the proportion of times that 3 heads, also 6 heads, turned up in each series, and in the combined series. Compare these results with the expected proportions.

  e. Subtract the mean of Series $A$ from that of Series $B$ (keep sign if negative). For the proportion of times 3 heads turned up, subtract the Series $A$ proportion from that for Series $B$ (keep sign).

  f. Bring all the results to class so that frequency distributions may be made for $M$s, $S$s, proportions, and differences between $M$s and between proportions.

**5.12.**  Do exercise 5.11, using 7 coins.

**5.13.**  If 42 of 60 rats turn to the right at the first choice point in a maze, would you conclude that rats, in general, prefer to turn to the right at this choice point?

**5.14.**  If at a particular time 50 per cent of all eligible voters favor the Democrats, how often would polls based on random samples of size 400 yield percentages of 55 or over as favoring Democrats?

**5.15.**  Items on an intelligence test of the Binet type are at times assigned an index of difficulty which is nothing more than the percentage passing the item. Given the following for an item: of 100 12-year-olds, 60 per cent passed; of 100 13-year-olds, 80 per cent passed. When possible sampling errors are considered, would you conclude from these two difficulty indices that the item is really more difficult for 12-year-olds? State the significance level associated with your conclusion.

**5.16.**  If a political issue is favored by 55 per cent of a sample of 200 Republicans, and by 46 per cent of a sample of 250 Democrats, would you conclude that the populations of Republicans and Democrats differ on the issue?

**5.17.**  a. Given the data in Table IV, do items $a$ and $b$ differ significantly in difficulty for the 4-year-olds? Ditto, the 5-year-olds?

  b. Is there a significant difference between 4- and 5-year-olds on item $a$? On item $b$?

**Table IV. Data for passing (P) and failing (F) items on the Stanford-Binet Test**

| 4-year-olds | | | | | | | | | | | | 5-year-olds | | | | | |
|---|---|---|---|---|---|---|---|---|---|---|---|---|---|---|---|---|---|

| | Item | | | Item | | | Item | | | Item | |
|---|---|---|---|---|---|---|---|---|---|---|---|
| Case | $a$ | $b$ | Case | $a$ | $b$ | Case | $a$ | $b$ | Case | $a$ | $b$ |
| 1 | F | F | 21 | P | F | 41 | P | P | 61 | P | P |
| 2 | P | F | 22 | P | F | 42 | P | P | 62 | P | P |
| 3 | P | P | 23 | P | F | 43 | P | F | 63 | P | F |
| 4 | F | P | 24 | P | P | 44 | F | F | 64 | P | P |
| 5 | P | F | 25 | P | F | 45 | P | P | 65 | F | F |
| 6 | F | F | 26 | P | P | 46 | P | P | 66 | P | P |
| 7 | F | F | 27 | P | F | 47 | F | F | 67 | F | F |
| 8 | P | F | 28 | F | F | 48 | P | P | 68 | P | P |
| 9 | P | F | 29 | P | P | 49 | P | P | 69 | P | P |
| 10 | F | F | 30 | P | F | 50 | P | P | 70 | F | F |
| 11 | P | P | 31 | P | P | 51 | P | P | 71 | P | P |
| 12 | P | P | 32 | P | F | 52 | P | P | 72 | F | F |
| 13 | F | F | 33 | P | F | 53 | P | P | 73 | P | F |
| 14 | P | P | 34 | F | F | 54 | P | P | 74 | P | F |
| 15 | P | F | 35 | P | P | 55 | P | F | 75 | F | F |
| 16 | P | F | 36 | P | F | 56 | P | F | 76 | F | F |
| 17 | P | P | 37 | P | P | 57 | P | P | 77 | P | F |
| 18 | F | F | 38 | F | F | 58 | P | P | 78 | P | F |
| 19 | F | F | 39 | F | F | 59 | P | F | 79 | F | P |
| 20 | P | P | 40 | P | F | 60 | P | F | 80 | P | F |

**5.18.** *a.* Would you conclude from the data of Table V that items $c$ and $d$ differ significantly in difficulty for the 6-year-olds? Ditto, the 7-year-olds?

*b.* Would you conclude from the data of Table V that, in general, 7-year-olds are more successful than 6-year-olds on item $c$? On item $d$?

**5.19.** In a student presidential election, Mr. Ralph received 2389, or 60 per cent, of the votes cast. Suppose that you had been able to poll a sample of 100 the day before the election. Assuming that such "last day" changes as took place were balanced so that neither candidate gained, how often would samples of 100 yield a majority? (i.e., what is the approximate probability that on the basis of a sample of 100 you would have predicted Mr. Ralph's election?).

**5.20.** A sample of $N = 100$ yields a percentage of 55 yeses to a question. Under what condition would you expect a large number of successive random samples to yield percentages of yeses that would average 55?

**Table V. Passing (P) and failing (F) information on two Binet Test items at two age levels**

| 6-year-olds | | | | | | 7-year-olds | | | | | |
|---|---|---|---|---|---|---|---|---|---|---|---|
| Item | | | Item | | | Item | | | Item | | |
| Case | c | d | Case | c | d | Case | c | d | Case | c | d |
| 1 | F | F | 21 | P | P | 41 | P | F | 61 | P | F |
| 2 | P | P | 22 | P | F | 42 | P | P | 62 | F | F |
| 3 | F | P | 23 | F | F | 43 | F | P | 63 | P | F |
| 4 | F | F | 24 | F | F | 44 | P | P | 64 | F | F |
| 5 | F | F | 25 | F | F | 45 | F | P | 65 | P | P |
| 6 | F | F | 26 | P | P | 46 | F | F | 66 | P | P |
| 7 | P | F | 27 | P | F | 47 | P | P | 67 | P | P |
| 8 | F | F | 28 | P | F | 48 | F | P | 68 | P | P |
| 9 | P | F | 29 | F | F | 49 | P | P | 69 | P | P |
| 10 | F | F | 30 | F | F | 50 | P | F | 70 | P | P |
| 11 | P | F | 31 | F | F | 51 | F | F | 71 | P | P |
| 12 | P | P | 32 | F | F | 52 | F | F | 72 | P | P |
| 13 | F | F | 33 | P | F | 53 | F | F | 73 | P | F |
| 14 | P | F | 34 | P | F | 54 | P | P | 74 | P | F |
| 15 | F | F | 35 | F | F | 55 | F | F | 75 | F | F |
| 16 | P | P | 36 | F | F | 56 | P | P | 76 | P | P |
| 17 | F | F | 37 | F | P | 57 | F | F | 77 | P | F |
| 18 | F | F | 38 | P | F | 58 | P | P | 78 | P | F |
| 19 | F | F | 39 | F | F | 59 | P | P | 79 | P | F |
| 20 | F | F | 40 | P | P | 60 | F | F | 80 | P | P |

**5.21.** Suppose a situation involving the difference between groups, which calls for a two-tailed test of significance and that we have decided on $P = .01$ as our level of significance.

    *a.* What is the probability of committing a type I error if the null hypothesis is really true?

    *b.* If the null hypothesis is not true, what is the probability of making a type I error?

    *c.* If the true difference were 3.3, what additional information would you need in order to figure out the probability of making a type II error?

**5.22.** Let beta stand for the probability of correctly rejecting the null hypothesis and let one minus beta stand for the probability of making the type II error. Under what condition could these two probabilities be equal?

**5.23.** For a sample of 100 it is found that 60 say yes and 40 say no when asked a certain question. For the difference, .60 − .40, between the two proportions why would it be incorrect to take the square root of $p_c q_c/100 + p_c q_c/100$ as the standard error of the difference?

**5.24.** Consider the setup for testing the difference between two nonindependent, or related, proportions via the square root of $(a + d)/N$. Although we have not explained the concept of correlation (or association), do you see a basis for saying that $(a + d)/N$ tends to be smaller the higher the correlation between the two sets of responses?

**5.25.** Suppose percentages of 37 and 39 are found for two samples, each of size 100, drawn from a defined population. Since a difference as large as 2 percentage points can easily, for the given $N$s, arise on a chance basis, it would seem safe to conclude that the samples are in very close agreement. From this degree of similarity in results, would you conclude that the sampling method has avoided bias? Explain or defend your answer.

**5.26.** Some textbooks have argued that whether or not a sample is representative (i.e., not biased) can be judged by splitting it (the sample) into random halves, and then claim representativeness if the proportions for the two halves are not significantly different. Any comment?

## CHAPTER 6

**6.1.** For a sample of 2970 cases, ages 2.5 to 18, the distribution of IQs on Form L of the 1937 Stanford-Binet yields:

$$\text{Mean} = 104.00 \quad \text{Skewness } (g_1) = .028$$
$$S = 17.03 \quad \text{Kurtosis } (g_2) = .346$$

In answering the following questions, indicate the steps in your computations.
  *a.* Would you conclude that the mean IQ of the population for these ages is 100 (the value expected for a properly constructed IQ test)?
  *b.* Is it reasonable to believe that the IQ distribution for the population, at these ages, has normal skewness?
  *c.* Would you conclude from the sample kurtosis that the kurtosis for the population differs from normal kurtosis?

**6.2.** Suppose that the mean IQ for the general population is 100 and the standard deviation is 17. If a sample of 289 cases were drawn at random, what would be the probability of obtaining a mean as great as 101? As low as 98?

**6.3.** Suppose it is known that the standard deviation of scores for a population is 20. How many cases would you need to draw in order that the standard error of
  *a.* a sample mean be 2 score points?
  *b.* a sample $S$ be 3 points?

**6.4.** Suppose that you are polling on an issue for which opinion seems about equally divided. How many cases (how large an $N$) would you need to be sure (at the .01 level of significance) that a sample deviation of 3 per cent from 50 per cent is nonchance?

**6.5.** One of the requirements of a good IQ test is that the mean IQ for un-selected cases of any school age group shall be 100, and that the distributions for the several age groups shall have the same standard deviations. Given the following for the 1937 Stanford-Binet Test:

| Age | 6 | 12 |
|---|---|---|
| N | 203 | 202 |
| M | 101.0 | 103.6 |
| S | 12.5 | 20.0 |

   a. Is it reasonable to believe that the test is yielding the desired mean when used with 12-year-olds?
   b. Would you judge from the results for these two age groups that the requirement of equal variability has been met?

**6.6.** The means and standard deviations for two groups of twins on spool packing are as follows:

| | Fraternals | Identicals |
|---|---|---|
| N | 92 | 94 |
| M | 761 | 741 |
| S | 79 | 66 |

Do these groups differ significantly in mean performance? In variability?

**6.7.** Two forms of a test, to be comparable, should yield similar means and similar standard deviations when given to a group. For 202 cases of age 7, we have the following data for the 1937 Stanford-Binet:

| | Form L | Form M |
|---|---|---|
| M | 101.8 | 103.5 |
| S | 16.2 | 15.6 |

In order to balance practice effect, one-half the group was tested on Form L, then on Form M, whereas the reverse order was used for the other half. The correlation between the two sets of IQs was .93. Is the obtained difference between means larger than one would expect on the basis of chance sampling? Ditto, the difference between the $S$s?

**6.8.** Measurements on 1000 of each sex at birth have been reported in the literature. The mean length of boys (in centimeters) was 50.51 and the $S$ was 2.99, and the values for the girls were 49.90 and 3.00. Is there evidence here for sex difference in length at birth?

**6.9.** Given a two-tailed test of the hypothesis that no change has taken place and that the standard error of the mean change is 3 points and that we use the .05 level for judging significance.

    *a.* If the true change is a loss of 6 points, the probability (beta) of correctly rejecting the null hypothesis is approximately ——— whereas if the true change is a gain of 12 points, the value of beta is approximately ———.

    *b.* If the true change is zero, the value of beta is ———.

    *c.* If, instead of a two-tailed test, a one-tailed test were used, the probability of making the type I error would be ———.

**6.10.** Given that a two-tailed test is appropriate, that we have chosen the .05 level of significance, that the obtained difference between two means is 3, and that the standard error of the difference is 2.

    *a.* Would we reject the null hypothesis?

    *b.* What is the probability that we will make a type I error?

    *c.* If the true or population difference were 6, what is the (approximate) probability that we would correctly reject the null hypothesis?

    *d.* If the true difference were zero, what can you say about the likelihood of making a type II error?

**6.11.** Given that a sample yields 98 per cent of yeses to a question and that the standard error of the percentage is 2. If we set the .99 confidence limits as 98 $\pm 2.58(2)$ we arrive at the absurdity of an upper limit in excess of 100 per cent. Why?

**6.12.** Suppose you draw a sample of size 3 (yielding scores of 90, 99, and 102; mean $= 97$) from a population which you know to have a mean of 100. What would be your best single estimate of the population variance?

**6.13.** Consider the following two statements: (*a*) the probability is .95 that sample means will not deviate more than $1.96S_M$ from the population mean, and (*b*) the probability is .99 that the population mean will not deviate more than $2.58S_M$ from a sample mean. Which statement is false? Why?

**6.14.** "The true mean has a 95 per cent chance of falling in the 95 per cent confidence interval for the true mean." This statement is incorrect. Why? Restate it in correct fashion.

**6.15.** If the standard deviation for a very large (infinite) population equals that for a small (finite) population, samples of the same size from the two populations will lead to confidence intervals for the population means that are the same or different in width? Why?

**6.16.** For normal distributions the sample mean and sample median are estimators of the same central value (location parameter). For fixed $N$, how will the .95 confidence limits differ when we use interval estimation based on the mean and also on the median? Why?

**6.17.** We did not discuss the sampling instability of percentiles. Do you think that for a distribution of 200 scores the 55th percentile point will be more or less stable in the sampling sense than the 95th percentile point? Why?

**6.18.** The text claims that, for samples drawn from a normally distributed population, the median is less stable in the sampling sense than the mean. Can you specify a type of distribution (as to shape) for which the median might be *more* stable than the mean? Explain.

# CHAPTER 7

**7.1.** Experimenter A randomly assigns 12 persons to an experimental group and 12 other persons to a control group, thus assuring independence for the two groups. Experimenter B assigns 12 persons at random to an experimental group but for the control group he selects 12 persons by matching (by pairs) them with the 12 individuals in the experimental group. Both experimenters evaluate the difference between the group means (their own groups) via the *t* test.
   *a.* How many degrees of freedom for A's *t* test?
   *b.* How many degrees of freedom for B's *t* test?
   *c.* Which experimenter do you think would have greater precision in results? Why?

**7.2.** When comparing means for two sets of scores belonging to two independent groups each of size $N = 8$, we find the number of degrees of freedom is ———, and when we are comparing means of two sets of scores belonging to two matched groups of 10 each the number of degrees of freedom is ———.

**7.3.** In connection with a topic not yet studied, the sums of squares of deviations about means may be combined for three groups in order to obtain a variance estimate. If the $N$s are 14, 8, and 10, the number of degrees of freedom will be what?

**7.4.** Given that the unbiased estimate of the standard error of a mean difference is 4, that the chosen level for judging significance is .01, that a one-tailed test is appropriate, and that for the $df = N - 1 = 24 - 1 = 23$ the value of $t$ must reach 2.50 for claiming significance at the adopted .01 level.
   *a.* Under these circumstances the probability of committing the type I error is what?
   *b.* How large would the mean difference for the population need to be in order to have the probability of making the type II error be exactly .50?

**7.5.** Given a mean of 50 based on 21 cases, with $s_M = 2$. Ascertain the .95 confidence interval; also the .99 confidence limits.

**7.6.** When setting the .95 confidence interval for a population mean, the procedure for small samples differs from that for large samples in what two respects?

**7.7.** An experimenter knows that the $\sigma$ for population A is 4 and the $\sigma$ for population B is 3. He draws a sample of size 10 from population A and a sample of size 10 from population B. In testing the significance of the difference between the means of the two samples he uses large sample techniques. Is he justified in doing this? Why?

**7.8.** An experimenter uses a sample of size 15. He wishes to test the hypothesis that the mean gain is nonzero. However, he knows nothing of small sample techniques. If he uses large sample techniques for his test, will he increase or decrease his probability of making a type II error over what it would be if he used small sample techniques? Why?

## CHAPTERS 8 AND 9

8–9.1.　*a.* Using the data of Table VI, make a scatter diagram with "Ex" on the *y* axis, intervals of size 5; and with "TMT" on the *x* axis, with *i* = 3 and the first interval taken as 105–107 (interval sizes are suggested in order to facilitate an exact check of the tallying and subsequent computations).

　　*b.* From the scatter diagram, compute the correlation between "Ex" and "TMT"; also compute the two means and the two standard deviations.

　　*c.* Write the regression equation for predicting "Ex" from "TMT". Draw the regression line on your scatter diagram.

　　*d.* Determine the error of estimate for predicting "Ex" from a knowledge of "TMT".

　　*e.* What percentage of the variance in "Ex" is due to or associated with variation in "TMT"?

8–9.2.　Do exercises 8–9.1 with "CM" substituted for "TMT" (an appropriate interval size for "CM" is rather obvious).

**Table VI. Data for 38 students in a course on mental tests ("Ex" stands for final examination scores; "TMT" stands for IQs based on Terman-McNemar Test of Mental Ability; "CM" stands for scores on the Terman Concept Mastery Test)**

| Ex | TMT | CM | Ex | TMT | CM | Ex | TMT | CM |
|---|---|---|---|---|---|---|---|---|
| 62 | 123 | 47 | 106 | 125 | 126 | 54 | 128 | 69 |
| 107 | 129 | 59 | 79 | 109 | 33 | 86 | 132 | 82 |
| 87 | 131 | 78 | 84 | 120 | 56 | 92 | 114 | 41 |
| 95 | 129 | 74 | 100 | 129 | 81 | 67 | 113 | 44 |
| 100 | 122 | 52 | 78 | 112 | 51 | 102 | 141 | 112 |
| 87 | 136 | 127 | 90 | 132 | 110 | 79 | 132 | 72 |
| 87 | 125 | 74 | 85 | 126 | 54 | 82 | 126 | 54 |
| 64 | 121 | 46 | 58 | 111 | 33 | 96 | 131 | 111 |
| 89 | 131 | 97 | 110 | 138 | 119 | 77 | 131 | 65 |
| 58 | 128 | 71 | 115 | 131 | 138 | 75 | 109 | 22 |
| 84 | 123 | 28 | 68 | 129 | 39 | 93 | 131 | 108 |
| 80 | 127 | 53 | 78 | 123 | 67 | 67 | 106 | 25 |
| 82 | 120 | 53 | 80 | 136 | 101 | | | |

**9.3.** The standard deviation of difference scores ($D$) based on $z_x$ and $z_y$ ($D = z_x - z_y$) is .40. What is the correlation between $X$ and $Y$? How did you get your answer?

**9.4.** Consider the general formula for the variance of a difference $S^2_{x-y} = S^2_x + S^2_y - 2r_{xy}S_xS_y$. Can you suggest a method for determining the $r$ between two sets of correlated scores?

**9.5.** Consider $Y$ as weight and $X$ as height and that $r$ has been computed and that $B$ and $A$ have been calculated for the regression equation $Y' = BX + A$. Now as regards the metric (or measurement units), $Y$ is in pounds and $X$ is in inches. What can you say about the units (or metric) for $A$? For $B$? For $r$?

**9.6.** Suppose we consider the regression lines, that for $Y$ on $X$ and that for $X$ on $Y$. If the assumption of linearity holds for both lines, under what condition will the two regression lines

 *a.* coincide?

 *b.* be at right angles to each other?

 *c.* both have negative slopes?

**9.7.** Occasionally we encounter a statement which goes something like this: the relationship seems to be higher for the low scorers than the high scorers. What does this imply?

**9.8.** Although it was argued that $S_{y,x}$ is preferable to the standard deviation of array distributions about their own means as a way of specifying errors when $Y$ is predicted from $X$, can you indicate two possible situations, distinctly different, for which the array sigmas would be much better than $S_{y,x}$?

**9.9.** Under what condition do three variances add to a total variance?

**9.10.** Given: $M_x = 40$, $M_y = 50$, $S_x = 8$, $S_y = 6$ and $r_{xy} = .00$. What will the mean and the standard deviation be for the sum score, $X + Y$?

**9.11.** In what ways can $X$ be manipulated without changing the correlation between $X$ and $Y$?

**9.12.** Suppose that the score on a first quiz is to be combined with the score on a second quiz so as to give equal weight to the two quizzes. Why might a simple addition of the two scores for an individual fail to accomplish the desired equal weighting?

**9.13.** We saw that $r^2$ was equal to the ratio of two variances, hence providing a percentage interpretation of correlation. Algebraically, by taking square roots we have $r$ as the ratio of two $S$s. Why cannot the latter ratio be safely interpreted in percentage terms?

**9.14.** Numerically an $r$ of .60 is twice an $r$ of .30; under what circumstances, as regards interpretation, can we regard

 *a.* .60 as *exactly* four times .30?

 *b.* .60 as nearly four times .30?

 *c.* .60 as being twice .30?

**9.15.** Test $Y$ has an $S = 10$. The $S$ of the predicted $Y$s from $X$ ($S_{y'}$) = 6. What is the standard error of estimate ($S_{y,x}$)? What is the correlation between $X$ and $Y$?

**9.16.** A critic of the text has said that $\Sigma(y - y')^2/N$ does not qualify as a variance unless it is first demonstrated that $\Sigma(y - y') = 0$. Can you supply a very, very simple algebraic proof that $\Sigma(y - y')$ does equal zero?

**9.17.** Given that the correlation between $X$ and $Y$ for Group A is .60 and for Group B also .60. Under what specific conditions would the correlation between $X$ and $Y$ for the two groups combined be much higher, say, .80? Ditto, much lower?

**9.18.** Suppose variables $X$ and $Y$ show no sex differences (equal means and $S$s). It is found that the correlation for boys and girls combined is .02, whereas for boys alone the correlation is $-.34$. How reconcile these two $r$s?

# CHAPTER 10

**10.1.** An $N$ of 101 will yield .10 as the standard error for near zero $r$s. If we have adopted the .05 level of significance and are using a two-tailed test:

    *a.* What is the probability of making a type I error if the population $r$ is zero?

    *b.* What is the probability of making a type I error if the population $r$ is $+.06$?

    *c.* What is the probability of making a type II error if the population $r$ is $-.20$?

    *d.* How often would we correctly reject the null hypothesis if the population $r$ were $+.20$?

**10.2.** For large samples the sampling distribution of low $r$s may be regarded as normal, with standard error of $1/\sqrt{N-1}$. Suppose in what follows that $N$ is 101, thus giving a $\sigma_r$ of .10.

    *a.* If the null hypothesis of no correlation for the population being sampled is true, the probability of a sample $r$ exceeding $+.20$ is approximately what?

    *b.* When the $r$ in the population is .10 and we are using a two-tailed test and the .05 level of significance, the probability of committing the type II error is approximately what?

    *c.* If a population $r$ is .26 and we are using a one-tailed test and the .01 level, the probability of correctly rejecting the null hypothesis is approximately what?

    *d.* If a sample $r$ is .15, the .95 confidence limits for the population $r$ are approximately what?

    *e.* For part *d*, the probability that the limits so set will not include the population value is what?

**10.3.** The classical formula, $(1 - r^2)/\sqrt{N}$, for the standard error of an $r$ implies that the degree of sampling stability for an $r = .90$ is greater than for an $r = .30$, yet by the $z$ transformation the standard errors for the two corresponding $z$s are the same. What can you say about the relative sampling stability of an $r$ of .30 and one of .90, each based on 103 cases?

**10.4.** Do you think that a sample $r$ of .80 could as readily be a chance deviation downward from a population value of .90 as it could be a chance deviation upward from a population value of .70? Explain.

**10.5.** It was argued that the degrees of freedom, $N - 2$, for the $t$ test of $r$ was logical because the two constants, $B$ and $A$, in the regression are calculated from the data hence two restrictions and hence 2 subtracted from $N$. What happens to the number of degrees of freedom when the regression equation is written in standard score form, or $z'_y = rz_x$?

**10.6.** Given that the population value for the correlation coefficient between two variables is $+1.00$ (possible only when errors of measurement are zero):

  *a.* What would you expect the correlation to be for a sample of 100 cases?

  *b.* What would be the value of the standard (sampling) error for the correlation based on 100 cases?

  *c.* If one variable were curtailed so as to lead to a 50 per cent reduction in its standard deviation, what would you expect as a sample value for $r$?

**10.7.** As regards the shape of their random sampling distributions, proportions and correlation coefficients have what in common under conditions A (specify) and what in common under conditions B (specify)?

**10.8.** For Group 1 the $r$ between two variables is found to be .25, which is significant at the .01 level, while for Group 2 the same two variables correlate .10, which is significant at the .16 level. Since the investigator has adopted the .05 level for judging significance, he concludes that (because one $r$ is and the other is not significant) the degree of correlation in the two populations being sampled is different. Any comment?

**10.9.** Suppose an $r$ of .90 based on 16 cases, and suppose that we establish the $P = .95$ confidence limits first by use of the classical standard error of $r$ and then by way of the $z$ transformation. Which method will yield the higher upper bound? Why? (No arithmetic called for.)

**10.10.** Distinguish between $S^2_x(1 - r^2)$ and $S^2_x(1 - r)$. What subscripts are needed for the $r$s?

**10.11.** Suppose you are to choose between two equally well-standardized tests of reading comprehension to use in a school system. Test A has a reliability coefficient of .95 and an $S_e$ of 4, whereas Test B has a reliability of .91 and an $S_e$ of 3. Which would you choose and why?

**10.12.** Test A yields an $S$ of 12 and an $S_e$ of 3, whereas Test B yields an $S$ of 20 and an $S_e$ of 4. Which test is the more reliable? Why?

**10.13.** It was argued that one result of measurement errors is a regressive effect. For example, those testing between 130 and 134 IQ points today will tend to test nearer average (100 points) tomorrow whereas those between 70 and 74 today will have regressed upward tomorrow. This would seem to imply that the group variability on tomorrow's testing will be less than the group variability today. Explain why this expected change in group variability does *not* take place even for very unreliable tests.

**10.14.** Suppose Form A and Form B are strictly parallel forms. This means that $M_a = M_b$, $S_a = S_b$ (both $= S_x$), and are similar in content; hence $r_{xx}$, the reliability coefficient, is simply $r_{ab}$. We showed that a true score, given an obtained score, $x_a$, is best estimated by the equation $x'_t = r_{xx}x_a = r_{ab}x_a$. The error of estimate is given by $S_{t \cdot a} = S_x \sqrt{r_{xx} - r^2_{xx}}$. It also follows from our discussion of $r$ and regression equations that the best estimate of a score on Form B, given an obtained $x_a$, is by $x'_b = r_{ab}x_a = r_{xx}x_a$ for which the standard error of estimate is $S_{b \cdot a} = S_x \sqrt{1 - r^2_{ab}}$. Obviously, the regression estimates of $x_t$ and $x_b$ lead to identical results, *but* the error of estimate for the latter is always larger than that for the former, as can be seen by examining the two error formulas. How would you account for the apparent paradox that two estimates which lead to precisely the same value have differing degrees of precision?

**10.15.** For $X = X_t + E$ we have $S^2_x = S^2_t + S^2_e$. Now when we solve $X = X_t + E$ for $X_t$ we get $X_t = X - E$, whence by the variance theorem we might write $S^2_t = S^2_x + S^2_e$ which is inconsistent with $S^2_x = S^2_t + S^2_e$. What is wrong?

**10.16.** Suppose two forms of a test, A and B, and that the form vs. form reliability $(r_{ab})$ is .91 and $S_a = S_b = 20$, thus leading to a standard error of measurement of 6.

      *a.* If we use scores obtained by summing the two form scores we will have scores with a reliability of how much? How did you arrive at this?

      *b.* Would the reliability coefficient be different if you averaged the two form scores instead of merely summing them? Why?

      *c.* Would the averaged values have a larger or smaller standard error of measurement than that for the summed values? Why?

**10.17.** It is sometimes said that intelligence can be held constant by choosing individuals (in forming a group) with the same IQ, or the same score, on an intelligence test. Any comment?

**10.18.** What logical connection do you see between the reliability of gain scores $(G = X_f - X_i)$ and $r_{if}$ corrected for attenuation?

**10.19.** For the Wechsler Adult Intelligence Scale the vocabulary test correlates .86 with the total score (including vocabulary). Why would you hesitate (or would you) to correct this $r$ for attenuation?

**10.20.** A student has discovered that when $R = .50$, $SD = 4$, and $sd = 2$ are substituted in formula (10.18), the value for $r$ becomes $-1.00$, which does not make sense. Do you see an explanation?

**10.21.** It can be shown that the correlation, $r_{t2}$, between $X_2$ and $X_T (= X_1 + X_2)$ will, under a specifiable condition, be .707 even when $r_{12} =$ zero. Under what condition would you expect $r_{t2}$ to exceed .707 when $r_{12}$ is zero?

**10.22.** Given the following for three variables and the total score on the three for 100 cases:

|   | $X_1$ | $X_2$ | $X_3$ | $X_T$ |
|---|---|---|---|---|
| $M$ | 60 | 80 | 50 | 190 |
| $S$ | 10 | 12 | 15 | 25 |

With only the foregoing information at your disposal, would you expect $X_2$ to correlate higher or lower than $X_3$ with $X_T$? Why?

**10.23.** Given the correlations between the four variables, $U$, $V$, $X$, and $Y$ as follows:

$$r_{uv} = .60, \quad r_{ux} = .00, \quad r_{uy} = .80$$
$$r_{vx} = .40, \quad r_{vy} = .50, \quad r_{xy} = -.30$$

Specify the numerical value of the following correlations:

- *a.* between $V$ and $Y$ with $U$ constant.
- *b.* between $X$ and $V$ with $U$ constant.
- *c.* between $V$ and $Y$ with the influence of $U$ partialed out of $Y$ only.
- *d.* between $X$ and $Y$ with $U$ constant.
- *e.* for $U$ and $V$ with $U$ constant.

**10.24.** Rorschach testers, facing the fact that the frequency (or score) for a category (say $X$) is influenced by the total number of responses $(R)$, have attempted to control $R$ by using a percentage scoring scheme: $100 \, X/R$. Can you suggest another (and perhaps better) scheme which would take care of individual differences in $R$ in the sense of yielding a score which is truly independent of $R$?

**10.25.** If 65 per cent of the variance in $Z$ is associated with variation in $X$ and 70 per cent of the variance in $Z$ is associated with variation in $Y$, what can be said about the correlation between $X$ and $Y$? Why?

**10.26.** If $r_{12} = .80$ and $r_{13} = .70$, how might we explain without recourse to p. 185 of the text that $r_{23}$ must be positive?

**10.27.** If $r_{12} = .60$, $r_{13} = .60$, and $r_{23} = .00$, the value of $r_{12 \cdot 3}$ becomes .75; that is the correlation between variables 1 and 2 goes up when variable 3 is "partialed out" even though 3 is uncorrelated with 2. How would you explain this?

**10.28.** We noted that the correlation coefficient is affected by heterogeneity with respect to one or both of the variables being correlated and with respect to a third variable, and we developed formulas which would correct for heterogeneity. Now suppose we have an $r_{xy}$ based on 100 boys and 100 girls ($N = 200$).

- *a.* Then by making separate sex distributions for $X$ we find a marked sex difference; how might our $r_{xy}$ be influenced by this sex difference? Why?
- *b.* Next it is discovered that there is also a sizable sex difference on $Y$. Considering now that both variables show sex differences, what can you say about the effect of such differences on $r_{xy}$? Again, why?
- *c.* Can you propose (in rough outline) a scheme for getting rid of the sex effect on $r_{xy}$?

## CHAPTER 11

**11.1.** How would you show that the choosing of multiple regression coefficients so as to minimize the error of prediction tends to maximize multiple $r$?

**11.2.** If a criterion measure has a reliability of only .60, what is the limiting percentage of the criterion's variance that is predictable by any single predictor or any combination of predictors?

**11.3.** Suppose in determining the beta coefficients for a 3-variable multiple correlation problem your calculations led to $\beta_2 = .70$ and $\beta_3 = .80$.

    *a.* Why might we suspect error in your computations?

    *b.* But under what conditions might your values be correct?

**11.4.** Consider the hypothetical multiple correlation situation involving a dependent variable, $X_1$, and two independent variables, $X_2$ and $X_3$, each of which correlates .707 with the dependent variable. If $r_{23} = 0$, and then a fourth variable is found which also correlates .707 with the dependent variable, what can you say regarding $r_{24}$ and $r_{34}$? Will both be zero or not? Why? (Negative hint: do not waste time substituting in formulas.)

**11.5.** Suppose a 26-variable multiple regression problem with each of the 25 possible predictor variables correlating .20 with the criterion variable and intercorrelating zero among themselves. Can you specify the value of the multiple $r$? How did you get your answer? (Obviously, you are not expected to answer this by using the Doolittle solution, so look for a "trick" solution.)

**11.6.** If each of $m$ independent and uncorrelated variables yields a correlation of .30 with a dependent or criterion variable, how many of them would you need in order to build up a multiple $r^2$ of .45? Of .90?

**11.7.** Given the following regression equation in raw score form:

$$X'_1 = 13.2X_2 + 39.1X_3 + 18.5$$

What are the possible factors that might be responsible for 39.1 being treble 13.2?

**11.8.** We learned that the $\beta$ weight for a suppressant variable tends to be negative. Suppose a 3-variable (one dependent and two independent) multiple regression equation in which $\beta_3$ is negative (say, $-.40$). Does it follow that variable 3 is a suppressant? Why?

**11.9.** When we come to the analysis of variance test of the significance of multiple $r$ we will learn that a multiple $r$ based on $N$ cases and $m$ independent (or predictor) variables will have associated with it a specifiable number of degrees of freedom. Can you anticipate what the $df$ will be? What is the basis for your answer?

**11.10.** Frequently the clinician will utilize a difference score $X - Y$, as a basis for predicting. Examples: in Rorschach this might be $M - C$; in the Babcock scheme for measuring mental deterioration the average standing on certain tests is subtracted from the score on a vocabulary test; on the Wechsler-Bellevue the score on the block design test may be subtracted from the information test score. Presuming that our clinical brethren use such difference scores for predicting a quantitative something of some kind, what pertinent comments can you make? (Note: there are at least two quite different considerations involved here.)

**11.11.** Given that for criterion measures the $S$ is 12 and $S_e$ is 8, and the report that a cross-validated multiple $r$ between the criterion and a battery of 9 selected tests is .80. Why might one suspect at least one of the given values?

## CHAPTER 12

**12.1.** Consider the fourfold table with entries *A*, *B*, *C*, and *D*, with the corresponding proportions *a*, *b*, *c*, and *d*. How might we go about setting up a measure of the success with which one variable can be predicted from another? (Note: to answer by saying compute an *r* of some kind is not enough.)

**12.2.** The coefficient of contingency is always a positive number. Does it follow that its use will never permit one to say a relationship is negative? Explain.

**12.3.** Sometimes in determining the interrelationship of items (scored as pass or fail) we encounter a fourfold table with zero frequency in either the upper-left or lower-right hand cell. Under what circumstances would you expect this to happen? (Your answer should be in terms of observables.)

**12.4.** What measure of correlation would you use to *describe* the relationship between sex (as near a point variable as one encounters in psychology) and passing or failing a test item? Why?

**12.5.** We discussed eta, the correlation ratio. What do you suppose "biserial eta" stands for? Write out a reasonable guess as to the formula for such a measure.

**12.6.** Occasionally we find instances in which the correlation between the percentile scores on two tests has been computed. Such a correlation coefficient resembles one type of correlation discussed somewhere in the text. Which? In what way?

**12.7.** Consider two variables, *X* and *Y*. Does saying that the two are uncorrelated (product-moment sense) necessarily mean that *Y* and *X* are independent? Explain.

**12.8.** Sometimes a skewed distribution of scores can be approximately normalized by a simple transformation, such as log *X* or the square root of *X*. Now suppose $r_{xy} = .60$ and that linearity of regression holds exactly. If the square root transformation is used on *X*, would you expect the correlation of these new *X* values and *Y* to be higher or lower than .60? Why? What change would you expect if for both the original and transformed *X*s we used the rank difference method in determining the correlation between the two variables?

**12.9.** Suppose we have a plot of the relationship between two variables and that the form of the relationship seems to be logarithmic. Accordingly, we write an equation for predicting *Y* from *X* as $Y' = B \log X + A$ with *B* and *A* so determined as to minimize $\Sigma(Y - Y')^2$. Obviously we can ascertain the variance about the curved line by computing $\Sigma(Y - Y')^2/N$, the square root of which would be analogous to the standard error of estimate. Next we define an index of relationship, say theta, as

$$\text{theta} = 1 - \frac{\Sigma(Y - Y')^2/N}{S^2_y}$$

We also compute *r* and eta. Problem: arrange theta, eta, and *r* as to magnitude. Basis?

**12.10.** Suppose a scattergram between $X$ and $Y$ (both measured in a graduated fashion, and both yielding symmetrical distributions) and that we compute $r$, eta, point biserial $r$, and the fourfold point $r$ (the last two by dichotomizing near the median or medians). Arrange these measures in order for expected magnitude.

**12.11.** One form of the formula for biserial $r$ contains $M_2 - M_1$, the other contains $M_2 - M_y$. Now suppose the number of cases in category 1 is so small that $M_1$ is very unstable in the sampling sense, hence the computed correlation must also be unstable. Why does not the use of the second form, which avoids the unstable mean, lead to a more stable $r$?

**12.12.** Suppose a discriminant-function weighting of six tests for differentiating delinquents from nondelinquents; and suppose that for those having a weighted (total) score between, say, 160 and 170 it is noted that 90 are nondelinquents and 10 are delinquents. What can you say about the predictive value of scores between 160 and 170? Any cautions?

## CHAPTER 13

**13.1.** Under what circumstance is the chief assumption underlying the chi square technique violated?

**13.2.** In their 1949 *Psychol. Bull.* article on "The use and misuse of the chi-square test," Lewis and Burke cite nine different types of errors frequently made in using chi square. One of these is "neglect of frequencies of non-occurrence." Can you illustrate what is meant by this?

**13.3.** Specify briefly the types of situations for which it is easy to substitute the simple binomial for chi square as a test of significance.

**13.4.** In the earlier days of chi square, Pearson and his followers claimed that the $df$ for a fourfold table was 2. R. A. Fisher disagreed, and finally won the ensuing argument by pointing out that the $df$ had to be 1 in order for a chi square $P$ to agree with another (well-established) significance test applied to the same fourfold table. Query: What other test? Be specific.

**13.5.** In what sense is chi square a "test of independence"?

**13.6.** The assumption that observations be independent (a requirement for chi square) seems to be violated in a true fourfold contingency table. How would you explain to the beginner that chi square for the fourfold table does not violate this assumption?

**13.7.** Occasionally in a contingency table a cell may have zero frequency. Should this lead to any changes in using the chi square technique for testing the hypothesis of independence? If so, what? If not so, why not?

**13.8.** Given pass (P) and fail (F) information for two test items in Table VII. Consider testing the two hypotheses: $H_i$ for independence and $H_d$ for difference in difficulty. For which would it be more important to use an exact probability test? Why? Be specific.

**Table VII**

|  |  | Item 2 | | |
|---|---|---|---|---|
|  |  | F | P | |
|  | P | 0 | 10 | 10 |
| Item 1 |  |  |  |  |
|  | F | 14 | 6 | 20 |
|  |  | 14 | 16 | 30 |

**13.9.** Specify (by example) a situation for which the $df$ for chi square is $k - 1$ for $k$ observed frequencies and then a situation for which the $df$ is also $k - 1$ but there are $2k$ observed frequencies.

**13.10.** Suppose a 3 by 3 contingency table, with $N = 100$, yields a chi square of 7.9 and that a 2 by 2 contingency table, with $N = 100$, also yields a chi square of 7.9. Thus the value of the contingency coefficient for each table becomes .27 (by substituting in the formula for $C$). Defend one or the other of the following statements regarding the statistical significance of the two $C$s.

    *a.* Both are of equal significance.

    *b.* They differ in significance.

**13.11.** A study is made to learn whether the items on the 1937 Stanford-Binet have the same difficulty values as during the standardization testing period, 1931–33. An item on which 5-, 6-, 7-, and 8-year-olds are tested might yield the difficulties (percentages for passing) given in Table VIII.

**Table VIII**

| Age | 5 | 6 | 7 | 8 | |
|---|---|---|---|---|---|
| N | 101 | 203 | 202 | 203 | |
|  |  |  |  |  | 1931–33 |
| % | 26 | 41 | 58 | 71 | |
| N | 90 | 120 | 130 | 115 | |
|  |  |  |  |  | 1955 |
| % | 20 | 37 | 49 | 68 | |

Needed: An over-all test for the significance of the differences between the two periods of testing. How would you do it? Indicate just how you would set up your work sheet for making the calculations.

**13.12.** Suppose an experiment involving two groups of rats, one of which (Group A) has been on a standard diet, the other (Group B) has been on a diet which is deficient in Vitamin A. Let $N_A = 30$ and $N_B = 20$. In order to see whether the vitamin deficient group will choose a food with high natural vitamin content (say food X) in preference to food $Y$ of low content, the members of each group are given a chance to choose on each of 5 successive days. In Group A there are 90 preferences for food X and in Group B there are 70 preferences for food X. Now our investigator has heard of the chi square test. He reasons that by chance 50 per cent of the choices in each group would be for food X, hence for Group A 75 of the 150 choices would be for X and in Group B 50 of the 100 choices would be for X. Thus he arrives at the following table:

|                | A  | B  |
|----------------|----|----|
| Chance for X   | 75 | 50 |
| Observed for X | 90 | 70 |

from which he calculates chi square as $(90 - 75)^2/75 + (70 - 50)^2/50 = 11$. He next worries about the number of degrees of freedom, and seeing no restrictions decides that $df = 2$. Hence he finds $P$ to be between the .01 and .001 levels of significance. In what ways, if any, can the statistical treatment be criticized?

**Table IX**

| | Item 2 | | | Item 4 | |
|--------|---|---|--------|---|---|
| | F | P | | F | P |
| P | 20 | 40 | P | 0 | 10 |
| Item 1 | | | Item 3 | | |
| F | 30 | 10 | F | 10 | 7 |

**13.13.** Given the fourfold tables in Table IX for passing (P) and failing (F) test items:

    *a.* Specify two meaningfull null hypotheses that are testable, via the chi square technique, with the 1–2 item table (formulas *not* called for).

    *b.* Why would it be unsafe to test, via chi square, similar hypotheses regarding items 3 and 4? (*N* small is not a very sophisticated answer.)

    *c.* Indicate alternative (and better) methods for handling the two hypotheses re the data for items 3 and 4.

**13.14.** Most of the items on the Strong Interest Inventory involve L (like), I (indifferent) and D (dislike) response categories.

    *a.* Suppose we have the responses from two independent groups on item *i* and we wish to test the hypothesis that the two populations from which our groups have been drawn are alike in their responses. Indicate the tabular setup and the techniques you would use in testing this hypothesis.

    *b.* Suppose item *i* and item *j* with responses from a single group and we wish to test the hypothesis that the responses to the two items are independent. Tabular setup? Technique?

    *c.* How would you go about testing the hypothesis that the "L" responses to item *i* *differed* from the "L" responses to item *j* for a single group?

**13.15.** The following statement may be found in the text chapter on chi square. "For $n$s larger than 30, the expression $\sqrt{2\chi^2} - \sqrt{2n-1}$ will have a sampling distribution which will follow very closely the unit normal curve. The probability is accordingly .05 that this expression will exceed $+1.64$, and .01 that it will exceed $+2.33$, by chance".

    *a.* To what does the $n$ refer?

    *b.* Why the concern with plus values only?

    *c.* How do you reconcile the use of 1.64 instead of 1.96 when you consider that 1.96 is for a two-tailed test, whereas chi square, under the appropriate condition, must equal 3.84 (or 1.96 squared) for a two-tailed significance level of .05?

**13.16.** In the earlier days chi square was typically regarded as a technique for testing the "goodness of fit" of curves. What kind of curves? Why not other kinds?

**13.17.** In general, would you expect chi square for a $2 \times 2$ table to be larger or smaller than for a $3 \times 3$ table? Why? Under what circumstance could the reverse be true?

**13.18.** Recently the author read a manuscript which involved data on a control group and three groups under three experimental conditions, a total of four groups with individuals assigned at random to the groups. The response data were dichotomous, hence chi square was used. A total of three fourfold tables (one for each experimental condition against control) were made, from which three chi squares were calculated, each with 1 *df*. Then the three chi squares (and the *df*s) were added for a total chi square based on 3 *df*. Any comment? (Note: no low expected values involved.) Be specific.

## CHAPTER 14

**14.1.** Suppose that in reading an older (circa 1920) study you find an $M_D$ and $S_{DD}$ (*S* of distribution of difference scores) based on $N = 10$. You wish to re-evaluate $M_D$ via the *t* technique. If no actual scores are reported, how would you proceed to get the needed unbiased estimate of the variance of the difference scores?

**14.2.** In what way is the concept of "degrees of freedom" similar for chi square applied to frequencies and for the variance estimation situation?

**14.3.** Typically, one tail (the right-hand side) of the chi square distribution is involved in tests of significance.

    *a.* Under what specific condition does this one tail provide a two-sided, or two-tailed, test?

    *b.* State two entirely different types of situations one of which *requires* using both tails of the chi square distribution and the other of which requires one tail and an alertness for the other.

**14.4.** How could you use $F$ to determine whether a variance of 25, based on a sample of $N$ cases, deviates significantly from a hypothetical value of 16?

**14.5.** In a certain textbook on statistical method you find the following data for $N = 30$ cases for two forms of a test having .93 as its form versus form reliability:

|        | Mean | $s^2$ |
|--------|------|-------|
| Form A | 44.4 | 193.2 |
| Form B | 42.8 | 146.3 |

To judge whether the two forms differ in variability, the author takes 193.2/146.3 as $F$. Any comment?

**14.6.** The distributions of scores on successive learning trials on a pursuit rotor typically increase in variance from trial to trial. Why is Bartlett's test not applicable for testing the differences among such variances?

**14.7.** Given for the 1937 Stanford-Binet the following standard deviations (all testing done during March 1956, and no siblings):

|        | Age 3 | | Age 4 | |
|--------|-------|-------|-------|-------|
|        | Boys  | Girls | Boys  | Girls |
| $N$s   | 50    | 48    | 49    | 51    |
| Form L | 17.5  | 18.0  | 16.8  | 16.5  |
| Form M | 16.2  | 17.2  | 16.3  | 17.6  |

    *a.* For the best test of the differences between $S$s, the difference
        $17.5 - 16.2$ would be tested by which formula?
        $18.0 - 16.5$ would be tested by which formula?
        $17.6 - 16.3$ would be tested by which formula?
    *b.* Specify (write down numerical values) a set of four $S$s that could be properly compared by Bartlett's test for homogeneity.

**14.8.** Suppose for a group of 60 college students it is found that their grades correlate .40 with a reading comprehension test and .50 with an intelligence test. If the grades (grade point ratios) are expressed in standard-score form with variance of unity, it is immediately seen that the two errors of estimate (or residual) variances are .84 and .75. The $r$ between the two tests is .70.
    *a.* Since .84 and .75 are both $S^2$ values, how would you proceed to convert them into $s^2$ values?
    *b.* Presuming that you have the appropriate $s^2$ values, do you have any suggestion as to how you might proceed to test the significance of the difference between the two estimated residual variances?

## CHAPTER 15

**15.1.** The tabled values of $F$ involve the ratio of two unbiased estimates of the same population variance. Why the requirement of "same" population variance? Is, or is not, this sameness concerned with the assumption of homogeneity of variances? Why?

**15.2.** When we use $F$ in the simple (one-way) analysis of variance to test the difference between, say, the means for four experimental conditions, the question (or requirement) of independence arises in at least three different places. Can you specify?

**15.3.** Cite two bits of evidence that the $F$ test as used in the analysis of variance *is* a two-tailed test despite the fact that the tabled value for the .01 level is actually the .02 level when a two-tailed test is appropriate for testing hypotheses regarding difference in variability of two groups.

**15.4.** Another textbook author says that the variance estimate based on the within groups sum of squares (one-way classification with $G$ groups) "corresponds to a standard error of a difference as used in the $t$ test." In what way might this statement be regarded as partially true, and how would it need to be modified to make it exactly true?

**15.5.** With reference to the variance estimates indicated in Table 15.4, why is it not permissible to take $F = s^2_r/s^2_w$ as a means of testing the hypothesis that the residual variance is greater than the within arrays variance?

**15.6.** What application of chi square might be used to test the tenability of which assumption in simple analysis of variance? Any cautions in arranging the data for such a test?

**15.7.** We had one $F$ test in which the numerator involved eta squared minus $r^2$, and another $F$ test involving the difference between two multiple $r$s squared. In what way are the two $F$s similar (or analogous)?

**15.8.** We learned of an $F$ test for $r_{1\cdot23}$ and for $r$. Do you see an easy way to test the significance of the point biserial $r$ by an $F$ test? How?

**15.9.** We have seen how the goodness of fit of a normal curve to frequencies can be checked (or tested) via the use of the chi square technique, and we have seen how the $F$ test can be used to test the goodness of fit of linear regression. Suppose we have height measurements on 100 children at each age, 1 to 18 inclusive (cross-sectional, i.e., we are not following the same children from 1 to 18). All measurements are within a week of a birthday. It is sometimes argued that growth follows the Gompertz curve, the equation of which is a double exponential function:

$$Y = vg^{h^X}$$

in which $v, g$, and $h$ are constants determinable from the data, $X$ is age, and $Y$ is height (of the children).

    *a.* Set up a variance table with appropriate symbols to indicate the breakdown of the sum of squares (for $Y$), the degrees of freedom (actual numerical values), and the variance estimates (in symbols) which you would use to test the goodness of fit of the data to the Gompertz curve. Specify symbolically the $F$ ratio you would use.

    *b.* In fitting a normal curve to a frequency distribution, we set up expected frequencies. What in the fitting of a Gompertz curve would you consider as analogous to "expected" frequency?

    *c.* Would your proposed scheme for testing the fit of the Gompertz curve be valid in case of a longitudinal study (i.e., we follow same children from ages 1 to 18)? Defend your answer briefly.

**15.10.** A recent (1967) textbook contains the following:

$$s^2{}_t = s^2{}_{ag} + s^2{}_{wg}$$

in which $s^2{}_t$ is "total variance," and $s^2{}_{ag}$ is "among-groups variance," and $s^2{}_{wg}$ is "within-groups variance." Any comment?

## CHAPTER 16

**16.1.** If you mistakenly used simple analysis of variance for testing the difference between $G$ means based on $N$ sets of matched individuals, would you expect a smaller or larger $P$ than would have resulted had you used the more appropriate two-way analysis of variance? Why?

**16.2.** Consider the ordinary $z$ test for the difference, via $M_D$, between correlated means. What in this $z$ test corresponds most closely to "interaction"? Be more specific than a mere statement that "it" is found in the error term or in the numerator or in the denominator or in the correlation between scores.

**16.3.** Given information regarding sex difference in weight at birth for large samples of American and of British babies. The data can obviously be placed in a two-way analysis of variance setup, which would permit testing for a nationality difference as well as for a sex difference. The sex by nationality interaction could also be tested by the analysis of variance method. Could this interaction have been tested prior to the invention of the $F$ technique? How?

**16.4.** During our discussion of chi square we did not mention "interaction." Suppose the following data for number of subjects who overcome "set" in the water-jar test:

    70 of 100 male science majors
    60 of 100 female science majors
    40 of 100 male history majors
    50 of 100 female history majors

Can you specify an interaction, and how would you proceed to test it for significance? (Note: this is not a small sample situation.)

**16.5.** The fourth edition of a certain textbook gives "an alternative test that is simple" for making further tests after finding, by $F$, significant differences among several means. The alternative test is indeed simple: Let $M$ stand for the group mean that deviates farthest from the mean, $M_t$, of all groups combined. Compute an estimate of the standard error of $M$ as $s_M = s_w/\sqrt{m}$ in which $m$ is the group size and $s_w$ is the usual within-groups variation. Then take

$$t = (M - M_t)/s_M \quad \text{with} \quad df = mG - G.$$

Any comments?

**16.6.** Suppose a two-way classification calling for the fixed constants model. Would you prefer to have $RC$ cases with measurement replication (thus leading to $m$ scores per cell) or would you prefer $mRC$ individuals (also leading to $m$ scores per cell)? Why?

**16.7.** In order to ascertain the effect of varying the color of the stimulus patch on critical flicker fusion you might plan to use 10 subjects, each measured 6 times under each of four color conditions: red, blue, yellow, and green (brightness controlled). Set up a schematic variance table, with sources, variance estimates, and numerical *df*s. Indicate legitimate *F*s (actual variance ratios, in symbols) that can be used to test hypotheses, and specify the generality of conclusions you would draw from possible significant *F*s.

**16.8.** Suppose you are consulted by a statistically naive individual who has the notion that he can plan a study having to do with the effects of height and weight on basal metabolism by using a factorial design with height and weight as the basis for a two-way classification.

    *a.* Any warnings to him regarding possible difficulties in such a design?
    *b.* What plan would you suggest instead?

**16.9.** Suppose researchers A and B both start with 12 litters of rats, 4 rats per litter. Both use identical T mazes. Researcher A splits each litter randomly so as to have four groups which are run under four different degrees of food deprivation. To test the between groups differences (deprivation effects), A computes an *F* with an interaction variance as "error." Researcher B runs all his rats under one condition, then calculates an *F* as the between litters variance estimate over the within litters variance estimate. (Each rat in both experiments has just one score.)

    *a.* Specify the degrees of freedom for A's *F* and for B's *F*.
    *b.* Why did A use an interaction variance estimate as "error" whereas B used a within variance estimate as his "error" term?

**16.10.** In a recent *Journal of Psychology* will be found the results, Table X, for an experiment designed to learn whether seeing the movie "Gentleman's Agreement" leads to changes in attitude toward the Jews. The researcher used the Levinson-Sanford Questionnaire on Anti-Semitism, an instrument with a reported reliability of .98. As usual in such studies, an experimental group was shown the movie, and a control group was not. Both groups were pretested with the questionnaire, and after the experimentals had seen the movie both groups were retested. The *N*s were 50 and 90.

**Table X. Summary statistics**

|  | Experimental | Control |
|---|---|---|
| Mean, pretest | 23.55, $S_M = 2.87$ | 26.54, $S_M = 2.26$ |
| Mean, posttest | 16.40, $S_M = 4.10$ | 27.06, $S_M = 2.79$ |
|  | Test-retest $r = .64$ | Test-retest $r = .84$ |
|  | Difference in means: 7.15 | Difference in means: .52 |
|  | $S_{D_M} = 3.16$, $z = 2.26$ | $S_{D_M} = 1.63$, $z = .33$ |

It was concluded that because the experimentals show a significant (at $P = .03$ level) difference, whereas the controls do not, the movie did lead to changes in attitude. Do you see anything wrong with his statistical treatment?

**16.11.** Because a smaller $F$ is needed for a prescribed level of significance as the $df$ for the denominator becomes larger, it has been argued frequently that we should strive to increase this $df$.

  *a.* Consider Design A with, say, 20 cases in each of two experimental groups having been assigned randomly and independently to the two groups in contrast to Design B in which we also have 20 cases per group, but the groups have been matched on a thought-to-be relevant variable by setting up 20 pairs of individuals. Which design provides the greater $df$ and why might it be unwise to use the design with the larger $df$?

  *b.* Consider the test for the significance of linear correlation. As we outlined the procedure, the $df$ for the denominator variance (residual) was $N - 2$. Now in some texts we can find that the denominator variance is taken as the within array variance about the array mean with $df = N - G$ where $G$ is the number of arrays. Obviously, the $df$s differ according to which variance is being used as "error." Why might the test based on $N - 2$ $df$ be no more apt to lead to significant $F$s than the one based on $N - G$ $df$?

  *c.* Part *b* involves a within array variance about the array means and a within array variance about a regression line. Why can't we test the difference between these two variances by taking $F$ as their ratio? Do you see a possible indirect method for making an inference regarding their difference?

**16.12.** Consider a two-way layout for the scores of 30 persons on $C = 3$ forms of a test.

  *a.* The remainder term will provide an estimate of error of measurement variance with how many (numerical value, please) degrees of freedom?

  *b.* If your statistical clerk regarded the data as simply 30 groups of scores with $m = 3$ per group, he (or she) would have a possibly different estimate of error of measurement variance in the within persons term, which would have (numerical) how many degrees of freedom?

  *c.* The "possibly different" above implies what possible major source of difference, aside from $df$s, in these two ways of estimating error variance?

  *d.* How would you test the difference between form means?

  *e.* How would you test the significance of the difference between the three form standard deviations? (Note: the available method may not be entirely satisfactory for a reason which you might specify.)

**16.13.** Lay out a series of possible plans for studying the effect of illumination and the effect of foveal versus peripheral vision on critical flicker fusion (CFF). Let us agree to use five levels of illumination and four areas of retinal stimulation (foveal plus three areas proceeding outward from the fovea toward the periphery). For each approach, indicate the sources of variation, the $df$s, the variance estimates, and appropriate $F$s. Evaluate the relative merits of your several plans. (Note: CFF is not affected by practice and can be measured in a couple of minutes.)

**16.14.** In an issue of the *Journal of Consulting Psychology* will be found a study of two groups (paranoid schizophrenics and normals), 27 cases per group. We are told that "the groups were equated for age, education, and intelligence." The dependent variable is a measure of "distortion" of responses to four stories. The authors present an analysis of variance table, given here as Table XI.

**Table XI. Analysis of variance**

| Source | df | SSqs | Var. Est. | F | P |
|--------|-----|--------|-----------|-------|------|
| Groups | 1 | 13.74 | 13.74 | 19.35 | <.01 |
| Stories | 3 | 17.81 | 5.94 | 8.37 | <.01 |
| Individuals | 53 | 144.87 | 2.73 | 3.85 | <.01 |
| G by S interaction | 3 | 4.56 | 1.52 | 2.14 | >.05 |
| Error | 144 | 101.73 | .71 | | |
| Total | 204 | 282.71 | | | |

What, if anything, is wrong with their statistical analysis?

**16.15.** In practically all applications of the $F$ test the hypothesis being tested determines the variance estimate to be placed in the numerator for the $F$ ratio, and usually we do not bother to compute $F$ if this numerator variance is smaller than the variance chosen as "error" for the denominator. Occasionally the numerator variance is so small relative to the chosen "error" variance that $F$ taken upside down is significant. In discussing this, another text says "The situations where the $F$ obtained in this manner is significant probably have no reasonable interpretation other than that they are the occasionally significant values which are to be expected in random sampling." Any comment?

**16.16.** A significant upside down $F$ has its counterpart in what kind of a chi square?

**16.17.** Do you think it possible, in a three-way fixed effects design, to have (*a*) three-way interaction without two-way interaction and (*b*) vice versa? Explain. (The use of diagrams may help here.)

**16.18.** It has been suggested that when an interaction is significant (fixed constants), we should proceed to a series of analyses of variance one lower in order. That is, a significant $R \times B \times C$ interaction should be followed by say, $B$ two-way analyses involving rows and columns; and a significant $R \times C$ interaction should lead to, say $R$ one-way analyses. What possible sense can you make of this suggestion?

**16.19.** The author of a recent letter criticizes this textbook for advocating, without qualifications, the use of, e.g., $s^2_{rc}$ for testing column effects in an $[a_r A_b A_c]$ design. He says that $s^2_{rc}$ should be used *only* when it is larger (not necessarily significantly larger) than $s^2_{rbc}$. Stated differently he says that if $s^2_{rc}$ is smaller than $s^2_{rbc}$ the latter should be used as "error." Obviously, his worry is about those situations for which $s^2_{rc}$ and $s^2_{rbc}$ do not differ significantly. He mentions "pooling" after making the point that "one could use either term," i.e., either $s^2_{rc}$ or $s^2_{rbc}$ as "error." Aside from "pooling," and in rather simple commonsense (but not statistically naive) terms,

      *a.* what argument do you think he set forth in favor of using $s^2_{rc}$ *only* when it is larger (though not significantly so) than $s^2_{rbc}$ as "error" for $F$?

      *b.* What argument would you use against his proposal? Note that this part can be answered even though you cannot answer part *a.*

**16.20.** Do you see any possible way of utilizing the analysis of variance technique for testing the hypothesis that the correlation between two variables is perfect within limits imposed by errors of measurement? This is the correction for attenuation problem under another guise. We intend this to be a hard question, and the only hint is to think in terms of standard scores for the two variables being correlated.

## CHAPTER 17

**17.1.** Occasionally, linear and quadratic trend analysis has been used in situations involving an independent "variable" that is actually qualitative but for which the investigator argues that in terms of the expected outcome for the dependent variable the "levels" on the independent variable can be ordered. Accordingly the qualitatively different levels are treated as though on a scale with equal spacings (distances) from level to level as a basis for trend analysis. Do you see any difficulty in this procedure?

**17.2.** In a recently published textbook is found an example in which five subjects are measured on three successive trials. The breakdown of the total sum of squares quite properly leads to sums of squares for trials ($df = 2$), for subjects ($df = 4$), and for subject by trial interaction ($df = 8$). The sum of squares for trials is 90.00, and the sum of squares for linear trend is also 90.00 ($df = 1$).

      *a.* What implication does this equality of the two sums of squares have for a possible quadratic component?

      *b.* The author failed to follow through with any implication regarding the consequence of having unequal *df*s (2 and 1) for the trial and the linear component sums of squares (of the same amount). What helpful remark could he have made?

**17.3.** Consider the situation for which we have just three levels and linear and quadratic components have been "taken out." What can be said, before you see the results, about the fit of a quadratic equation (curve) to the three means? Does it follow that the quadratic component must be statistically significant?

**17.4.** No use of individual slopes was made prior to p. 400 of the text. How could we use individual slopes in the setup involving a single linear trend based on correlated observations?

**17.5.** Suppose the $r$ between $Y$ and $X$ is .32 and the correlation ratio (eta) for $Y$ on $X$ is .40, both computed for a sample of $N = 100$ with $G = 14$ intervals on the $x$ axis. The $F$ for testing eta will have 13 and 86 degrees of freedom, whereas the $F$ for testing $r$ will have 1 and 98 degrees of freedom. When the $F$s are computed and the significance levels determined, we have $F_{eta} = 1.26$, $P$ about .30, and $F_r = 13.76$, $P$ about .001. How do you account for $r$, though smaller than eta, being the more significant? (Do not forget that the larger the $n_1$ the smaller the $F$ required for significance.)

**17.6.** When we have the Case XV setup, we can test the differences among the possible $B$ linear trends by following the procedure given on pp. 399–400. Suppose we wish to test the differences between the quadratic components of the $B$ trend lines. How would you do this? (Note: You might first take out the linear components or you might proceed directly to the quadratic part.)

## CHAPTER 18

**18.1.** Line 5 of Table 18.2 indicates the possibility of computing an $r$ based on between sums. Why would such an $r$ be meaningless when $G = 2$?

**18.2.** Suppose an appreciable negative correlation between $X$ and an uncontrolled variable $Y$. Defend either the proposition that the covariance adjustment can be used or the proposition that it cannot be used.

**18.3.** The correlation between two variables based on combined sex groups will be distorted by possible sex differences on either or both variables. How might you proceed to obtain a single $r$ that is not distorted by the sex difference?

**18.4.** An assumption underlying the covariance adjustment technique is homogeneity of regression from group to group. Does the text provide a method for testing this assumption? Where (or what test)?

## CHAPTER 19

**19.1.** Suppose you have 22 cases measured under normal (control) conditions, then measured under a prescribed experimental condition. You are interested in evaluating the changes, but because of marked skewness (in which scores?) you are skeptical of the tenability of the $t$ test. Do you see a way of testing the significance of the changes by a method for which you might use chi square as an approximation?

**19.2.** A possible extension of the general idea of the median test for two or more groups would be to classify the scores of each group into four categories according to their position relative to $Q_3$, $Q_2$, and $Q_1$ based on the combined groups. This would lead to a 4 by $k$ table when we have $k$ groups, from which a chi square with $3(k - 1)$ $df$ could be computed. Aside from difficulties when samples are very small and the loss of efficiency caused by grouping, do you see any possible problem in connection with the meaning of a significant chi square from such a setup?

**19.3.** While reviewing a manuscript submitted to *Psychological Monographs*, the author encountered a two-way fixed effects design in which rows stood for eight different "treatments" and columns stood for two groups (pilots and nonpilots), with 16 (independent) cases per cell. Apparently the writer of the manuscript was a devotee of nonparametric methods: instead of testing the $T$ by $G$ interaction by the conventional $F$ test with the within cells as the error term, he used Kendall's tau. If, his argument goes, the tau for the eight sets of means is significantly negative he would conclude that the interaction is significant. This tau is, of course, the correlation between a rank-ordering of the means in the first column and a rank-ordering of the means in the second column, with $n = 8$ pairs of ranks. For $n = 8$, tau must reach (negative) .49 for significance at the chosen level. Now there are three distinctly different bugs in this, every one of which nullifies his procedure. OK, try your critical powers.

## CHAPTER 20

**20.1.** When experimental and control groups are set up on the basis of individuals paired on two control variables, the gain in precision (or the error reduction) depends upon what fact(s), presumably available before carrying out the experiment?

**20.2.** Suppose you wish to do an experiment in which the cost per experimental subject is far greater than that for a control. Accordingly you decide to take $N_C$ as $4N_E$. What scheme, other than randomization, would you use to assure comparability for the two groups? And how would you proceed to test for significance the difference between the means of the two groups?

**20.3.** In a recent study of sex differences in problem solving ($X$), the fact of differences in general intelligence as measured by a college aptitude test ($Y$) was taken into account by comparing males and females who had been paired on $Y$.

    *a.* What statistical procedure do you think was used in testing the null hypothesis of no sex difference on $X$?

    *b.* Can you suggest an alternative experimental-statistical plan for getting at sex difference on $X$ with $Y$ controlled?

**20.4.** For the large sample situation, the sampling variance of a mean based on a sample stratified on variable $U$ is given by $S^2_{\bar{X}} = S^2_x(1 - r^2_{xu})/N$ and for the sampling variance of the difference between means based on groups matched as to distribution on control variable $Y$ (not individual pairing), we have

$$S^2_D = S^2_{x1}(1 - r^2_{xy1})/N_1 + S^2_{x2}(1 - r^2_{xy2})/N_2$$

Perhaps you will have noted the similarity of sampling variance for the stratified sampling situation and the matched distributions situation. Do you see a connection between the foregoing formulas and the analysis of variance technique?

**20.5.** When attempting to evaluate the relative effect of two movies on attitudes (measured in a continuous fashion), we may form two groups by random assignment of individuals and then we may follow either (*a*) the procedure of a pretest, show movie, posttest, with the statistical analysis based on a comparison of the two mean changes or (*b*) the "after only" plan in which one movie is shown to each group after which both groups are tested and the difference between the resulting "after" means is tested for significance. This second, or "after only," method is frequently more feasible than the first procedure. In general, which design would you expect to be more precise? Why? Can you specify a condition which might make the other design more precise? (Hint: presume that all four possible standard deviations are equal.)

# *APPENDIX*
## TABLES A TO G

## Table A. Normal curve functions

| z or x/σ | Area: m to z | Area: q Smaller | h or Ordinate |
|---|---|---|---|
| .00 | .00000 | .50000 | .3989 |
| .05 | .01994 | .48006 | .3984 |
| .10 | .03983 | .46017 | .3970 |
| .15 | .05962 | .44038 | .3945 |
| .20 | .07926 | .42074 | .3910 |
| .25 | .09871 | .40129 | .3867 |
| .30 | .11791 | .38209 | .3814 |
| .35 | .13683 | .36317 | .3752 |
| .40 | .15542 | .34458 | .3683 |
| .45 | .17364 | .32636 | .3605 |
| .50 | .19146 | .30854 | .3521 |
| .55 | .20884 | .29116 | .3429 |
| .60 | .22575 | .27425 | .3332 |
| .65 | .24215 | .25785 | .3230 |
| .70 | .25804 | .24196 | .3123 |
| .75 | .27337 | .22663 | .3011 |
| .80 | .28814 | .21186 | .2897 |
| .85 | .30234 | .19766 | .2780 |
| .90 | .31594 | .18406 | .2661 |
| .95 | .32894 | .17106 | .2541 |
| 1.00 | .34134 | .15866 | .2420 |
| 1.05 | .35314 | .14686 | .2299 |
| 1.10 | .36433 | .13567 | .2179 |
| 1.15 | .37493 | .12507 | .2059 |
| 1.20 | .38493 | .11507 | .1942 |
| 1.25 | .39435 | .10565 | .1826 |
| 1.30 | .40320 | .09680 | .1714 |
| 1.35 | .41149 | .08851 | .1604 |
| 1.40 | .41924 | .08076 | .1497 |
| 1.45 | .42647 | .07353 | .1394 |
| 1.50 | .43319 | .06681 | .1295 |
| 1.55 | .43943 | .06057 | .1200 |
| 1.60 | .44520 | .05480 | .1109 |
| 1.65 | .45053 | .04947 | .1023 |
| 1.70 | .45543 | .04457 | .0940 |

**Table A. Normal curve functions** (*continued*)

| $z$ or $x/\sigma$ | Area: $m$ to $z$ | Area: $q$ Smaller | $h$ or Ordinate |
|---|---|---|---|
| 1.75 | .45994 | .04006 | .0863 |
| 1.80 | .46407 | .03593 | .0790 |
| 1.85 | .46784 | .03216 | .0721 |
| 1.90 | .47128 | .02872 | .0656 |
| 1.95 | .47441 | .02559 | .0596 |
| 2.00 | .47725 | .02275 | .0540 |
| 2.05 | .47982 | .02018 | .0488 |
| 2.10 | .48214 | .01786 | .0440 |
| 2.15 | .48422 | .01578 | .0396 |
| 2.20 | .48610 | .01390 | .0355 |
| 2.25 | .48778 | .01222 | .0317 |
| 2.30 | .48928 | .01072 | .0283 |
| 2.35 | .49061 | .00939 | .0252 |
| 2.40 | .49180 | .00820 | .0224 |
| 2.45 | .49286 | .00714 | .0198 |
| 2.50 | .49379 | .00621 | .0175 |
| 2.55 | .49461 | .00539 | .0154 |
| 2.60 | .49534 | .00466 | .0136 |
| 2.65 | .49598 | .00402 | .0119 |
| 2.70 | .49653 | .00347 | .0104 |
| 2.75 | .49702 | .00298 | .0091 |
| 2.80 | .49744 | .00256 | .0079 |
| 2.85 | .49781 | .00219 | .0069 |
| 2.90 | .49813 | .00187 | .0060 |
| 2.95 | .49841 | .00159 | .0051 |
| 3.00 | .49865 | .00135 | .0044 |
| 3.25 | .49942 | .00058 | .0020 |
| 3.50 | .49977 | .00023 | .0009 |
| 3.75 | .49991 | .00009 | .0004 |
| 4.00 | .49997 | .00003 | .0001 |

**Table B. Transformation of $r$ to $z$**

| $r$ | $z$ | $r$ | $z$ | $r$ | $z$ |
|-----|------|-----|------|-----|-------|
| .01 | .010 | .34 | .354 | .67 | .811 |
| .02 | .020 | .35 | .366 | .68 | .829 |
| .03 | .030 | .36 | .377 | .69 | .848 |
| .04 | .040 | .37 | .389 | .70 | .867 |
| .05 | .050 | .38 | .400 | .71 | .887 |
| .06 | .060 | .39 | .412 | .72 | .908 |
| .07 | .070 | .40 | .424 | .73 | .929 |
| .08 | .080 | .41 | .436 | .74 | .950 |
| .09 | .090 | .42 | .448 | .75 | .973 |
| .10 | .100 | .43 | .460 | .76 | .996 |
| .11 | .110 | .44 | .472 | .77 | 1.020 |
| .12 | .121 | .45 | .485 | .78 | 1.045 |
| .13 | .131 | .46 | .497 | .79 | 1.071 |
| .14 | .141 | .47 | .510 | .80 | 1.099 |
| .15 | .151 | .48 | .523 | .81 | 1.127 |
| .16 | .161 | .49 | .536 | .82 | 1.157 |
| .17 | .172 | .50 | .549 | .83 | 1.188 |
| .18 | .181 | .51 | .563 | .84 | 1.221 |
| .19 | .192 | .52 | .577 | .85 | 1.256 |
| .20 | .203 | .53 | .590 | .86 | 1.293 |
| .21 | .214 | .54 | .604 | .87 | 1.333 |
| .22 | .224 | .55 | .618 | .88 | 1.376 |
| .23 | .234 | .56 | .633 | .89 | 1.422 |
| .24 | .245 | .57 | .648 | .90 | 1.472 |
| .25 | .256 | .58 | .663 | .91 | 1.528 |
| .26 | .266 | .59 | .678 | .92 | 1.589 |
| .27 | .277 | .60 | .693 | .93 | 1.658 |
| .28 | .288 | .61 | .709 | .94 | 1.738 |
| .29 | .299 | .62 | .725 | .95 | 1.832 |
| .30 | .309 | .63 | .741 | .96 | 1.946 |
| .31 | .321 | .64 | .758 | .97 | 2.092 |
| .32 | .332 | .65 | .775 | .98 | 2.298 |
| .33 | .343 | .66 | .793 | .99 | 2.647 |

## Table C. Transformation of $z$ to $r$*

| $z$ | .00 | .01 | .02 | .03 | .04 | .05 | .06 | .07 | .08 | .09 |
|---|---|---|---|---|---|---|---|---|---|---|
| .0 | .0000 | .0100 | .0200 | .0300 | .0400 | .0500 | .0599 | .0699 | .0798 | .0898 |
| .1 | .0997 | .1096 | .1194 | .1293 | .1391 | .1489 | .1586 | .1684 | .1781 | .1877 |
| .2 | .1974 | .2070 | .2165 | .2260 | .2355 | .2449 | .2543 | .2636 | .2729 | .2821 |
| .3 | .2913 | .3004 | .3095 | .3185 | .3275 | .3364 | .3452 | .3540 | .3627 | .3714 |
| .4 | .3800 | .3885 | .3969 | .4053 | .4136 | .4219 | .4301 | .4382 | .4462 | .4542 |
| .5 | .4621 | .4699 | .4777 | .4854 | .4930 | .5005 | .5080 | .5154 | .5227 | .5299 |
| .6 | .5370 | .5441 | .5511 | .5580 | .5649 | .5717 | .5784 | .5850 | .5915 | .5980 |
| .7 | .6044 | .6107 | .6169 | .6231 | .6291 | .6351 | .6411 | .6469 | .6527 | .6584 |
| .8 | .6640 | .6696 | .6751 | .6805 | .6858 | .6911 | .6963 | .7014 | .7064 | .7114 |
| .9 | .7163 | .7211 | .7259 | .7306 | .7352 | .7398 | .7443 | .7487 | .7531 | .7574 |
| 1.0 | .7616 | .7658 | .7699 | .7739 | .7779 | .7818 | .7857 | .7895 | .7932 | .7969 |
| 1.1 | .8005 | .8041 | .8076 | .8110 | .8144 | .8178 | .8210 | .8243 | .8275 | .8306 |
| 1.2 | .8337 | .8367 | .8397 | .8426 | .8455 | .8483 | .8511 | .8538 | .8565 | .8591 |
| 1.3 | .8617 | .8643 | .8668 | .8692 | .8717 | .8741 | .8764 | .8787 | .8810 | .8832 |
| 1.4 | .8854 | .8875 | .8896 | .8917 | .8937 | .8957 | .8977 | .8996 | .9015 | .9033 |
| 1.5 | .9051 | .9069 | .9087 | .9104 | .9121 | .9138 | .9154 | .9170 | .9186 | .9201 |
| 1.6 | .9217 | .9232 | .9246 | .9261 | .9275 | .9289 | .9302 | .9316 | .9329 | .9341 |
| 1.7 | .9354 | .9366 | .9379 | .9391 | .9402 | .9414 | .9425 | .9436 | .9447 | .9458 |
| 1.8 | .9468 | .9478 | .9488 | .9498 | .9508 | .9518 | .9527 | .9536 | .9545 | .9554 |
| 1.9 | .9562 | .9571 | .9579 | .9587 | .9595 | .9603 | .9611 | .9618 | .9626 | .9633 |
| 2.0 | .9640 | .9647 | .9654 | .9661 | .9668 | .9674 | .9680 | .9686 | .9693 | .9699 |
| 2.1 | .9704 | .9710 | .9716 | .9722 | .9727 | .9732 | .9738 | .9743 | .9748 | .9753 |
| 2.2 | .9757 | .9762 | .9767 | .9771 | .9776 | .9780 | .9785 | .9789 | .9793 | .9797 |
| 2.3 | .9801 | .9805 | .9809 | .9812 | .9816 | .9820 | .9823 | .9827 | .9830 | .9834 |
| 2.4 | .9837 | .9840 | .9843 | .9846 | .9849 | .9852 | .9855 | .9858 | .9861 | .9864 |
| 2.5 | .9866 | .9869 | .9871 | .9874 | .9876 | .9879 | .9881 | .9884 | .9886 | .9888 |
| 2.6 | .9890 | .9892 | .9894 | .9897 | .9899 | .9901 | .9903 | .9904 | .9906 | .9908 |
| 2.7 | .9910 | .9912 | .9914 | .9915 | .9917 | .9919 | .9920 | .9922 | .9923 | .9925 |
| 2.8 | .9926 | .9928 | .9929 | .9931 | .9932 | .9933 | .9935 | .9936 | .9937 | .9938 |
| 2.9 | .9940 | .9941 | .9942 | .9943 | .9944 | .9945 | .9946 | .9948 | .9948 | .9950 |

* Table C is abridged from Table VII of Fisher and Yates: *Statistical tables for biological, agricultural and medical research*, Oliver and Boyd, Ltd., Edinburgh, by permission of the authors and publishers.

## Table D. Distribution of $\chi^2$*

| n | P = .99 | .98 | .95 | .90 | .80 | .70 | .50 |
|---|---------|-----|-----|-----|-----|-----|-----|
| 1 | .00016 | .00063 | .0039 | .016 | .064 | .15 | .46 |
| 2 | .02 | .04 | .10 | .21 | .45 | .71 | 1.39 |
| 3 | .12 | .18 | .35 | .58 | 1.00 | 1.42 | 2.37 |
| 4 | .30 | .43 | .71 | 1.06 | 1.65 | 2.20 | 3.36 |
| 5 | .55 | .75 | 1.14 | 1.61 | 2.34 | 3.00 | 4.35 |
| 6 | .87 | 1.13 | 1.64 | 2.20 | 3.07 | 3.83 | 5.35 |
| 7 | 1.24 | 1.56 | 2.17 | 2.83 | 3.82 | 4.67 | 6.35 |
| 8 | 1.65 | 2.03 | 2.73 | 3.49 | 4.59 | 5.53 | 7.34 |
| 9 | 2.09 | 2.53 | 3.32 | 4.17 | 5.38 | 6.39 | 8.34 |
| 10 | 2.56 | 3.06 | 3.94 | 4.86 | 6.18 | 7.27 | 9.34 |
| 11 | 3.05 | 3.61 | 4.58 | 5.58 | 6.99 | 8.15 | 10.34 |
| 12 | 3.57 | 4.18 | 5.23 | 6.30 | 7.81 | 9.03 | 11.34 |
| 13 | 4.11 | 4.76 | 5.89 | 7.04 | 8.63 | 9.93 | 12.34 |
| 14 | 4.66 | 5.37 | 6.57 | 7.79 | 9.47 | 10.82 | 13.34 |
| 15 | 5.23 | 5.98 | 7.26 | 8.55 | 10.31 | 11.72 | 14.34 |
| 16 | 5.81 | 6.61 | 7.96 | 9.31 | 11.15 | 12.62 | 15.34 |
| 17 | 6.41 | 7.26 | 8.67 | 10.08 | 12.00 | 13.53 | 16.34 |
| 18 | 7.02 | 7.91 | 9.39 | 10.86 | 12.86 | 14.44 | 17.34 |
| 19 | 7.63 | 8.57 | 10.12 | 11.65 | 13.72 | 15.35 | 18.34 |
| 20 | 8.26 | 9.24 | 10.85 | 12.44 | 14.58 | 16.27 | 19.34 |
| 21 | 8.90 | 9.92 | 11.59 | 13.24 | 15.44 | 17.18 | 20.34 |
| 22 | 9.54 | 10.60 | 12.34 | 14.04 | 16.31 | 18.10 | 21.34 |
| 23 | 10.20 | 11.29 | 13.09 | 14.85 | 17.19 | 19.02 | 22.34 |
| 24 | 10.86 | 11.99 | 13.85 | 15.66 | 18.06 | 19.94 | 23.34 |
| 25 | 11.52 | 12.70 | 14.61 | 16.47 | 18.94 | 20.87 | 24.34 |
| 26 | 12.20 | 13.41 | 15.38 | 17.29 | 19.82 | 21.79 | 25.34 |
| 27 | 12.88 | 14.12 | 16.15 | 18.11 | 20.70 | 22.72 | 26.34 |
| 28 | 13.56 | 14.85 | 16.93 | 18.94 | 21.59 | 23.65 | 27.34 |
| 29 | 14.26 | 15.57 | 17.71 | 19.77 | 22.48 | 24.58 | 28.34 |
| 30 | 14.95 | 16.31 | 18.49 | 20.60 | 23.36 | 25.51 | 29.34 |

* Table D is abridged from Table IV of Fisher and Yates: *Statistical tables for biological, agricultural and medical research*, Oliver and Boyd, Ltd., Edinburgh, by permission of the authors and publishers.

## Table D. Distribution of $\chi^{2*}$ (*continued*)

| $n$ | .30 | .20 | .10 | .05 | .02 | .01 | .001 |
|---|---|---|---|---|---|---|---|
| 1 | 1.07 | 1.64 | 2.71 | 3.84 | 5.41 | 6.64 | 10.83 |
| 2 | 2.41 | 3.22 | 4.60 | 5.99 | 7.82 | 9.21 | 13.82 |
| 3 | 3.66 | 4.64 | 6.25 | 7.82 | 9.84 | 11.34 | 16.27 |
| 4 | 4.88 | 5.99 | 7.78 | 9.49 | 11.67 | 13.28 | 18.46 |
| 5 | 6.06 | 7.29 | 9.24 | 11.07 | 13.39 | 15.09 | 20.52 |
| 6 | 7.23 | 8.56 | 10.64 | 12.59 | 15.03 | 16.81 | 22.46 |
| 7 | 8.38 | 9.80 | 12.02 | 14.07 | 16.62 | 18.48 | 24.32 |
| 8 | 9.52 | 11.03 | 13.36 | 15.51 | 18.17 | 20.09 | 26.12 |
| 9 | 10.66 | 12.24 | 14.68 | 16.92 | 19.68 | 21.67 | 27.88 |
| 10 | 11.78 | 13.44 | 15.99 | 18.31 | 21.16 | 23.21 | 29.59 |
| 11 | 12.90 | 14.63 | 17.28 | 19.68 | 22.62 | 24.72 | 31.26 |
| 12 | 14.01 | 15.81 | 18.55 | 21.03 | 24.05 | 26.22 | 32.91 |
| 13 | 15.12 | 16.98 | 19.81 | 22.36 | 25.47 | 27.69 | 34.53 |
| 14 | 16.22 | 18.15 | 21.06 | 23.68 | 26.87 | 29.14 | 36.12 |
| 15 | 17.32 | 19.31 | 22.31 | 25.00 | 28.26 | 30.58 | 37.70 |
| 16 | 18.42 | 20.46 | 23.54 | 26.30 | 29.63 | 32.00 | 39.25 |
| 17 | 19.51 | 21.62 | 24.77 | 27.59 | 31.00 | 33.41 | 40.79 |
| 18 | 20.60 | 22.76 | 25.99 | 28.87 | 32.35 | 34.80 | 42.31 |
| 19 | 21.69 | 23.90 | 27.20 | 30.14 | 33.69 | 36.19 | 43.82 |
| 20 | 22.78 | 25.04 | 28.41 | 31.41 | 35.02 | 37.57 | 45.32 |
| 21 | 23.86 | 26.17 | 29.62 | 32.67 | 36.34 | 38.93 | 46.80 |
| 22 | 24.94 | 27.30 | 30.81 | 33.92 | 37.66 | 40.29 | 48.27 |
| 23 | 26.02 | 28.43 | 32.01 | 35.17 | 38.97 | 41.64 | 49.73 |
| 24 | 27.10 | 29.55 | 33.20 | 36.42 | 40.27 | 42.98 | 51.18 |
| 25 | 28.17 | 30.68 | 34.38 | 37.65 | 41.57 | 44.31 | 52.62 |
| 26 | 29.25 | 31.80 | 35.56 | 38.88 | 42.86 | 45.64 | 54.05 |
| 27 | 30.32 | 32.91 | 36.74 | 40.11 | 44.14 | 46.96 | 55.48 |
| 28 | 31.39 | 34.03 | 37.92 | 41.34 | 45.42 | 48.28 | 56.89 |
| 29 | 32.46 | 35.14 | 39.09 | 42.56 | 46.69 | 49.59 | 58.30 |
| 30 | 33.53 | 36.25 | 40.26 | 43.77 | 47.96 | 50.89 | 59.70 |

* Table D is abridged from Table IV of Fisher and Yates: *Statistical tables for biological, agricultural and medical research*, Oliver and Boyd, Ltd., Edinburgh, by permission of the authors and publishers.

## Table E. Distribution of $t$*

| $n$ | $P = .1$ | .05 | .02 | .01 | .001 |
|---|---|---|---|---|---|
| 1 | 6.314 | 12.706 | 31.821 | 63.657 | 636.619 |
| 2 | 2.920 | 4.303 | 6.965 | 9.925 | 31.598 |
| 3 | 2.353 | 3.182 | 4.541 | 5.841 | 12.941 |
| 4 | 2.132 | 2.776 | 3.747 | 4.604 | 8.610 |
| 5 | 2.015 | 2.571 | 3.365 | 4.032 | 6.859 |
| 6 | 1.943 | 2.447 | 3.143 | 3.707 | 5.959 |
| 7 | 1.895 | 2.365 | 2.998 | 3.499 | 5.405 |
| 8 | 1.860 | 2.306 | 2.896 | 3.355 | 5.041 |
| 9 | 1.833 | 2.262 | 2.821 | 3.250 | 4.781 |
| 10 | 1.812 | 2.228 | 2.764 | 3.169 | 4.587 |
| 11 | 1.796 | 2.201 | 2.718 | 3.106 | 4.437 |
| 12 | 1.782 | 2.179 | 2.681 | 3.055 | 4.318 |
| 13 | 1.771 | 2.160 | 2.650 | 3.012 | 4.221 |
| 14 | 1.761 | 2.145 | 2.624 | 2.977 | 4.140 |
| 15 | 1.753 | 2.131 | 2.602 | 2.947 | 4.073 |
| 16 | 1.746 | 2.120 | 2.583 | 2.921 | 4.015 |
| 17 | 1.740 | 2.110 | 2.567 | 2.898 | 3.965 |
| 18 | 1.734 | 2.101 | 2.552 | 2.878 | 3.922 |
| 19 | 1.729 | 2.093 | 2.539 | 2.861 | 3.883 |
| 20 | 1.725 | 2.086 | 2.528 | 2.845 | 3.850 |
| 21 | 1.721 | 2.080 | 2.518 | 2.831 | 3.819 |
| 22 | 1.717 | 2.074 | 2.508 | 2.819 | 3.792 |
| 23 | 1.714 | 2.069 | 2.500 | 2.807 | 3.767 |
| 24 | 1.711 | 2.064 | 2.492 | 2.797 | 3.745 |
| 25 | 1.708 | 2.060 | 2.485 | 2.787 | 3.725 |
| 26 | 1.706 | 2.056 | 2.479 | 2.779 | 3.707 |
| 27 | 1.703 | 2.052 | 2.473 | 2.771 | 3.690 |
| 28 | 1.701 | 2.048 | 2.467 | 2.763 | 3.674 |
| 29 | 1.699 | 2.045 | 2.462 | 2.756 | 3.659 |
| 30 | 1.697 | 2.042 | 2.457 | 2.750 | 3.646 |
| 40 | 1.684 | 2.021 | 2.423 | 2.704 | 3.551 |
| 60 | 1.671 | 2.000 | 2.390 | 2.660 | 3.460 |
| 120 | 1.658 | 1.980 | 2.358 | 2.617 | 3.373 |
| ∞ | 1.645 | 1.960 | 2.326 | 2.576 | 3.291 |

* Table E is abridged from Table III of Fisher and Yates: *Statistical tables for biological, agricultural and medical research*, Oliver and Boyd, Ltd., Edinburgh, by permission of the authors and publishers.

## Table F. Table of $F$ for .05 (roman), .01 (*italic*), and .001 (**bold face**) levels of significance*

| $n_2$ \ $n_1$ | 1 | 2 | 3 | 4 | 5 | 6 | 8 | 12 | 24 | ∞ |
|---|---|---|---|---|---|---|---|---|---|---|
| 1 | 161 | 200 | 216 | 225 | 230 | 234 | 239 | 244 | 249 | 254 |
| | *4052* | *4999* | *5403* | *5625* | *5724* | *5859* | *5981* | *6106* | *6234* | *6366* |
| | **405284** | **500000** | **540379** | **562500** | **576405** | **585937** | **598144** | **610667** | **623497** | **636619** |
| 2 | 18.51 | 19.00 | 19.16 | 19.25 | 19.30 | 19.33 | 19.37 | 19.41 | 19.45 | 19.50 |
| | *98.49* | *99.01* | *99.17* | *99.25* | *99.30* | *99.33* | *99.36* | *99.42* | *99.46* | *99.50* |
| | **998.5** | **999.0** | **999.2** | **999.2** | **999.3** | **999.3** | **999.4** | **999.4** | **999.5** | **999.5** |
| 3 | 10.13 | 9.55 | 9.28 | 9.12 | 9.01 | 8.94 | 8.84 | 8.74 | 8.64 | 8.53 |
| | *34.12* | *30.81* | *29.46* | *28.71* | *28.24* | *27.91* | *27.49* | *27.05* | *26.60* | *26.12* |
| | **167.5** | **148.5** | **141.1** | **137.1** | **134.6** | **132.8** | **130.6** | **128.3** | **125.9** | **123.5** |
| 4 | 7.71 | 6.94 | 6.59 | 6.39 | 6.26 | 6.16 | 6.04 | 5.91 | 5.77 | 5.63 |
| | *21.20* | *18.00* | *16.69* | *15.98* | *15.52* | *15.21* | *14.80* | *14.37* | *13.93* | *13.46* |
| | **74.14** | **61.25** | **56.18** | **53.44** | **51.71** | **50.53** | **49.00** | **47.41** | **45.77** | **44.05** |
| 5 | 6.61 | 5.79 | 5.41 | 5.19 | 5.05 | 4.95 | 4.82 | 4.68 | 4.53 | 4.36 |
| | *16.26* | *13.27* | *12.06* | *11.39* | *10.97* | *10.67* | *10.27* | *9.89* | *9.47* | *9.02* |
| | **47.04** | **36.61** | **33.20** | **31.09** | **29.75** | **28.84** | **27.64** | **26.42** | **25.14** | **23.78** |
| 6 | 5.99 | 5.14 | 4.76 | 4.53 | 4.39 | 4.28 | 4.15 | 4.00 | 3.84 | 3.67 |
| | *13.74* | *10.92* | *9.78* | *9.15* | *8.75* | *8.47* | *8.10* | *7.72* | *7.31* | *6.88* |
| | **35.51** | **27.00** | **23.70** | **21.90** | **20.81** | **20.03** | **19.03** | **17.99** | **16.89** | **15.75** |
| 7 | 5.59 | 4.74 | 4.35 | 4.12 | 3.97 | 3.87 | 3.73 | 3.57 | 3.41 | 3.23 |
| | *12.25* | *9.55* | *8.45* | *7.85* | *7.46* | *7.19* | *6.84* | *6.47* | *6.07* | *5.65* |
| | **29.22** | **21.69** | **18.77** | **17.19** | **16.21** | **15.52** | **14.63** | **13.71** | **12.73** | **11.69** |
| 8 | 5.32 | 4.46 | 4.07 | 3.84 | 3.69 | 3.58 | 3.44 | 3.28 | 3.12 | 2.93 |
| | *11.26* | *8.65* | *7.59* | *7.01* | *6.63* | *6.37* | *6.03* | *5.67* | *5.28* | *4.86* |
| | **25.42** | **18.49** | **15.83** | **14.39** | **13.49** | **12.86** | **12.04** | **11.19** | **10.30** | **9.34** |
| 9 | 5.12 | 4.26 | 3.86 | 3.63 | 3.48 | 3.37 | 3.23 | 3.07 | 2.90 | 2.71 |
| | *10.56* | *8.02* | *6.99* | *6.42* | *6.06* | *5.80* | *5.47* | *5.11* | *4.73* | *4.31* |
| | **22.86** | **16.39** | **13.90** | **12.56** | **11.71** | **11.13** | **10.37** | **9.57** | **8.72** | **7.81** |
| 10 | 4.96 | 4.10 | 3.71 | 3.48 | 3.33 | 3.22 | 3.07 | 2.91 | 2.74 | 2.54 |
| | *10.04* | *7.56* | *6.55* | *5.99* | *5.64* | *5.39* | *5.06* | *4.71* | *4.33* | *3.91* |
| | **21.04** | **14.91** | **12.55** | **11.28** | **10.48** | **9.92** | **9.20** | **8.45** | **7.64** | **6.76** |
| 11 | 4.84 | 3.98 | 3.59 | 3.36 | 3.20 | 3.09 | 2.95 | 2.79 | 2.61 | 2.40 |
| | *9.65* | *7.20* | *6.22* | *5.67* | *5.32* | *5.07* | *4.74* | *4.40* | *4.02* | *3.60* |
| | **19.69** | **13.81** | **11.56** | **10.35** | **9.58** | **9.05** | **8.35** | **7.63** | **6.85** | **6.00** |
| 12 | 4.75 | 3.88 | 3.49 | 3.26 | 3.11 | 3.00 | 2.85 | 2.69 | 2.50 | 2.30 |
| | *9.33* | *6.93* | *5.95* | *5.41* | *5.06* | *4.82* | *4.50* | *4.16* | *3.78* | *3.36* |
| | **18.64** | **12.97** | **10.80** | **9.63** | **8.89** | **8.38** | **7.71** | **7.00** | **6.25** | **5.42** |

* Table F is reprinted, in rearranged form, from Table V of Fisher and Yates: *Statistical tables for biological, agricultural and medical research*, Oliver and Boyd, Ltd., Edinburgh, by permission of the authors and publishers.

**Table F. Table of $F$ for .05 (roman), .01 (*italic*), and .001 (bold face) levels of significance\* (*continued*)**

| $n_1$ / $n_2$ | 1 | 2 | 3 | 4 | 5 | 6 | 8 | 12 | 24 | ∞ |
|---|---|---|---|---|---|---|---|---|---|---|
| 13 | 4.67 | 3.80 | 3.41 | 3.18 | 3.02 | 2.92 | 2.77 | 2.60 | 2.42 | 2.21 |
|  | 9.07 | 6.70 | 5.74 | 5.20 | 4.86 | 4.62 | 4.30 | 3.96 | 3.59 | 3.16 |
|  | **17.81** | **12.31** | **10.21** | **9.07** | **8.35** | **7.86** | **7.21** | **6.52** | **5.78** | **4.97** |
| 14 | 4.60 | 3.74 | 3.34 | 3.11 | 2.96 | 2.85 | 2.70 | 2.53 | 2.35 | 2.13 |
|  | 8.86 | 6.51 | 5.56 | 5.03 | 4.69 | 4.46 | 4.14 | 3.80 | 3.43 | 3.00 |
|  | **17.14** | **11.78** | **9.73** | **8.62** | **7.92** | **7.43** | **6.80** | **6.13** | **5.41** | **4.60** |
| 15 | 4.54 | 3.68 | 3.29 | 3.06 | 2.90 | 2.79 | 2.64 | 2.48 | 2.29 | 2.07 |
|  | 8.68 | 6.36 | 5.42 | 4.89 | 4.56 | 4.32 | 4.00 | 3.67 | 3.29 | 2.87 |
|  | **16.59** | **11.34** | **9.34** | **8.25** | **7.57** | **7.09** | **6.47** | **5.81** | **5.10** | **4.31** |
| 16 | 4.49 | 3.63 | 3.24 | 3.01 | 2.85 | 2.74 | 2.59 | 2.42 | 2.24 | 2.01 |
|  | 8.53 | 6.23 | 5.29 | 4.77 | 4.44 | 4.20 | 3.89 | 3.55 | 3.18 | 2.75 |
|  | **16.12** | **10.97** | **9.00** | **7.94** | **7.27** | **6.81** | **6.19** | **5.55** | **4.85** | **4.06** |
| 17 | 4.45 | 3.59 | 3.20 | 2.96 | 2.81 | 2.70 | 2.55 | 2.38 | 2.19 | 1.96 |
|  | 8.40 | 6.11 | 5.18 | 4.67 | 4.34 | 4.10 | 3.79 | 3.45 | 3.08 | 2.65 |
|  | **15.72** | **10.66** | **8.73** | **7.68** | **7.02** | **6.56** | **5.96** | **5.32** | **4.63** | **3.85** |
| 18 | 4.41 | 3.55 | 3.16 | 2.93 | 2.77 | 2.66 | 2.51 | 2.34 | 2.15 | 1.92 |
|  | 8.28 | 6.01 | 5.09 | 4.58 | 4.25 | 4.01 | 3.71 | 3.37 | 3.00 | 2.57 |
|  | **15.38** | **10.39** | **8.49** | **7.46** | **6.81** | **6.35** | **5.76** | **5.13** | **4.45** | **3.67** |
| 19 | 4.38 | 3.52 | 3.13 | 2.90 | 2.74 | 2.63 | 2.48 | 2.31 | 2.11 | 1.88 |
|  | 8.18 | 5.93 | 5.01 | 4.50 | 4.17 | 3.94 | 3.63 | 3.30 | 2.92 | 2.49 |
|  | **15.08** | **10.16** | **8.28** | **7.26** | **6.61** | **6.18** | **5.59** | **4.97** | **4.29** | **3.52** |
| 20 | 4.35 | 3.49 | 3.10 | 2.87 | 2.71 | 2.60 | 2.45 | 2.28 | 2.08 | 1.84 |
|  | 8.10 | 5.85 | 4.94 | 4.43 | 4.10 | 3.87 | 3.56 | 3.23 | 2.86 | 2.42 |
|  | **14.82** | **9.95** | **8.10** | **7.10** | **6.46** | **6.02** | **5.44** | **4.82** | **4.15** | **3.38** |
| 21 | 4.32 | 3.47 | 3.07 | 2.84 | 2.68 | 2.57 | 2.42 | 2.25 | 2.05 | 1.81 |
|  | 8.02 | 5.78 | 4.87 | 4.37 | 4.04 | 3.81 | 3.51 | 3.17 | 2.80 | 2.36 |
|  | **14.59** | **9.77** | **7.94** | **6.95** | **6.32** | **5.88** | **5.31** | **4.70** | **4.03** | **3.26** |
| 22 | 4.30 | 3.44 | 3.05 | 2.82 | 2.66 | 2.55 | 2.40 | 2.23 | 2.03 | 1.78 |
|  | 7.94 | 5.72 | 4.82 | 4.31 | 3.99 | 3.76 | 3.45 | 3.12 | 2.75 | 2.31 |
|  | **14.38** | **9.61** | **7.80** | **6.81** | **6.19** | **5.76** | **5.19** | **4.58** | **3.92** | **3.15** |
| 23 | 4.28 | 3.42 | 3.03 | 2.80 | 2.64 | 2.53 | 2.38 | 2.20 | 2.00 | 1.76 |
|  | 7.88 | 5.66 | 4.76 | 4.26 | 3.94 | 3.71 | 3.41 | 3.07 | 2.70 | 2.26 |
|  | **14.19** | **9.47** | **7.67** | **6.69** | **6.08** | **5.65** | **5.09** | **4.48** | **3.82** | **3.05** |
| 24 | 4.26 | 3.40 | 3.01 | 2.78 | 2.62 | 2.51 | 2.36 | 2.18 | 1.98 | 1.73 |
|  | 7.82 | 5.61 | 4.72 | 4.22 | 3.90 | 3.67 | 3.36 | 3.03 | 2.66 | 2.21 |
|  | **14.03** | **9.34** | **7.55** | **6.59** | **5.98** | **5.55** | **4.99** | **4.39** | **3.74** | **2.97** |

\* Table F is reprinted, in rearranged form, from Table V of Fisher and Yates: *Statistical tables for biological, agricultural and medical research*, Oliver and Boyd, Ltd., Edinburgh, by permission of the authors and publishers.

**Table F. Table of _F_ for .05 (roman), .01 (_italic_), and .001 (bold face) levels of significance\* (_continued_)**

| $n_2$ \ $n_1$ | 1 | 2 | 3 | 4 | 5 | 6 | 8 | 12 | 24 | ∞ |
|---|---|---|---|---|---|---|---|---|---|---|
| 25 | 4.24 | 3.38 | 2.99 | 2.76 | 2.60 | 2.49 | 2.34 | 2.16 | 1.96 | 1.71 |
|  | _7.77_ | _5.57_ | _4.68_ | _4.18_ | _3.86_ | _3.63_ | _3.32_ | _2.99_ | _2.62_ | _2.17_ |
|  | **13.88** | **9.22** | **7.45** | **6.49** | **5.88** | **5.46** | **4.91** | **4.31** | **3.66** | **2.89** |
| 26 | 4.22 | 3.37 | 2.98 | 2.74 | 2.59 | 2.47 | 2.32 | 2.15 | 1.95 | 1.69 |
|  | _7.72_ | _5.53_ | _4.64_ | _4.14_ | _3.82_ | _3.59_ | _3.29_ | _2.96_ | _2.58_ | _2.13_ |
|  | **13.74** | **9.12** | **7.36** | **6.41** | **5.80** | **5.38** | **4.83** | **4.24** | **3.59** | **2.82** |
| 27 | 4.21 | 3.35 | 2.96 | 2.73 | 2.57 | 2.46 | 2.30 | 2.13 | 1.93 | 1.67 |
|  | _7.68_ | _5.49_ | _4.60_ | _4.11_ | _3.78_ | _3.56_ | _3.26_ | _2.93_ | _2.55_ | _2.10_ |
|  | **13.61** | **9.02** | **7.27** | **6.33** | **5.73** | **5.31** | **4.76** | **4.17** | **3.52** | **2.75** |
| 28 | 4.20 | 3.34 | 2.95 | 2.71 | 2.56 | 2.44 | 2.29 | 2.12 | 1.91 | 1.65 |
|  | _7.64_ | _5.45_ | _4.57_ | _4.07_ | _3.75_ | _3.53_ | _3.23_ | _2.90_ | _2.52_ | _2.06_ |
|  | **13.50** | **8.93** | **7.19** | **6.25** | **5.66** | **5.24** | **4.69** | **4.11** | **3.46** | **2.70** |
| 29 | 4.18 | 3.33 | 2.93 | 2.70 | 2.54 | 2.43 | 2.28 | 2.10 | 1.90 | 1.64 |
|  | _7.60_ | _5.42_ | _4.54_ | _4.04_ | _3.73_ | _3.50_ | _3.20_ | _2.87_ | _2.49_ | _2.03_ |
|  | **13.39** | **8.85** | **7.12** | **6.19** | **5.59** | **5.18** | **4.64** | **4.05** | **3.41** | **2.64** |
| 30 | 4.17 | 3.32 | 2.92 | 2.69 | 2.53 | 2.42 | 2.27 | 2.09 | 1.89 | 1.62 |
|  | _7.56_ | _5.39_ | _4.51_ | _4.02_ | _3.70_ | _3.47_ | _3.17_ | _2.84_ | _2.47_ | _2.01_ |
|  | **13.29** | **8.77** | **7.05** | **6.12** | **5.53** | **5.12** | **4.58** | **4.00** | **3.36** | **2.59** |
| 40 | 4.08 | 3.23 | 2.84 | 2.61 | 2.45 | 2.34 | 2.18 | 2.00 | 1.79 | 1.51 |
|  | _7.31_ | _5.18_ | _4.31_ | _3.83_ | _3.51_ | _3.29_ | _2.99_ | _2.66_ | _2.29_ | _1.80_ |
|  | **12.61** | **8.25** | **6.60** | **5.70** | **5.13** | **4.73** | **4.21** | **3.64** | **3.01** | **2.23** |
| 60 | 4.00 | 3.15 | 2.76 | 2.52 | 2.37 | 2.25 | 2.10 | 1.92 | 1.70 | 1.39 |
|  | _7.08_ | _4.98_ | _4.13_ | _3.65_ | _3.34_ | _3.12_ | _2.82_ | _2.50_ | _2.12_ | _1.60_ |
|  | **11.97** | **7.76** | **6.17** | **5.31** | **4.76** | **4.37** | **3.87** | **3.31** | **2.69** | **1.90** |
| 120 | 3.92 | 3.07 | 2.68 | 2.45 | 2.29 | 2.17 | 2.02 | 1.83 | 1.61 | 1.25 |
|  | _6.85_ | _4.79_ | _3.95_ | _3.48_ | _3.17_ | _2.96_ | _2.66_ | _2.34_ | _1.95_ | _1.38_ |
|  | **11.38** | **7.31** | **5.79** | **4.95** | **4.42** | **4.04** | **3.55** | **3.02** | **2.40** | **1.56** |
| ∞ | 3.84 | 2.99 | 2.60 | 2.37 | 2.21 | 2.09 | 1.94 | 1.75 | 1.52 | 1.00 |
|  | _6.64_ | _4.60_ | _3.78_ | _3.32_ | _3.02_ | _2.80_ | _2.51_ | _2.18_ | _1.79_ | _1.00_ |
|  | **10.83** | **6.91** | **5.42** | **4.62** | **4.10** | **3.74** | **3.27** | **2.74** | **2.13** | **1.00** |

\* Table F is reprinted, in rearranged form, from Table V of Fisher and Yates: _Statistical tables for biological, agricultural and medical research_, Oliver and Boyd, Ltd., Edinburgh, by permission of the authors and publishers.

PSYCHOLOGICAL STATISTICS

## Table G. Squares and square roots

| N | N² | √N | √10N | N | N² | √N | √10N |
|---|---|---|---|---|---|---|---|
| **1.00** | 1.0000 | 1.00000 | 3.16228 | **1.50** | 2.2500 | 1.22474 | 3.87298 |
| 1.01 | 1.0201 | 1.00499 | 3.17805 | 1.51 | 2.2801 | 1.22882 | 3.88587 |
| 1.02 | 1.0404 | 1.00995 | 3.19374 | 1.52 | 2.3104 | 1.23288 | 3.89872 |
| 1.03 | 1.0609 | 1.01489 | 3.20936 | 1.53 | 2.3409 | 1.23693 | 3.91152 |
| 1.04 | 1.0816 | 1.01980 | 3.22490 | 1.54 | 2.3716 | 1.24097 | 3.92428 |
| 1.05 | 1.1025 | 1.02470 | 3.24037 | 1.55 | 2.4025 | 1.24499 | 3.93700 |
| 1.06 | 1.1236 | 1.02956 | 3.25576 | 1.56 | 2.4336 | 1.24900 | 3.94968 |
| 1.07 | 1.1449 | 1.03441 | 3.27109 | 1.57 | 2.4649 | 1.25300 | 3.96232 |
| 1.08 | 1.1664 | 1.03923 | 3.28634 | 1.58 | 2.4964 | 1.25698 | 3.97492 |
| 1.09 | 1.1881 | 1.04403 | 3.30151 | 1.59 | 2.5281 | 1.26095 | 3.98748 |
| **1.10** | 1.2100 | 1.04881 | 3.31662 | **1.60** | 2.5600 | 1.26491 | 4.00000 |
| 1.11 | 1.2321 | 1.05357 | 3.33167 | 1.61 | 2.5921 | 1.26886 | 4.01248 |
| 1.12 | 1.2544 | 1.05830 | 3.34664 | 1.62 | 2.6244 | 1.27279 | 4.02492 |
| 1.13 | 1.2769 | 1.06301 | 3.36155 | 1.63 | 2.6569 | 1.27671 | 4.03733 |
| 1.14 | 1.2996 | 1.06771 | 3.37639 | 1.64 | 2.6896 | 1.28062 | 4.04969 |
| 1.15 | 1.3225 | 1.07238 | 3.39116 | 1.65 | 2.7225 | 1.28452 | 4.06202 |
| 1.16 | 1.3456 | 1.07703 | 3.40588 | 1.66 | 2.7556 | 1.28841 | 4.07431 |
| 1.17 | 1.3689 | 1.08167 | 3.42053 | 1.67 | 2.7889 | 1.29228 | 4.08656 |
| 1.18 | 1.3924 | 1.08628 | 3.43511 | 1.68 | 2.8224 | 1.29615 | 4.09878 |
| 1.19 | 1.4161 | 1.09087 | 3.44964 | 1.69 | 2.8561 | 1.30000 | 4.11096 |
| **1.20** | 1.4400 | 1.09545 | 3.46410 | **1.70** | 2.8900 | 1.30384 | 4.12311 |
| 1.21 | 1.4641 | 1.10000 | 3.47851 | 1.71 | 2.9241 | 1.30767 | 4.13521 |
| 1.22 | 1.4884 | 1.10454 | 3.49285 | 1.72 | 2.9584 | 1.31149 | 4.14729 |
| 1.23 | 1.5129 | 1.10905 | 3.50714 | 1.73 | 2.9929 | 1.31529 | 4.15933 |
| 1.24 | 1.5376 | 1.11355 | 3.52136 | 1.74 | 3.0276 | 1.31909 | 4.17133 |
| 1.25 | 1.5625 | 1.11803 | 3.53553 | 1.75 | 3.0625 | 1.32288 | 4.18330 |
| 1.26 | 1.5876 | 1.12250 | 3.54965 | 1.76 | 3.0976 | 1.32665 | 4.19524 |
| 1.27 | 1.6129 | 1.12694 | 3.56371 | 1.77 | 3.1329 | 1.33041 | 4.20714 |
| 1.28 | 1.6384 | 1.13137 | 3.57771 | 1.78 | 3.1684 | 1.33417 | 4.21900 |
| 1.29 | 1.6641 | 1.13578 | 3.59166 | 1.79 | 3.2041 | 1.33791 | 4.23084 |
| **1.30** | 1.6900 | 1.14018 | 3.60555 | **1.80** | 3.2400 | 1.34164 | 4.24264 |
| 1.31 | 1.7161 | 1.14455 | 3.61939 | 1.81 | 3.2761 | 1.34536 | 4.25441 |
| 1.32 | 1.7424 | 1.14891 | 3.63318 | 1.82 | 3.3124 | 1.34907 | 4.26615 |
| 1.33 | 1.7689 | 1.15326 | 3.64692 | 1.83 | 3.3489 | 1.35277 | 4.27785 |
| 1.34 | 1.7956 | 1.15758 | 3.66060 | 1.84 | 3.3856 | 1.35647 | 4.28952 |
| 1.35 | 1.8225 | 1.16190 | 3.67423 | 1.85 | 3.4225 | 1.36015 | 4.30116 |
| 1.36 | 1.8496 | 1.16619 | 3.68782 | 1.86 | 3.4596 | 1.36382 | 4.31277 |
| 1.37 | 1.8769 | 1.17047 | 3.70135 | 1.87 | 3.4969 | 1.36748 | 4.32435 |
| 1.38 | 1.9044 | 1.17473 | 3.71484 | 1.88 | 3.5344 | 1.37113 | 4.33590 |
| 1.39 | 1.9321 | 1.17898 | 3.72827 | 1.89 | 3.5721 | 1.37477 | 4.34741 |
| **1.40** | 1.9600 | 1.18322 | 3.74166 | **1.90** | 3.6100 | 1.37840 | 4.35890 |
| 1.41 | 1.9881 | 1.18743 | 3.75500 | 1.91 | 3.6481 | 1.38203 | 4.37035 |
| 1.42 | 2.0164 | 1.19164 | 3.76829 | 1.92 | 3.6864 | 1.38564 | 4.38178 |
| 1.43 | 2.0449 | 1.19583 | 3.78153 | 1.93 | 3.7249 | 1.38924 | 4.39318 |
| 1.44 | 2.0736 | 1.20000 | 3.79473 | 1.94 | 3.7636 | 1.39284 | 4.40454 |
| 1.45 | 2.1025 | 1.20416 | 3.80789 | 1.95 | 3.8025 | 1.39642 | 4.41588 |
| 1.46 | 2.1316 | 1.20830 | 3.82099 | 1.96 | 3.8416 | 1.40000 | 4.42719 |
| 1.47 | 2.1609 | 1.21244 | 3.83406 | 1.97 | 3.8809 | 1.40357 | 4.43847 |
| 1.48 | 2.1904 | 1.21655 | 3.84708 | 1.98 | 3.9204 | 1.40712 | 4.44972 |
| 1.49 | 2.2201 | 1.22066 | 3.86005 | 1.99 | 3.9601 | 1.41067 | 4.46094 |
| **1.50** | 2.2500 | 1.22474 | 3.87298 | **2.00** | 4.0000 | 1.41421 | 4.47214 |
| N | N² | √N | √10N | N | N² | √N | √10N |

## Table G. Squares and square roots (*continued*)

| N | N² | √N | √10N | N | N² | √N | √10N |
|---|----|----|------|---|----|----|------|
| **2.00** | 4.0000 | 1.41421 | 4.47214 | **2.50** | 6.2500 | 1.58114 | 5.00000 |
| 2.01 | 4.0401 | 1.41774 | 4.48330 | 2.51 | 6.3001 | 1.58430 | 5.00999 |
| 2.02 | 4.0804 | 1.42127 | 4.49444 | 2.52 | 6.3504 | 1.58745 | 5.01996 |
| 2.03 | 4.1209 | 1.42478 | 4.50555 | 2.53 | 6.4009 | 1.59060 | 5.02991 |
| 2.04 | 4.1616 | 1.42829 | 4.51664 | 2.54 | 6.4516 | 1.59374 | 5.03984 |
| 2.05 | 4.2025 | 1.43178 | 4.52769 | 2.55 | 6.5025 | 1.59687 | 5.04975 |
| 2.06 | 4.2436 | 1.43527 | 4.53872 | 2.56 | 6.5536 | 1.60000 | 5.05964 |
| 2.07 | 4.2849 | 1.43875 | 4.54973 | 2.57 | 6.6049 | 1.60312 | 5.06952 |
| 2.08 | 4.3264 | 1.44222 | 4.56070 | 2.58 | 6.6564 | 1.60624 | 5.07937 |
| 2.09 | 4.3681 | 1.44568 | 4.57165 | 2.59 | 6.7081 | 1.60935 | 5.08920 |
| **2.10** | 4.4100 | 1.44914 | 4.58258 | **2.60** | 6.7600 | 1.61245 | 5.09902 |
| 2.11 | 4.4521 | 1.45258 | 4.59347 | 2.61 | 6.8121 | 1.61555 | 5.10882 |
| 2.12 | 4.4944 | 1.45602 | 4.60435 | 2.62 | 6.8644 | 1.61864 | 5.11859 |
| 2.13 | 4.5369 | 1.45945 | 4.61519 | 2.63 | 6.9169 | 1.62173 | 5.12835 |
| 2.14 | 4.5796 | 1.46287 | 4.62601 | 2.64 | 6.9696 | 1.62481 | 5.13809 |
| 2.15 | 4.6225 | 1.46629 | 4.63681 | 2.65 | 7.0225 | 1.62788 | 5.14782 |
| 2.16 | 4.6656 | 1.46969 | 4.64758 | 2.66 | 7.0756 | 1.63095 | 5.15752 |
| 2.17 | 4.7089 | 1.47309 | 4.65833 | 2.67 | 7.1289 | 1.63401 | 5.16720 |
| 2.18 | 4.7524 | 1.47648 | 4.66905 | 2.68 | 7.1824 | 1.63707 | 5.17687 |
| 2.19 | 4.7961 | 1.47986 | 4.67974 | 2.69 | 7.2361 | 1.64012 | 5.18652 |
| **2.20** | 4.8400 | 1.48324 | 4.69042 | **2.70** | 7.2900 | 1.64317 | 5.19615 |
| 2.21 | 4.8841 | 1.48661 | 4.70106 | 2.71 | 7.3441 | 1.64621 | 5.20577 |
| 2.22 | 4.9284 | 1.48997 | 4.71169 | 2.72 | 7.3984 | 1.64924 | 5.21536 |
| 2.23 | 4.9729 | 1.49332 | 4.72229 | 2.73 | 7.4529 | 1.65227 | 5.22494 |
| 2.24 | 5.0176 | 1.49666 | 4.73286 | 2.74 | 7.5076 | 1.65529 | 5.23450 |
| 2.25 | 5.0625 | 1.50000 | 4.74342 | 2.75 | 7.5625 | 1.65831 | 5.24404 |
| 2.26 | 5.1076 | 1.50333 | 4.75395 | 2.76 | 7.6176 | 1.66132 | 5.25357 |
| 2.27 | 5.1529 | 1.50665 | 4.76445 | 2.77 | 7.6729 | 1.66433 | 5.26308 |
| 2.28 | 5.1984 | 1.50997 | 4.77493 | 2.78 | 7.7284 | 1.66733 | 5.27257 |
| 2.29 | 5.2441 | 1.51327 | 4.78539 | 2.79 | 7.7841 | 1.67033 | 5.28205 |
| **2.30** | 5.2900 | 1.51658 | 4.79583 | **2.80** | 7.8400 | 1.67332 | 5.29150 |
| 2.31 | 5.3361 | 1.51987 | 4.80625 | 2.81 | 7.8961 | 1.67631 | 5.30094 |
| 2.32 | 5.3824 | 1.52315 | 4.81664 | 2.82 | 7.9524 | 1.67929 | 5.31037 |
| 2.33 | 5.4289 | 1.52643 | 4.82701 | 2.83 | 8.0089 | 1.68226 | 5.31977 |
| 2.34 | 5.4756 | 1.52971 | 4.83735 | 2.84 | 8.0656 | 1.68523 | 5.32917 |
| 2.35 | 5.5225 | 1.53297 | 4.84768 | 2.85 | 8.1225 | 1.68819 | 5.33854 |
| 2.36 | 5.5696 | 1.53623 | 4.85798 | 2.86 | 8.1796 | 1.69115 | 5.34790 |
| 2.37 | 5.6169 | 1.53948 | 4.86826 | 2.87 | 8.2369 | 1.69411 | 5.35724 |
| 2.38 | 5.6644 | 1.54272 | 4.87852 | 2.88 | 8.2944 | 1.69706 | 5.36656 |
| 2.39 | 5.7121 | 1.54596 | 4.88876 | 2.89 | 8.3521 | 1.70000 | 5.37587 |
| **2.40** | 5.7600 | 1.54919 | 4.89898 | **2.90** | 8.4100 | 1.70294 | 5.38516 |
| 2.41 | 5.8081 | 1.55242 | 4.90918 | 2.91 | 8.4681 | 1.70587 | 5.39444 |
| 2.42 | 5.8564 | 1.55563 | 4.91935 | 2.92 | 8.5264 | 1.70880 | 5.40370 |
| 2.43 | 5.9049 | 1.55885 | 4.92950 | 2.93 | 8.5849 | 1.71172 | 5.41295 |
| 2.44 | 5.9536 | 1.56205 | 4.93964 | 2.94 | 8.6436 | 1.71464 | 5.42218 |
| 2.45 | 6.0025 | 1.56525 | 4.94975 | 2.95 | 8.7025 | 1.71756 | 5.43139 |
| 2.46 | 6.0516 | 1.56844 | 4.95984 | 2.96 | 8.7616 | 1.72047 | 5.44059 |
| 2.47 | 6.1009 | 1.57162 | 4.96991 | 2.97 | 8.8209 | 1.72337 | 5.44977 |
| 2.48 | 6.1504 | 1.57480 | 4.97996 | 2.98 | 8.8804 | 1.72627 | 5.45894 |
| 2.49 | 6.2001 | 1.57797 | 4.98999 | 2.99 | 8.9401 | 1.72916 | 5.46809 |
| **2.50** | 6.2500 | 1.58114 | 5.00000 | **3.00** | 9.0000 | 1.73205 | 5.47723 |
| N | N² | √N | √10N | N | N² | √N | √10N |

## Table G. Squares and square roots (*continued*)

| N | N² | √N̄ | √10N̄ | N | N² | √N̄ | √10N̄ |
|---|---|---|---|---|---|---|---|
| **3.00** | 9.0000 | 1.73205 | 5.47723 | **3.50** | 12.2500 | 1.87083 | 5.91608 |
| 3.01 | 9.0601 | 1.73494 | 5.48635 | 3.51 | 12.3201 | 1.87350 | 5.92453 |
| 3.02 | 9.1204 | 1.73781 | 5.49545 | 3.52 | 12.3904 | 1.87617 | 5.93296 |
| 3.03 | 9.1809 | 1.74069 | 5.50454 | 3.53 | 12.4609 | 1.87883 | 5.94138 |
| 3.04 | 9.2416 | 1.74356 | 5.51362 | 3.54 | 12.5316 | 1.88149 | 5.94979 |
| 3.05 | 9.3025 | 1.74642 | 5.52268 | 3.55 | 12.6025 | 1.88414 | 5.95819 |
| 3.06 | 9.3636 | 1.74929 | 5.53173 | 3.56 | 12.6736 | 1.88680 | 5.96657 |
| 3.07 | 9.4249 | 1.75214 | 5.54076 | 3.57 | 12.7449 | 1.88944 | 5.97495 |
| 3.08 | 9.4864 | 1.75499 | 5.54977 | 3.58 | 12.8164 | 1.89209 | 5.98331 |
| 3.09 | 9.5481 | 1.75784 | 5.55878 | 3.59 | 12.8881 | 1.89473 | 5.99166 |
| **3.10** | 9.6100 | 1.76068 | 5.56776 | **3.60** | 12.9600 | 1.89737 | 6.00000 |
| 3.11 | 9.6721 | 1.76352 | 5.57674 | 3.61 | 13.0321 | 1.90000 | 6.00833 |
| 3.12 | 9.7344 | 1.76635 | 5.58570 | 3.62 | 13.1044 | 1.90263 | 6.01664 |
| 3.13 | 9.7969 | 1.76918 | 5.59464 | 3.63 | 13.1769 | 1.90526 | 6.02495 |
| 3.14 | 9.8596 | 1.77200 | 5.60357 | 3.64 | 13.2496 | 1.90788 | 6.03324 |
| 3.15 | 9.9225 | 1.77482 | 5.61249 | 3.65 | 13.3225 | 1.91050 | 6.04152 |
| 3.16 | 9.9856 | 1.77764 | 5.62139 | 3.66 | 13.3956 | 1.91311 | 6.04979 |
| 3.17 | 10.0489 | 1.78045 | 5.63028 | 3.67 | 13.4689 | 1.91572 | 6.05805 |
| 3.18 | 10.1124 | 1.78326 | 5.63915 | 3.68 | 13.5424 | 1.91833 | 6.06630 |
| 3.19 | 10.1761 | 1.78606 | 5.64801 | 3.69 | 13.6161 | 1.92094 | 6.07454 |
| **3.20** | 10.2400 | 1.78885 | 5.65685 | **3.70** | 13.6900 | 1.92354 | 6.08276 |
| 3.21 | 10.3041 | 1.79165 | 5.66569 | 3.71 | 13.7641 | 1.92614 | 6.09098 |
| 3.22 | 10.3684 | 1.79444 | 5.67450 | 3.72 | 13.8384 | 1.92873 | 6.09918 |
| 3.23 | 10.4329 | 1.79722 | 5.68331 | 3.73 | 13.9129 | 1.93132 | 6.10737 |
| 3.24 | 10.4976 | 1.80000 | 5.69210 | 3.74 | 13.9876 | 1.93391 | 6.11555 |
| 3.25 | 10.5625 | 1.80278 | 5.70088 | 3.75 | 14.0625 | 1.93649 | 6.12372 |
| 3.26 | 10.6276 | 1.80555 | 5.70964 | 3.76 | 14.1376 | 1.93907 | 6.13188 |
| 3.27 | 10.6929 | 1.80831 | 5.71839 | 3.77 | 14.2129 | 1.94165 | 6.14003 |
| 3.28 | 10.7584 | 1.81108 | 5.72713 | 3.78 | 14.2884 | 1.94422 | 6.14817 |
| 3.29 | 10.8241 | 1.81384 | 5.73585 | 3.79 | 14.3641 | 1.94679 | 6.15630 |
| **3.30** | 10.8900 | 1.81659 | 5.74456 | **3.80** | 14.4400 | 1.94936 | 6.16441 |
| 3.31 | 10.9561 | 1.81934 | 5.75326 | 3.81 | 14.5161 | 1.95192 | 6.17252 |
| 3.32 | 11.0224 | 1.82209 | 5.76194 | 3.82 | 14.5924 | 1.95448 | 6.18061 |
| 3.33 | 11.0889 | 1.82483 | 5.77062 | 3.83 | 14.6689 | 1.95704 | 6.18870 |
| 3.34 | 11.1556 | 1.82757 | 5.77927 | 3.84 | 14.7456 | 1.95959 | 6.19677 |
| 3.35 | 11.2225 | 1.83030 | 5.78792 | 3.85 | 14.8225 | 1.96214 | 6.20484 |
| 3.36 | 11.2896 | 1.83303 | 5.79655 | 3.86 | 14.8996 | 1.96469 | 6.21289 |
| 3.37 | 11.3569 | 1.83576 | 5.80517 | 3.87 | 14.9769 | 1.96723 | 6.22093 |
| 3.38 | 11.4244 | 1.83848 | 5.81378 | 3.88 | 15.0544 | 1.96977 | 6.22896 |
| 3.39 | 11.4921 | 1.84120 | 5.82237 | 3.89 | 15.1321 | 1.97231 | 6.23699 |
| **3.40** | 11.5600 | 1.84391 | 5.83095 | **3.90** | 15.2100 | 1.97484 | 6.24500 |
| 3.41 | 11.6281 | 1.84662 | 5.83952 | 3.91 | 15.2881 | 1.97737 | 6.25300 |
| 3.42 | 11.6964 | 1.84932 | 5.84808 | 3.92 | 15.3664 | 1.97990 | 6.26099 |
| 3.43 | 11.7649 | 1.85203 | 5.85662 | 3.93 | 15.4449 | 1.98242 | 6.26897 |
| 3.44 | 11.8336 | 1.85472 | 5.86515 | 3.94 | 15.5236 | 1.98494 | 6.27694 |
| 3.45 | 11.9025 | 1.85742 | 5.87367 | 3.95 | 15.6025 | 1.98746 | 6.28490 |
| 3.46 | 11.9716 | 1.86011 | 5.88218 | 3.96 | 15.6816 | 1.98997 | 6.29285 |
| 3.47 | 12.0409 | 1.86279 | 5.89067 | 3.97 | 15.7609 | 1.99249 | 6.30079 |
| 3.48 | 12.1104 | 1.86548 | 5.89915 | 3.98 | 15.8404 | 1.99499 | 6.30872 |
| 3.49 | 12.1801 | 1.86815 | 5.90762 | 3.99 | 15.9201 | 1.99750 | 6.31664 |
| **3.50** | 12.2500 | 1.87083 | 5.91608 | **4.00** | 16.0000 | 2.00000 | 6.32456 |
| **N** | **N²** | **√N̄** | **√10N̄** | **N** | **N²** | **√N̄** | **√10N̄** |

Table G. Squares and square roots (*continued*)

| N | N² | √N | √10N | N | N² | √N | √10N |
|---|---|---|---|---|---|---|---|
| **4.00** | 16.0000 | 2.00000 | 6.32456 | **4.50** | 20.2500 | 2.12132 | 6.70820 |
| 4.01 | 16.0801 | 2.00250 | 6.33246 | 4.51 | 20.3401 | 2.12368 | 6.71565 |
| 4.02 | 16.1604 | 2.00499 | 6.34035 | 4.52 | 20.4304 | 2.12603 | 6.72309 |
| 4.03 | 16.2409 | 2.00749 | 6.34823 | 4.53 | 20.5209 | 2.12838 | 6.73053 |
| 4.04 | 16.3216 | 2.00998 | 6.35610 | 4.54 | 20.6116 | 2.13073 | 6.73795 |
| 4.05 | 16.4025 | 2.01246 | 6.36396 | 4.55 | 20.7025 | 2.13307 | 6.74537 |
| 4.06 | 16.4836 | 2.01494 | 6.37181 | 4.56 | 20.7936 | 2.13542 | 6.75278 |
| 4.07 | 16.5649 | 2.01742 | 6.37966 | 4.57 | 20.8849 | 2.13776 | 6.76018 |
| 4.08 | 16.6464 | 2.01990 | 6.38749 | 4.58 | 20.9764 | 2.14009 | 6.76757 |
| 4.09 | 16.7281 | 2.02237 | 6.39531 | 4.59 | 21.0681 | 2.14243 | 6.77495 |
| **4.10** | 16.8100 | 2.02485 | 6.40312 | **4.60** | 21.1600 | 2.14476 | 6.78233 |
| 4.11 | 16.8921 | 2.02731 | 6.41093 | 4.61 | 21.2521 | 2.14709 | 6.78970 |
| 4.12 | 16.9744 | 2.02978 | 6.41872 | 4.62 | 21.3444 | 2.14942 | 6.79706 |
| 4.13 | 17.0569 | 2.03224 | 6.42651 | 4.63 | 21.4369 | 2.15174 | 6.80441 |
| 4.14 | 17.1396 | 2.03470 | 6.43428 | 4.64 | 21.5296 | 2.15407 | 6.81175 |
| 4.15 | 17.2225 | 2.03715 | 6.44205 | 4.65 | 21.6225 | 2.15639 | 6.81909 |
| 4.16 | 17.3056 | 2.03961 | 6.44981 | 4.66 | 21.7156 | 2.15870 | 6.82642 |
| 4.17 | 17.3889 | 2.04206 | 6.45755 | 4.67 | 21.8089 | 2.16102 | 6.83374 |
| 4.18 | 17.4724 | 2.04450 | 6.46529 | 4.68 | 21.9024 | 2.16333 | 6.84105 |
| 4.19 | 17.5561 | 2.04695 | 6.47302 | 4.69 | 21.9961 | 2.16564 | 6.84836 |
| **4.20** | 17.6400 | 2.04939 | 6.48074 | **4.70** | 22.0900 | 2.16795 | 6.85565 |
| 4.21 | 17.7241 | 2.05183 | 6.48845 | 4.71 | 22.1841 | 2.17025 | 6.86294 |
| 4.22 | 17.8084 | 2.05426 | 6.49615 | 4.72 | 22.2784 | 2.17256 | 6.87023 |
| 4.23 | 17.8929 | 2.05670 | 6.50384 | 4.73 | 22.3729 | 2.17486 | 6.87750 |
| 4.24 | 17.9776 | 2.05913 | 6.51153 | 4.74 | 22.4676 | 2.17715 | 6.88477 |
| 4.25 | 18.0625 | 2.06155 | 6.51920 | 4.75 | 22.5625 | 2.17945 | 6.89202 |
| 4.26 | 18.1476 | 2.06398 | 6.52687 | 4.76 | 22.6576 | 2.18174 | 6.89928 |
| 4.27 | 18.2329 | 2.06640 | 6.53452 | 4.77 | 22.7529 | 2.18403 | 6.90652 |
| 4.28 | 18.3184 | 2.06882 | 6.54217 | 4.78 | 22.8484 | 2.18632 | 6.91375 |
| 4.29 | 18.4041 | 2.07123 | 6.54981 | 4.79 | 22.9441 | 2.18861 | 6.92098 |
| **4.30** | 18.4900 | 2.07364 | 6.55744 | **4.80** | 23.0400 | 2.19089 | 6.92820 |
| 4.31 | 18.5761 | 2.07605 | 6.56506 | 4.81 | 23.1361 | 2.19317 | 6.93542 |
| 4.32 | 18.6624 | 2.07846 | 6.57267 | 4.82 | 23.2324 | 2.19545 | 6.94262 |
| 4.33 | 18.7489 | 2.08087 | 6.58027 | 4.83 | 23.3289 | 2.19773 | 6.94982 |
| 4.34 | 18.8356 | 2.08327 | 6.58787 | 4.84 | 23.4256 | 2.20000 | 6.95701 |
| 4.35 | 18.9225 | 2.08567 | 6.59545 | 4.85 | 23.5225 | 2.20227 | 6.96419 |
| 4.36 | 19.0096 | 2.08806 | 6.60303 | 4.86 | 23.6196 | 2.20454 | 6.97137 |
| 4.37 | 19.0969 | 2.09045 | 6.61060 | 4.87 | 23.7169 | 2.20681 | 6.97854 |
| 4.38 | 19.1844 | 2.09284 | 6.61816 | 4.88 | 23.8144 | 2.20907 | 6.98570 |
| 4.39 | 19.2721 | 2.09523 | 6.62571 | 4.89 | 23.9121 | 2.21133 | 6.99285 |
| **4.40** | 19.3600 | 2.09762 | 6.63325 | **4.90** | 24.0100 | 2.21359 | 7.00000 |
| 4.41 | 19.4481 | 2.10000 | 6.64078 | 4.91 | 24.1081 | 2.21585 | 7.00714 |
| 4.42 | 19.5364 | 2.10238 | 6.64831 | 4.92 | 24.2064 | 2.21811 | 7.01427 |
| 4.43 | 19.6249 | 2.10476 | 6.65582 | 4.93 | 24.3049 | 2.22036 | 7.02140 |
| 4.44 | 19.7136 | 2.10713 | 6.66333 | 4.94 | 24.4036 | 2.22261 | 7.02851 |
| 4.45 | 19.8025 | 2.10950 | 6.67083 | 4.95 | 24.5025 | 2.22486 | 7.03562 |
| 4.46 | 19.8916 | 2.11187 | 6.67832 | 4.96 | 24.6016 | 2.22711 | 7.04273 |
| 4.47 | 19.9809 | 2.11424 | 6.68581 | 4.97 | 24.7009 | 2.22935 | 7.04982 |
| 4.48 | 20.0704 | 2.11660 | 6.69328 | 4.98 | 24.8004 | 2.23159 | 7.05691 |
| 4.49 | 20.1601 | 2.11896 | 6.70075 | 4.99 | 24.9001 | 2.23383 | 7.06399 |
| **4.50** | 20.2500 | 2.12132 | 6.70820 | **5.00** | 25.0000 | 2.23607 | 7.07107 |
| N | N² | √N | √10N | N | N² | √N | √10N |

## Table G. Squares and square roots (*continued*)

| N | N² | √N | √10N | N | N² | √N | √10N |
|---|---|---|---|---|---|---|---|
| **5.00** | 25.0000 | 2.23607 | 7.07107 | **5.50** | 30.2500 | 2.34521 | 7.41620 |
| 5.01 | 25.1001 | 2.23830 | 7.07814 | 5.51 | 30.3601 | 2.34734 | 7.42294 |
| 5.02 | 25.2004 | 2.24054 | 7.08520 | 5.52 | 30.4704 | 2.34947 | 7.42967 |
| 5.03 | 25.3009 | 2.24277 | 7.09225 | 5.53 | 30.5809 | 2.35160 | 7.43640 |
| 5.04 | 25.4016 | 2.24499 | 7.09930 | 5.54 | 30.6916 | 2.35372 | 7.44312 |
| 5.05 | 25.5025 | 2.24722 | 7.10634 | 5.55 | 30.8025 | 2.35584 | 7.44983 |
| 5.06 | 25.6036 | 2.24944 | 7.11337 | 5.56 | 30.9136 | 2.35797 | 7.45654 |
| 5.07 | 25.7049 | 2.25167 | 7.12039 | 5.57 | 31.0249 | 2.36008 | 7.46324 |
| 5.08 | 25.8064 | 2.25389 | 7.12741 | 5.58 | 31.1364 | 2.36220 | 7.46994 |
| 5.09 | 25.9081 | 2.25610 | 7.13442 | 5.59 | 31.2481 | 2.36432 | 7.47663 |
| **5.10** | 26.0100 | 2.25832 | 7.14143 | **5.60** | 31.3600 | 2.36643 | 7.48331 |
| 5.11 | 26.1121 | 2.26053 | 7.14843 | 5.61 | 31.4721 | 2.36854 | 7.48999 |
| 5.12 | 26.2144 | 2.26274 | 7.15542 | 5.62 | 31.5844 | 2.37065 | 7.49667 |
| 5.13 | 26.3169 | 2.26495 | 7.16240 | 5.63 | 31.6969 | 2.37276 | 7.50333 |
| 5.14 | 26.4196 | 2.26716 | 7.16938 | 5.64 | 31.8096 | 2.37487 | 7.50999 |
| 5.15 | 26.5225 | 2.26936 | 7.17635 | 5.65 | 31.9225 | 2.37697 | 7.51665 |
| 5.16 | 26.6256 | 2.27156 | 7.18331 | 5.66 | 32.0356 | 2.37908 | 7.52330 |
| 5.17 | 26.7289 | 2.27376 | 7.19027 | 5.67 | 32.1489 | 2.38118 | 7.52994 |
| 5.18 | 26.8324 | 2.27596 | 7.19722 | 5.68 | 32.2624 | 2.38328 | 7.53658 |
| 5.19 | 26.9361 | 2.27816 | 7.20417 | 5.69 | 32.3761 | 2.38537 | 7.54321 |
| **5.20** | 27.0400 | 2.28035 | 7.21110 | **5.70** | 32.4900 | 2.38747 | 7.54983 |
| 5.21 | 27.1441 | 2.28254 | 7.21803 | 5.71 | 32.6041 | 2.38956 | 7.55645 |
| 5.22 | 27.2484 | 2.28473 | 7.22496 | 5.72 | 32.7184 | 2.39165 | 7.56307 |
| 5.23 | 27.3529 | 2.28692 | 7.23187 | 5.73 | 32.8329 | 2.39374 | 7.56968 |
| 5.24 | 27.4576 | 2.28910 | 7.23878 | 5.74 | 32.9476 | 2.39583 | 7.57628 |
| 5.25 | 27.5625 | 2.29129 | 7.24569 | 5.75 | 33.0625 | 2.39792 | 7.58288 |
| 5.26 | 27.6676 | 2.29347 | 7.25259 | 5.76 | 33.1776 | 2.40000 | 7.58947 |
| 5.27 | 27.7729 | 2.29565 | 7.25948 | 5.77 | 33.2929 | 2.40208 | 7.59605 |
| 5.28 | 27.8784 | 2.29783 | 7.26636 | 5.78 | 33.4084 | 2.40416 | 7.60263 |
| 5.29 | 27.9841 | 2.30000 | 7.27324 | 5.79 | 33.5241 | 2.40624 | 7.60920 |
| **5.30** | 28.0900 | 2.30217 | 7.28011 | **5.80** | 33.6400 | 2.40832 | 7.61577 |
| 5.31 | 28.1961 | 2.30434 | 7.28697 | 5.81 | 33.7561 | 2.41039 | 7.62234 |
| 5.32 | 28.3024 | 2.30651 | 7.29383 | 5.82 | 33.8724 | 2.41247 | 7.62889 |
| 5.33 | 28.4089 | 2.30868 | 7.30068 | 5.83 | 33.9889 | 2.41454 | 7.63544 |
| 5.34 | 28.5156 | 2.31084 | 7.30753 | 5.84 | 34.1056 | 2.41661 | 7.64199 |
| 5.35 | 28.6225 | 2.31301 | 7.31437 | 5.85 | 34.2225 | 2.41868 | 7.64853 |
| 5.36 | 28.7296 | 2.31517 | 7.32120 | 5.86 | 34.3396 | 2.42074 | 7.65506 |
| 5.37 | 28.8369 | 2.31733 | 7.32803 | 5.87 | 34.4569 | 2.42281 | 7.66159 |
| 5.38 | 28.9444 | 2.31948 | 7.33485 | 5.88 | 34.5744 | 2.42487 | 7.66812 |
| 5.39 | 29.0521 | 2.32164 | 7.34166 | 5.89 | 34.6921 | 2.42693 | 7.67463 |
| **5.40** | 29.1600 | 2.32379 | 7.34847 | **5.90** | 34.8100 | 2.42899 | 7.68115 |
| 5.41 | 29.2681 | 2.32594 | 7.35527 | 5.91 | 34.9281 | 2.43105 | 7.68765 |
| 5.42 | 29.3764 | 2.32809 | 7.36206 | 5.92 | 35.0464 | 2.43311 | 7.69415 |
| 5.43 | 29.4849 | 2.33024 | 7.36885 | 5.93 | 35.1649 | 2.43516 | 7.70065 |
| 5.44 | 29.5936 | 2.33238 | 7.37564 | 5.94 | 35.2836 | 2.43721 | 7.70714 |
| 5.45 | 29.7025 | 2.33452 | 7.38241 | 5.95 | 35.4025 | 2.43926 | 7.71362 |
| 5.46 | 29.8116 | 2.33666 | 7.38918 | 5.96 | 35.5216 | 2.44131 | 7.72010 |
| 5.47 | 29.9209 | 2.33880 | 7.39594 | 5.97 | 35.6409 | 2.44336 | 7.72658 |
| 5.48 | 30.0304 | 2.34094 | 7.40270 | 5.98 | 35.7604 | 2.44540 | 7.73305 |
| 5.49 | 30.1401 | 2.34307 | 7.40945 | 5.99 | 35.8801 | 2.44745 | 7.73951 |
| **5.50** | 30.2500 | 2.34521 | 7.41620 | **6.00** | 36.0000 | 2.44949 | 7.74597 |
| N | N² | √N | √10N | N | N² | √N | √10N |

## Table G. Squares and square roots (*continued*)

| N | N² | √N | √10N | N | N² | √N | √10N |
|---|---|---|---|---|---|---|---|
| **6.00** | 36.0000 | 2.44949 | 7.74597 | **6.50** | 42.2500 | 2.54951 | 8.06226 |
| 6.01 | 36.1201 | 2.45153 | 7.75242 | 6.51 | 42.3801 | 2.55147 | 8.06846 |
| 6.02 | 36.2404 | 2.45357 | 7.75887 | 6.52 | 42.5104 | 2.55343 | 8.07465 |
| 6.03 | 36.3609 | 2.45561 | 7.76531 | 6.53 | 42.6409 | 2.55539 | 8.08084 |
| 6.04 | 36.4816 | 2.45764 | 7.77174 | 6.54 | 42.7716 | 2.55734 | 8.08703 |
| 6.05 | 36.6025 | 2.45967 | 7.77817 | 6.55 | 42.9025 | 2.55930 | 8.09321 |
| 6.06 | 36.7236 | 2.46171 | 7.78460 | 6.56 | 43.0336 | 2.56125 | 8.09938 |
| 6.07 | 36.8449 | 2.46374 | 7.79102 | 6.57 | 43.1649 | 2.56320 | 8.10555 |
| 6.08 | 36.9664 | 2.46577 | 7.79744 | 6.58 | 43.2964 | 2.56515 | 8.11172 |
| 6.09 | 37.0881 | 2.46779 | 7.80385 | 6.59 | 43.4281 | 2.56710 | 8.11788 |
| **6.10** | 37.2100 | 2.46982 | 7.81025 | **6.60** | 43.5600 | 2.56905 | 8.12404 |
| 6.11 | 37.3321 | 2.47184 | 7.81665 | 6.61 | 43.6921 | 2.57099 | 8.13019 |
| 6.12 | 37.4544 | 2.47386 | 7.82304 | 6.62 | 43.8244 | 2.57294 | 8.13634 |
| 6.13 | 37.5769 | 2.47588 | 7.82943 | 6.63 | 43.9569 | 2.57488 | 8.14248 |
| 6.14 | 37.6996 | 2.47790 | 7.83582 | 6.64 | 44.0896 | 2.57682 | 8.14862 |
| 6.15 | 37.8225 | 2.47992 | 7.84219 | 6.65 | 44.2225 | 2.57876 | 8.15475 |
| 6.16 | 37.9456 | 2.48193 | 7.84857 | 6.66 | 44.3556 | 2.58070 | 8.16088 |
| 6.17 | 38.0689 | 2.48395 | 7.85493 | 6.67 | 44.4889 | 2.58263 | 8.16701 |
| 6.18 | 38.1924 | 2.48596 | 7.86130 | 6.68 | 44.6224 | 2.58457 | 8.17313 |
| 6.19 | 38.3161 | 2.48797 | 7.86766 | 6.69 | 44.7561 | 2.58650 | 8.17924 |
| **6.20** | 38.4400 | 2.48998 | 7.87401 | **6.70** | 44.8900 | 2.58844 | 8.18535 |
| 6.21 | 38.5641 | 2.49199 | 7.88036 | 6.71 | 45.0241 | 2.59037 | 8.19146 |
| 6.22 | 38.6884 | 2.49399 | 7.88670 | 6.72 | 45.1584 | 2.59230 | 8.19756 |
| 6.23 | 38.8129 | 2.49600 | 7.89303 | 6.73 | 45.2929 | 2.59422 | 8.20366 |
| 6.24 | 38.9376 | 2.49800 | 7.89937 | 6.74 | 45.4276 | 2.59615 | 8.20975 |
| 6.25 | 39.0625 | 2.50000 | 7.90569 | 6.75 | 45.5625 | 2.59808 | 8.21584 |
| 6.26 | 39.1876 | 2.50200 | 7.91202 | 6.76 | 45.6976 | 2.60000 | 8.22192 |
| 6.27 | 39.3129 | 2.50400 | 7.91833 | 6.77 | 45.8329 | 2.60192 | 8.22800 |
| 6.28 | 39.4384 | 2.50599 | 7.92465 | 6.78 | 45.9684 | 2.60384 | 8.23408 |
| 6.29 | 39.5641 | 2.50799 | 7.93095 | 6.79 | 46.1041 | 2.60576 | 8.24015 |
| **6.30** | 39.6900 | 2.50998 | 7.93725 | **6.80** | 46.2400 | 2.60768 | 8.24621 |
| 6.31 | 39.8161 | 2.51197 | 7.94355 | 6.81 | 46.3761 | 2.60960 | 8.25227 |
| 6.32 | 39.9424 | 2.51396 | 7.94984 | 6.82 | 46.5124 | 2.61151 | 8.25833 |
| 6.33 | 40.0689 | 2.51595 | 7.95613 | 6.83 | 46.6489 | 2.61343 | 8.26438 |
| 6.34 | 40.1956 | 2.51794 | 7.96241 | 6.84 | 46.7856 | 2.61534 | 8.27043 |
| 6.35 | 40.3225 | 2.51992 | 7.96869 | 6.85 | 46.9225 | 2.61725 | 8.27647 |
| 6.36 | 40.4496 | 2.52190 | 7.97496 | 6.86 | 47.0596 | 2.61916 | 8.28251 |
| 6.37 | 40.5769 | 2.52389 | 7.98123 | 6.87 | 47.1969 | 2.62107 | 8.28855 |
| 6.38 | 40.7044 | 2.52587 | 7.98749 | 6.88 | 47.3344 | 2.62298 | 8.29458 |
| 6.39 | 40.8321 | 2.52784 | 7.99375 | 6.89 | 47.4721 | 2.62488 | 8.30060 |
| **6.40** | 40.9600 | 2.52982 | 8.00000 | **6.90** | 47.6100 | 2.62679 | 8.30662 |
| 6.41 | 41.0881 | 2.53180 | 8.00625 | 6.91 | 47.7481 | 2.62869 | 8.31264 |
| 6.42 | 41.2164 | 2.53377 | 8.01249 | 6.92 | 47.8864 | 2.63059 | 8.31865 |
| 6.43 | 41.3449 | 2.53574 | 8.01873 | 6.93 | 48.0249 | 2.63249 | 8.32466 |
| 6.44 | 41.4736 | 2.53772 | 8.02496 | 6.94 | 48.1636 | 2.63439 | 8.33067 |
| 6.45 | 41.6025 | 2.53969 | 8.03119 | 6.95 | 48.3025 | 2.63629 | 8.33667 |
| 6.46 | 41.7316 | 2.54165 | 8.03741 | 6.96 | 48.4416 | 2.63818 | 8.34266 |
| 6.47 | 41.8609 | 2.54362 | 8.04363 | 6.97 | 48.5809 | 2.64008 | 8.34865 |
| 6.48 | 41.9904 | 2.54558 | 8.04984 | 6.98 | 48.7204 | 2.64197 | 8.35464 |
| 6.49 | 42.1201 | 2.54755 | 8.05605 | 6.99 | 48.8601 | 2.64386 | 8.36062 |
| **6.50** | 42.2500 | 2.54951 | 8.06226 | **7.00** | 49.0000 | 2.64575 | 8.36660 |
| N | N² | √N | √10N | N | N² | √N | √10N |

## Table G. Squares and square roots (*continued*)

| N | N² | √N̄ | √10N̄ | N | N² | √N̄ | √10N̄ |
|---|---|---|---|---|---|---|---|
| **7.00** | 49.0000 | 2.64575 | 8.36660 | **7.50** | 56.2500 | 2.73861 | 8.66025 |
| 7.01 | 49.1401 | 2.64764 | 8.37257 | 7.51 | 56.4001 | 2.74044 | 8.66603 |
| 7.02 | 49.2804 | 2.64953 | 8.37854 | 7.52 | 56.5504 | 2.74226 | 8.67179 |
| 7.03 | 49.4209 | 2.65141 | 8.38451 | 7.53 | 56.7009 | 2.74408 | 8.67756 |
| 7.04 | 49.5616 | 2.65330 | 8.39047 | 7.54 | 56.8516 | 2.74591 | 8.68332 |
| 7.05 | 49.7025 | 2.65518 | 8.39643 | 7.55 | 57.0025 | 2.74773 | 8.68907 |
| 7.06 | 49.8436 | 2.65707 | 8.40238 | 7.56 | 57.1536 | 2.74955 | 8.69483 |
| 7.07 | 49.9849 | 2.65895 | 8.40833 | 7.57 | 57.3049 | 2.75136 | 8.70057 |
| 7.08 | 50.1264 | 2.66083 | 8.41427 | 7.58 | 57.4564 | 2.75318 | 8.70632 |
| 7.09 | 50.2681 | 2.66271 | 8.42021 | 7.59 | 57.6081 | 2.75500 | 8.71206 |
| **7.10** | 50.4100 | 2.66458 | 8.42615 | **7.60** | 57.7600 | 2.75681 | 8.71780 |
| 7.11 | 50.5521 | 2.66646 | 8.43208 | 7.61 | 57.9121 | 2.75862 | 8.72353 |
| 7.12 | 50.6944 | 2.66833 | 8.43801 | 7.62 | 58.0644 | 2.76043 | 8.72926 |
| 7.13 | 50.8369 | 2.67021 | 8.44393 | 7.63 | 58.2169 | 2.76225 | 8.73499 |
| 7.14 | 50.9796 | 2.67208 | 8.44985 | 7.64 | 58.3696 | 2.76405 | 8.74071 |
| 7.15 | 51.1225 | 2.67395 | 8.45577 | 7.65 | 58.5225 | 2.76586 | 8.74643 |
| 7.16 | 51.2656 | 2.67582 | 8.46168 | 7.66 | 58.6756 | 2.76767 | 8.75214 |
| 7.17 | 51.4089 | 2.67769 | 8.46759 | 7.67 | 58.8289 | 2.76948 | 8.75785 |
| 7.18 | 51.5524 | 2.67955 | 8.47349 | 7.68 | 58.9824 | 2.77128 | 8.76356 |
| 7.19 | 51.6961 | 2.68142 | 8.47939 | 7.69 | 59.1361 | 2.77308 | 8.76926 |
| **7.20** | 51.8400 | 2.68328 | 8.48528 | **7.70** | 59.2900 | 2.77489 | 8.77496 |
| 7.21 | 51.9841 | 2.68514 | 8.49117 | 7.71 | 59.4441 | 2.77669 | 8.78066 |
| 7.22 | 52.1284 | 2.68701 | 8.49706 | 7.72 | 59.5984 | 2.77849 | 8.78635 |
| 7.23 | 52.2729 | 2.68887 | 8.50294 | 7.73 | 59.7529 | 2.78029 | 8.79204 |
| 7.24 | 52.4176 | 2.69072 | 8.50882 | 7.74 | 59.9076 | 2.78209 | 8.79773 |
| 7.25 | 52.5625 | 2.69258 | 8.51469 | 7.75 | 60.0625 | 2.78388 | 8.80341 |
| 7.26 | 52.7076 | 2.69444 | 8.52056 | 7.76 | 60.2176 | 2.78568 | 8.80909 |
| 7.27 | 52.8529 | 2.69629 | 8.52643 | 7.77 | 60.3729 | 2.78747 | 8.81476 |
| 7.28 | 52.9984 | 2.69815 | 8.53229 | 7.78 | 60.5284 | 2.78927 | 8.82043 |
| 7.29 | 53.1441 | 2.70000 | 8.53815 | 7.79 | 60.6841 | 2.79106 | 8.82610 |
| **7.30** | 53.2900 | 2.70185 | 8.54400 | **7.80** | 60.8400 | 2.79285 | 8.83176 |
| 7.31 | 53.4361 | 2.70370 | 8.54985 | 7.81 | 60.9961 | 2.79464 | 8.83742 |
| 7.32 | 53.5824 | 2.70555 | 8.55570 | 7.82 | 61.1524 | 2.79643 | 8.84308 |
| 7.33 | 53.7289 | 2.70740 | 8.56154 | 7.83 | 61.3089 | 2.79821 | 8.84873 |
| 7.34 | 53.8756 | 2.70924 | 8.56738 | 7.84 | 61.4656 | 2.80000 | 8.85438 |
| 7.35 | 54.0225 | 2.71109 | 8.57321 | 7.85 | 61.6225 | 2.80179 | 8.86002 |
| 7.36 | 54.1696 | 2.71293 | 8.57904 | 7.86 | 61.7796 | 2.80357 | 8.86566 |
| 7.37 | 54.3169 | 2.71477 | 8.58487 | 7.87 | 61.9369 | 2.80535 | 8.87130 |
| 7.38 | 54.4644 | 2.71662 | 8.59069 | 7.88 | 62.0944 | 2.80713 | 8.87694 |
| 7.39 | 54.6121 | 2.71846 | 8.59651 | 7.89 | 62.2521 | 2.80891 | 8.88257 |
| **7.40** | 54.7600 | 2.72029 | 8.60233 | **7.90** | 62.4100 | 2.81069 | 8.88819 |
| 7.41 | 54.9081 | 2.72213 | 8.60814 | 7.91 | 62.5681 | 2.81247 | 8.89382 |
| 7.42 | 55.0564 | 2.72397 | 8.61394 | 7.92 | 62.7264 | 2.81425 | 8.89944 |
| 7.43 | 55.2049 | 2.72580 | 8.61974 | 7.93 | 62.8849 | 2:81603 | 8.90505 |
| 7.44 | 55.3536 | 2.72764 | 8.62554 | 7.94 | 63.0436 | 2.81780 | 8.91067 |
| 7.45 | 55.5025 | 2.72947 | 8.63134 | 7.95 | 63.2025 | 2.81957 | 8.91628 |
| 7.46 | 55.6516 | 2.73130 | 8.63713 | 7.96 | 63.3616 | 2.82135 | 8.92188 |
| 7.47 | 55.8009 | 2.73313 | 8.64292 | 7.97 | 63.5209 | 2.82312 | 8.92749 |
| 7.48 | 55.9504 | 2.73496 | 8.64870 | 7.98 | 63.6804 | 2.82489 | 8.93308 |
| 7.49 | 56.1001 | 2.73679 | 8.65448 | 7.99 | 63.8401 | 2.82666 | 8.93868 |
| **7.50** | 56.2500 | 2.73861 | 8.66025 | **8.00** | 64.0000 | 2.82843 | 8.94427 |
| **N** | **N²** | **√N̄** | **√10N̄** | **N** | **N²** | **√N̄** | **√10N̄** |

## Table G. Squares and square roots (*continued*)

| N | N² | √N | √10N | N | N² | √N | √10N |
|---|---|---|---|---|---|---|---|
| **8.00** | 64.0000 | 2.82843 | 8.94427 | **8.50** | 72.2500 | 2.91548 | 9.21954 |
| 8.01 | 64.1601 | 2.83019 | 8.94986 | 8.51 | 72.4201 | 2.91719 | 9.22497 |
| 8.02 | 64.3204 | 2.83196 | 8.95545 | 8.52 | 72.5904 | 2.91890 | 9.23038 |
| 8.03 | 64.4809 | 2.83373 | 8.96103 | 8.53 | 72.7609 | 2.92062 | 9.23580 |
| 8.04 | 64.6416 | 2.83549 | 8.96660 | 8.54 | 72.9316 | 2.92233 | 9.24121 |
| 8.05 | 64.8025 | 2.83725 | 8.97218 | 8.55 | 73.1025 | 2.92404 | 9.24662 |
| 8.06 | 64.9636 | 2.83901 | 8.97775 | 8.56 | 73.2736 | 2.92575 | 9.25203 |
| 8.07 | 65.1249 | 2.84077 | 8.98332 | 8.57 | 73.4449 | 2.92746 | 9.25743 |
| 8.08 | 65.2864 | 2.84253 | 8.98888 | 8.58 | 73.6164 | 2.92916 | 9.26283 |
| 8.09 | 65.4481 | 2.84429 | 8.99444 | 8.59 | 73.7881 | 2.93087 | 9.26823 |
| **8.10** | 65.6100 | 2.84605 | 9.00000 | **8.60** | 73.9600 | 2.93258 | 9.27362 |
| 8.11 | 65.7721 | 2.84781 | 9.00555 | 8.61 | 74.1321 | 2.93428 | 9.27901 |
| 8.12 | 65.9344 | 2.84956 | 9.01110 | 8.62 | 74.3044 | 2.93598 | 9.28440 |
| 8.13 | 66.0969 | 2.85132 | 9.01665 | 8.63 | 74.4769 | 2.93769 | 9.28978 |
| 8.14 | 66.2596 | 2.85307 | 9.02219 | 8.64 | 74.6496 | 2.93939 | 9.29516 |
| 8.15 | 66.4225 | 2.85482 | 9.02774 | 8.65 | 74.8225 | 2.94109 | 9.30054 |
| 8.16 | 66.5856 | 2.85657 | 9.03327 | 8.66 | 74.9956 | 2.94279 | 9.30591 |
| 8.17 | 66.7489 | 2.85832 | 9.03881 | 8.67 | 75.1689 | 2.94449 | 9.31128 |
| 8.18 | 66.9124 | 2.86007 | 9.04434 | 8.68 | 75.3424 | 2.94618 | 9.31665 |
| 8.19 | 67.0761 | 2.86182 | 9.04986 | 8.69 | 75.5161 | 2.94788 | 9.32202 |
| **8.20** | 67.2400 | 2.86356 | 9.05539 | **8.70** | 75.6900 | 2.94958 | 9.32738 |
| 8.21 | 67.4041 | 2.86531 | 9.06091 | 8.71 | 75.8641 | 2.95127 | 9.33274 |
| 8.22 | 67.5684 | 2.86705 | 9.06642 | 8.72 | 76.0384 | 2.95296 | 9.33809 |
| 8.23 | 67.7329 | 2.86880 | 9.07193 | 8.73 | 76.2129 | 2.95466 | 9.34345 |
| 8.24 | 67.8976 | 2.87054 | 9.07744 | 8.74 | 76.3876 | 2.95635 | 9.34880 |
| 8.25 | 68.0625 | 2.87228 | 9.08295 | 8.75 | 76.5625 | 2.95804 | 9.35414 |
| 8.26 | 68.2276 | 2.87402 | 9.08845 | 8.76 | 76.7376 | 2.95973 | 9.35949 |
| 8.27 | 68.3929 | 2.87576 | 9.09395 | 8.77 | 76.9129 | 2.96142 | 9.36483 |
| 8.28 | 68.5584 | 2.87750 | 9.09945 | 8.78 | 77.0884 | 2.96311 | 9.37017 |
| 8.29 | 68.7241 | 2.87924 | 9.10494 | 8.79 | 77.2641 | 2.96479 | 9.37550 |
| **8.30** | 68.8900 | 2.88097 | 9.11043 | **8.80** | 77.4400 | 2.96648 | 9.38083 |
| 8.31 | 69.0561 | 2.88271 | 9.11592 | 8.81 | 77.6161 | 2.96816 | 9.38616 |
| 8.32 | 69.2224 | 2.88444 | 9.12140 | 8.82 | 77.7924 | 2.96985 | 9.39149 |
| 8.33 | 69.3889 | 2.88617 | 9.12688 | 8.83 | 77.9689 | 2.97153 | 9.39681 |
| 8.34 | 69.5556 | 2.88791 | 9.13236 | 8.84 | 78.1456 | 2.97321 | 9.40213 |
| 8.35 | 69.7225 | 2.88964 | 9.13783 | 8.85 | 78.3225 | 2.97489 | 9.40744 |
| 8.36 | 69.8896 | 2.89137 | 9.14330 | 8.86 | 78.4996 | 2.97658 | 9.41276 |
| 8.37 | 70.0569 | 2.89310 | 9.14877 | 8.87 | 78.6769 | 2.97825 | 9.41807 |
| 8.38 | 70.2244 | 2.89482 | 9.15423 | 8.88 | 78.8544 | 2.97993 | 9.42338 |
| 8.39 | 70.3921 | 2.89655 | 9.15969 | 8.89 | 79.0321 | 2.98161 | 9.42868 |
| **8.40** | 70.5600 | 2.89828 | 9.16515 | **8.90** | 79.2100 | 2.98329 | 9.43398 |
| 8.41 | 70.7281 | 2.90000 | 9.17061 | 8.91 | 79.3881 | 2.98496 | 9.43928 |
| 8.42 | 70.8964 | 2.90172 | 9.17606 | 8.92 | 79.5664 | 2.98664 | 9.44458 |
| 8.43 | 71.0649 | 2.90345 | 9.18150 | 8.93 | 79.7449 | 2.98831 | 9.44987 |
| 8.44 | 71.2336 | 2.90517 | 9.18695 | 8.94 | 79.9236 | 2.98998 | 9.45516 |
| 8.45 | 71.4025 | 2.90689 | 9.19239 | 8.95 | 80.1025 | 2.99166 | 9.46044 |
| 8.46 | 71.5716 | 2.90861 | 9.19783 | 8.96 | 80.2816 | 2.99333 | 9.46573 |
| 8.47 | 71.7409 | 2.91033 | 9.20326 | 8.97 | 80.4609 | 2.99500 | 9.47101 |
| 8.48 | 71.9104 | 2.91204 | 9.20869 | 8.98 | 80.6404 | 2.99666 | 9.47629 |
| 8.49 | 72.0801 | 2.91376 | 9.21412 | 8.99 | 80.8201 | 2.99833 | 9.48156 |
| **8.50** | 72.2500 | 2.91548 | 9.21954 | **9.00** | 81.0000 | 3.00000 | 9.48683 |
| N | N² | √N | √10N | N | N² | √N | √10N |

## Table G. Squares and square roots (*continued*)

| N | N² | √N̄ | √10N | N | N² | √N̄ | √10N |
|---|---|---|---|---|---|---|---|
| **9.00** | 81.0000 | 3.00000 | 9.48683 | **9.50** | 90.2500 | 3.08221 | 9.74679 |
| 9.01 | 81.1801 | 3.00167 | 9.49210 | 9.51 | 90.4401 | 3.08383 | 9.75192 |
| 9.02 | 81.3604 | 3.00333 | 9.49737 | 9.52 | 90.6304 | 3.08545 | 9.75705 |
| 9.03 | 81.5409 | 3.00500 | 9.50263 | 9.53 | 90.8209 | 3.08707 | 9.76217 |
| 9.04 | 81.7216 | 3.00666 | 9.50789 | 9.54 | 91.0116 | 3.08869 | 9.76729 |
| 9.05 | 81.9025 | 3.00832 | 9.51315 | 9.55 | 91.2025 | 3.09031 | 9.77241 |
| 9.06 | 82.0836 | 3.00998 | 9.51840 | 9.56 | 91.3936 | 3.09192 | 9.77753 |
| 9.07 | 82.2649 | 3.01164 | 9.52365 | 9.57 | 91.5849 | 3.09354 | 9.78264 |
| 9.08 | 82.4464 | 3.01330 | 9.52890 | 9.58 | 91.7764 | 3.09516 | 9.78775 |
| 9.09 | 82.6281 | 3.01496 | 9.53415 | 9.59 | 91.9681 | 3.09677 | 9.79285 |
| **9.10** | 82.8100 | 3.01662 | 9.53939 | **9.60** | 92.1600 | 3.09839 | 9.79796 |
| 9.11 | 82.9921 | 3.01828 | 9.54463 | 9.61 | 92.3521 | 3.10000 | 9.80306 |
| 9.12 | 83.1744 | 3.01993 | 9.54987 | 9.62 | 92.5444 | 3.10161 | 9.80816 |
| 9.13 | 83.3569 | 3.02159 | 9.55510 | 9.63 | 92.7369 | 3.10322 | 9.81326 |
| 9.14 | 83.5396 | 3.02324 | 9.56033 | 9.64 | 92.9296 | 3.10483 | 9.81835 |
| 9.15 | 83.7225 | 3.02490 | 9.56556 | 9.65 | 93.1225 | 3.10644 | 9.82344 |
| 9.16 | 83.9056 | 3.02655 | 9.57079 | 9.66 | 93.3156 | 3.10805 | 9.82853 |
| 9.17 | 84.0889 | 3.02820 | 9.57601 | 9.67 | 93.5089 | 3.10966 | 9.83362 |
| 9.18 | 84.2724 | 3.02985 | 9.58123 | 9.68 | 93.7024 | 3.11127 | 9.83870 |
| 9.19 | 84.4561 | 3.03150 | 9.58645 | 9.69 | 93.8961 | 3.11288 | 9.84378 |
| **9.20** | 84.6400 | 3.03315 | 9.59166 | **9.70** | 94.0900 | 3.11448 | 9.84886 |
| 9.21 | 84.8241 | 3.03480 | 9.59687 | 9.71 | 94.2841 | 3.11609 | 9.85393 |
| 9.22 | 85.0084 | 3.03645 | 9.60208 | 9.72 | 94.4784 | 3.11769 | 9.85901 |
| 9.23 | 85.1929 | 3.03809 | 9.60729 | 9.73 | 94.6729 | 3.11929 | 9.86408 |
| 9.24 | 85.3776 | 3.03974 | 9.61249 | 9.74 | 94.8676 | 3.12090 | 9.86914 |
| 9.25 | 85.5625 | 3.04138 | 9.61769 | 9.75 | 95.0625 | 3.12250 | 9.87421 |
| 9.26 | 85.7476 | 3.04302 | 9.62289 | 9.76 | 95.2576 | 3.12410 | 9.87927 |
| 9.27 | 85.9329 | 3.04467 | 9.62808 | 9.77 | 95.4529 | 3.12570 | 9.88433 |
| 9.28 | 86.1184 | 3.04631 | 9.63328 | 9.78 | 95.6484 | 3.12730 | 9.88939 |
| 9.29 | 86.3041 | 3.04795 | 9.63846 | 9.79 | 95.8441 | 3.12890 | 9.89444 |
| **9.30** | 86.4900 | 3.04959 | 9.64365 | **9.80** | 96.0400 | 3.13050 | 9.89949 |
| 9.31 | 86.6761 | 3.05123 | 9.64883 | 9.81 | 96.2361 | 3.13209 | 9.90454 |
| 9.32 | 86.8624 | 3.05287 | 9.65401 | 9.82 | 96.4324 | 3.13369 | 9.90959 |
| 9.33 | 87.0489 | 3.05450 | 9.65919 | 9.83 | 96.6289 | 3.13528 | 9.91464 |
| 9.34 | 87.2356 | 3.05614 | 9.66437 | 9.84 | 96.8256 | 3.13688 | 9.91968 |
| 9.35 | 87.4225 | 3.05778 | 9.66954 | 9.85 | 97.0225 | 3.13847 | 9.92472 |
| 9.36 | 87.6096 | 3.05941 | 9.67471 | 9.86 | 97.2196 | 3.14006 | 9.92975 |
| 9.37 | 87.7969 | 3.06105 | 9.67988 | 9.87 | 97.4169 | 3.14166 | 9.93479 |
| 9.38 | 87.9844 | 3.06268 | 9.68504 | 9.88 | 97.6144 | 3.14325 | 9.93982 |
| 9.39 | 88.1721 | 3.06431 | 9.69020 | 9.89 | 97.8121 | 3.14484 | 9.94485 |
| **9.40** | 88.3600 | 3.06594 | 9.69536 | **9.90** | 98.0100 | 3.14643 | 9.94987 |
| 9.41 | 88.5481 | 3.06757 | 9.70052 | 9.91 | 98.2081 | 3.14802 | 9.95490 |
| 9.42 | 88.7364 | 3.06920 | 9.70567 | 9.92 | 98.4064 | 3.14960 | 9.95992 |
| 9.43 | 88.9249 | 3.07083 | 9.71082 | 9.93 | 98.6049 | 3.15119 | 9.96494 |
| 9.44 | 89.1136 | 3.07246 | 9.71597 | 9.94 | 98.8036 | 3.15278 | 9.96995 |
| 9.45 | 89.3025 | 3.07409 | 9.72111 | 9.95 | 99.0025 | 3.15436 | 9.97497 |
| 9.46 | 89.4916 | 3.07571 | 9.72625 | 9.96 | 99.2016 | 3.15595 | 9.97998 |
| 9.47 | 89.6809 | 3.07734 | 9.73139 | 9.97 | 99.4009 | 3.15753 | 9.98499 |
| 9.48 | 89.8704 | 3.07896 | 9.73653 | 9.98 | 99.6004 | 3.15911 | 9.98999 |
| 9.49 | 90.0601 | 3.08058 | 9.74166 | 9.99 | 99.8001 | 3.16070 | 9.99500 |
| **9.50** | 90.2500 | 3.08221 | 9.74679 | **10.00** | 100.000 | 3.16228 | 10.0000 |
| N | N² | √N̄ | √10N | N | N² | √N̄ | √10N |

# INDEX